Laser Engineering

Laser Engineering

Edited by **Juan Landers**

CWILLFORD PRESS

New York

Published by Willford Press,
118-35 Queens Blvd., Suite 400,
Forest Hills, NY 11375, USA
www.willfordpress.com

Laser Engineering
Edited by Juan Landers

International Standard Book Number: 978-1-68285-081-7 (Hardback)

The publisher's policy is to use permanent paper from mills that operate a sustainable forestry policy. Furthermore, the publisher ensures that the text paper and cover boards used have met acceptable environmental accreditation standards.

Trademark Notice: Registered trademark of products or corporate names are used only for explanation and identification without intent to infringe.

Printed in the United States of America.

Contents

Permissions

List of Contributors

Preface

The applications of laser engineering can be found across a large number of industries such as laser printing, welding materials, etc. This book traces the progress of laser engineering and highlights some of its key concepts through detailed elaborations of topics like types of lasers, industrial uses of lasers, properties of lasers, optical fiber technology, etc. The ever growing need for advanced technology is the reason that has fueled the research in the field of laser engineering in recent times. The book is appropriate for students seeking detailed information in this area as well as for experts engaged in research in this discipline.

After months of intensive research and writing, this book is the end result of all who devoted their time and efforts in the initiation and progress of this book. It will surely be a source of reference in enhancing the required knowledge of the new developments in the area. During the course of developing this book, certain measures such as accuracy, authenticity and research focused analytical studies were given preference in order to produce a comprehensive book in the area of study.

This book would not have been possible without the efforts of the authors and the publisher. I extend my sincere thanks to them. Secondly, I express my gratitude to my family and well-wishers. And most importantly, I thank my students for constantly expressing their willingness and curiosity in enhancing their knowledge in the field, which encourages me to take up further research projects for the advancement of the area.

Editor

Density measurements of laser interaction with ordered structured 'snow' targets

E. Schleifer, M. Botton, E. Nahum, S. Eisenman, A. Zigler, and Z. Henis
Racah Institute Of Physics, Hebrew University, Jerusalem, Israel

Abstract
This paper presents a new method to control the position of a micro-column snow target. This target enables the measurement of the mean electron density of the pre-plasma created by a pre-pulse with different time delays. This research will allow a better understanding of the generation of fast protons from the interaction between a structured pre-plasma and a high intensity laser.

Keywords: high intensity lasers; plasma; proton acceleration

Laser driven proton acceleration is an active field of research due to its high potential in reducing the size and cost of conventional accelerators. The acceleration scheme consists of a very high intensity ($>10^{18}$ W cm^{-2}), high contrast (10^{-10}) laser pulse[1, 2]. The laser pulse interacts with a solid target, commonly a thin foil, from which high energy ions and protons are accelerated. Enhancement of the energy of the generated protons using a compact laser source (moderate intensity, 10^{17}–10^{19} W cm^{-2}) is a challenging but rewarding task due to the fact that the proton energy in traditional schemes scales as the electric field employed on the target[3]. One of the promising ways to increase the ion energy is the use of structured targets[4], mainly nanotargets[5]. In this kind of target the energy conversion efficiency from the laser to the proton beam is increased[6].

Recently, we demonstrated that by using frozen H_2O micro-column targets, which were grown on a sapphire substrate, significantly improved absorption of the laser energy by the H_2O micro-column target took place. This study showed that more than 90% of the incident energy was absorbed by the target[7]. Moreover, we have shown that the emitted proton energy scales up by a factor of 10 compared with standard laser driven protons schemes[8–11]. The maximal proton energy as a function of laser power is presented in Figure 1. Numerical particle in cell (PIC) simulations indicate that the enhancement of proton energy is attributed to the structure of the snow target and the pre-plasma formed by the pre-pulse and the density gradients formed[10].

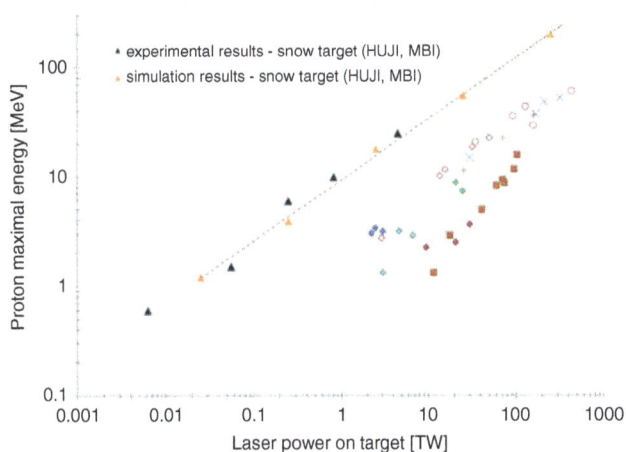

Figure 1. Proton maximal energy as a function of laser power the on target (triangles: snow target experimental and simulation results)[8–10]; other shapes: different targets and laser facilities)[11].

Characterization and control of the morphology of the micro-structured targets is necessary to gain a better understanding of the interaction process and for optimization of the proton acceleration. As a first step, we present here a method to control the position of the micro-columns which is based on seeding of nucleation centers on the sapphire substrate. The nucleation centers were implanted on the sapphire substrate by spattering of aluminum. The size and shape of the seeding dots were determined by laser writing lithography. Selective growth of the micro-columns on the nucleation centers was achieved. The obtained structured target is characterized by enhanced order of each micro-column. The detailed shape of the column is determined by

Correspondence to: Elad Schleifer, High Intensity Laser Lab (HIL), Racah Institute of Physics, Hebrew University of Jerusalem, Givat Ram, Jerusalem, Israel. Email: elad.schleifer@mail.huji.ac.il

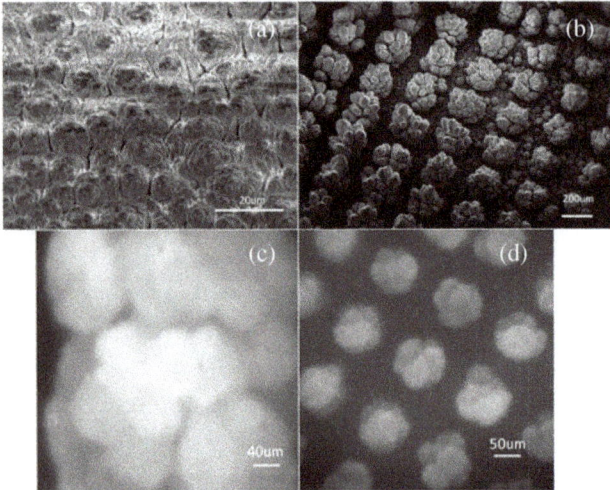

Figure 2. Snow morphology images. Images (a) and (c) are SEM and optical pictures of snow micro-columns without nucleation centers. Images (b) and (d) are SEM and optical pictures of snow micro-columns with nucleation centers.

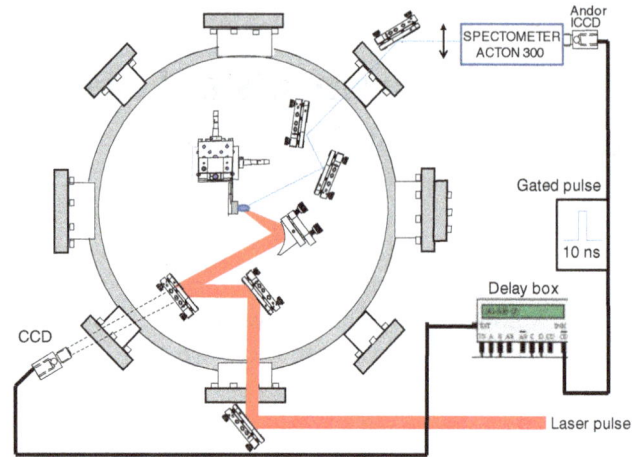

Figure 3. Schematics of the experimental setup.

the growth parameters, which will be described elsewhere. Scanning electron microscopy (SEM) and optical images of the structured target are shown in Figure 2. Figures 2(a) and (c) are of non-structured targets whereas Figures 2(b) and (d) are of the structured targets. The difference is self-evident.

Previous experiments and PIC simulations have shown the important role of the pre-plasma on the proton acceleration mechanism[10]. In this paper we present a measurement of the mean plasma density for different time delays after the pre-pulse interacts with the target. This density measurement profile will be integrated into our PIC simulation and will allow us to better understand the acceleration mechanism. The structured targets were used to characterize the plasma generated by the laser pre-pulse. The pre-pulse originates from the regenerative amplifier, thus the pre-pulse comes 10 ns before the short main pulse. Both the pulse and the pre-pulse are 40 fs (FWHM) with better than 10^{-5} contrast up to 1 ps. In this case the plasma forms 10 ns before the main pulse and during this time the plasma can freely expand and a highly nonuniform plasma cloud is formed. By the time of its arrival, the main pulse meets this highly structured dynamic plasma cloud and interacts with it to produce the accelerated protons[10]. The ordered structured targets enabled measurement of the electron density of the pre-plasma created by a pre-pulse. Due to the increased spacing between the columns, the laser beam interacts with only one of them each shot, hence the other columns do not block the optical path of the plasma formed by the interaction of the laser with the target. The plasma density was derived by spectroscopic measurements. It was calculated using the line broadening of the hydrogen due to the Stark effect[12]. Regular interferometry and Thompson scattering could not be utilized due to the morphology of the target and the

impossibility of perpendicular line of sight. The spectroscopic method gives a fairly good measurement of the mean density, but lacks in spatial resolution. The method presented here could realize the detection of the density profile for different times in usual target normal sheath acceleration (TNSA) experiments with the use of different spectroscopic line broadenings that were chosen for our experiment. The mean plasma electron density in our experiment was derived from the line broadening caused by the linear Stark effect in the Balmer hydrogen series at 656.28 nm. This line was chosen due to the richness of hydrogen in our target and the mean density expected in our experiment.

The experiments were performed at the Hebrew University High Intensity Laser facility (HUJI). A schematic representation of the experimental setup is shown in Figure 3. The laser used for the experiments was a chirped pulse amplification Ti:sapphire laser, which delivered a 20 mJ in 40 fs (FWHM) pulse at a central wavelength of 800 nm with 10 Hz repetition rate. The intensity contrast ratio between the main laser pulse and the pre-pulse was 10^{-3} and the time difference was 10 ns; both pulses had less than 10^{-5} pedestal up to 1 ps before the corresponding pulse as measured by a combination of a fast photodiode and oscilloscope. The laser beam was focused using an off-axis parabola having an F value of 3.3 with an incident angle of $60°$ to the target surface normal. A deformable mirror was used to overcome the deformations in the wave front, and in order to reach a focal spot area (FWHM) of 100 μm^2, reaching an estimated intensity of 5×10^{17} W cm^{-2} in the main pulse. In order to reproduce the same conditions as the interaction of the pre-pulse with the target, a filter with an optical density (OD) of 3 was used on the main beam. The radiation emitted by the plasma was collected using a lens and was imaged into a spectrometer entrance slit (0.3 m optical length, 1800 lines/mm grating) coupled to a fast gated ICCD. The gating was varied between 10 and 90 ns to within a few ns jitter with respect to the ablating laser. The spectral

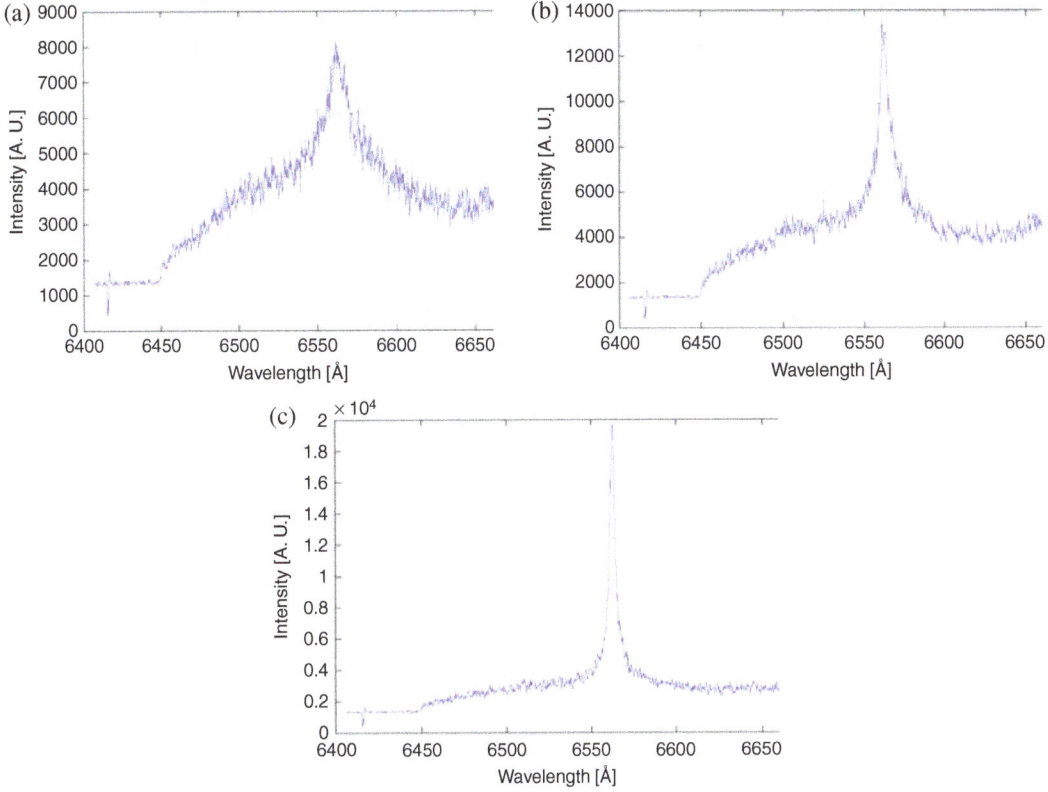

Figure 4. Line broadening of hydrogen for different time delays between the pre-pulse and the main pulse: (a) $t = 10$ ns, (b) $t = 40$ ns, (c) $t = 90$ ns.

resolution was calibrated using the doublet of sodium at 589.59 nm and was better than 3 Å.

In Figure 4 we present the broadening of the hydrogen line (at 656.28 nm) for different time delays corresponding to the ablating laser.

The mean electron density was calculated using[12]

$$N_e(10^{18} \text{ cm}^{-3}) = \left(\frac{\Delta\lambda_s(\text{nm})}{5.4} \right)^{3/2}.$$

The mean electron density is presented as a function of the delay in Figure 5. It was found that the mean electron density of the pre-plasma after 10 ns was $N_e \cong 2 \times 10^{17}$ cm^{-3}. This mean density is the pre-plasma density which our main pulse is interacting with.

Although the mean density agrees with our estimates, a better spatial description of the plasma cloud is sought. We are currently in the process of developing a system in which spatial measurements of the plasma cloud can be measured. It should be noted that these measurements are not trivial as the average density is low but the gradients are not. This measurement will contribute to a better understanding of the interaction of a high intensity laser with our targets. The spatial electron density measured will be integrated into our PIC code and will allow a better understanding of the acceleration mechanism.

In conclusion, we have achieved the manufacture of ordered snow micro-column targets using nucleation sites

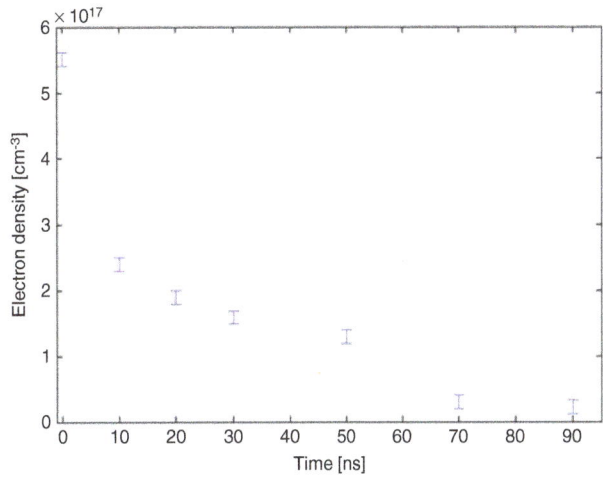

Figure 5. Mean electron density as a function of the delay after the interaction of the pre-pulse with the snow target.

deposited on a sapphire substrate. This target allowed the measurement of the plasma density generated by the pre-pulse for different time delays after the interaction between the pre-pulse and the snow column target. This research will allow a better understanding of the generation of fast protons from the interaction between structured pre-plasma and high intensity lasers.

References

1. S. C. Wilks, A. B. Langdon, T. E. Cowan, M. Roth, M. Singh, S. Hatchett, M. H. Key, D. Pennington, A. MacKinnon, and R. A. Snavely, Phys. Plasmas **8**, 542 (2001).

2. S. P. Hatchett, C. G. Brown, T. E. Cowan, E. A. Henry, J. S. Johnson, M. H. Key, J. A. Koch, A. B. Langdon, B. F. Lasinski, R. W. Lee, A. J. Mackinnon, D. M. Pennington, M. D. Perry, T. W. Phillips, M. Roth, T. C. Sangster, M. S. Singh, R. A. Snavely, M. A. Stoyer, S. C. Wilks, and K. Yasuike, Phys. Plasmas **7**, 2076 (2000).

3. S. C. Wilks and W. L. Kruer, IEEE J. Quantum Electron. **33**, 1954 (1997).

4. S. Kawata, T. Nagashima, M. Takano, T. Izumiyama, D. Kamiyama, D. Barada, Q. Kong, Y. J. Gu, P. X. Wang, Y. Y. Ma, W. M. Wang, W. Zhang, J. Xie, H. Zhang, and D. Dai, High Power Laser Sci. Eng. **2**, e4 (2014).

5. D. Margarone, O. Klimo, I. J. Kim, J. Prokůpek, J. Limpouch, T. M. Jeong, T. Mocek, J. Pšikal, H. T. Kim, J. Proška, K. H. Nam, L. Štolcová, I. W. Choi, S. K. Lee, J. H. Sung, T. J. Yu, and G. Korn, Phys. Rev. Lett. **109**, 234801 (2012).

6. Y. Nodera, S. Kawata, N. Onuma, J. Limpouch, O. Klimo, and T. Kikuchi, Phys. Rev. E. **78**, 046401 (2008).

7. T. Plachan, S. Pecker, Z. Henis, S. Eisenmann, and A. Zigler, Appl. Phys. Lett. **90**, 041501 (2007).

8. T. Palchan, Z. Henis, A. Y. Faenov, A. I. Magunov, S. A. Pikuz, S. V. Gasilov, I. Yu. Skobelev, and A. Zigler, Appl. Phys. Lett. **91**, 251501 (2007).

9. A. Zigler, T. Palchan, N. Bruner, E. Schleifer, S. Eisenmann, M. Botton, Z. Henis, S. A. Pikuz, A. Y. Faenov, Jr., D. Gordon, and P. Sprangle, Phys. Rev. Lett. **106**, 134801 (2011).

10. A. Zigler, S. Eisenman, M. Botton, E. Nahum, E. Schleifer, A. Baspaly, I. Pomerantz, F. Abicht, J. Branzel, G. Priebe, S. Steinke, A. Andreev, M. Schnuerer, W. Sandner, D. Gordon, P. Sprangle, and K. W. D. Ledingham, Phys. Rev. Lett. **110**, 215004 (2013).

11. K. Zeil, S. D. Kraft, S. Bock, M. Bussmann, T. E. Cowan, T. Kluge, J. Metzkes, T. Richter, R. Sauerbrey, and U. Schramm, New J. Phys. **12**, 045015 (2010).

12. H. R. Griem, J. Halenka, and W. Olchawa, J. Phys. B At. Mol. Opt. Phys. **38**, 975 (2005).

Large aperture N31 neodymium phosphate laser glass for use in a high power laser facility

Lili Hu, Shubin Chen, Jingping Tang, Biao Wang, Tao Meng, Wei Chen, Lei Wen, Junjiang Hu, Shunguang Li, Yongchun Xu, Yasi Jiang, Junzhou Zhang, and Zhonghong Jiang

Shanghai Institute of Optics and Fine Mechanics, Chinese Academy of Sciences, Shanghai 201800, China

Abstract

Large aperture Nd:phosphate laser glass is a key optical element for an inertial confinement fusion (ICF) facility. N31, one type of neodymium doped phosphate glasses, was developed for high peak power laser facility applications in China. The composition and main properties of N31 glass are given, together with those of LHG-8, LG-770, and KGSS-0180 Nd:phosphate laser glasses, from Hoya and Schott, and from Russia. The technologies of pot melting, continuous melting, and edge cladding of large size N31 phosphate laser glass are briefly described. The small signal gain profiles of N31 glass slabs from both pot melting and continuous melting at various values of the pumping energy of the xenon lamp are presented. N31 glass is characterized by a stimulated emission cross section of 3.8×10^{-20} cm^2 at 1053 nm, an absorption coefficient of 0.10–0.15% cm^{-1} at laser wavelength, small residual stress around the interface between the cladding glass and the laser glass, optical homogeneity of $\sim 2 \times 10^{-6}$ in a 400 mm aperture, and laser damage threshold larger than 42 J/cm^2 for a 3 ns pulse width at 1064 nm wavelength.

Keywords: neodymium phosphate laser glass; large aperture glass; ICF facility

1. Introduction

Large aperture Nd:phosphate laser glass is at the heart of a high power laser system. For high peak power inertial confinement fusion (ICF) facility application, there are many strict technical requirements on laser glass, such as high gain, low nonlinear refractive index, low attenuation at laser wavelength, excellent optical homogeneity, and large laser damage threshold. Large aperture Nd:phosphate laser glasses have been successfully applied in the NIF facility in the United States, and an over 2 MJ ultraviolet laser has already been realized in this largest laser facility[1, 2]. Phosphate glass has good solubility to rare earth ions, medium phonon energy, low nonlinear refractive index, strong ability to resolve platinum particles, and good spectroscopic properties for Nd^{3+} ions compared with silicate glass. Since the late 1970s, Nd:phosphate laser glass has been developed for high peak power laser facility use. LHG-8 and LG-770 Nd:phosphate laser glasses, which are used in the NIF facility in the United States, were developed and fabricated by Hoya and Schott, with continuous melting technology[3]. KGSS-0180 Nd:phosphate glass was developed in Russia, and it has been used in the four-channel

laser system "Luch"[4]. In China, N21 and N31, two types of Nd:phosphate laser glass, have been used in high power laser facilities. N21 Nd:phosphate laser glass was developed in the 1980s in Shanghai Institute of Optics and Fine Mechanics (SIOM), and it was used in Shen Guang I and Shen Guang II high peak power laser facilities[5, 6]. N31 Nd:phosphate laser glass was developed in Shanghai Institute of Optics and Fine Mechanics in the 1990s. It has been applied in the Shen Guang series of high power laser facilities in China for more than ten years[7]. In recent years, different technologies concerning the mass fabrication of large aperture N31 laser glass have been explored[8]. In this paper we present the composition, main properties, and fabrication techniques of large aperture N31 glass.

2. Compositions and properties of N31 phosphate laser glass

Similar to other neodymium phosphate laser glasses used in an ICF facility, N31 glass is a kind of metaphosphate glass composed of P$_2$O$_5$-R$_2$O-MO-T$_2$O$_3$. R$_2$O represents an alkali oxide. MO represents an alkaline oxide. T$_2$O$_3$ represents mixtures of Al$_2$O$_3$, La$_2$O$_3$, and Nd$_2$O$_3$. Up to 5wt% Nd$_2$O$_3$ can be easily doped in N31 glass without obvious change of properties besides density and refractive

Correspondence to: Lili Hu, Shanghai Institute of Optics and Fine Mechanics, Chinese Academy of Sciences, Shanghai 201800, China. Email: hulili@siom.ac.cn

Table 1. Main parameters of neodymium phosphate laser glasses from Hoya[1], Schott[1], Russia (GOI)[4, 9], and SIOM.

Parameters	N31	LHG-8	LG-770	KGSS-0180
$\sigma/10^{-20}$ cm^2	3.8	3.6	3.9	3.6
* τ_{rad}/μs	351	365	351	
$\Delta\lambda_{eff}$/nm	25.8	26.5	25.4	
*d/g/cm^3	2.87	2.83	2.59	2.83
* n_d	1.540	1.5296	1.5067	
$n_{1053\,nm}$	1.533	1.5201	1.4991	
Abbe number	65.8	66.5	68.4	
$n_2/10^{-13}$esu	1.18	1.12	1.01	1.1
Tg/°C	450	485	460	460
$\alpha/10^{-7}$/K(20–100°C)	115	115	116	116
dn/dT/10^{-7}/K	−43	−53	−47	−40
dS/dT/10^{-7}/K	14	6	11	
k/W/m K	0.56	0.58	0.57	
E/Gpa	56.4	50.1	47.3	59.0

* Parameters which vary with Nd$_2$O$_3$ concentration.

index. Its composition is satisfied for the requirements of mass fabrication and laser facility applications.

A comparison of the main basic properties of N31 with those of LHG-8, LG-770, and KGSS-0180 glasses is given in table 1. The most important property of neodymium laser glass is gain coefficient. It is expressed in eq. (1). In order to get optimal gain properties and energy extraction efficiency from neodymium phosphate laser glass, the stimulated emission cross section at emission peak wavelength is required to be in the range 3.5 to 4.0 $\times 10^{-20}$ cm^2 for ICF applications.

$$g = [\sigma(\lambda)N^* - \alpha], \qquad (1)$$

where σ is the emission cross section at wavelength λ. For Nd:phosphate glass, the maximum emission cross section in the near infrared region is at 1053 nm. N* is the Nd^{3+} inversion density in the $^4F_{3/2}$ state. It is determined by the fluorescent lifetime and fluorescence effective bandwidth of this state. α is the attenuation at laser wavelength. In order to obtain a high gain coefficient, large stimulated emission cross section and long fluorescent lifetime are preferred.

The stimulated emission cross section of N31 glass is higher than that of LHG-8 glass and close to that of LG-770 glass. In order to suppress the damage from self focus due to the optical nonlinear effect at high peak energy fluence, the nonlinear refractive index n$_2$ of neodymium phosphate laser glass should be controlled. From Table 1 it is seen that n$_2$ is below 1.2 $\times 10^{-13}$ esu for these Nd:phosphate glasses.

The laser gain curves of N3122 and N3130 glasses, with 2.2wt% and 3.0wt% Nd$_2$O$_3$ doping concentration, were detected at various xenon lamp pumping voltages. The fluorescence lifetime and optical loss at 1053 nm will have an important influence on the small signal gain. Only samples with an optical loss of 0.1–0.15% cm^{-1} and a lifetime of 340 μs (for 2.2wt%) or 320 μs (for 3.0wt%) were chosen for measurement. The results, shown in Figure 1, were detected by the same experimental setup and under the same pumping conditions. It clearly indicates that the small signal gain of N31 neodymium phosphate laser glass increases with Nd$_2$O$_3$ concentration.

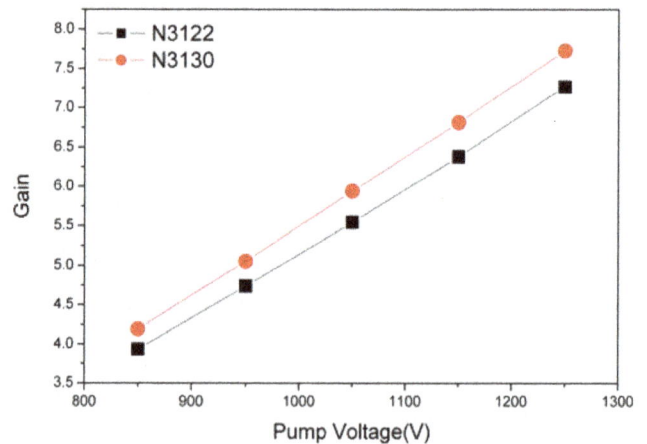

Figure 1. Gain profiles of N3122 and N3130 glass rods of size Φ 8 × 162 mm at various xenon lamp pumping voltages. N3122 and N3130 correspond to the glasses with Nd$_2$O$_3$ concentration of 2.2 wt% and 3.0 wt%.

3. Melting technologies of 400 mm large aperture N31 phosphate laser glass

It is well known that most of the key parameters of laser glass such as fluorescent lifetime, number of platinum inclusions, bubble and optical homogeneity, birefringence, optical attenuation at lasing wavelength, residual OH$^-$, and absorption at 400 nm are determined by the fabrication technology. The fabrication process of a laser glass slab includes melting, forming, rough annealing, fine annealing, and edge cladding. The fabrication technology, especially the melting technology, is very important in ensuring the quality of laser glass. An N31 glass rod with diameter 90 mm and a slab with a clear aperture of 400 mm have been fabricated in SIOM.

The melting technology of N31 glass has been explored since the mid 1990s, and several fabrication technologies concerning pot melting of N31 laser glass were developed in early 2000. A patented pot melting technology has been established instead of traditional two-step melting[10]. The glass melt was directly flowed into a platinum crucible to refine and homogenize after the primary melting and dehydration in a refractory crucible. This patented pot melting method ensures a lower degree of contamination by impurities and high efficiency of dehydration.

High purity raw materials with Fe, Cu, Cr, Ni, and V trace elements (less than 3 ppm) have been domestically fabricated. Through the controlling of the purity of raw materials and melting processing, the total amount of transition metal oxides is less than 10 ppm in N31 glass. Research has been done on the effect of Fe and Cu impurities on the optical attenuation of N31 glass[11, 12]. It is found that the most harmful trace element is copper ions, which can seriously affect the optical absorption at laser wavelength even at an amount of only several tens of ppb in glass[12]. The high purity raw materials and patented pot melting technology make the attenuation at lasing wavelength lower than 0.15% cm^{-1}. Using the reactive atmosphere processing (RAP) dehydration method[6, 13], the absorption at 3000 cm^{-1} can be controlled to be less than 1 cm^{-1} in N31 glass produced by pot melting. Platinum inclusions were removed by controlling the redox condition of glass melting in order to ionize the metallic platinum[14].

The pot melting efficiency is too low to manufacture thousands of laser glass slabs. Since 2006, research on continuous melting technology of N31 glass has been carried out. The continuous melting technology of large aperture laser glass is more complicated than that of traditional optical glass due to its special technical parameters. Based on the matured laser glass fabrication technology of pot melting[10] as well as theoretical and small scale experimental modeling[15, 16], we successfully developed a continuous melting technology for N31 laser glass[17].

The continuous melting line of N31 laser glass consists of an interconnected melter, conditioner, refiner, homogenizer, forming, and annealing lehr. Figure 2 shows the N31 glass running from the annealing lehr in a continuous melting line. Large size N31 glass with thickness greater than 50 mm and width of about 500 mm was formed through a platinum pipe into a special mold. A series of technical issues has been overcome in which the greatest challenges are melting batched raw materials with low contamination from refractory materials, dehydration, forming, and fracture in the annealing lehr. Now 400 mm large aperture N31 laser glass can be fabricated by both pot melting and continuous melting in SIOM.

Table 2 lists the main parameters of mass produced N31 glass with 3.5wt% Nd_2O_3 (denoted N3135). The absorption coefficients at 3333 nm (3000 cm^{-1}) and 400 nm were measured with a commercial Nicolet 6700 FTIR Infrared Spectrometer and Lambda UV/VIS/NIR 1050Spectrophotometers with a glass sample of size 10 × 10 × 20 mm^3.

Figure 2. Large size N31 glass running from the annealing lehr of a continuous melting line.

The loss at lasing wavelength was measured with a self-made double beam spectrometer using a Nd:YAG laser using a glass rod of size $\Phi 8 × 160$ mm^3. The fluorescence lifetime was measured with an Edinburgh FLSP920 time resolved fluorescence spectrometer with a glass sample of size 10 × 10 × 10 mm^3. The refractive index of the glass at a wavelength of 587.6 nm was measured with a Shimadzu KPR-2000 Refractometer. The optical homogeneity of the glass slab was measured and calculated using a Zygo 24″ MST interferometer. A 650 nm laser diode was used to measure the platinum inclusions of full size 400 mm aperture N31 glass slab. A TEM$_{00}$ Nd:YAG laser with a pulse width of 12 ns was used to detect the bulk laser damage threshold of N3135 glass.

In the measurement of the single shot laser induced bulk damage threshold (bulk LIDT), ISO 11254-1 was taken as a standard[18]. Because the polished surface is the weakest part of the glass sample under laser radiation, the laser beam was focused into the sample by a lens with 100 mm focal length to evaluate the bulk LIDT. The spot size was measured with a beam analyzer, and the effective spot area was about 0.02 mm^2. The test error of the spot size is about 25% because of the highly convergent beam. The laser energy on the target was controlled by an attenuator, and the pulse energy was recorded for each shot by an energy meter from a split-off portion of the beam. The fluctuation of output energy was 3%. A He–Ne laser was used to monitor the glass sample, and high sensitivity on-line damage detection was performed by using a microscopy system. Under the condition of 12 ns pulse width, the damage threshold is around 85.2–86.2 J/cm^2. Accordingly, the calculated threshold for N3135 neodymium phosphate laser glass is around 42.6-43.2 J/cm^2 under a 3 ns laser radiation.

Figure 3 shows a comparison of the gain coefficient of N3135 glasses produced by pot melting and continuous melting. A 4 × 2 × 3 amplifier module, where the $n × m × q$ designation denotes the slab number

Table 2. Parameters of mass production N3135 glass.

Parameters	Data
Attenuation at 1053 nm (cm^{-1})	0.10–0.14%
Fluorescent lifetime (μs)	310–315
n_d	1.540 ± 0.001
Absorption coefficient at 400 nm (cm^{-1})	0.12–0.23
Absorption coefficient at 3000 cm^{-1} (cm^{-1})	0.5-1.8
Optical homogeneity	$\sim 2 \times 10^{-6}$
Platinum inclusion	No platinum for more than 80% glass slabs
Damage threshold at 1064 nm, 3 ns	No bulk damage at 42 J/cm^2 energy influence

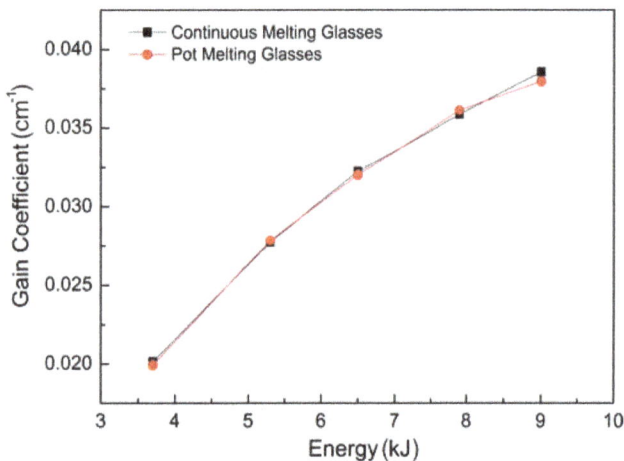

Figure 3. Gain coefficient of 400 mm aperture N3135 glass slabs from pot melting and continuous melting.

(height × width × length) of parallel amplifying channels, was used to measure the small signal gain coefficient. The tested rectangular glass slabs of 810 × 460 × 40 mm^3 were placed in the amplifier module and oriented at the Brewster angle with respect to the input beam with aperture of 380 × 380 mm^2, respectively. The central flashlamps cassettes with eight flashlamps pump glass slabs in both directions, while two sided flashlamp cassettes, each with six flashlamps and large silver reflectors, pump glass slabs in one direction. These flashlamps are configured as 20 circuits, with each circuit having two flashlamps in series. The operating voltage is from 15 to 23.5 kV, the capacitance is 308 μF per circuit, and the duration of discharge is 360 μs. When the charge voltage is 23.5 kV, the power condition module will deliver nominally 92 kJ of energy to each flashlamp pair during the main discharge pulse. In Figure 3, it can been seen that the N3135 laser glasses produced by the pot melting and continuous melting techniques provide nearly the same gain performance.

A typical 633 nm transmitted wavefront of 400 mm aperture N3135 laser glass, which was measured by Zygo interferometer with a test aperture of 600 mm, is shown in Figure 4. To measure the optical homogeneity of one 810 × 460 × 43.7 mm^3 glass slab with a wedge angle of

$1'$–$5'$, a four-step procedure [19] was done on the digital phase-shift Fizeau interferometer, for both the left and the right part of the slab. Figure 4 shows the 633 nm transmitted wavefront of the left and the right parts of the same glass slab. The peak and valley (PV) data shown in Figure 4 are the results obtained by subtracting the PV of two reflection surfaces of the slab, as well as PV of the empty cavity of the interferometer, by the same 400 mm aperture testing laser beam normal to the slab surface. In the light of the PV values shown in Figures 4a and 4b, the optical homogeneity Δn of the slab can be estimated by $\Delta n = PV * 632.8\,\text{nm} / t$, where t is the thickness of the slab. Thus, the optical homogeneity of the refractive index is 2.6×10^{-6} and $2.5. \times 10^{-6}$, for the left and right 400 × 400 mm^2 aperture of the N3135 slab, respectively. The testing precision of optical homogeneity by this four-step procedure method is about 1×10^{-6}.

4. Edge cladding of N31 phosphate laser glass

Edge cladding is an important technology to suppress the amplified stimulated emission ASE and to ensure the gain properties of large size Nd:phosphate glass. Cu^{2+} ion doped phosphate glass with a precise refractive index match to N31 laser glass has been designed as a cladding glass. A kind of self-developed epoxy adhesive agent with precise refractive index match to both the laser glass and the cladding glass is used to bond these two glasses. It provides an adhesive strength of 18 MPa. This adhesive agent has been tested to be highly resistant to high intensity pump and laser power as well as humid environments in the polishing process.

The Cu^{2+} doping level is limited by the temperature rise at the interface between the cladding glass and the laser glass. This temperature rise is due to strong absorption of ASE energy of a laser pulse. The temperature rise ΔT after a laser shot can be expressed by

$$\Delta T(x) = \frac{\beta I_O}{\rho \cdot C_P} \exp(-\beta x), \qquad (2)$$

where x is the distance from the interface to the inside of laser glass. I_0 is the pumping energy fluence. ρ is the density of the cladding glass. C_p is heat capacity. β is absorption coefficient of cladding glass at 1 μm. The absorption

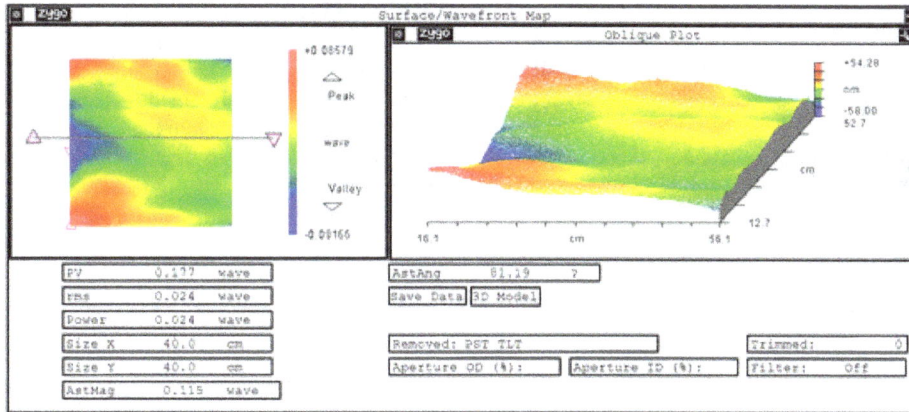

(a) the left part of 400×400 mm^2

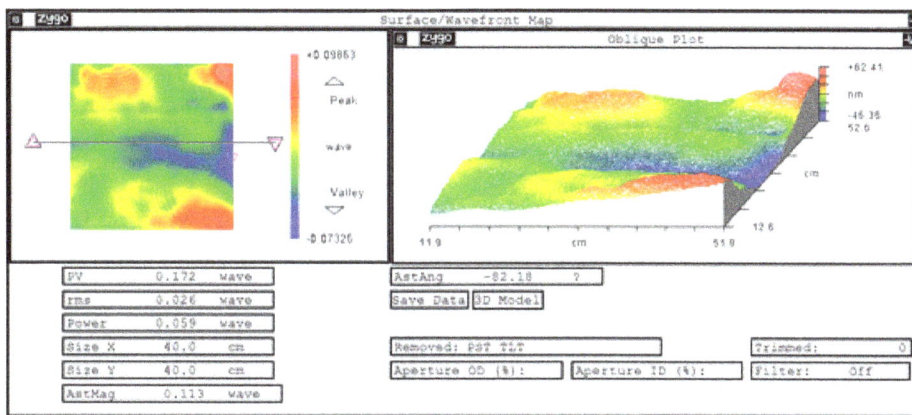

(b) the right part of 400×400 mm^2

Figure 4. 633 nm transmitted wavefront of a N3135 glass slab of size $810 \times 460 \times 43.7$ mm^3. (a) The left part of 400×400 mm^2 and (b) the right part of 400×400 mm^2.

coefficient of Cu^{2+} doped cladding glass is usually 3–5 cm^{-1} at 1053 nm to suppress the temperature rise between the interface of the cladding and laser glasses. A small temperature rise can limit the thermal stress around the interface between the cladding glass and the laser glass after a laser shot.

A patented edge cladding technique has been developed for large aperture N31 laser glass[20]. It has been applied in the fabrication of large aperture N31 laser glass slab. More than 10,000 shots of high energy fluence have been applied in our edge cladding N31 glass slab in the Shen Guang facility.

The residual stress caused by edge cladding is kept small by the proper choice of adhesive agent and its curing parameters. Figure 5 shows the averaged stress distribution within 50 mm distance to the interface of the cladding for a 400 mm aperture N31 laser glass slab before and after cladding, which was scanned with a Strain Optics DIAS 1600 stress analyzer. The test area is 70×100 mm^2 for one time scanning. In the measurement, the polarized beam is normal to the surface of the slab. The averaged result indicates that this patented cladding technique does not induce additional stress

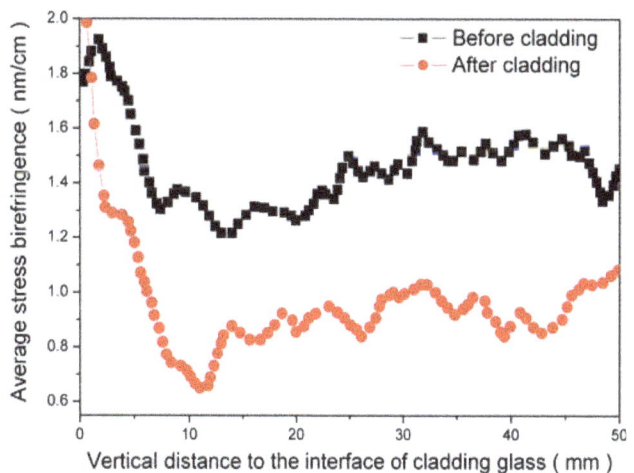

Figure 5. The stress distribution of 400 mm aperture N31 glass slab before and after cladding.

inside the neodymium glass 12 mm away from the interface between the neodymium glass and the cladding glass.

Figure 6. 400 mm aperture N31 glass slabs before installation.

Figure 6 shows 400 mm aperture N31 laser glass slabs with edge cladding after fine polishing and cleaning. These glass slabs have been successfully applied in the Shen Guang facility.

5. Conclusions

The main composition and properties of N31 Nd-doped phosphate laser glass are reported. Three key techniques for laser glass fabrication (pot melting, continuous melting, and edge cladding) have been developed in Shanghai Institute of Optics and Fine Mechanics. The glass parameters and laser gain of N3135 phosphate glass produced by continuous melting are almost the same as those obtained by pot melting. 400 mm clear aperture N31 glass slabs with high quality have been fabricated for building the Shen Guang high peak power laser facilities in China.

References

1. J. H. Campbell, and T. I. Suratwala, J. Non-Cryst. Solids **263&264**, 318 (2000).

2. E. Hand, Nature, doi:10.1038/nature.2012.10269.

3. J. H. Campbell, T. I. Suratwalaa, C. B. Thorsnessa, J. S. Haydenb, A. J. Thorneb, J. M. Ciminob, A. J. Marker IIIb, K. Takeuchic, M. Smolleyc, and G. F. Ficini-Dornd, J. Non-Cryst. Solids **263&264**, 342 (2000).

4. V. I. Arbuzov, Y. K. Fyodorov, S. I. Kramarev, S. G. Lunter, S. I. Nikitina, A. N. Pozharskii, A. V. Shashkin, A. D. Semyonov, V. E. Ter-Nersesyants, A. V. Charukhchev, V. S. Sirazetdinov, S. G. Garanin, and S. A. Sukharev, Glass Technol. **46**, 67 (2005).

5. Z. Jiang, Chin. J. Lasers **33**, 1265 (2006) (in Chinese).

6. Y. Jiang, J. Zhang, W. Xu, Z. Ma, X. Ying, H. Mao, S. Mao, and J. Li, J. Non-Cryst. Solids **80**, 623 (1986).

7. L. Hu, and Z. Jiang, Bull. Chin. Ceramics Society **25**, 125 (2005) (in Chinese).

8. L. Hu, S. Chen, T. Meng, W. Chen, J. Tang, B. Wang, J. Hu, L. Wen, S. Li, Y. Jiang, J. Zhang, and Z. Jiang, High power laser and particle beams **23**, 2560 (2011) (in Chinese).

9. V. I. Arbuzov, Yu. K. Fedorov, S. I. Kramarev, and A. V. Shashkin, J. Opt. Technol. **80**, 321 (2013).

10. J. Zhang, H. Mao, S. Chen, X. Ying, B. Wang, B. Zhou, H. Zhu, and C. Jin, Chinese patent ZL 00127617.4.

11. Y. Xu, S. Li, L. Hu, and W. Chen, Chin. Opt. Lett. **3**, 701 (2005).

12. Y. Xu, S. Li, L. Hu, and W. Chen, J. Rare Earths **29**, 614 (2011).

13. D. Zhuo, W. Xu, and Y. Jiang, Chin. J. Laser **12**, 173 (1985).

14. B. Zhou, L. Hu, Y. Jiang, H. Mao, and J. Zhang, Chin. J. Lasers **28A**, 837 (2001).

15. J. Tang, L. Hu, and F. Gan, J. Wuhan Univ. Technol. **29**, Suppl.1, 210 (2007).

16. J. Tang, L. Hu, X. Ying, C. Jin, and F. Gan, Glass & Enamel (Monograph) 30 (2007).

17. J. Tang, B. Wang, and L. Hu, Annual Reports on Inertial Confinement Fusion of China Academy of Engineering Physics (2011) 1.

18. ISO11254-1. 2000. Lasers and laser-related equipment — Determination of laser-induced damage threshold of optical surfaces - Part 1: 1-on-1 test.

19. J. Schwider, R. Burow, K.-E. Elssner, and R. Spolaczyk, J. Grzanna, Appl. Opt. **24**, 3059 (1985).

20. T. Meng, J. Tang, J. Hu, L. Wen, L. Chen, W. Chen, and L. Hu, Chinese Patent, ZL 2010 1 0273819.7.

3

Design of a kJ-class HiLASE laser as a driver for inertial fusion energy

Antonio Lucianetti[1], Magdalena Sawicka[1], Ondrej Slezak[1], Martin Divoky[1], Jan Pilar[1],
Venkatesan Jambunathan[1], Stefano Bonora[1,2], Roman Antipenkov[3], and Tomas Mocek[1]

[1] HiLASE Project, Institute of Physics AS CR, Na Slovance 2, 18221, Prague, Czech Republic

[2] CNR-IFN, Via Trasea 7, 35131, Padova, Italy

[3] ELI Beamlines Project, Institute of Physics AS CR, Na Slovance 2, 18221, Prague, Czech Republic

Abstract

We present the results of performance modeling of a diode-pumped solid-state HiLASE laser designed for use in inertial fusion energy power plants. The main amplifier concept is based on a He-gas-cooled multi-slab architecture similar to that employed in Mercury laser system. Our modeling quantifies the reduction of thermally induced phase aberrations and average depolarization in Yb^{3+}:YAG slabs by a combination of helium cryogenic cooling and properly designed (doping/width) cladding materials.

Keywords: ASE; birefringence; cryogenic cooling; slab lasers; thermooptic effects

1. Introduction

Laser-driven inertial fusion energy is one of the most promising approaches for the sustainable generation of electrical power. Research on laser-driven inertial confinement fusion (ICF) has resulted in the world's largest laser systems, such as NIF and LMJ [1, 2]. While single-shot facilities can be used to study the basic physics and technology of laser fusion, they are not applicable for the continuously operated power plants of the future. Recent studies have shown that diode-pumped solid-state lasers (DPSSLs) are the most promising laser systems to reach the requirements for such a driver, namely multi-100 kJ energy of ns pulses, multi-Hz repetition rates and high wall-plug efficiencies between 10% and 15%[3, 4]. The HiLASE team is developing an Yb^{3+}:YAG gain medium based concept for a 100 J/10 Hz DPSSL amplifier that could potentially be scaled to the kJ regime[5–7]. While there are several projects around the world that are trying to achieve the same goal[8–11], HiLASE is expected to be completed in May 2015 and it will be the world's highest pulse energy short pulse (2–10 ns) DPSSL at 100 J and 1–10 Hz.

In this paper, we examine the predicted performance of a kJ-class HiLASE laser which is based on a gas-cooled slab-stack architecture. It uses multiple thin slabs of Yb^{3+}:YAG gain medium, face-cooled with high-pressure streaming helium gas[12, 13]. The goal of this paper is to provide an amplifier design for the kJ-class laser architecture that minimizes accumulated nonlinear phase (B integral), thermally induced wavefront aberrations and stress birefringence. In Section 2, we describe our model for the calculation of output energy. In Section 3, we describe the HiLASE laser concept and show calculations performed using our energetics modeling. In Section 4, we present the results from the MIRO model used to determine the temporal profile of the output beam and the evolution of the amplified beam as it propagates through the optical system. In Section 5, we show thermal modeling results, including optimization of optical path difference (OPD) and thermally induced stress birefringence. In Section 6, we present the optimized structure parameters of deformable mirrors (DMs) for wavefront correction. Finally, we present our modeling results for frequency conversion in Section 7.

2. Energetics modeling

To quantitatively assess energy storage and amplified spontaneous emission (ASE) losses within the laser active material in multi-slab geometry, a numerical model has been developed[5]. Figure 1 shows a flow chart of the code. At the beginning, the parameters of the amplifier head are specified including the temperature of operation. The population of each laser level is calculated from the Boltzmann

Correspondence to: Antonio Lucianetti, Institute of Physics, AS CR, Na Slovance 2, 182 21 Prague, Czech Republic. Email: lucianetti@fzu.cz

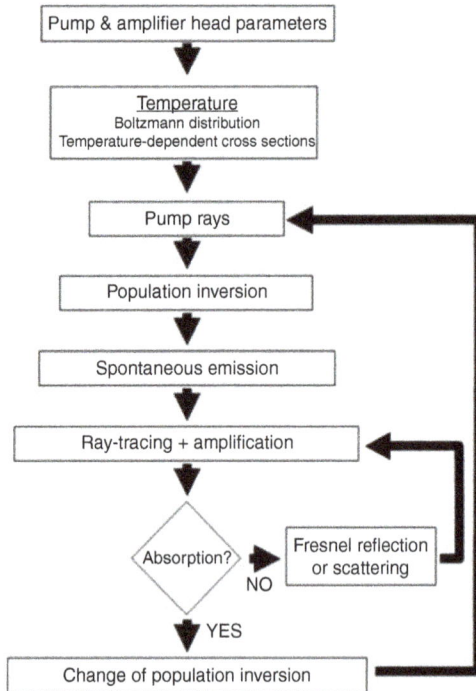

Figure 1. A schematic flow diagram of the model.

distribution. For a given temperature, wavelength-dependent emission and absorption cross sections are uploaded from the database. Each slab is divided into pixels, which contain information about the total number of active ions. The information about the excited fraction of ions is also stored in the memory. The slabs will absorb polychromatic pump radiation proportionally to the number of non-excited active ions and the wavelength resolved absorption cross-section.

Based on the excited ion density, the spontaneously emitted photons are generated in the form of rays. Each ray is traced through the slab and the number of photons is changed after the ray undergoes amplification or absorption. At the same time, the excited ion density in appropriate cells is also recalculated. Photons that are not absorbed in the cladding

propagate up to the side edge of the slab. The probability of reflection at the edge is calculated from the Fresnel equations for the incident angle of the ray. If photons are reflected back, they pass through the absorptive layer again and, if they are not absorbed, they propagate back to the gain medium and decrease the stored energy[6].

3. Amplifier concept and simulations

For our kJ-class HiLASE laser, we have chosen to model an amplifier with the characteristics derived from the previous ASE and thermal study[6, 7], i.e., eight 1 cm thick Yb^{3+}:YAG slabs with transverse dimensions of 14 cm $\times$ 14 cm. A 2 cm wide Cr^{4+}:YAG absorptive cladding ($a = 1.1$ cm^{-1} at 1030 nm) around the edge of each slab was also included to further suppress ASE and prevent unwanted parasitic oscillations. The edges of the cladding were modeled as roughened surfaces which scatter rays. The optimized doping concentrations for the eight-slab amplifier were 0.29, 0.38, 0.56, and 0.85 at.%. The super-Gaussian pump dimensions were kept at 14 cm $\times$ 14 cm which corresponded to a total pump area of 196 cm^2. The operating temperature was allowed to vary between 160 and 240 K. Two different beamline concepts are considered here. In the first concept, the output from a low-energy preamplifier (0.2–5 J) is sent to the main kJ amplifier consisting of two identical heads (Figure 2).

In the second design, the laser beam is preamplified up to a 100 J level and then sent to a single kJ laser head (Figure 3).

To find out how the energy is stored in the amplifier, a time-resolved calculation was conducted. In this case, the two-head design was considered. Figures 4–6 show the extractable energy as a function of time for different pump intensities at temperatures of 160, 200, and 240 K.

Large values of the ASE saturate the gain and the pump duration of 1 ms is too long for effective energy storage at low temperatures (see Figure 4). It is noted that an extractable energy of more than 1 kJ can be obtained for pump durations of 1 ms and temperatures greater than

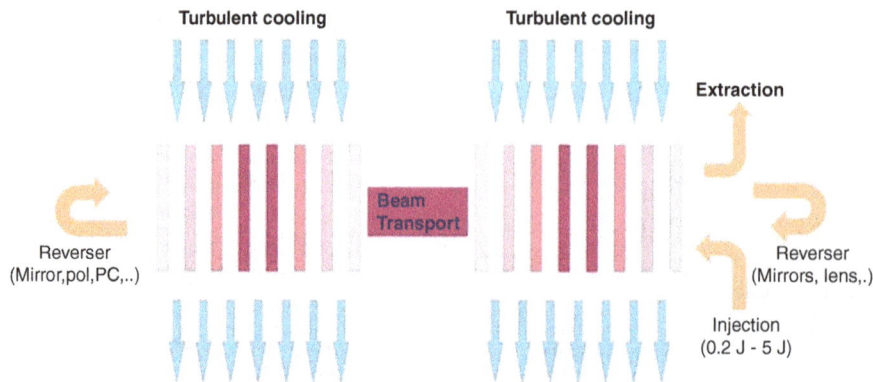

Figure 2. Block diagram of the HiLASE kJ laser (two-head configuration).

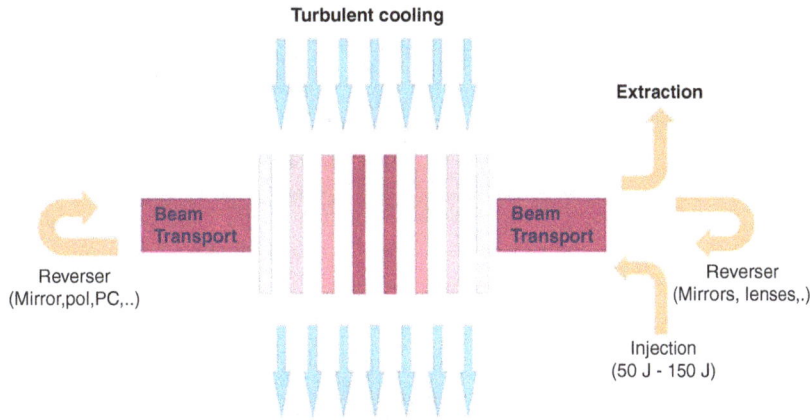

Figure 3. Block diagram of the HiLASE kJ laser (single-head configuration).

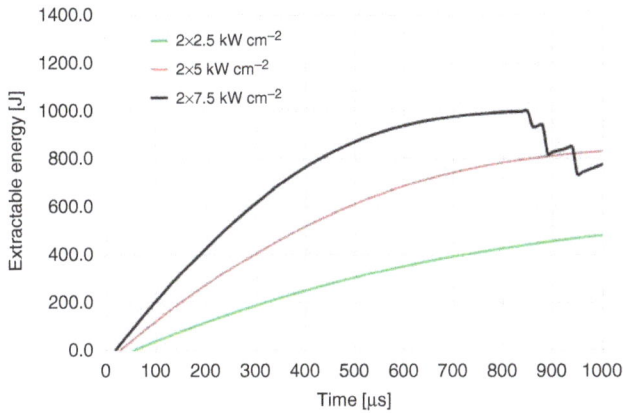

Figure 4. The time-resolved extractable energy in the HiLASE slab for different pump intensities ($T = 160$ K).

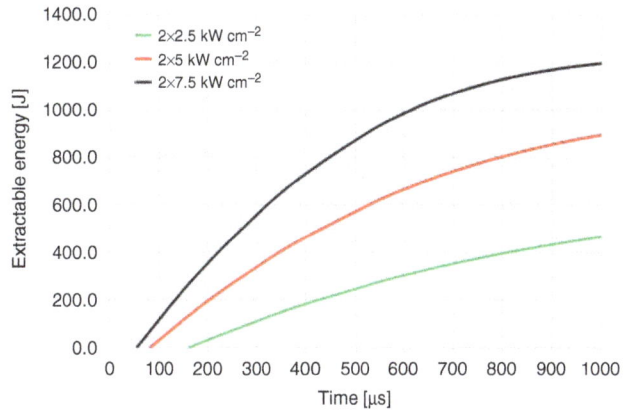

Figure 5. The time-resolved extractable energy in the HiLASE slab for different pump intensities ($T = 200$ K).

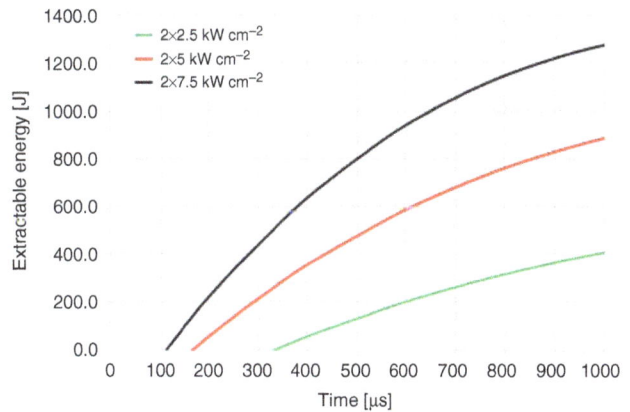

Figure 6. The time-resolved extractable energy in the HiLASE slab for different pump intensities ($T = 240$ K).

200 K. In addition, there is an optimal temperature for which the extractable energy is the highest. For higher pump intensities, the excitation of the gain medium is higher and the temperature-dependent emission cross section plays a significant role. If the temperature is increasing, the emission cross section of the Yb:YAG is decreasing and therefore the ASE losses are also decreasing. Therefore, the optimal temperature is shifting towards higher temperatures with increasing pump intensity. The extractable energy is the stored energy minus the energy bound to the lower laser level due to the three-level nature of the Yb^{3+}:YAG (energy needed for the gain medium to stay transparent). The extractable energy as a function of temperature was calculated for different pump intensities for a single amplifier head (Figure 7).

The storage efficiency was calculated as the ratio between the absorbed energy and the extractable energy (Figure 8). Higher pump intensity causes higher ASE losses. Therefore, at lower temperatures the storage efficiency decreases with increase of pump intensity.

Based on the previous results, an operating temperature of 200 K and a pump duration of 1 ms were selected. The following figures give an indication of the evolution of the energy generated after each pass through the pair of amplifiers (Figures 9 and 10) or single amplifier (Figures 11 and 12) at 200 K.

It is noted that an output energy of 0.92 kJ can be reached in the single-head design with reasonably low values of optical losses (10%) and an operating temperature of 200 K.

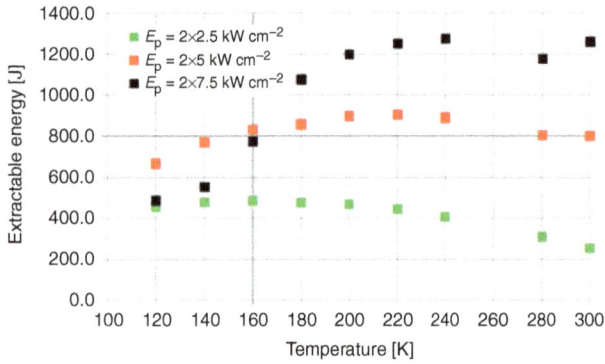

Figure 7. The extractable energy as a function of the operating temperature for different pump intensities.

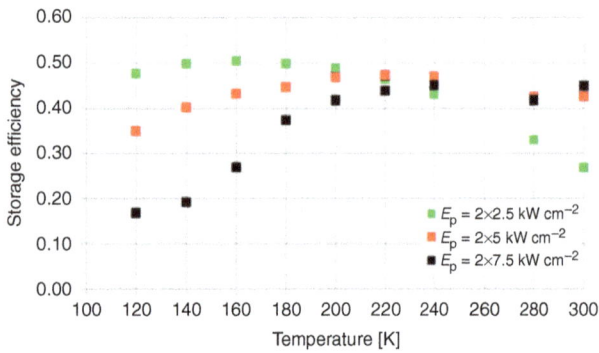

Figure 8. The storage efficiency as a function of the operating temperature for different pump intensities.

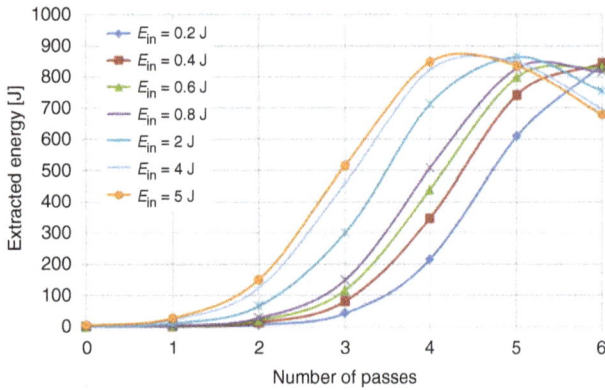

Figure 9. The evolution of the extracted energy for different input energies at 200 K (two heads, 20% optical losses per round trip pass). The total pump intensity was 2×10 kW cm^{-2}.

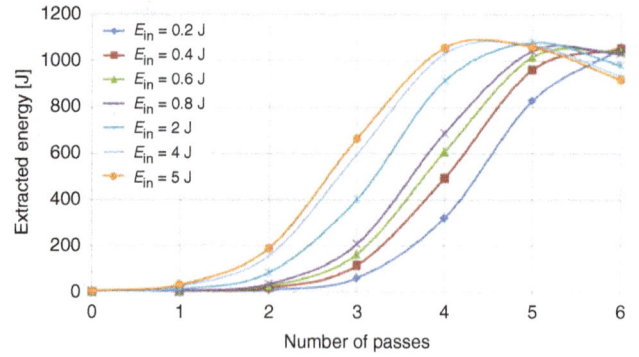

Figure 10. The evolution of the extracted energy for different input energies at 200 K (two heads, 16% optical losses per round trip pass). The total pump intensity was 2×10 kW cm^{-2}.

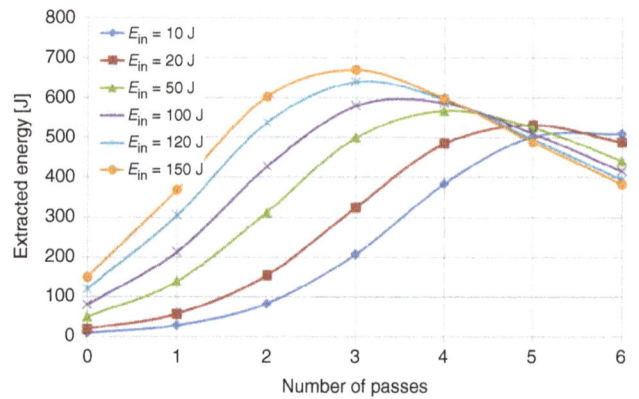

Figure 11. The evolution of the extracted energy for different input energies at 200 K (one head, 18% optical losses per round trip pass). The total pump intensity was 15 kW cm^{-2}.

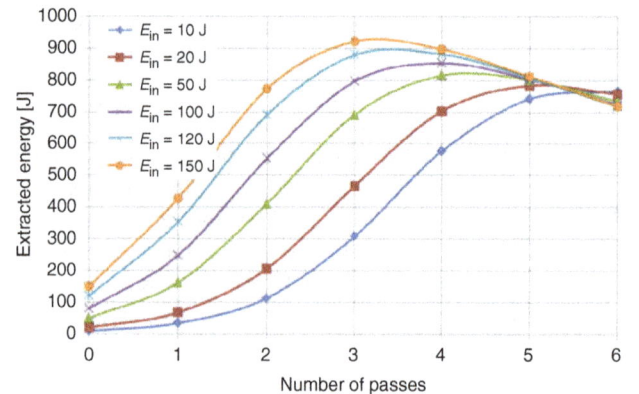

Figure 12. The evolution of the extracted energy for different input energies at 200 K (one head, 10% optical losses per round trip pass). The total pump intensity was 15 kW cm^{-2}.

In this case, the input beam can be provided by next-generation high-energy-class (HEC)-DPSSL facilities with output energies in the range of 100–150 J[8, 12, 13].

4. Beam propagation

In order to carry out fundamental calculations on how a HiLASE kJ laser would operate, a MIRO model has been constructed for the two-head configuration. These calculations are based on previous modeling results for the two-head design for a 100 J-class HiLASE amplifier[12]. We assumed that all surfaces are anti-reflection coated with a reflectivity of 0.5%, and all optical elements are made of

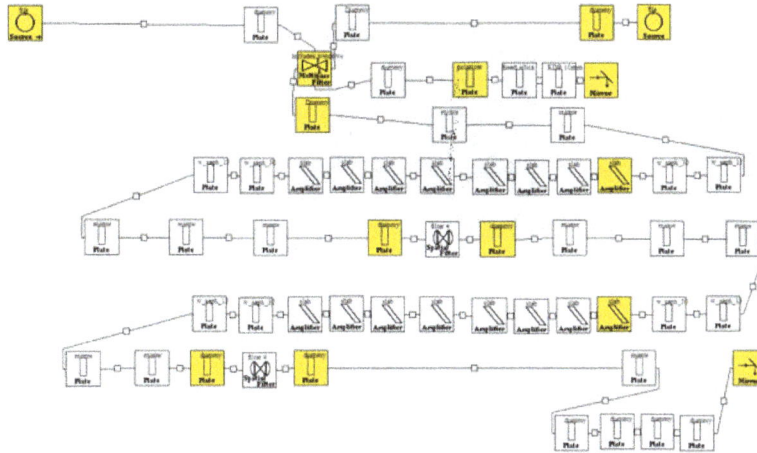

Figure 13. The MIRO model used to calculate the temporal shape, spatial shape, and B integral of the HiLASE kJ laser.

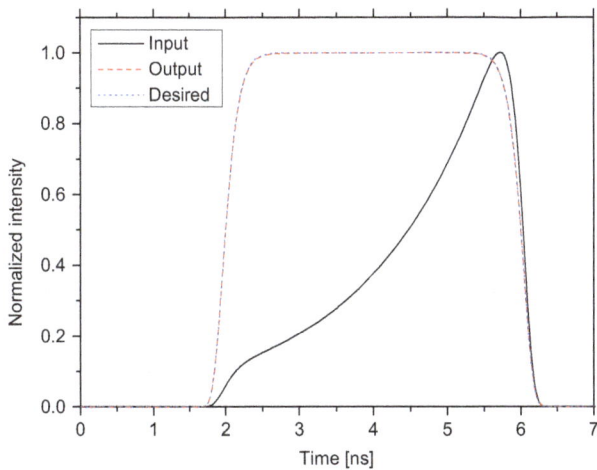

Figure 14. Input, output, and desired temporal profiles of the MIRO model for the HiLASE kJ laser.

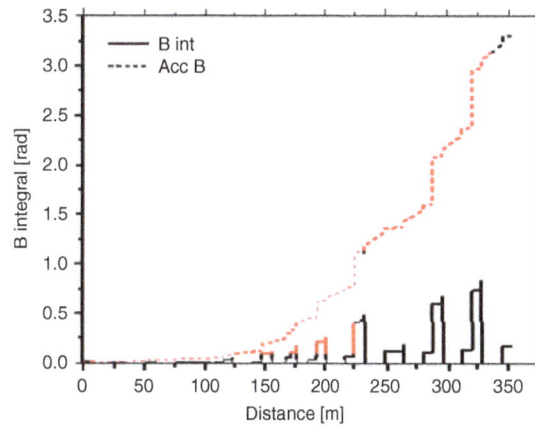

Figure 15. The evolution of the B integral and accumulated B integral upon beam propagation in the HiLASE kJ laser.

appropriate material, i.e., fused silica ($n_2 = 3e^{-20}$ m^2 W^{-1}) for the lenses and windows of the spatial filters, DKDP ($n_2 = 10e^{-20}$ m^2 W^{-1}) for the Pockels cell, Yb^{3+}:YAG ($n_2 = 7e^{-20}$ m^2 W^{-1}) for the laser slabs, and sapphire ($n_2 = 3e^{-20}$ m^2 W^{-1}) for the amplifier head windows. The internal transmission is assumed to be 100% for the lenses, windows, and slabs and 99% for the DKDP Pockels cells. The graphical representation of the MIRO model is shown in Figure 13.

The quality of each component corresponded to $\lambda/10$ at 1030 nm. The calculation was run with a spatial resolution of 50 μm and a temporal resolution of 30 ps. The MIRO model included only thermal aberrations. The aim of the calculation was to optimize the input pulse shape as well as to evaluate diffraction and non-linear phase accumulation (breakup integral) in the beam. The initial modeling concentrated on achieving a top-hat temporal output pulse profile. The input pulse profile was obtained by an iteration method. At each

step, the input pulse was multiplied by the ratio between the output pulse and the desired output pulse. The optimized input and output temporal profiles of the MIRO model are shown in Figure 14.

Advanced temporal pulse shaping is therefore required on the front end seed source to achieve the desired top-hat pulse profile. Two B integral types are calculated with MIRÓ. The first B integral is reset to zero in each spatial filter where the beam passes through. The second is the accumulated B integral which is the sum of B integrals in all sections between spatial filters. For an input pulse energy of 4 J and a pump intensity of 10 kW cm^{-2}, the evolution of the B integral and the accumulated B integral is shown in Figure 15 after four passes. It is noted that the maximum values of the B integrals are calculated to be 0.75 and 3.3, respectively.

Figures 16(a)–(c) show the beam profile, the phase and the phase after subtraction of defocus and tilt in the MIRO model after four passes at 1.1 kJ.

The diffraction pattern in the beam is caused by shift of the beam due to large aberrations in the system and clipping of

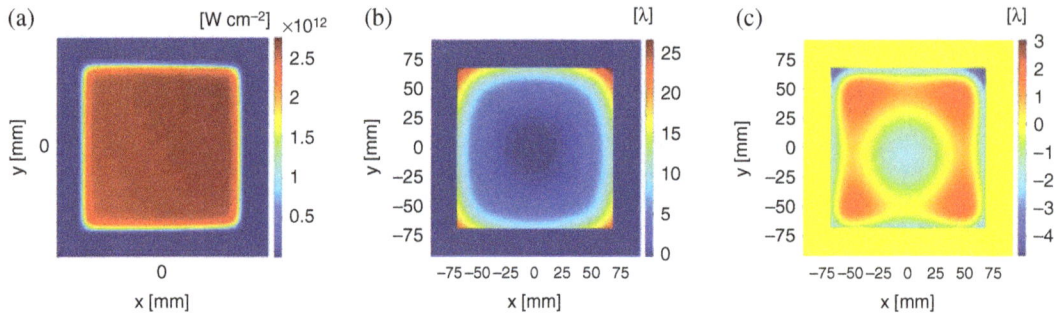

Figure 16. (a) Beam profile, (b) and (c) phase after subtraction of defocus and tilt of the output beam.

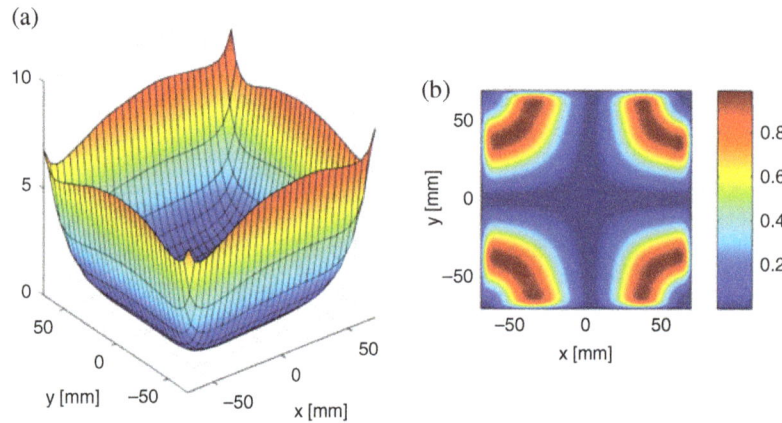

Figure 17. (a) The stress- and temperature-induced OPD after a single pass through the laser head (after one pass through eight slabs). (b) The depolarization of the beam after a single pass through the head caused by stress-induced birefringence. The Cr^{4+}:YAG cladding thickness was 20 mm.

the beam. This can be remedied by decreasing the beam size considerably or by adding wavefront correction after each pass by implementing a DM in the laser chain (like NIF and LMJ). An optimized MIRO model which also includes the manufacturing defects of the optical components will be required in the future to provide a detailed understanding of the operational parameters of an optimized HiLASE laser design.

5. Thermal modeling

Substantial analysis has been performed in order to optimize the Yb:YAG gain medium size, coolant flow rates, arrangement of pumped and unpumped regions, and absorbing materials for properly designed (doping/width) cladding. The laser slabs are cooled by forced flow of He gas with a temperature of 190 K and a pressure of 5 bar. The model assumes that the energy deposition is the same for all slabs, which is a fair approximation given the stepped doping profile. A 3D finite-element method (FEM) using Comsol Multiphysics software was chosen to model the thermal effects in the amplifiers and the fluid dynamics of the helium flow. The calculation takes into account the

temperature changes of the material parameters[14–16], the turbulent flow of the He and the resulting spatial variation of the heat transfer coefficient. The model is then used to calculate all mechanical stresses in the laser slab and birefringence depolarization losses for eight slabs according to the approach described in [17]. The stress- and temperature-induced OPD and depolarization after one pass through eight slabs (i.e. one laser head) are shown in Figure 17. This figure shows a maximum OPD value of 8.9 waves and an average depolarization of 36.6%.

It is found that the thermal OPD and average depolarization can be substantially reduced by increasing the size of the cladding, and even more drastically by inserting a thin layer of undoped YAG around the gain medium[7] with the two-stepped doping profile of the Cr^{4+}:YAG cladding. The geometry and zone layout used for heat deposition modeling in the HiLASE amplifier (single and double clad configurations) are shown in Figure 18.

Table 1 summarizes the results of the three different layouts under consideration, i.e., single, enlarged single and double clad geometries. The enlarged single geometry had a cladding width of 53 mm, i.e., the same total cladding width as the double clad geometry consisting of a 3 mm

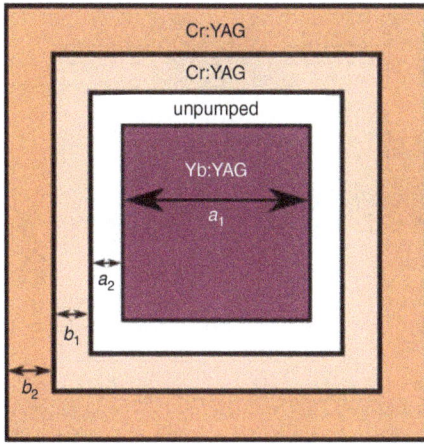

Figure 18. The geometry and zone layout used for heat deposition modeling in the HiLASE amplifier slab.

Table 1. The Gain Medium and Cladding Dimensions used for Simulation of HiLASE Square Amplifiers.

Geometry	Yb^{3+}:YAG	Undoped/ unpumped	Cr^{4+}:YAG	Absorption coefficient
	a_1 (mm)	a_2 (mm)	b_1/b_2 (mm)	α_1/α_2 (cm^{-1})
Single clad	140	0	20/0	1.15/0
Enlarged single clad	140	0	53/0	0.43/0
Double clad	140	3	25/25	0.24/0.72

Table 2. Thermal Results for HiLASE Square Amplifiers (Single, Enlarged Single, and Double Clad).

	Single clad	Enlarged single clad	Double clad
T_{max} (K)	213.4	203.5	199.3
$\langle T \rangle$ (K)	202.1	197.5	196.1
γ (%) – eight slabs	36.3	3.20	0.079
OPD (λ) – eight slabs	8.96	2.97	1.54
OPD w/o TD (λ) – eight slabs	6.85	2.31	0.49

layer of undoped YAG and two 25 mm layers of cladding with different doping levels.

Figure 19 shows the total OPD and depolarization due to eight slabs in the double clad geometry.

Table 2 summarizes the thermal results for the single, enlarged single and double clad geometries. These results include the maximal temperature reached within the slab volume, T_{max}, the average temperature, $\langle T \rangle$, the total depolarization loss, γ, due to eight slabs (one amplifier head), the peak-to-valley (P-V) value of the OPD due to eight slabs, and the total eight-slab OPD with tilt and defocus subtracted.

The double clad geometry showed the minimum depolarization losses and OPD values. It should be noted that the tilted OPD profile in the double clad geometry is due to the helium gas flow which generates a transverse dependence of the heat transfer coefficient. However, the magnitude of the aberrations is well within the correction capability of the current DMs used for adaptive optics (AO)[18]. These results suggest that a cryogenic helium gas approach coupled with properly designed (doping/width) cladding materials should

be able to provide sufficient cooling capacity while introducing minimal optical distortions and thermal depolarization for operation of a kJ-class amplifier.

6. Wavefront correction

The calculated wavefront of the beam at the output of the laser system (see Figure 16) was corrected by a numerical model of a DM. We consider the worst condition for the amplifier of placing the DM after the last pass. The reliability of the wavefront correction code was verified experimentally in a slab simulator[18]. The simulation was performed for a variable number of actuators from 5×5 to 8×8. The numerical model for wavefront correction calculates influence functions from a plate equation describing bending of the thin face sheet for each individual actuator of the DM. Figure 20 shows the actuator layout of the DM, where 'c'

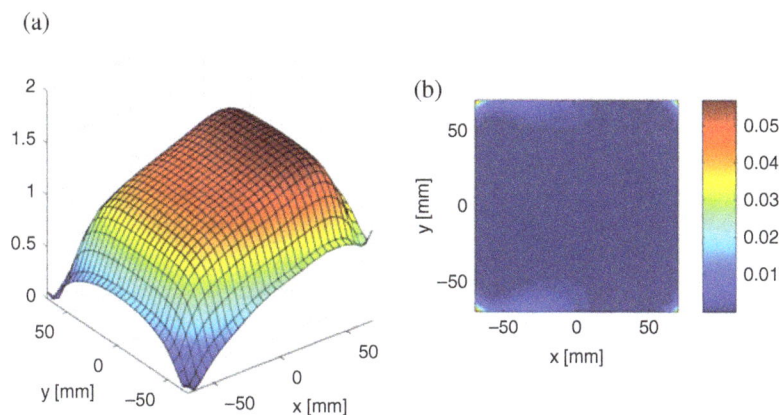

Figure 19. (a) The calculated OPD and (b) the depolarization loss due to eight slabs. A 3 mm layer of undoped YAG and two 25 mm Cr:YAG layers of cladding with different doping levels were added around the gain medium.

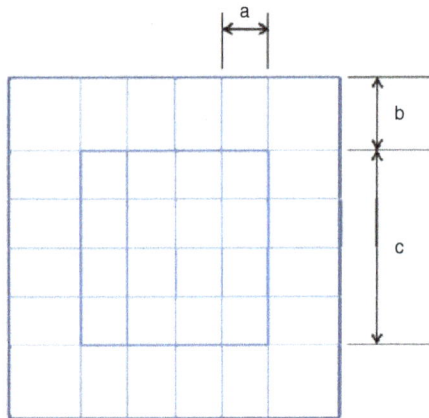

Figure 20. The actuator layout of the DM.

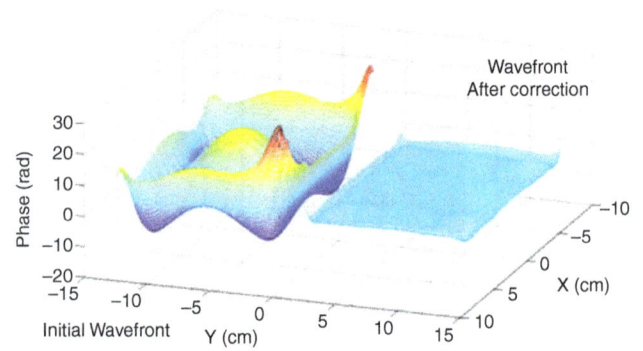

Figure 22. (a) The output wavefront calculated in MIRO and shown in Figure 16(a) after subtraction of tilt and defocus. (b) The residual wavefront after correction by the DM with 8 × 8 actuators (b/c = 0.43, stroke = 12 μm).

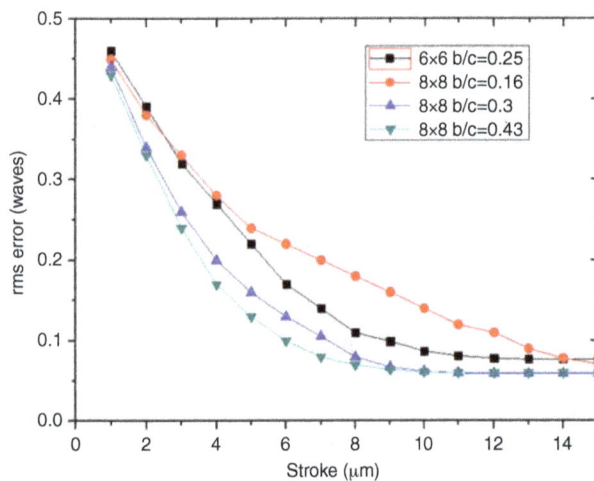

Figure 21. Residual rms values of the OPD as a function of the stroke after correction by the DM.

is the size of the active region (or beam size) which is also characterized by the size 'a' of the actuators. The results have been calculated for different values of the b/c ratio and stroke of the DM.

The deformation of the mirror is computed as a superposition of influence functions and the algorithm minimizes the OPD rms value. Figure 21 shows the residual rms values of the OPD as a function of the stroke after correction by the DM with 6 × 6 and 8 × 8 actuators.

It is noted that the residual rms value after correction is more influenced by the b/c ratio than by the actuator density. Therefore, optimization of the DM requires large-area actuators outside the active region, i.e. large b/c values rather than high actuator density. It should be noted that the larger the stroke is, the smaller the dynamic range used by the AO system will be. The DM is able to achieve numerical correction of the initial aberrations, as shown in Figure 22 (b/c = 0.43, stroke = 12 μm). The OPD after subtraction of defocus and tilt and the OPD corrected by the DM are shown

in Figure 22. The rms value of the OPD was reduced from 0.56 waves down to 0.070 waves.

The corresponding far-field images before and after wavefront correction are shown in Figures 23(b) and (c). Figure 23(a) shows the far field with an ideal flat wavefront. The Strehl ratios of the aberrated and corrected beams are 0.083 and 0.971, respectively.

7. Frequency conversion

Frequency conversion of the fundamental wavelength at 1030 nm to the second and third harmonic wavelengths of 515 and 343 nm allows for more efficient absorption of laser energy by the deuterium–tritium target. In this section, we present the modeling results for the second harmonic generation (SHG) and third harmonic generation (THG) conversion efficiency. It is proposed to use LBO crystal for frequency conversion due to its excellent nonlinear properties and recently demonstrated large crystal sizes[19]. RTP is also an interesting material for frequency conversion[20]. To the best of our knowledge, however, no multi-J/few-Hz operation of large-size RTP crystals has been reported so far. In addition, green absorption in RTP is much stronger than in LBO. Although the size of the LBO crystals that has been used in our calculations exceeds the currently available size, we believe that the required apertures will become available in the near future. The numerical simulations were performed using home-written code based on a three-wave interaction model, in which the symmetrized split-step method was used for calculations. A grid of 128 × 128 × 128 points was used for the 3D simulations. The spatial profile was assumed to be square shaped with dimensions of 140 mm × 140 mm. The intensity profile in the simulations was assumed to be ideal, i.e., to be defined by a super-Gaussian function of the 20th order for the spatial profile and of the eighth order for the temporal profile. To accommodate such a laser beam comfortably, the aperture of the LBO crystal

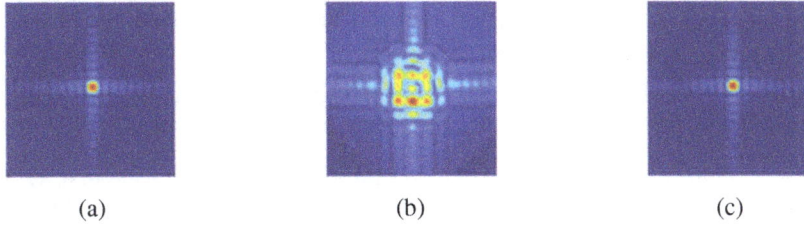

Figure 23. (a) Far field with ideal flat wavefront. (b) Far-field image before correction by the DM. (c) Far-field image after correction with 8×8 actuators (b/c = 0.43, stroke = 12 μm).

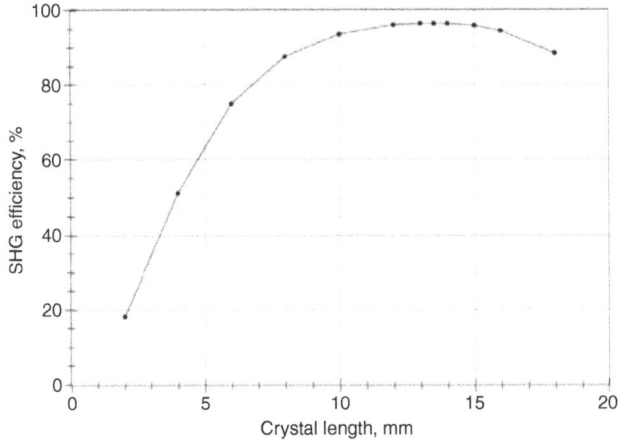

Figure 24. The SHG efficiency for different LBO thickness values.

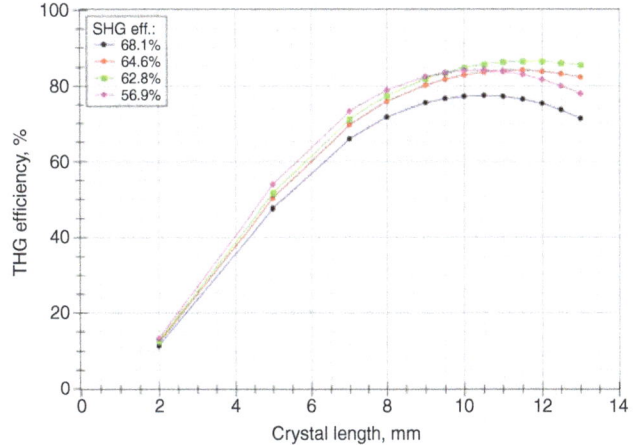

Figure 25. The THG efficiency for different LBO thickness values.

was set to 160 mm × 160 mm. Since the shape of the pulse ideally ensures uniform intensity distribution in time and space, very high conversion efficiencies are easily achievable if significant wavefront distortions are absent. Figure 24 shows that SHG can achieve up to 90% conversion efficiency and even higher. In this calculation, the LBO crystal was oriented for type-I SHG, XY-plane, $\varphi = 13.78°$. We have assumed a fundamental wavelength pulse energy of 1 kJ, which corresponds to 1.32 GW cm^{-2} peak intensity.

As can be seen from Figure 24, the optimal LBO crystal thickness is close to 13 mm, with a theoretical conversion efficiency of more than 95%. For the case of THG, we investigated the most common setup where the second harmonic is first generated in LBO crystal (type-I, ~5 mm) with ~60% efficiency. Then the pulses of both the first harmonic (FH) and the second harmonic are mixed in a second LBO crystal (type-II, o + e = o) for sum-frequency generation (SFG). The second LBO crystal was oriented for type-II nonlinear interaction, YZ plane, $\theta = 50.1°$. The total THG conversion efficiency is shown in Figure 25. Several curves, representing different SHG conversion efficiencies in the first stage of the setup, are presented. In the optimized case, the maximum third harmonic output is obtained at ~63% SHG conversion.

Angular and temperature acceptance are important parameters in the case of high energy and high average power

operation. In the case of LBO type-I SHG at 1030 nm, we have calculated the angular acceptance using the Sellmeier equations to be 3.42 mrad cm (which yields 2.63 mrad for 13 mm crystal). For the THG via SFG in the given conditions, the angular acceptance was calculated to be 3.09 mrad cm (2.38 mrad for 13 mm crystal). For the temperature acceptance calculation, we currently rely on values provided by SNLO software, recalculated for the full width at half maximum (FWHM) of the sinc$^2(\Delta kL/2)$ function: 6.36 K cm for SHG and 3.19 K cm for THG (for 13 mm crystal 4.89 and 2.45 K accordingly). The acceptance values calculated above are useful for the comparison of different nonlinear crystals. However, in order to evaluate the actual angular tolerance of the crystal for a given pulse, we have also performed a numerical full 3D simulation for a set of angles in the vicinity of optimal phase matching. The simulation shows that the crystal detuning tolerance in this case is 0.93 mrad. We define this number as the FWHM level for the conversion efficiency function. The same simulation was performed for the second LBO crystal of 11.5 mm length, in which the third harmonic was generated via SFG, and it was shown that the crystal detuning tolerance in this case was 1.57 mrad. These simulations show that the output efficiency is more sensitive to angular detuning in the case of high efficiency of the frequency conversion processes and fundamental pulse depletion, in comparison to angular acceptance values calculated from the Sellmeier equations. It is noted that when phase and amplitude distortions become significant, the efficiency of the nonlinear process is expected to decrease. A more extensive model of the frequency convertor which includes amplitude or phase noise will be developed.

Acknowledgements

This work benefitted from the support of the Czech Republic's Ministry of Education, Youth and Sports to the HiLASE (CZ.1.05/2.1.00/01.0027), DPSSLasers (CZ.1.07/2.3.00/20.0143), and Postdok (CZ.1.07/2.3.00/30.0057) projects, co-financed by the European Regional Development Fund. This research was supported by grant RVO 68407700. The authors would like to thank Dr. Audrius Zaukevicius from Vilnius Unversity for kindly providing his code for nonlinear interaction simulations.

References

1. C. A. Haynam, P. J. Wegner, J. M. Auerbach, M. W. Bowers, S. N. Dixit, G. V. Erbert, G. M. Heestand, M. A. Henesian, M. R. Hermann, K. S. Jancaitis, K. R. Manes, C. D. Marshall, N. C. Mehta, J. Menapace, E. Moses, J. R. Murray, M. C. Nostrand, C. D. Orth, R. Patterson, R. A. Sacks, M. J. Shaw, M. Spaeth, S. B. Sutton, W. H. Williams, C. C. Widmayer, R. K. White, S. T. Yang, and B. M. Van Wonterghem, Appl. Opt. **46**, 3276 (2007).

2. J. Ebrardt and J. M. Chaput, J. Phys. Conf. Ser. **244**, 032017 (2010).

3. J.-C. Chanteloup, D. Albach, A. Lucianetti, K. Ertel, S. Banerjee, P. D. Mason, C. Hernandez-Gomez, J. L. Collier, J. Hein, M. Wolf, J. Körner, and B. J. L. Garrec, J. Phys. Conf. Ser. **244**, 012010 (2010).

4. A. C. Erlandson, S. M. Aceves, A. J. Bayramian, A. L. Bullington, R. J. Beach, C. D. Boley, J. A. Caird, R. J. Deri, A. M. Dunne, D. L. Flowers, M. A. Henesian, K. R. Manes, E. I. Moses, S. I. Rana, K. I. Schaffers, M. L. Spaeth, C. J. Stolz, and S. J. Telford, Opt. Mat. Express **1**, 1341 (2011).

5. M. Sawicka, M. Divoky, J. Novak, A. Lucianetti, B. Rus, and T. Mocek, J. Opt. Soc. Am. B **29**, 1270 (2012).

6. M. Sawicka, M. Divoky, A. Lucianetti, and T. Mocek, Laser Part. Beams **31**, 553 (2013).

7. O. Slezak, A. Lucianetti, M. Divoky, M. Sawicka, and T. Mocek, IEEE J. Quant. Electron. **49**, 960 (2013).

8. T. Novo, D. Albach, B. Vincent, M. Arzakantsyan, and J.-C. Chanteloup, Opt. Express **21**, 855 (2013).

9. S. Banerjee, K. Ertel, P. D. Mason, P. J. Phillips, M. Siebold, M. Loeser, C. Hernandez-Gomez, and J. L. Collier, Opt. Lett. **37**, 2175 (2012).

10. T. Kawashima, T. Ikegawa, J. Kawanaka, N. Miyanaga, M. Nakatsuka, Y. Izawa, O. Matsumoto, R. Yasuhara, T. Kurita, T. Sekine, M. Miyamoto, H. Kan, H. Furukawa, S. Motokoshi, and T. Kanabe, J. Phys. IV France **133**, 615 (2006).

11. J. Hein, M. C. Kaluza, R. Bodefeld, M. Siebold, S. Podleska, and R. Sauerbrey, In *Lasers and Nuclei*, H. Schwoerer, J. Magill and B. Beleites, eds. (Springer, Berlin, 2006), p. 47.

12. M. Divoky, P. Sikocinski, J. Pilar, A. Lucianetti, M. Sawicka, O. Slezak, and T. Mocek, Opt. Eng. **52**, 064201 (2013).

13. K. Ertel, S. Banerjee, P. D. Mason, P. J. Phillips, M. Siebold, C. Hernandez-Gomez, and J. C. Collier, Opt. Express **19**, 26610 (2011).

14. R. L. Aggarwal, D. J. Ripin, J. R. Ochoa, and T. Y. Fan, J. Appl. Phys. **98**, 103514 (2005).

15. H. Yagi, K. Takaichi, K. Ueda, Y. Yamasaki, T. Yanagitani, and A. A. Kaminskii, Las. Phys. **15**, 1338 (2005).

16. K. Ueda, J.-F. Bisson, H. Yagi, K. Takaichi, A. Shirakawa, T. Yanagitani, and A. A. Kaminskii, Las. Phys. **15**, 927 (2005).

17. A. L. Bullington, S. B. Sutton, A. J. Bayramian, J. A. Caird, R. J. Deri, A. C. Erlandson, and M. A. Henesian, Proc. SPIE **7916**, 79160V (2011).

18. J. Pilar, M. Divoky, P. Sikocinski, V. Kmetik, O. Slezak, A. Lucianetti, S. Bonora, and T. Mocek, Proc. SPIE **8780**, 878011 (2013).

19. Z. Hu, Y. Zhao, Y. Yue, and X. Hu, J. Cryst. Growth **335**, 133 (2011).

20. Y. S. Oseledchik, A. I. Pisarevzsy, A. L. Prosvirnin, V. V. Starshenko, and N. V. Svitanko, Opt. Mater. **3**, 237 (1994).

4

Rapid growth of a large-scale (600 mm aperture) KDP crystal and its optical quality

Guohui Li, Guozong Zheng, Yingkun Qi, Peixiu Yin, En Tang, Fei Li, Jing Xu, Taiming Lei, Xiuqin Lin, Min Zhang, Junye Lu, Jinbo Ma, Youping He, and Yuangen Yao

Fujian Institute of Research on the Structure of Matter, Chinese Academy of Sciences, Fuzhou 350002, China

Abstract

Potassium dihydrogen phosphate (KDP) single crystals are the only nonlinear crystals currently used for electro-optic switches and frequency converters in inertial confinement fusion research, due to their large dimension and exclusive physical properties. Based on the traditional solution-growth process, large bulk KDP crystals, usually with sizes up to 600×600 mm^2 so as to make a frequency doubler for the facility requirement loading highly flux of power laser, can be grown in standard Holden-type crystallizers, without spontaneous nucleation and visible defects, one to two orders of magnitude faster than by conventional methods. Pure water and KDP raw material with a few ion impurities such as Fe, Cr, and Al (less than 0.1 ppm) were used. The rapid-growth method includes extreme conditions such as temperature range from 60 to 35°C, overcooling up to 5°C, growth rates exceeding 10 mm/day, and crystal size up to 600 mm. The optical parameters of KDP crystals were determined. The optical properties of crystals determined indicate that they are of favorable quality for application in the facility.

Keywords: large-scale KDP crystals; rapid growth; optical quality

1. Introduction

Laser fusion systems such as the National Ignition Facility (NIF) require large-aperture crystals (410×410 mm) of KDP for optical switching and frequency conversion. To achieve economically useful yields, crystals grown for the NIF should exceed 53 cm in all three dimensions. The growth period of KDP crystal is about one year by the conventional lowering temperature method, which results in higher cost due to long period of growth. For this reason, new fast-growth techniques have been developed recently[1, 2].

Conventional crystal growth uses a temperature reduction method, but the growth rate is only 1–2 mm/d, and large vessels with capacity more than 1000 l are needed for large-size crystals. The rapid growth technique has been developed greatly in recent years, and the growth rates can be more than an order of magnitude larger than those obtained with the traditional techniques. Initiated in the early 1980s by the need for large-aperture single-crystal plates in Nova and Nova's successor, NIF, Zaitseva *et al.* of LLNL grew large-scale (40–55 cm) KDP crystals at rates of 10–20 mm/d, the

rapid growth method is based on the use of a 'point seed'[3]. Nakatsuka *et al.* used external energy to grow KDP crystals of 60 mm in size at high rates, in excess of 50 mm/d[4]. It is shown that rapidly grown crystals can have the same high optical quality as of those typically grown by the traditional technique[5, 6].

The 'point seed' in the Z cutting direction, with a size of $1 \times 1 \times 1$ cm^3, grows uniformly along the three directions (X, Y, and Z axes) under the condition of high supersaturation. The rapid crystal growth needs a higher-level stable solution without spontaneous crystallization. The level of solution stability depends on many different factors such as highly pure raw material, ultra fine filtration, overheating at high temperature, rate of temperature reduction, hydrodynamics of the crystallizer, and impurity content of solution, which actively keep the solution from spontaneous nucleation at high overcooling.

This paper describes a successful design for a large crystallizer of a metric tonnage volume at high supersaturation; the growth parameters of a real growth process in the crystallizer are summarized. This paper reports the spectral transmittance and the laser damage threshold of the rapidly grown crystal. The optical qualities of the KDP crystals grown rapidly are not significantly different from those of KDP crystals grown by the traditional method.

Correspondence to: Guohui Li, Fujian Institute of Research on the Structure of Matter, Chinese Academy of Sciences, Fuzhou 350002, China. Email: igh@fjirsm.ac.cn

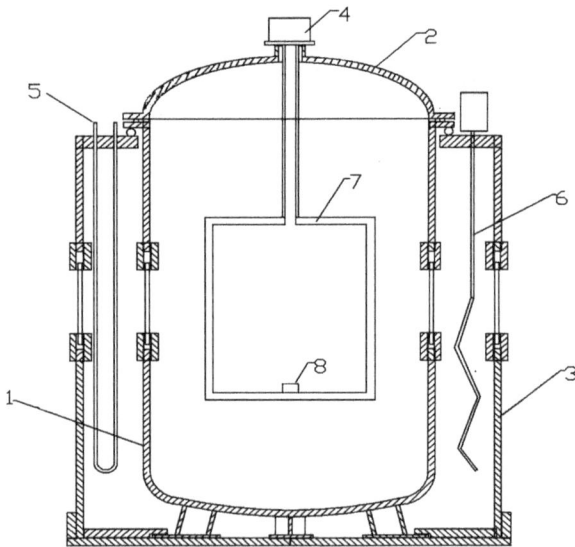

Figure 1. 2000 L crystallizer. 1-stainless steel crystallizer; 2-vessel cover; 3-thermostat; 4-rotational motor; 5-heater; 6-stirrer; 7-aluminum platform; 8-point seed.

2. Experimental

2.1. Apparatus

This experiment was carried out in a 2000 L crystallizer (Figure 1) that was placed in a water bath of temperature fluctuation less than $\pm 0.02°C$. The device of crystal growth can be similarly used for conventional crystal growth. The crystallizer is made from stainless steel and the inside of the crystallizer is a film of polyvinyl-fluoride of F_{46}. An SYL-50 rare-earth permanent magnet DC torque motor was used to make the crystal rotation smooth and continuous. The crystal was rotated alternately in one direction and then the other on a symmetrically programmed schedule with controlled acceleration, deceleration, and rotation rates.

The acrylic or aluminum platform is of Holden type. Two brackets connect the disc plate up and down. On the centre of the platform, there is a hole, which is of suitable size for planting the crystal seed.

2.2. Crystal growth

The impurity content of KDP raw material, mainly metal ions (Fe^{3+}, Cr^{3+}, Al^{3+}), was controlled to be less than 0.1 ppm by using ultra pure product provided by a domestic manufacturer. This kind of material can reduce the width of the 'dead zone' [7]. Extrapure water produced by a Milli-Q filter system with a resistivity of 18.2 M.$\Omega \cdot$ cm and was used. The temperature of the saturated solution was 60°C. The solution was filtered by two filters with pore diameter of 0.22 and 0.1 μm successively. The KDP solution had been overheated at 70°C for 24 hours. The seed was planted on

the platform through a tube inserted from the top that would direct the seed into the hole after the overheating procedure. The solution temperature was lowered to 55°C instantly and the point seed began to rotate at about 15 rpm and to grow in both the [001] and [100] directions. The size of the seed crystal was about 1 cubic centimeter. The overcooling was controlled between 4.5 and 5°C. The grown crystals shown in Figure 2 are examples of successful growth. The average growth rate was between 10–15 mm/day.

Some crystals had no visible defects, though in many crystals solution inclusions appeared randomly on both prismatic and pyramidal faces. These defects typically were slight on the prismatic {100} faces and much more pronounced on the top of the pyramidal {101} faces. The nature of these defects is not yet clear and is a subject of current investigations.

3. Characterizations of KDP crystals

3.1. Optical transmission

The KDP crystals grown in the experiments with a size of about $20 \times 20 \times 10$ mm were polished on the (001) faces for optical measurements. The samples of the rapidly grown crystal and the conventionally grown crystal were z-cut from the pyramidal sector. The transmission spectra, measured with a PE-lambda35 spectrometer, are shown in Figure 3. These crystals were transparent throughout both the visible and the ultraviolet (UV) regions. In the UV region, the rapidly grown KDP crystal had lower optical absorption than the KDP crystal grown by the conventional method. Additive UV absorption was found in slow-growth KDP, caused by incorporating metal-ion impurities such as Fe, Al, and Cr[8].

3.2. Laser damage threshold

The laser damage thresholds (LDTs) were measured using well-polished z-cut samples obtained in the above experiments. The measurements were carried out with a damage tester, including a Nd:YAG laser, operating at 10 Hz and pulse length 10 ns at 1064 nm. Test samples were exposed to a Gaussian beam with effective cross section of 0.79 mm². The damage tests are of R/1 type[9]. Table 1 shows the results of laser-induced damage threshold measurements in the prismatic and pyramidal sectors of rapidly grown KDP crystals. This standard test showed that the laser damage threshold remains at the same level in both sectors. The damage resistance meets the facility requirement.

3.3. Refractive-index nonuniformities

Efficient frequency doubling and tripling are critical for the successful operation of inertial confinement fusion laser systems. High frequency conversion efficiency is strongly

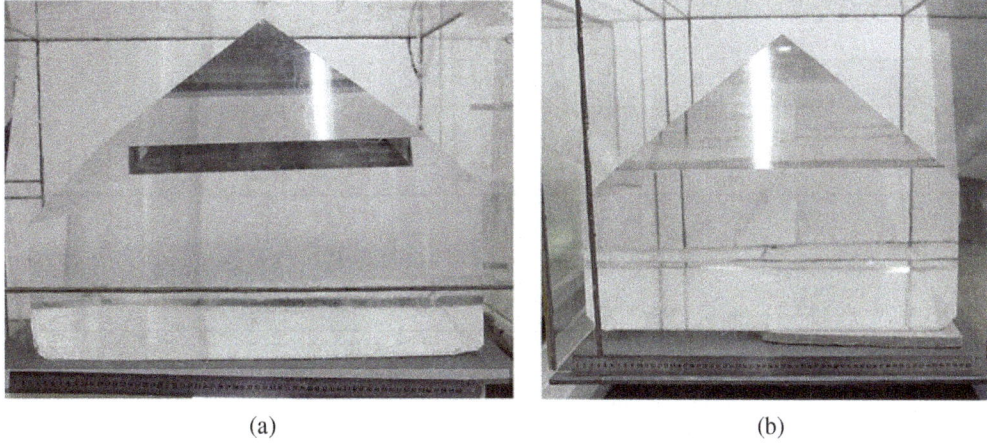

(a)   (b)

Figure 2. The grown KDP crystals with sizes of (a) 680 × 580 × 540 mm³ (growth period: 150 days) and (b) 650 × 550 × 540 mm³ (growth period: 140 days) (b).

Table 1. Laser damage thresholds of Z-cut KDP crystals.

Sample	Sector	R-on-1 (J/cm², 10 ns)	LDT(Scale to 3 ns)
KDP-1	prismatic	25.4	15.1
KDP-2		22.0	13.1
KDP-1	pyramidal	19.5	11.6
KDP-2		19.3	11.5

dependent on attainment of the phase-matching condition. In an ideal converter crystal, one can obtain the phase-matching condition throughout by angle tuning or temperature tuning of the crystal as a whole. In real crystals, imperfections in the crystal structure prohibit the attainment of phase matching at all locations in the crystal[10]. We have modeled frequency doubling and tripling with a quantitative measure of this departure from phase matching in real crystals, which we refer to as orthogonal polarization interferometry (OPI).

Wei *et al.*[11] used the ZYGO MST large-aperture interferometer to attain the refraction index inhomogeneity distribution of a large-size KDP crystal with precision as high as 10^{-7}. The diameter of the test aperture is 32″ and wavelength of the used interferometer is 633 nm. The resulting $\Delta(n_e - n_o)$ distribution for the rapid-growth KDP crystal is shown in Figure 4. It shows that there are two stripes on the upper part of the crystal. One slopes up from left to right, and another slopes down from right to left. The two stripes cut across each other at a midway position of the upper part, forming a boundary which represents the discontinuity of the refractive induced distribution. The sharp feature in Figure 4 is the boundary between 101 (pyramidal) and 100 (prismatic) growth regions in the rapid-growth KDP crystal. Conventional growth material is all pyramidal. The refractive-index nonuniformities existing in the rapid-growth KDP doubler will prohibit the attainment of perfect phase matching across the beam.

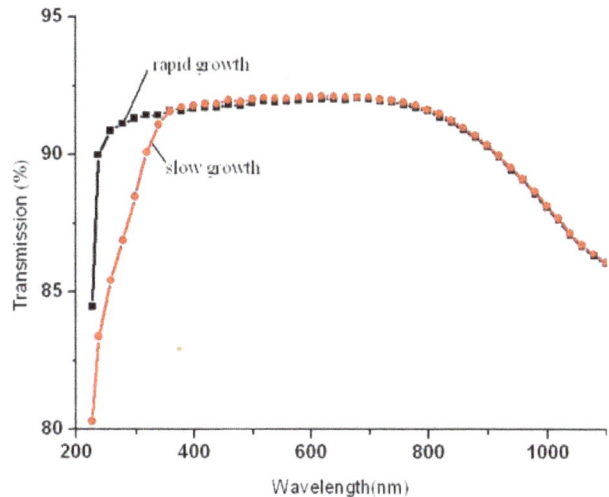

Figure 3. Transmission spectra of KDP crystals.

We carried out third-harmonic conversion experiments on a large-scale laser facility. The crystals are 33 cm and 43 cm plates cut from a KDP boule such that the surface normal is at a specified angle relative to the crystal optical axis. The angle at which the plates are cut from the boule and oriented in the beam is chosen so as to minimize the phase mismatch factor. The result is shown in Table 2. The RMS values of rapidly grown KDP doublers are less than 1×10^{-6}. The phase mismatch factor due to the

Table 2. The comparison of refractive-index nonuniformities.

Sample	Size	Refractive-index nonuniformities RMS ($\times 10^{-6}$)
KDP-1 (doubler-cut)	330×330 mm	0.843
KDP-2 (doubler-cut)	430×430 mm	0.572
KDP-3 (doubler-cut)	430×430 mm	0.727

Table 3. The comparison of refractive-index nonuniformities.

Sample	Size	RMS ($\times 10^{-6}$) (before annealing)	RMS ($\times 10^{-6}$) (after annealing)
KDP doubler-1 (rapid growth)	330×330 mm	0.843	0.824
KDP doubler-2 (rapid growth)	430×430 mm	0.572	0.699
KDP doubler-3 (rapid growth)	430×430 mm	0.727	0.763
KDP doubler-4 (rapid growth)	45×45 mm	0.771	1.23
KDP doubler-5 (slow growth)	45×45 mm	1.02	1.13
KDP tripler-1 (slow growth)	45×45 mm	0.469	0.736

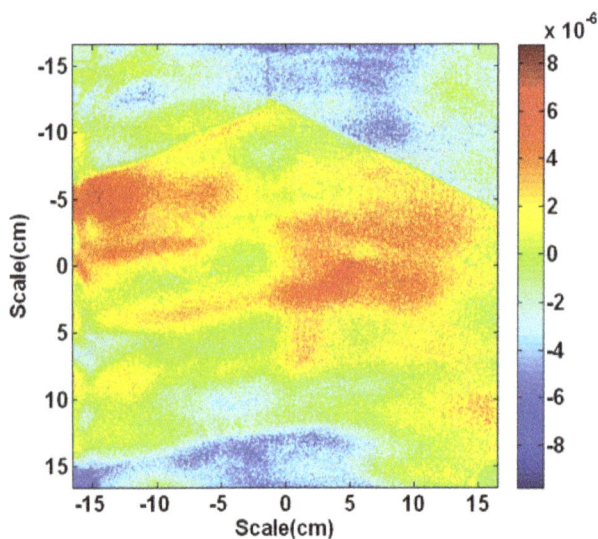

Figure 4. Measured $\Delta(n_e - n_o)$ distribution for the rapid-growth KDP doubler.

refractive-index nonuniformities in the rapid-growth KDP doubler will not lead to a remarkable reduction of the third-harmonic generation (THG) conversion efficiency at high input fundamental intensity. The property of the rapid growth KDP crystal can also meet the requirement of high conversion efficiency and good beam quality.

3.4. Thermal conditioning

Potential thermal conditioning has been reported to increase the optically induced damage threshold[12]. As confirmation, the thermal conditioning was tested to see if it could improve the laser damage threshold of rapidly grown crystals. The same samples as used for the laser damage threshold experiment were used. An electric oven heated the samples to a temperature of 160°C, just below the destructive tetragonal/monoclinic phase transition temperature, 180°C. The rising and lowering rate of temperature was 5°C/h, to prevent breaking of the samples. After baking, the refractive-index nonuniformities and the laser damage threshold were measured under the same conditions as those described above. The results of the measurement are shown in Tables 3 and 4. Thermal annealing cannot obviously improve refractive-index nonuniformities of rapidly grown KDP crystals, but thermal conditioning at 160°C for 24 h improved the damage threshold of rapidly grown KDP crystal to 27.7 J/cm^2. The thermal conditioning of the rapidly grown crystals was effective in reducing the strain in the crystals and improving the laser damage threshold.

3.5. THG conversion efficiency

In order to study the influence of refractive-index nonuniformities on THG conversion efficiency and the near field of the third-harmonic beam, Han et al.[13] carried out third-harmonic conversion experiments on the prototype of the SG-III laser facility which can deliver 2.4 kJ per beam at a wavelength of 1053 nm with a narrow bandwidth, a 1 ns pulse duration, and a 29 cm $\times$ 29 cm beam size. The fundamental beam from the main amplifier of the prototype was delivered into the final optical assembly, which consists of a vacuum window, a frequency converter consisting of a KDP doubler and a KDP tripler, and a fused silica focusing lens. The THG configuration we used in the experiment is

Table 4. The comparison of damage threshold.

Sample	Wavelength	Pulse width	R-on-1 (J/cm^2)
KDP(before annealing)	1064 nm	10 ns	20.2
KDP(after annealing)	1064 nm	10 ns	27.7

an 11 mm thick type I rapid growth KDP doubler followed by a 9 mm thick type II traditional growth KDP tripler. The aperture of both crystals was 330 mm × 330 mm. After the target, a sampling mirror was used for third-harmonic beam energy and nearfield measurements. A self-made MSC-type energy meter was used to determine the efficiency with sensitivity of 10 μv/mJ and measuring energy range 200 mJ–20 J.

The highest conversion efficiencies at the input fundamental intensities of 1.2 GW/cm^2 and 2.5 GW/cm^2 are 58.7% and 58.4%, respectively. In the experiment, the optimum detuning angle of the doubler is about 130 rad, smaller than the theoretical value. This is because other factors, such as the fundamental beam wavefront, refractive-index nonuniformities, surface figure of the doubler, and so on, reduced the doubling efficiency. The phase mismatch factor due to the refractive-index nonuniformities in rapid-growth KDP doubler will not lead to a remarkable reduction of the THG conversion efficiency at high input fundamental intensity.

4. Conclusions

The result obtained from the present work was on the basis of the development of the rapid-growth technique. Our results show that large KDP single crystals for laser fusion systems can be grown rapidly using high supersaturations. In our experiments we managed to accelerate the KDP crystal growth rate by aggrandizing the supersaturation of the KDP solution, improving the purity of the KDP solution, and making the KDP crystal rotation reasonable. We have successfully grown a KDP crystal with a size of 680 × 580 × 540 mm^3 at an average rapid-growth rate of 10 mm/day without a complicated continuous filtration system. The optical quality of the KDP crystals was characterized. We have developed a stable technique for obtaining KDP up to 60 cm in size having high optical quality including a high laser damage threshold.

According to the experimental results, the impact of refractive-index nonuniformities on the THG conversion efficiency can be deemed acceptable, but efforts should be made to eliminate refractive index discontinuities in rapid-growth KDP crystals.

Acknowledgements

This work was performed under the auspices of the State High Technology Program for Inertial Confinement Fusion, which is gratefully acknowledged.

References

1. G. M. Loiacono, J. J. Zola, and G. Kostescky, J. Crystal Growth **62**, 543 (1983).
2. N. P. Zaitseva, L. N. Rashkovich, and S. V. Bogatyreva, J. Crystal Growth **148**, 276 (1995).
3. N. P. Zaitseva, J. J. De Yoreo, M. R. Dehaven, R. L. Vital, K. E. Montgomery, M. Richardson, and L. J. Atherton, J. Crystal Growth **180**, 255 (1997).
4. M. Nakatsuka, K. Fujioka, T. Kanabe, and H. Fujita, J. Crystal Growth **171**, 531 (1997).
5. V. L. Bespalov, V. I. Bredikhim, V. P. Ershov, V. I. Katsman, and L. A. Lavrov, J. Crystal Growth **82**, 776 (1987).
6. J. J. De Yoreo, Z. U. Rek, N. P. Zaitseva, T. A. Land, and B. W. Woods, J. Crystal Growth **166**, 291 (1996).
7. L. N. Rashkovich, 1991 *KDP Family of Crystals*. Adam-Hilger, New York.
8. Terry A. Land, Tracie L. Martin, and Sergey Potapenko, Nature **399**, 442 (1999).
9. Alan K. Burnham, Michael Runkel, and Michael D. Feit, Appl. Opt. **42**, 5483 (2003).
10. Jerome M. Auerbach, Paul J. Wegner, and Scott A. Couture, Appl. Optics **40**, 20 (2001).
11. X. Wei, L. Chai, B. Gao, and Q. Li, J. Appl. Opt. **34**, 300 (2013).
12. J. E. Swain, S. E. Stokowski, and D. Milam, Appl. Phys. Lett. **41**, 12 (1982).
13. W. Han, L. Zhou, F. Wang, F. Li, K. Li, L. Wang, B. Feng, Q. Zhu, J. Su, and G. Mali, Optik **24**, 6506 (2013).

5

Bragg accelerator optimization

Adi Hanuka and Levi Schächter

Department of Electrical Engineering, Technion – Israel Institute of Technology, Haifa 32000, Israel

Abstract

We present the first steps of a design of the optimal parameters for a full Bragg X-Ray free electron laser (BX-FEL). Aiming towards a future source of coherent X-ray radiation, operating in the strong Compton regime, we envisage the system to be the seed for an advanced light source or compact medical X-ray source. Here we focus on the design of the accelerator parameters: maximum gradient, optimal accelerated charge, maximum efficiency, and 'wake coefficient', which relates to the decelerating electric field generated due to the motion of a charged-line or train of charged-lines. Specifically, we demonstrate that the maximum efficiency has optimal value and given the fluence of the materials, the maximum accelerated charge in the train is constant. These two results might be important in any future design.

Keywords: Bragg; wake-field; Compton scattering; energy conversion; efficiency; fluence; light source; medical accelerator

1. Introduction

X-ray sources based on Compton scattering of a laser from a relativistic counter-propagating electron beam (e-beam) have recently drawn increasing interest due to several potential advantages over magnetostatic free electron lasers (FELs), such as compact size, low-cost operation, and reduced e-beam energy requirements. Recent work[1] demonstrated that X-ray radiation emitted by relativistic electrons scattered by a counter-propagating laser pulse guided by an adequate Bragg structure (spontaneous emission) surpassed by about two orders of magnitude the intensity generated by a conventional free-space Gaussian-beam configuration, given the same e-beam and injected laser power in both configurations.

Based on this configuration, we proposed a Bragg configuration based X-FEL operating in the collective regime. The full system consists of three main components (Figure 1): an *optical injector* bunches the electrons to the *accelerator Bragg structure* which supports a TM_{01} mode. The co-propagating laser which accelerates the electrons is dumped at the end. Next, the electrons are transported into a second Bragg waveguide. This structure supports a TEM laser mode in the vacuum core (TM mode at the Bragg layers) counter-propagating to the e-beam – the latter acts as an *electromagnetic (EM) wiggler*. The scattering of free electrons with a counter-propagating TEM-like laser mode generates X-ray radiation – inverse Compton scattering.

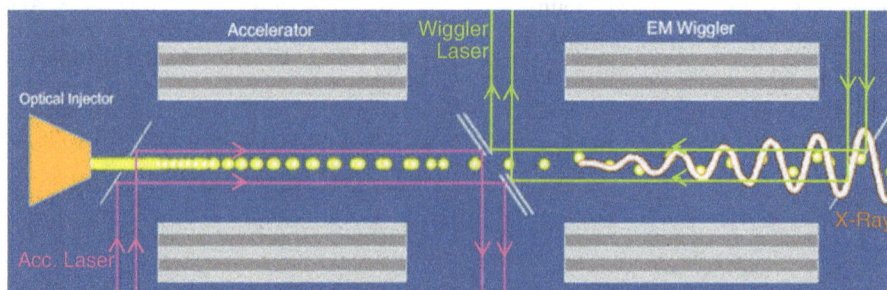

Figure 1. Schematic of an all-Bragg system. On the left, the Bragg accelerator supports a co-propagating TM_{01} mode which accelerates the e-beam. The latter is injected into another Bragg structure which supports a TEM mode (inside the vacuum core) counter-propagating to the electrons, which as a result generates X-ray radiation.

Correspondence to: Adi Hanuka, HARAVA 5 Nesher, 36863, Israel.
Email: Hanukaadi@gmail.com

Figure 2. Planar Bragg waveguide with a vacuum region of width $2D_{int}$.

In this study we focus on the Bragg accelerator. Structure-based laser-driven linear accelerators have been the subject of intense investigation, primarily for use with relativistic particles, where obtaining high energies in compact geometries is desired.

To accelerate particles efficiently, EM waves must be guided or confined to the region in which the particles travel. An electric field component in the direction of desired acceleration is strictly necessary. Traditionally, EM waves have been confined to a vacuum channel surrounded by metallic structures. Field confinement can also be achieved through surrounding dielectric layers where reflections from different layers interfere constructively (Bragg reflection).

A planar Bragg waveguide[2] consists of dielectric layers surrounding a sub-wavelength vacuum region which is symmetrical relative to the central plane (Figure 2). The clearance is a vacuum region of width $2D_{int}$, surrounding alternating periodic layers ($\varepsilon_2 = 4$, $\varepsilon_3 = 2.1$).

The layers are made of two lossless dielectric materials; the first layer has a relative dielectric coefficient ε_1. For single-mode operation: $D_{int} = 0.25\lambda_L \rightarrow 0.55\lambda_L$. For an optical accelerator having a *vacuum core*, the surrounding layers must have, at the operating wavelength, an effective dielectric coefficient smaller than unity, i.e., $\varepsilon_{eff} < 1$, thus creating the need for a Bragg structure with a matching layer[3].

The present study is organized as follows: A self-consistent solution for maximum gradient is presented in Section 2. In Section 3 we present the optimal charge of the e-beam injected into the acceleration module from a perspective of high efficiency. A charged-line moving in a planar Bragg acceleration structure generates a reaction field which, by virtue of linearity of Maxwell's equations, is related through the so-called, wake coefficient. We show a detailed evaluation of the wake coefficient in Section 4 for a single bunch and in Section 5 for train of micro-bunches.

2. Accelerating gradient

For assessment of the accelerating gradient, we examine the maximum energy flux that can be sustained by the structure; typically, this occurs at the vacuum–dielectric interface,

where $S_{z,max}(D_{int}^+) = \frac{1}{2}H_y^*(D_{int}^+)E_x(D_{int}^+)$. The material is characterized by the fluence (F), which represents the energy per unit surface before breakdown occurs. Consequently, the threshold for a pulse of duration τ_p is determined by

$$S_{max}(D_{int}^+) = \frac{1}{2\eta_0\varepsilon_1}\left[\frac{\omega_L}{c}D_{int}G_0\right]^2 = \frac{F}{\tau_P}, \qquad (1)$$

where G_0 is the accelerating field gradient, ω_L is the laser frequency and η_0 is the vacuum impedance. We rely on the three EM field components of the accelerating mode in the vacuum region

$$E_z = G_0 \exp\left(-j\frac{\omega_L}{c}z\right),$$
$$E_x(x) = j\frac{\omega_L}{c}xG_0\exp\left(-j\frac{\omega_L}{c}z\right), \qquad (2)$$
$$H_y(x) = j\frac{\omega_L}{c}x\frac{G_0}{\eta_0}\exp\left(-j\frac{\omega_L}{c}z\right).$$

Dielectric damage involves heating of conduction band electrons by the incident radiation and transfer of this energy to the lattice; damage occurs via conventional heat deposition. Thus, we consider the threshold fluence (energy/area) of the material, which, in turn, depends on the pulse duration. An empirical fluence threshold of fused silica has been published[4] by the LLNL group

$$F\left(\frac{J}{cm^2}\right) = \begin{cases} 1.44\tau_p^{1/2} & \tau_p \text{ (ps)} > 10 \\ 2.51\tau_p^{1/4} & 0.4 < \tau_p \text{ (ps)} < 10 \\ 2 & \tau_p \text{ (ps)} < 0.4. \end{cases} \qquad (3)$$

It should be pointed out that the above empirical expression assumes a TEM mode impinging *perpendicular* to the dielectric surface, whereas in our case the energy flows in the *parallel* direction. It is assumed that since the electric field is vertical to the surface the probability of flash-over is reduced – therefore, adopting this criterion (Equation (3)) is an underestimate of the fluence our structure can sustain.

In RF-based accelerators, the radiation wavelength is typically more than 10 cm in length, thus we may use 10^{10} electrons in a 100 μm bunch. Assuming similar dimensions in the transverse directions, we realize that the density of electrons is on the order of $n \sim 10^{23}$ m^{-3}. If the density is kept the same in the optical regime, say $\lambda_L \sim 1$ μm, the micro-bunch needs to be on the order of 30 nm long, 100 nm high and, assuming a sheet-beam about 10 μm wide, then the number of electrons in one optical micro-bunch is 300. In spite of this clearly being a very rough estimate, in order to accelerate a significant number of electrons, and keep the electron density as in an RF machine, we need to use a multiple number of periods of the accelerating mode, thus the accelerated bunch is actually a *train* of micro-bunches.

Moreover, it is strictly necessary to use a pre-bunched beam at optical wavelengths before injection into the optical

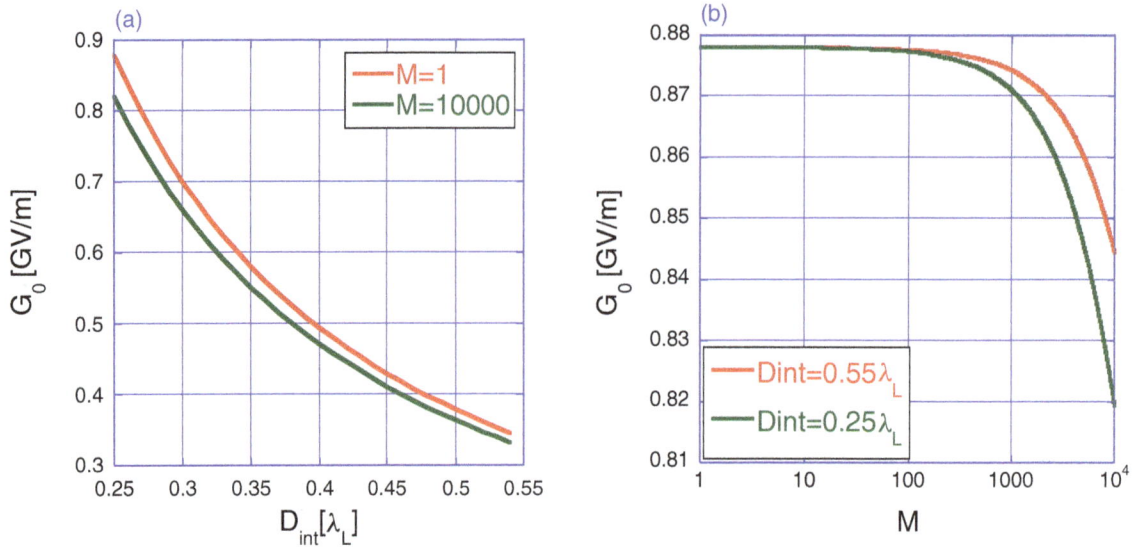

Figure 3. (a) Gradient versus clearance of the accelerator structure. Red line for a single bunch and green line for $M = 10^4$. As the clearance is increased, the gradient drops. (b) Gradient versus number of micro-bunches in the train. Red line for $D_{int} = 0.25\lambda_L$; green line for $D_{int} = 0.55\lambda_L$, multiplied by a factor (2.377) such that at $M = 1$ both curves coincide for G_0 ($D_{int} = 0.25\lambda_L$). There is a critical value at approximately $M = 1000$.

accelerator. Without the pre-bunching, the beam energy spread is too large to be useful. Energy modulation converted into a density modulation may also lead to increased efficiency in the accelerator. A configuration of modulator and chicane may be used as a pre-buncher injector to increase the number of electrons in the optimal phase of the accelerating laser. An efficient method for bunching the beam at optical wavelengths was suggested in[5] and it was demonstrated that a pre-bunched beam at optical wavelengths indeed reduced the beam energy spread in laser accelerators[6]. We are in the process of analysing a novel injector that generates density-modulated beams at optical wavelengths, but this is beyond the scope of this study.

An assessment of the laser pulse duration requires one to take into consideration that the EM wave propagates at the group/energy velocity whereas the electrons propagate virtually at the speed of light in vacuum. For full overlap of the two pulses, the duration of the EM pulse is

$$\tau_p = \frac{L_{geo}}{c}\left(\frac{1}{\beta_{gr}} - 1\right) + \frac{L_{beam}}{c}, \quad (4)$$

where $L_{beam} = (M - 1)\lambda_L$ is the train length, M represents the number of micro-bunches in the train, and λ_L is the laser wavelength in vacuum. The geometrical length of the structure is $L_{geo} = \gamma m_e c^2 / eG_0$, where $\gamma = \sqrt{\lambda_L/4\lambda_r}$ is the relativistic factor and is set by the resonance condition in an EM wiggler. As an example, for a laser wavelength of 1 micron and a radiation wavelength of X-rays $\lambda_X = 0.1$ nm we get $\gamma \simeq 50$.

Due to the constraint imposed by the fluence and for the specified energy ($\gamma \simeq 50$), the maximum accelerating gradient depends on two parameters: the clearance of the

Table 1. Typical Values of the Parameters for $D_{int} = 0.3\lambda_L$.

	G_0 (GV m^{-1})	$\langle P_L \rangle$ (kW)	τ_{beam} (ps)	τ_p (ps)	F (J cm^{-2})	β_{gr}	L_{geo} (m)
$M = 1$	0.7	35.56	~ 0	171	18.83	0.424	0.038
$M = 10^4$	0.66	117.77	35	216	21.2	0.424	0.04

structure (D_{int}) and the number of micro-bunches in the train (M). Note the interdependence between the various parameters requires a self-consistent solution: the laser field depends on the laser pulse duration, which in turn depends on the train's total charge and the gradient itself.

A self-consistent solution is illustrated in Figure 3. For $M < 10^3$ micro-bunches, the gradient is virtually independent of M (see Figure 3(b)). For larger values of M the gradient decreases for the same clearance. Figure 3(a) shows that it is advantageous to operate with the smallest possible vacuum tunnel – leading to a maximum gradient of less than 0.9 GV m^{-1}. Further simulations show it is better to use a lower dielectric coefficient for the first layer, and in order to achieve a gradient of 1 GV m^{-1} we need to replace the silica with a material whose typical fluence is higher by a factor of 1.5 – assuming the pulse dependence is the same. Typical values of the parameters for $D_{int} = 0.3\lambda_L$ are given in Table 1.

At this point it warrants making a comment regarding a more realistic scenario: The wake generated by the accelerated bunches tends to reduce the laser field – this is the well known beam-loading effect. Consequently, in the presence of the electrons, the field experienced by the structure is thus reduced accordingly for a given fluence, the applied gradient may be significantly higher. In the context of the fluence effect we consider the worst case scenario and ignore this process. Taking it into consideration, we estimate that the accelerating gradient will be enhanced.

3. Optimal charge

Given the accelerating gradient G_0, which is evaluated self-consistently, we now calculate the optimal number of electrons in a micro-bunch injected into the acceleration module. Optimum charge occurs for maximum efficiency of the acceleration process. We can interpret the reason for this optimum as follows: for a given accelerating gradient, if the accelerated charge is small, the energy transferred is negligibly small (zero). At the other extreme, when the charge is large, beam loading may suppress the effective accelerating gradient to zero; therefore, again, the transferred energy is minuscule. Between these two 'zeros' the function of energy transferred is expected to have a maximum.

3.1. Single bunch

In the single bunch case ($M = 1$), the efficiency of the acceleration process may be determined as

$$\eta = \frac{\Delta U_{\text{kinetic}}}{U_{\text{EM}}} = 4\eta_{\max} \frac{q}{q_0} \left(1 - \frac{q}{q_0}\right), \quad (5)$$

where $q_0 = G_0/\kappa$ is the charge for which the wake generated by the bunch balances the laser gradient – in other words, there is no net acceleration. The maximum value of efficiency, occurring for $q_{\text{opt}} = q_0/2$, is determined by the projection of the total deceleration, represented by κ, on the fundamental mode, represented in turn by κ_1 – explicitly, $\eta_{\max} = \kappa_1/\kappa \equiv W_1$; W_1 is the weight function of the first mode[7]. The maximum efficiency is dependent on the clearance (Stupakov & Bane[8]):

$$\eta_{\max}(D_{\text{int}}) = \frac{1}{4\varepsilon_0 \lambda_L^2 \kappa} \frac{\beta_{gr}}{1 - \beta_{gr}} \frac{Z_{\text{int}}}{\eta_0}, \quad (6)$$

where κ, β_{gr}, and Z_{int} are dependent on the vacuum clearance[2]. In the case of a dielectric planar Bragg waveguide with a vacuum tunnel of $2D_{\text{int}}$ along which a charged-line propagates, the wake coefficient associated with the decelerating field is $\kappa = E^{(\text{dec})}/q = 1/(4\varepsilon_0 D_{\text{int}})$ (see Section 4).

The maximum efficiency itself has an optimum at $D_{\text{int}} = 0.35\lambda_L$ (Figure 4). The reason for this optimum is a combination of two facts: the wake coefficient and the interaction impedance drop as the clearance is widened, whereas the group velocity increases (Figure 5).

3.2. Train of micro-bunches

For the case of train of M micro-bunches, the beam-loading causes different micro-bunches to experience different effec-

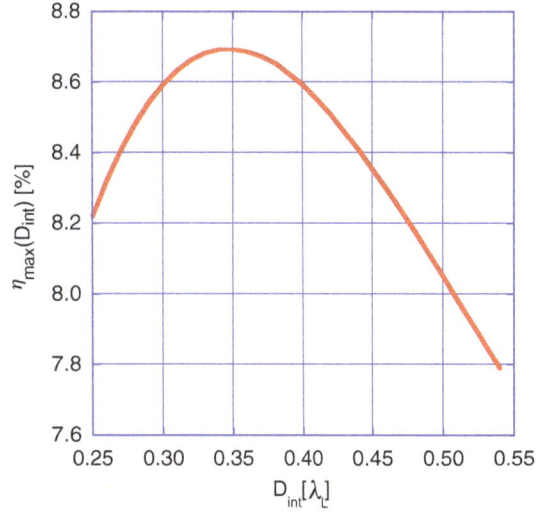

Figure 4. Maximum efficiency for a single bunch versus half clearance width (Equation (6)).

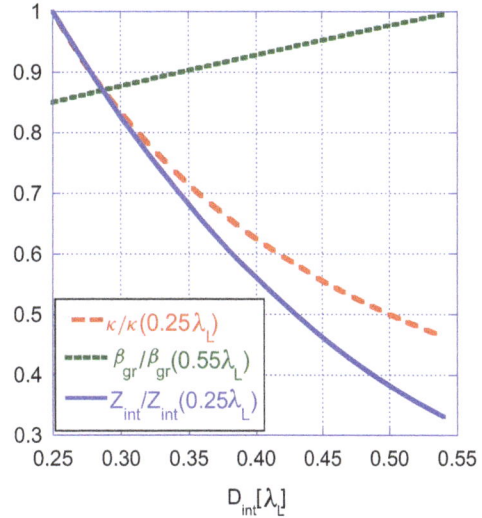

Figure 5. Wake coefficient, interaction impedance, and group velocity versus half clearance width. Each of the parameters is normalized to its maximum value.

tive accelerating gradients. In order to eliminate this effect, the laser pulse must be tapered according to

$$G(t) = G_0 + \frac{t}{\tau_P} \kappa q_{\text{mb}}(M - 1), \quad (7)$$

where q_{mb} is the charge in one micro-bunch; tacitly assuming that all micro-bunches are identical. It is assumed that for a sufficiently large number of M the weight function of the first mode is dominant, i.e., $\kappa_1 \sim \kappa$. κ is the wake coefficient, which depends on the structure and number of accelerated micro-bunches. The wake coefficients for a single bunch and train of micro-bunches are different; however, since the laser is tapered, we consider the wake coefficient for a single bunch and a single discontinuity.

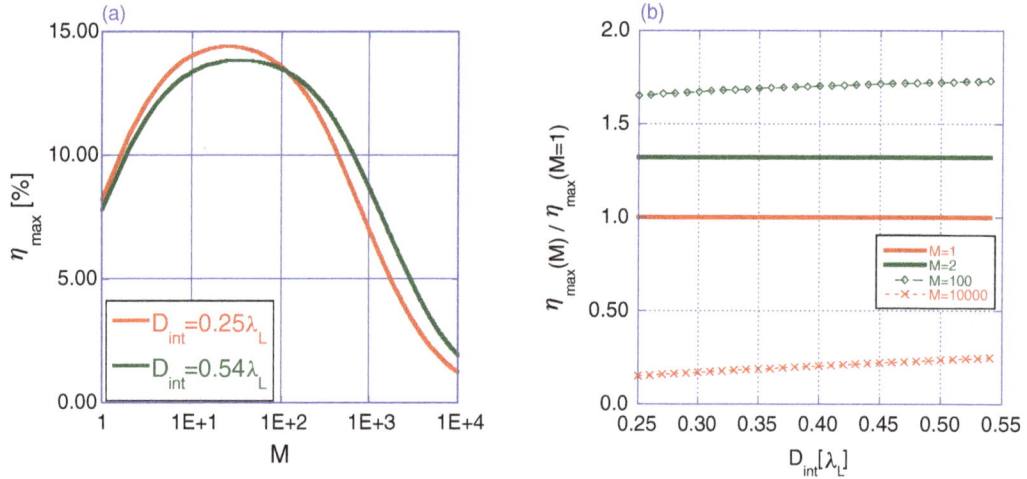

Figure 6. (a) Maximum efficiency versus number of micro-bunches in the train. Red line for $D_{int} = 0.25\lambda_L$ and green line for $D_{int} = 0.54\lambda_L$. The optimum value is 15% for $M = 30$. (b) Maximum efficiency normalized to the single bunch case versus clearance of the accelerator structure. For each clearance there is an optimal value for M.

The EM energy injected into the system may be readily calculated using

$$U_{EM} = \int_0^{\tau_p} dt\, P_L(t) = \int_0^{\tau_p} dt\, \frac{|\lambda_L|^2}{Z_{int}} G^2(t), \qquad (8)$$

together with Equation (4). The efficiency in this case is given by

$$\eta = \frac{\Delta U_{kinetic}}{U_{EM}} = \eta_{max}(M=1)\frac{\frac{12M\bar{q}(1-\bar{q})}{[3+3\bar{q}(M-1)+\bar{q}^2(M-1)^2]}}{1+\frac{\beta_{gr}}{1-\beta_{gr}}\frac{L_{beam}}{L_{geo}}}. \quad (9)$$

By neglecting the geometric length dependence on the charge, since it does not change significantly, the optimal charge is

$$q_{opt}(M) = q_0 \frac{-3+\sqrt{9+3(M-1)(M+2)}}{(M-1)(M+2)} \equiv q_0\xi(M). \tag{10}$$

Several facts are evident: (i) The maximum efficiency value depends on two parameters – the clearance and the number of micro-bunches in a train. Figure 6(a) illustrates the efficiency and its maximum. (ii) For the case of a single bunch ($M = 1$) we get $\xi(M = 1) = 0.5$, and get the maximum efficiency which was calculated explicitly for a single bunch. (iii) Comparing to the latter case, the efficiency more than doubled for $M \sim 50$. (iv) Figure 6(b) shows a weak dependence of the maximum efficiency on the vacuum clearance and a strong dependence on the number of micro-bunches.

Another perspective of the energy conversion efficiency is the total amount of charge accelerated and its distribution among the various numbers of micro-bunches. The number of electrons in a train as a function of the number of micro-bunches is almost constant $\sim 1 \times 10^6$. Thus, for larger values of M, the number of electrons in a micro-bunch drops.

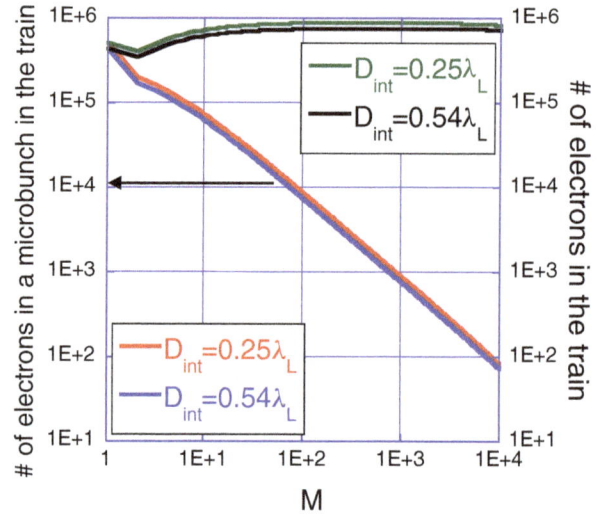

Figure 7. Number of electrons in a microbunch (left y-axis) and number of electrons in the train (right y-axis) versus the number of micro-bunches for $D_{int} = 0.25\lambda_L, 0.54\lambda_L$.

Figure 7 reveals a weak dependence of q_{mb} and Mq_{mb} on the vacuum clearance.

The average power per unit length (Δ_y) of the tapered laser is given by

$$\langle P_L \rangle = \frac{U_{EM}}{\tau_P} = \frac{1}{\tau_P}\int_0^{\tau_P} dt\, P_L(t)$$
$$= \Delta_y \frac{|\lambda_L|^2}{Z_{int}}\left[G_0^2 + \frac{\alpha^2}{3} + \alpha G_0\right], \qquad (11)$$

where $\alpha = \kappa q_{mb}(M-1)$. It has a maximum for $M \sim 700$ (Figure 8). This maximum is the result of two contradicting trends: the accelerating gradient increases for a reduction of

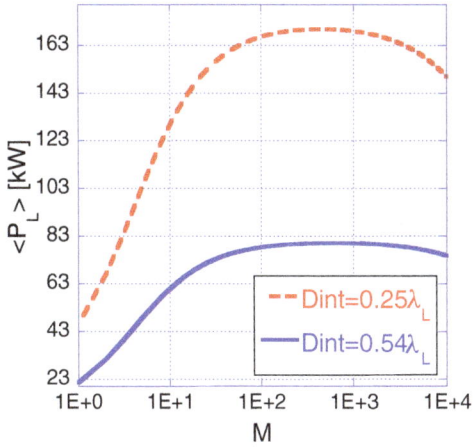

Figure 8. Average laser power for $D_{\text{int}} = 0.25\lambda_L$, $0.54\lambda_L$. Its maximum (170 kW) occurs for $D_{\text{int}} = 0.25\lambda_L$, $M = 700$.

M, whereas α decreases. As an example, for $\Delta_y = 10\ \mu\text{m}$ and 10^3 micro-bunches in the train, the average tapered laser power is 117 kW, whereas for $\Delta_y = 1\ \mu\text{m}$ the latter is only 11 kW. It should be pointed out that pulse shaping can be done using several methods, such as spatial light modulators, adaptive beam shaping and fixed masks[9, 10].

4. Wake coefficient – single bunch

In this section we determine and investigate the wake coefficient (κ) for a single bunch (in the following section we repeat this for a train of micro-bunches). Both are essential for establishing the optimal charge in the micro-bunch and determining the beam loading effect.

A laser pulse accelerates a point charge q_{mb} moving in a vacuum tunnel of planar Bragg acceleration structure and generates an EM wake (Cerenkov radiation). Associated with this wake there is a decelerating electric field which, by virtue of the linearity of Maxwell's equations, is proportional to the charge, namely $E_{\text{dec}} = \kappa q$, where the wake coefficient κ depends on the structure.

The vacuum–dielectric discontinuity generates a reflected wave that can affect the point charge. Any reflection occurring further away from the first discontinuity reaches the structure's axis only after the point charge has passed– thus it may affect only trailing micro-bunches.

In the absence of reflections, the wake coefficient is determined by the structure (Appendix A) and given as

$$\kappa = \frac{1}{4\varepsilon_0 D_{\text{int}}} \left(\frac{\Omega}{\text{ms}} \right). \tag{12}$$

In the presence of reflections, for a single bunch, only the first discontinuity affects a line-charge. However, we demonstrate that quantitatively using a previously defined formulation[2]. The effective wake coefficient on the first

bunch ($\bar{\tau} = 0$) is

$$E_{\text{dec}} \triangleq E_z^{(s)}(\bar{\tau} = 0) = \left[\frac{\kappa}{\pi} \int_{-\infty}^{\infty} d\bar{\omega}\ \frac{1}{\frac{1-R(\omega)}{1+R(\omega)} + j\bar{\omega}} \right] \bar{\lambda} \triangleq \bar{\kappa}\bar{\lambda},$$
$$\tag{13}$$

where the $R(\omega)$ is the reflection coefficient of the structure – including the effect of the first (matching) layer as well as the 'Bragg layers'. Numerical evaluation of Equation (13) reveals that $\bar{\kappa} = \kappa$, with an error of less than 0.1% – i.e., the wake coefficient for a single bunch including reflections is almost equal to the wake coefficient with no reflections. In fact, it is possible to demonstrate analytically that since reflection reaches the axis behind the charged-line micro-bunch, $\bar{\kappa} = \kappa$.

5. Wake coefficient – train of bunches

In the case of a train of micro-bunches the field spatial distribution trailing the particle is strongly affected. The Bragg dielectric structure allows energy to escape from the structure through the layers, and the trailing bunches are less affected by the wake field – except eigenmodes of the structure.

For a train of M micro-bunches, of length L_{mb} each, separated by one wavelength λ_L (Figure 9), the wake coefficient in the case of a train of micro-bunches is strongly dependent on the structure

$$\tilde{\kappa} = \kappa \int_{-\infty}^{\infty} d\bar{\omega}\ W(\bar{\omega}) \text{sinc}^2 \left(\frac{\chi}{2} \frac{\bar{\omega}}{\bar{\omega}_L} \right) \frac{\text{sinc}^2 \left(\pi \frac{\bar{\omega}}{\bar{\omega}_L} M \right)}{\text{sinc}^2 \left(\pi \frac{\bar{\omega}}{\bar{\omega}_L} \right)}, \tag{14}$$

where $\chi = 2\pi L_{\text{mb}}/\lambda_L$ is the relative length of the micro-bunch and $W(\bar{\omega}) \equiv \left[\frac{1-R(\omega)}{1+R(\omega)} + j\bar{\omega} \right]^{-1}$ are the weights[2].

Several observations are evident: $\tilde{\kappa}$ decays as $1/M^2$ and the 'sinc' function acts as a low-pass filter (Figure 10); higher frequencies than the fundamental are suppressed (shorter wavelengths). M has an effect on the non-fundamental modes, i.e., $\tilde{\kappa} \simeq \kappa_1$. κ_1 is independent of the number of micro-bunches (M) and thus remains the same. The projection of the wake on the fundamental mode (W_1-weight of the first mode) increases with the number of micro-bunches in the train (since higher frequencies are suppressed). Therefore, we should be able to enhance to some extent the efficiency for $M \gg 1$, and increase the amount of charge accelerated.

Our simplified model allows all $n \cdot \omega_L$ harmonics, since the dielectric coefficient is independent of frequency. In practice, the dielectric function is frequency dependent and, as a result, the higher harmonics are also suppressed.

6. Conclusions

The maximum accelerating gradient is evaluated self-consistently based on the constraints imposed by the pulse

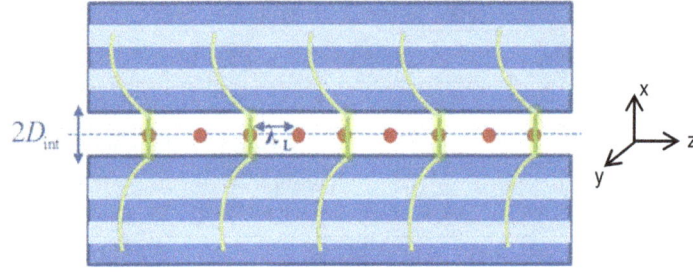

Figure 9. Planar waveguide acceleration module with a vacuum region of width $2D_{int}$. The e-beam is accelerated by a co-propagating TM_{01} laser mode. The macrobunch consists of a train of M line charges, separated by a laser wavelength.

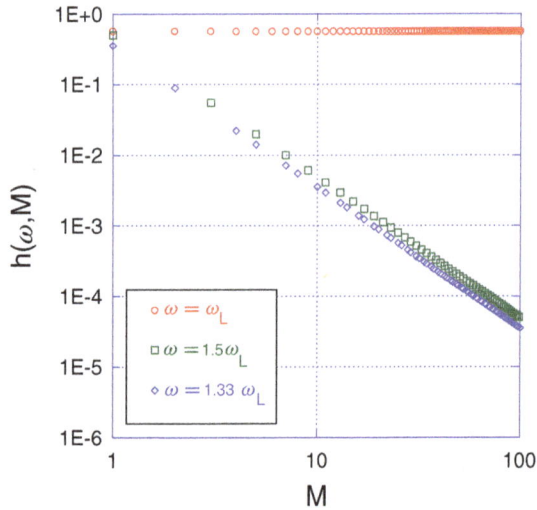

Figure 10. Spectrum of the decelerating field multiplied by the 'sinc' function of the number of micro-bunches: $h(\omega, M) = w(\omega)\text{sinc}^2(\pi\omega M/\omega_L)/\text{sinc}^2(\pi\omega/\omega_L)$ on a log–log scale. Frequencies other than the fundamental are suppressed as the number of micro-bunches in the train increases.

duration and fluence. It depends on two parameters: the clearance of the structure (D_{int}) and the number of micro-bunches in the train (M). In the worst case scenario, it may reach levels of 1 GV m^{-1}.

Optimum charge occurs for maximum efficiency of the acceleration process. For a train of micro-bunches, two constraints must be satisfied: the laser pulse duration must be longer than the macro bunch length and the laser's envelope must be tapered to compensate for the beam loading, ensuring uniform gradient acceleration of all micro-bunches. The maximum efficiency has an optimal value ($\sim$15%) which depends on two parameters: a weak dependence on the vacuum clearance and a strong dependence on the number of micro-bunches.

The optimal number of electrons to be accelerated is determined by the laser field and the maximum efficiency requirement. For $M = 1000$, the number electrons in a micro-bunch is $\sim$1150, while the total number of electrons

in the train is almost constant ($\sim$10^6). There is weak dependence of q_{mb} and Mq_{mb} on the vacuum clearance.

The optimal charge in the micro-bunch and the beam loading effect are also determined by the wake coefficient. The latter is a property of the structure and refers to the decelerating field. The maximum efficiency increases with the number of micro-bunches in the train since higher frequencies are suppressed.

Acknowledgements

This study was supported by the BiNational US–Israel Science Foundation and the Israel Ministry of Science.

Appendix A

We investigate the wake field, separating the wake into two components, as developed in Refs. [5] and [7]. For the primary we consider a line charge, infinite in the y direction, moving with a constant velocity v in the z-direction inside the vacuum core of the planar Bragg acceleration. All field components are excited by the current density $J_z = -\bar{\lambda}v\delta(x)\delta(z - vt)$, where $\bar{\lambda} = q/\Delta_y$ is the charge per unit length.

The EM field is derived from the nonhomogeneous wave equation of the magnetic vector potential, subject to the Lorentz gauge $\nabla \cdot A + (\varepsilon/c^2)\partial_t\phi = 0$, thus

$$\left[\nabla^2 - \frac{\varepsilon}{c^2}\frac{\partial^2}{\partial t^2}\right]A_z = -\mu_0 J_z. \tag{A1}$$

On the one hand this primary field is generated by a charged-line in free space, whereas on the other hand the secondary field is the reaction to the presence of the surrounding structure. It is this secondary field which is responsible for the decelerating force which acts on the charged-line.

Using the time Fourier transform defined by $A_z(x, \tau = t - z/v) = \int_{-\infty}^{+\infty} d\omega\, e^{j\omega\tau} A_z(x, \omega)$ and its corresponding source term $J_z(x, \omega) = \bar{\lambda}\delta(x)/2\pi$, Equation (A1) reads

$$[\partial_x^2 - (\omega/c\beta\gamma)^2]A_z(x,\omega) = \mu_0\bar{\lambda}\delta(x)/2\pi, \qquad (A\,2)$$

and the general solution has the form

$$A_z^{(p)}(x,\omega) = \begin{cases} a\exp\left(-\dfrac{|\omega|x}{c\beta\gamma}\right) & x > 0 \\[2mm] b\exp\left(+\dfrac{|\omega|x}{c\beta\gamma}\right) & x < 0. \end{cases} \qquad (A\,3)$$

By integrating both sides of the equation $\partial_x A_z(0^+,\omega) - \partial_x A_z(0^-,\omega) = \mu_0\bar{\lambda}/2\pi$, the constants are $a = b = -\mu_0\bar{\lambda}c\beta\gamma/4\pi|\omega|$. Finally, the magnetic vector potential for the primary field is

$$A_z^{(p)}(x,z,t) = -\frac{\mu_0\bar{\lambda}}{4\pi}\int_{-\infty}^{+\infty} d\omega\,\frac{\gamma v}{|\omega|}$$
$$\times \exp\left[-j\frac{\omega}{v}(z-vt) - \frac{1}{v\gamma}|\omega x|\right]. \quad (A\,4)$$

The magnetic vector potential for the secondary field is determined using the dielectric coefficient of the first layer adjacent to the vacuum core ($\varepsilon_1 > 1$). This is the condition for Cerenkov radiation. Moreover, we assume that the charge does not experience any reflection from higher layers.

$$A_z^{(s)}(x,\tau) = -\frac{\mu_0\bar{\lambda}}{4\pi}\int_{-\infty}^{+\infty} d\omega\,\frac{\gamma v}{|\omega|}\exp(j\omega\tau)$$
$$\times \begin{cases} A_1\exp[-\Gamma(x - D_{\text{int}})] & x > D_{\text{int}} \\[2mm] A_0\cosh\left(\dfrac{|\omega|}{\gamma v}x\right) & |x| < D_{\text{int}} \\[2mm] A_2\exp[\Gamma(x - D_{\text{int}})] & x < -D_{\text{int}}, \end{cases}$$
$$(A\,5)$$

where $\Gamma = |\omega|\sqrt{1 - \beta^2\varepsilon}/v$, and due to symmetry ($A_2 = A_1$) we may solve the equation only for the half-space $x > 0$. Based on the Lorentz gauge, the electric scalar potential is

$$\partial_z A_z + \frac{j\omega\varepsilon}{c^2}\varphi = j\frac{\omega}{v}A_z + \frac{j\omega\varepsilon}{c^2}\varphi = 0 \rightarrow \varphi = \frac{c^2}{\varepsilon v}A_z, \quad (A\,6)$$

thus

$$\begin{cases} E_z = -\partial_z\varphi - j\omega A_z = -j\omega\left(1 - \dfrac{1}{\varepsilon\beta^2}\right)A_z \\[2mm] H_y = -\dfrac{1}{\mu_0}\partial_x A_z, \end{cases} \quad (A\,7)$$

or, explicitly,

$$E_z^{(p)} = \frac{\mu_0\bar{\lambda}}{4\pi}\int_{-\infty}^{+\infty} d\omega\,\frac{j\omega\gamma v}{|\omega|}\left(1 - \frac{1}{\beta^2}\right)$$
$$\times \exp\left[j\omega\tau - \frac{1}{v\gamma}|\omega x|\right]$$

$$E_z^{(s)} = \frac{\mu_0\bar{\lambda}}{4\pi}\int_{-\infty}^{+\infty} d\omega\,\frac{j\omega\gamma v}{|\omega|}\exp(j\omega\tau)$$
$$\times \begin{cases} \left(1 - \dfrac{1}{\varepsilon\beta^2}\right)A_1\exp[-\Gamma(x - D_{\text{int}})] \\ \qquad x > D_{\text{int}} \\[2mm] \left(1 - \dfrac{1}{\beta^2}\right)A_0\cosh\left(\dfrac{|\omega|}{\gamma v}x\right) \\ \qquad |x| < D_{\text{int}}. \end{cases} \quad (A\,8)$$

Similarly, the magnetic field satisfies

$$H_y^{(p)} = -\frac{\mu_0\bar{\lambda}}{4\pi}\int_{-\infty}^{+\infty} d\omega\,\frac{1}{\mu_0}\exp\left[j\omega\tau - \frac{1}{v\gamma}|\omega x|\right]\text{sign}(x)$$

$$H_y^{(s)} = \frac{\mu_0\bar{\lambda}}{4\pi}\int_{-\infty}^{+\infty} d\omega\,\frac{1}{\mu_0}\exp(j\omega\tau)$$
$$\times \begin{cases} -\gamma\sqrt{1 - \beta^2\varepsilon}\,A_1\exp[-\Gamma(x - D_{\text{int}})] \\ \qquad x > D_{\text{int}} \\[2mm] A_0\sinh\left(\dfrac{|\omega|}{\gamma v}x\right) & |x| < D_{\text{int}}. \end{cases} \quad (A\,9)$$

Imposing boundary conditions, from the continuity of E_z on the vacuum-dielectric interface $E_z^{(p)}(x = D_{\text{int}}^-) + E_z^{(s)}(x = D_{\text{int}}^-) = E_z^{(s)}(x = D_{\text{int}}^+)$ we obtain

$$\exp\left(-\frac{|\omega|}{v\gamma}D_{\text{int}}\right) + A_0\cosh\left(\frac{|\omega|}{\gamma v}D_{\text{int}}\right) = \frac{\varepsilon\beta^2 - 1}{\varepsilon(\beta^2 - 1)}A_1, \quad (A\,10)$$

whereas the continuity of H_y entails $H_y^{(p)}(x = D_{\text{int}}^-) + H_y^{(s)}(x = D_{\text{int}}^-) = H_y^{(s)}(x = D_{\text{int}}^+)$

$$-\exp\left(-\frac{1}{v\gamma}|\omega|D_{\text{int}}\right) + A_0\sinh\left(\frac{|\omega|}{\gamma v}D_{\text{int}}\right)$$
$$= -\gamma\sqrt{1 - \beta^2\varepsilon}\,A_1. \quad (A\,11)$$

The solution for the amplitude in the vacuum region from the above two equations is

$$A_0 = \frac{2}{\frac{\Omega - 1}{\Omega + 1}\exp\left(2\frac{|\omega|}{\gamma v}D_{\text{int}}\right) - 1}, \quad (A\,12)$$

where $\Omega = -\sqrt{1 - \beta^2\varepsilon}/[\gamma\varepsilon(1 - \beta^2)] = -\gamma\sqrt{1 - \beta^2\varepsilon}/\varepsilon$. Accordingly, the decelerating field is the secondary field acting on the charged particle in the vacuum core

$$E_{\text{dec}} \triangleq E_z^{(s)}(x = 0, z = vt, t)$$
$$= \frac{\mu_0\bar{\lambda}}{4\pi}\int_{-\infty}^{+\infty} d\omega\,\frac{j\omega\gamma v}{|\omega|}\left(1 - \frac{1}{\beta^2}\right)A_0$$
$$= -\frac{\mu_0\bar{\lambda}}{4\pi}2Re\left\{\int_0^{+\infty} d\omega\,\frac{jv}{\gamma\beta^2}\frac{2}{\frac{\Omega-1}{\Omega+1}\exp\left(2\frac{|\omega|}{\gamma v}D_{\text{int}}\right) - 1}\right\}$$
$$= -\frac{\mu_0\bar{\lambda}}{4\pi}2Re\left\{\frac{2\gamma v}{2D_{\text{int}}}\frac{jv}{\gamma\beta^2}\ln\left(\frac{1 - \Omega}{2}\right)\right\}. \quad (A\,13)$$

Having in mind that

$$\int_0^{+\infty} d\omega \, \frac{1}{u_1 \exp(u_2|\omega|) - 1}$$
$$= \frac{1}{u_2} \ln\left(1 - \frac{\exp(-u_2|\omega|)}{u_1}\right)\Big|_0^\infty$$
$$= \frac{1}{u_2} \ln\left(\frac{u_1}{u_1 - 1}\right)$$

and tacitly assuming that Ω is independent of ω, as well as taking the relativistic limit ($\gamma \gg 1$) for confinement, we may write

$$1 - \Omega = 1 + \gamma\sqrt{1 - \beta^2\varepsilon}/\varepsilon \simeq 1 + j\gamma\sqrt{\varepsilon - 1/\varepsilon^2}.$$

With Taylor expansion for $x \to \infty \ln(1 + jx) = -\ln(1/x) + j/x$, together with $\ln(-j\varsigma) = \ln(\varsigma) - j\pi/2$, the deceleration field is

$$E_{\text{dec}} = -\frac{\bar{\lambda}}{2\pi\varepsilon_0 D_{\text{int}}} Re\left\{j\ln\left(1 + j\gamma\sqrt{\frac{\varepsilon - 1}{\varepsilon^2}}\right)\right\}$$
$$\overset{\gamma \to \infty}{\simeq} \frac{\bar{\lambda}}{2\pi\varepsilon_0 D_{\text{int}}}\frac{\pi}{2}$$
$$= \frac{\bar{\lambda}}{4\varepsilon_0 D_{\text{int}}}, \tag{A 14}$$

implying that the decelerating field for a given charge is E_{dec} (V/m) $= \kappa$ (Ω/ms) q (C) and, as a result, the wake coefficient is

$$\kappa = \frac{1}{4\varepsilon_0 D_{\text{int}}}\left(\frac{\Omega}{\text{ms}}\right). \tag{A 15}$$

References

1. V. Karagodsky, D. Schieber, and L. Schächter, Phys. Rev. Lett. **104**, 024801 (2010).
2. A. Mizrahi and L. Schächter, Phys. Rev. E **70**, 016505 (2004).
3. A. Mizrahi and L. Schächter, Opt. Express **12**, 3156 (2004).
4. B. C. Stuart, M. D. Feit, A. M. Rubenchik, B. W. Shore, and M. D. Perry, Phys. Rev. Lett. **74**, 2248 (1995).
5. E. Hemsing and D. Xiang, Phys. Rev. Spec. Top. Accel. Beams **16**, 010706 (2013).
6. M. Dunning, E. Hemsing, C. Hast, T. O. Raubenheimer, S. Weathersby, D. Xiang, and F. Fu, Phys. Rev. Lett. **110**, 244801 (2013).
7. L. Schächter, R. L. Byer, and R. H. Siemann, AAC 23 (2002).
8. K. Bane and G. Stupakov, Phys. Rev. Spec. Top. Accel. Beams **6**, 024401 (2003).
9. A. M. Weiner, Rev. Sci. Instrum. **71**, 1929 (2000).
10. C. Chang, J. Liang, D. Hei, M. F. Becker, K. Tang, Y. Feng, V. Yakimenko, C. Pellegrini, and J. Wu, Opt. Express **21**, 32013 (2013).

6

Wavefront measurement techniques used in high power lasers

Haiyan Wang[1], Cheng Liu[1], Xiaoliang He[2], Xingchen Pan[1], Shenlei Zhou[1], Rong Wu[1], and Jianqiang Zhu[1]

[1] *Shanghai Institute of Optics and Fine Mechanics, Chinese Academy of Sciences, Shanghai 201800, China*

[2] *College of Sciences, Jiangnan University, Wuxi 214122, China*

Abstract

The properties of a series of phase measurement techniques, including interferometry, the Hartmann–Shack wavefront sensor, the knife-edge technique, and coherent diffraction imaging, are summarized and their performance in high power laser applications is compared. The advantages, disadvantages, and application ranges of each technique are discussed.

Keywords: wavefront measurement; phase retrieval; focus prediction; Ptychographical Iterative Engine

1. Introduction

High power solid-state laser facilities for Inertial Confinement Fusion (ICF) employ thousands of large optical components, including amplifiers, polarizing films, electro-optical switches, lenses, and mirrors[1–3]. These components usually possess large sizes and weights; some of them have diameters larger than half a meter and weigh up to hundreds of kilogram. The performance of these components is easily influenced by material non-uniformity, manufacturing errors, assembly stresses, and temperature changes[4]. For example, while the surfaces of large-aperture lenses can be accurately manufactured, their large weight and size cause deformations due to gravity, and assembly stress formed during installation may introduce remarkable changes in their shapes[5]. These variations, in turn, introduce significant aberrations to the passing laser beam. Wavefront distortion causes irregular malformation of far-field beams and reduction of the through-hole efficiency of spatial filtering; the frequency doubling efficiency is also substantially decreased, leading to serious degradation in the performance of the entire facility[6]. Low frequency wavefront distortion changes the spatial distribution of the focal spot[4], whereas high frequency wavefront distortion, which results from the noise of phase perturbation, may lead to self-focusing inside the optical components and generate serious material damage[7]. Extremely accurate techniques are necessary in laser beam sensing and optical component measurement to

satisfy the requirements of ICF systems for laser beam wavefronts. High precision measurements of optical components and laser beams are important endeavors in ICF research.

Wavefronts in high power lasers possess two distinguishing features. First, the laser beam to be measured is pulsed, which makes extraction of phase information using conventional phase-shifting techniques difficult. Second, the space for pulse synchronization and wave shaping optics inside the laser driver system is insufficient. An ideal phase measurement technique for an ICF system should feature high spatial resolution, high accuracy, simple setup, and rapid data acquisition. Given that most of the commonly used devices for wavefront measurement are unable to satisfy all of these requirements simultaneously and that the number of phase measurement techniques applicable to an ICF system is limited, several measurement techniques must be used in combination to obtain the required accuracy.

First, the current paper presents a general discussion on the wavefront measurement techniques applicable in ICF systems. Second, three traditional measurement methods, namely the Hartmann–Shack wavefront sensor, interferometry, and the knife-edge test, as well as their advantages and disadvantages, are discussed. Third, the development of Coherent Diffraction Imaging (CDI) and its application in high power laser systems is introduced in detail. Fourth, some newly developed techniques including the Ptychographical Iterative Engine (PIE) and Phase Modulation (PM) are introduced; some of their applications and potential applications in the field of high power laser systems are also demonstrated.

Correspondence to: Liu Cheng, Shanghai Institute of Optics and Fine Mechanics, Chinese Academy of Sciences, No. 390, Qinghe Road, Jiading District, Shanghai 201800, China. Email: cheng.liu@hotmail.co.uk

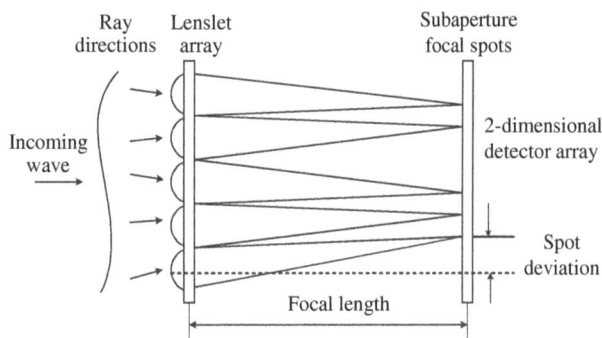

Figure 1. Schematic of a Hartmann–Shack sensor.

2. Traditional methods of phase measurement

Commonly used phase measurement techniques for ICF systems include the Hartmann–Shack wavefront sensor, interferometry, and the knife-edge technique, all of which are well-developed and commercialized techniques. The Hartmann–Shack wavefront sensor is used to detect online wavefronts and realize feedback control because of its high data acquisition speed and rapid computational processing. Interferometry is used to evaluate the properties of the optical elements. The knife-edge technique, which features the simplest structure among the three phase measurement techniques, is used to evaluate the characteristics of optics qualitatively. Apart from these traditional measurement techniques, the CDI method, which was developed for X-ray and electron beam imaging, is also used in some special cases and has shown significant advantages in high power lasers.

2.1. Hartmann–Shack wavefront sensor

The Hartmann–Shack wavefront sensor is mainly composed of a micro-lens array and a CCD camera (Figure 1). The focal spot array of an ideal plane wave is used as the reference pattern when a deformed wavefront is measured,

and the focal spots of the deformed wavefront shift from that of the reference beam. The spot deviation of Δy in the y direction is related to the slope of the deformed wavefront:

$$\frac{\Delta y}{f} = \frac{\partial W(x, y)}{\partial y}, \quad (1)$$

where f is the focal length of the lenslet. In this way, the wavefront distortion is converted to the spot offsets at the CCD sensor plane, and the phase map of the wavefront can be generated by integrating the calculated slope.

The most remarkable advantages of the Hartmann–Shack sensor are its simple structure and rapid data processing; these merits allow the use of the sensor in measurement of dynamic wavefronts[8], evaluation of laser beam quality, and realization of closed-loop wavefront control in combination with adaptive optics[9]. However, given its limited number of micro-lenses, low resolution is an inherent disadvantage of the Hartmann–Shack sensor. Most commercial Hartmann–Shack sensors have a limited number of micro-lenses. As such, only the low frequency components of the wavefront can be measured, which results in low accuracy measurements.

In OMEGA EP system, a Hartmann–Shack sensor of 133×133 lenslets is applied to set up a focal spot diagnostic (FSD) system to measure a lower energy sample of the main beam that is attenuated and down-collimated to a more convenient beam size (12 mm × 12 mm). Stretched pulses (250 mJ, 8 nm square spectrum, 5 Hz) are amplified using a multipass Nd:glass amplifier and compressed by a tiled-grating compressor, then 99.5% of the compressed pulse energy is reflected by a diagnostic pickoff mirror and the remainder is transmitted as a sample beam for the laser diagnostics package. The Hartmann–Shack sensor is positioned at an image plane conjugate to the fourth compressor grating, as shown in Figure 2. A local wavefront gradient as high as 15 mrad can be measured[10]. The FSD is qualified using a sequence of experiments designed to compare measurements made by the FSD and focal-spot microscope (FSM). Figure 3 shows some of the measurement

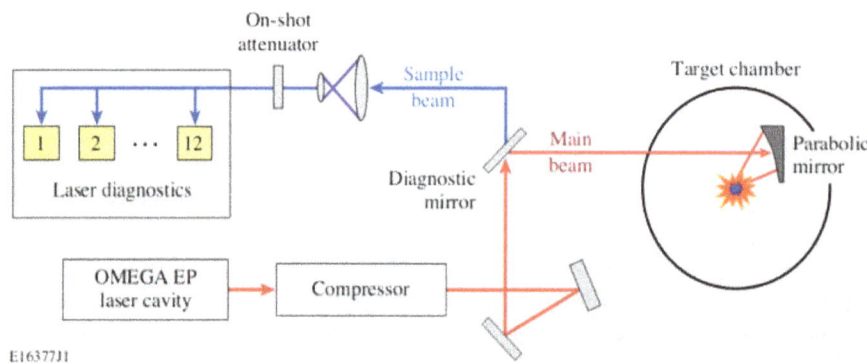

Figure 2. Overview of an OMEGA EP, showing the relative location of the main laser beam and the sample beam used by diagnostics for on-shot measurement of the laser properties. The FSD wavefront sensor is one of many laser diagnostics that characterize the sample beam (from Ref. [10]).

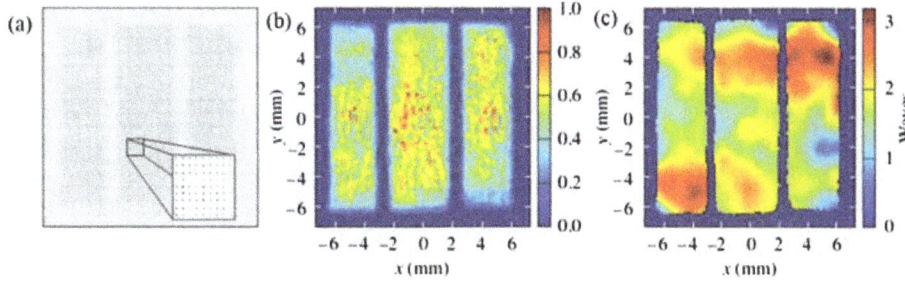

Figure 3. FSD measurements using the OPCPA front end (by Bromage). (a) Raw Hartmann–Shack image with inset showing the spots formed by each lenslet, (b) fluence (normalized), and (c) wavefront (in units of waves) (from Ref. [10]).

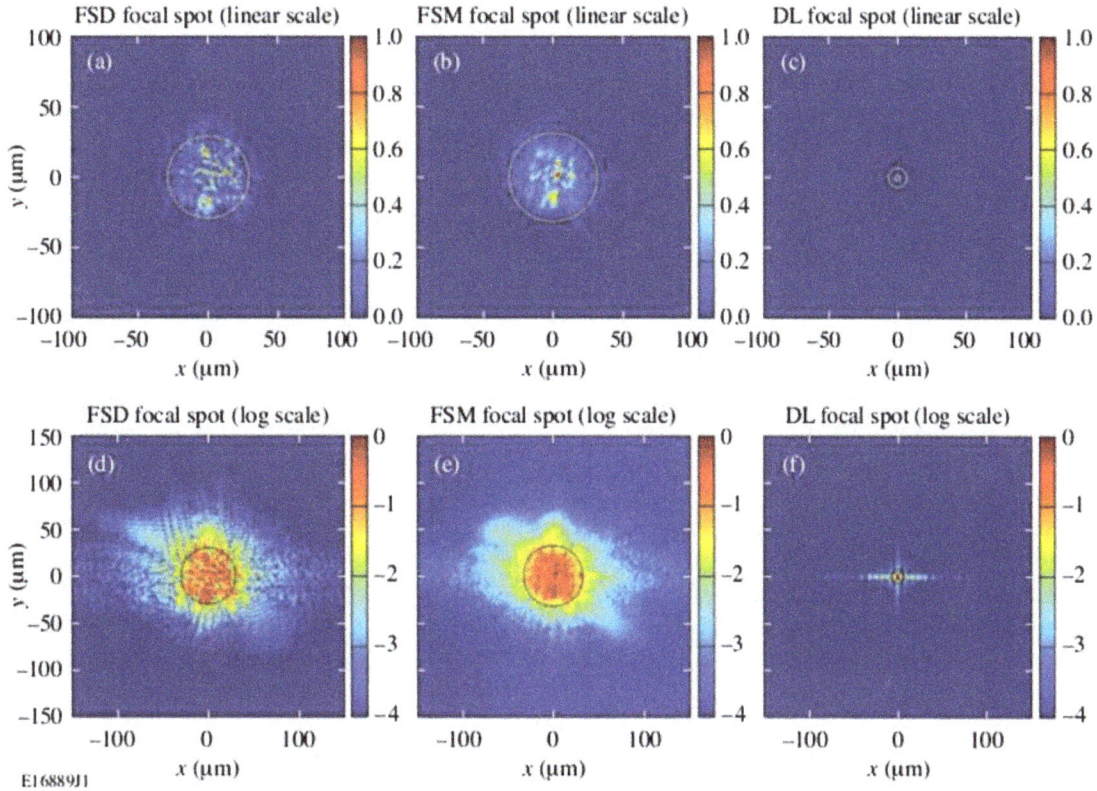

Figure 4. Same-shot measurements of a focal spot using the FSD and FSM (by Bromage) and the diffraction-limited (DL) spot, which are calculated by setting the wavefront error to zero. (a–c) Linear scale plots; (d–f) logarithmic scale plots. The circles contain 80% of the energy (from Ref. [10]).

results of the FSD wavefront sensor in these experiments; wave-plate throttles are set so that $400\,\mu\text{J}$ of the 100 mJ front end is focused in the target chamber, which provides enough energy for the FSD wavefront sensor and is not too high for the FSM. The image plane for this sensor is the last of the four tiled gratings inside the compressor. Figure 3(a) illustrates the raw data of the Hartmann–Shack sensor, Figure 3(b) shows the measured fluence of the near-field, and Figure 3(c) shows the corresponding measured wavefront. The performance of this system is evaluated by comparing its measurement results with those of the FSM system, which can be assumed to be correct. Figures 4(a) and (b) indicate the distributions of the focal spot measured using the FSD and FSM systems, respectively. Given

the finite spatial resolution of the FSD wavefront sensor (133×133 lenslets), the wavefront cannot be captured accurately and the fine details of the focal spot cannot be faithfully represented; however, the encircled energies of these two measurements show good agreement. Figures 4(d) and (e) show the spot distributions in logarithmic scales to demonstrate their difference more clearly.

The Hartmann–Shack sensor is also used at National Ignition Facility (NIF)[2] for phase measurements, where the beam passes though a micro-lens array and is focused on a CCD. A hexagonal lens array of 77 micro-lenses is used to form the focal spot array and a measurement accuracy of 0.1λ is achieved at 1.053 mm in the offline test.

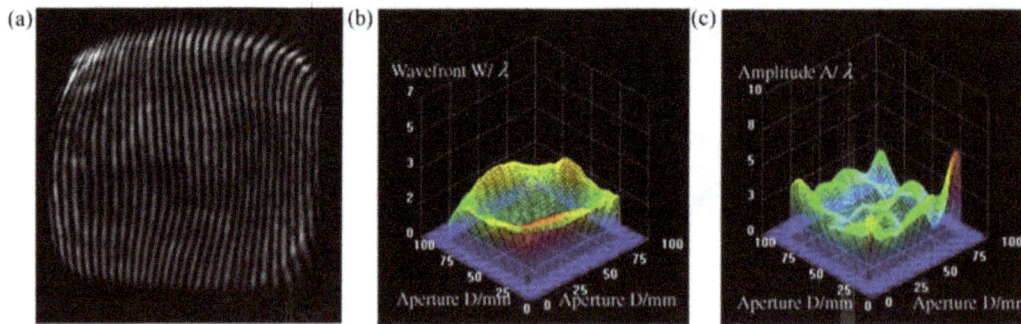

Figure 5. Testing results of an ICF system. (a) Radial shearing interferogram with spatial PM, (b) phase of the tested laser wavefront, and (c) amplitude of the tested laser wavefront (from Ref. [14]).

In most cases, measurements obtained using the Hartmann–Shack sensor present the closest real wavefront, but the resolution and accuracy of this technique are significantly lower than those obtained from interferometry.

2.2. Interferometry

2.2.1. Traditional interferometry

Interferometry is a classic method to determine phase distributions and detect wavefronts. The light wave to be measured initially interferes with a regular spherical or planar reference wave to generate interference fringe patterns, and the resultant phase distribution is extracted from the recorded fringes using various methods, including phase shifting and fringe carriers. Given that interferometry is remarkably sensitive to environmental turbulence, such as mechanical vibration and air fluence, the interferometer must be installed on a vibration-isolated table inside a room at constant temperature and humidity to achieve high accuracy. Using the phase-shift technique and high resolution CCD, the highest measurement accuracy of interferometry can reach $\lambda/1000$ in theory. Although the accuracy of interferometry is significantly higher than that of other techniques, its setup complexity and high requirement on its working environment limit its application in high power laser systems.

Interferometers used in high power lasers must possess large apertures because the diameters of large optical elements can reach up to half a meter. Given the difficulties associated with fabricating large reference standard mirrors and the related optical components, the cost of an interferometer increases drastically with its diameter, and its requirements on the working environment become more difficult to satisfy. To detect the machining quality of optical elements used at NIF, VEECO (USA) developed seven large Fizeau interferometers with diameters of up to 610 mm[11]; the highest measurement precision of these instruments reached $\lambda/10$ (PV) and the measurement repeatability depended on the test surface gradient. The interferometer applied by the French Atomic Energy Commission (CEA) has a diameter of 800 mm and is the largest interferometer ever reported.

In addition to optical element measurements, the Fizeau laser interferometer is also used in adjusting the optical alignment[4].

Interferometers are capable of performing phase measurement with high accuracy and can meet the requirements of most high power laser applications in terms of resolution and aperture size. However, the use of interferometers to detect wavefronts online is difficult because of their complex structure and complicated environmental requirements.

2.2.2. Shearing interferometry

The reference standard mirror used in classic interferometry must be fabricated at an accuracy one order higher than that for a common optical element. Fabrication of a large reference mirror that is suitable for use in high power lasers remains challenging. Shearing interferometers do not need an accurate reference standard mirror; they possess simple structures, strong anti-interference abilities, and stable stripes. In theory, shearing interferometers are a good alternative to traditional interferometers for measuring the quality of optical elements and light beam wavefronts. Shearing interferometers can be classified into lateral and radial shearing interferometers. Given that two orthogonal shearing interferograms at precise displacements in the x and y directions are necessary for lateral shearing interferometers, the wavefront retrieval process is markedly complex and error prone. Although the cross-grating lateral shearing interferometer has been proposed to test the density distribution of deuterium–tritium (DT) ice in ICF experiments and the root-mean-square error has been found to be $<\lambda/15$[12] in theory, no experimental result has yet been reported. The radial shearing interferometer introduces radially symmetric shearing by interfering with the wavefront with an expanded copy of itself[13]. The wavefronts are measured with high precision without setting special reference light beams. The radial shearing interferometer is also sensitive to mechanical vibrations and disturbances in the environment. The accuracy of radial shearing interferometers is better than that of Zygo interferometers in the comparison made by Liu[14], yielding an error of $<1/1000\lambda$ in computer simulations. Figure 5 shows practical testing results from a 1064 nm

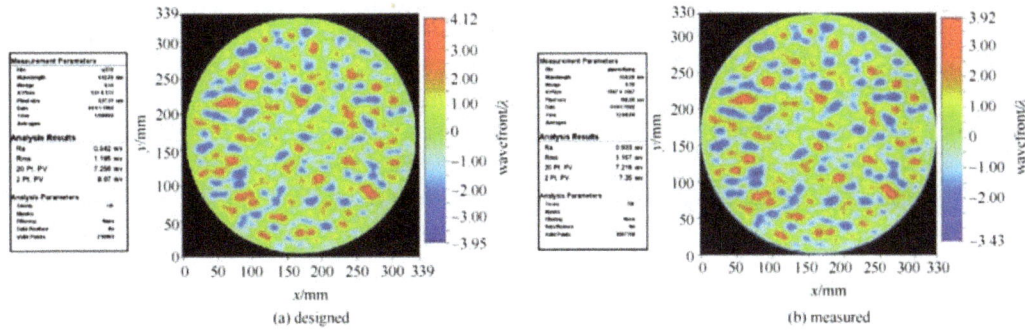

Figure 6. Designed and measured surfaces of a CPP with 380 mm diameter (from Ref. [18]).

pulse laser using a radial shearing interferometer in the ICF system.

Given that shearing interferometers measure the phase variance and that the measured phase map indicates the differential distribution of the wavefront, these instruments are not as intuitive as classic interferometers. Complicated imaging processing and wavefront reconstruction algorithms are necessary to retrieve the wavefront from the differential data, thereby limiting the applications of shearing interferometers.

2.2.3. Subaperture stitching method

In classical interferometry, the elements to be measured are imaged by a CCD camera. Thus, the resolution is L/N, where L is the size of the element to be measured and N is the pixel number of the CCD camera. The spatial resolution of a 400 mm diameter element and a CCD with 2048 pixels $\times$ 2048 pixels is approximately 195 µm, which is too low for most high power laser applications. Subaperture stitching interferometers can measure the large optical components of various subapertures, after which the measurements are stitched together to obtain a high resolution. This measurement method has been used to measure large optical elements with sizes of up to 800 mm $\times$ 400 mm at Laser Méga Joule and NIF[15–17]. Wen[18] successfully measured a continuous phase plate (CPP) using the subaperture stitching method, the global least squares method, and the image fusion technique. Figure 6 shows the designed and measured CPP surfaces (380 mm diameter). The original stitching method only detects low order components of the surface by fitting polynomials to non-overlapping subapertures[19, 20]; by contrast, the stitching method of overlapping subapertures and connecting interferograms minimizes the error of connection down to $\lambda/20$[21].

The use of subaperture stitching interferometers significantly reduces cost and increases spatial resolution[22] and measurement accuracy[23]. However, the subaperture stitching method possesses inherent disadvantages. First, this method requires a high quality standard mirror, which is difficult to fabricate. Second, error transfer and the unstable solution are unavoidable sources of errors, which make it difficult to obtain sufficiently high measurement accuracy. In addition, the subaperture stitching interferometer detection process is time consuming, and the requirements of environmental stability are remarkably complicated.

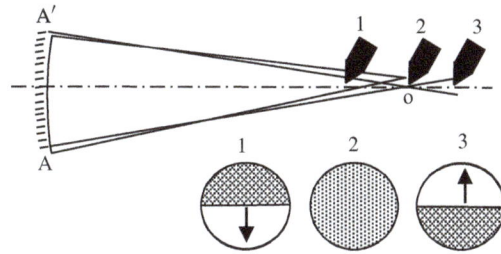

Figure 7. Geometric principle of the knife-edge test.

2.3. Knife-edge test method

2.3.1. Traditional Foucault method

The knife-edge method was proposed by Foucault in 1858[13]. The geometric principle of the knife-edge test is shown in Figure 7, where AA' is an ideal spherical surface with center o. When a point source is placed at o, the light from o to AA' coincides with the normal direction of the corresponding surface element and the reflected light returns to o. When the knife-edge cuts the reflected beam from right to left, the mirror is observed behind the knife-edge. When the knife-edge is located at position 1, the shadow on AA' darkens from right to left, i.e., the shadow and knife-edge move in the same direction. The entire field is bright if the knife-edge has not met point o and when the knife-edge is located at position 2, which is at the center of the sphere; AA' darkens immediately when the knife-edge first cuts through point o. When the knife-edge cuts the light at position 3, the shadow on AA' darkens from left to right, which is opposite to the moving direction of the knife-edge. This method is sensitive to slopes rather than heights, and only one direction can be measured at a single orientation of the knife-edge.

Given that only qualitative measurements can be realized by the traditional Foucault test method, a digitized Foucault tester (Figure 8) is used for quantitative measurements.

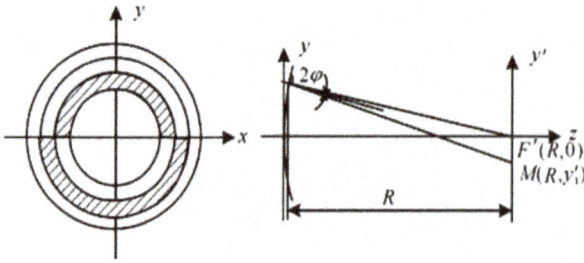

Figure 8. Principle of a digitized Foucault tester.

When a point source is located at the center of a spherical mirror, the returning beam focuses on the same position as the point source if the mirror under testing possesses an ideal spherical surface. When defects are present on the surface of the mirror, the returning beam deviates from the focal point and reaches a new location. The angle by which it departs from the ideal position can be calculated as follows:

$$\varphi_x = -\frac{1}{n}\left[\frac{\partial \Delta\omega(x, y)}{\partial x}\right], \qquad (2)$$

where n is the refractive index of air and $\Delta\omega(x, y)$ is the wave aberration. The deviation of the returning beam from the ideal position E_x and E_y can be calculated from the location of the knife-edge, and the wave aberration of the mirror can be obtained using

$$\Delta\text{OPD}_{\text{waves}} = \iint E(x, y)\text{d}x\text{d}y, \qquad (3)$$

where $E_x = R\varphi_x$, $E_y = R\varphi_y$, and R is the radius of curvature.

The performance of the digitized Foucault test method has been compared with that of interferometry[24, 25]; when the air turbulence is totally eliminated, the precision of the former is comparable with that of the latter. Foucault testers have been used in nonlinear measurement of wavefront sensorless adaptive optics systems[26] as well as in measurements of mid-frequency surface errors in ICF[27]. However, given that the Foucault tester has strict testing environment requirements, even slight turbulence in the optical path may cause remarkable image distortion and affect the final test results. Thus, significant improvements are needed to enable the use of this method in high power laser applications.

3. CDI technology and its applications

3.1. Development of CDI

CDI is a phase-retrieval method based on computer iterative calculations. The original purpose of CDI is imaging of the wave phase using X-rays and electron beams when high quality optics are unavailable. Given its outstanding advantages, which include simple setup, compact structure,

and low environmental requirements, CDI has been used in high power laser systems to measure wavefronts and predict focus.

The principle of CDI was first developed by Hoppe in the 1970s and then improved by Fienup. The main theories include the Gerchberg–Saxton (G–S) algorithm, the error-reduction algorithm, and the input–output algorithm[28]. These algorithms have been widely used in X-ray and electron imaging[29, 30]. The traditional CDI method suffers from disadvantages with respect to viewing field, convergence speed, and reliability. To overcome these disadvantages, Rodenburg[31] proposed an improved phase retrieval algorithm called PIE, which used a Wegener filter-like algorithm to reconstruct images iteratively from a set of diffraction patterns and extended it to an extended PIE (ePIE) algorithm to obtain an accurate model of the illumination and specimen functions simultaneously. PIE and ePIE are promising algorithms for imaging using X-rays[32, 33], electron beams[34, 35], and visible light[36, 37]. Given that the CDI algorithms can directly measure the phase distribution of a laser beam from the recorded diffraction intensity, they are often used in various high power laser applications, including wavefront detection, large optical element measurement, and FSDs.

3.2. Applications of traditional CDI

The wavefront of a high power laser system is easily distorted because of the high complexity of the system, which contains thousands of optical elements. The wavefront of the laser beam is difficult to control because it involves routing tasks of all of the ICF facilities. However, most of the commonly used measurement techniques do not satisfy the requirements for accurate online measurement because of the compact structure and limited inner space of CDI. In 2000, CDI was first used to measure the phase of high power laser beams[38]. The intensity of the laser beam at two different planes vertical to the optical axis were recorded, and the Fresnel phase-retrieval algorithm based on the G–S algorithm and the Fienup phase-retrieval algorithm was used to reconstruct the complex amplitudes of these two recording planes. Figure 9 shows the experimental results, where the first two images from the left of each row indicate the laser intensities at the two recording planes; the third image provides the measured phase distribution of the laser beam.

In 2006 Brady and Fienup[39] measured a concave spherical mirror using the CDI method (setup shown in Figure 10), where a He–Ne laser was filtered by a microscope objective and a pinhole placed near the center of curvature of the concave spherical mirror was used as the illumination beam; the intensity distributions of the resulting diffraction spots were measured using a CCD mounted on a computer-controlled translation stage, which could accurately shift

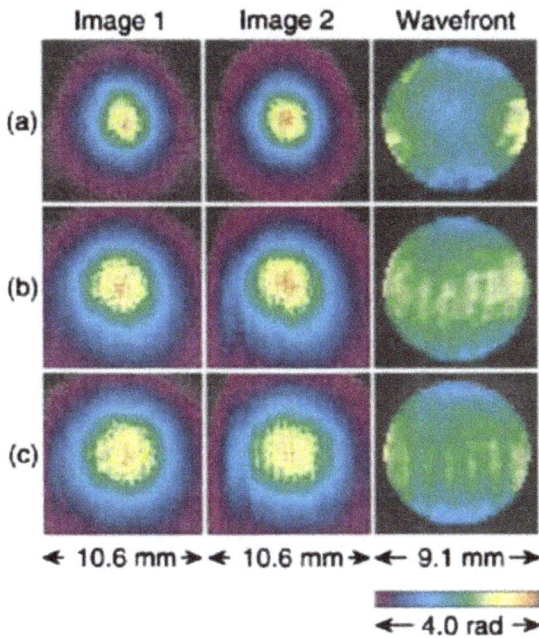

Figure 9. Measured intensity distributions (image 1 and image 2) and reconstructed wavefronts of 100 fs pulses at different output power levels (from Ref. [38]).

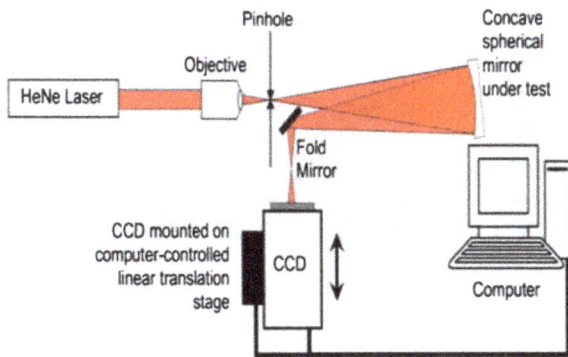

Figure 10. Experimental arrangement used for phase retrieval measurements (from Ref. [39]).

along the optical axis. The measurement results are shown in Figure 11, where the images in the top row are the recorded diffraction intensities and the images in the bottom row are the obtained surface shapes. However, the accuracy of these results is not ensured because the element is not measured independently using an instrument with sufficient accuracy. The diameter of the optical elements used in the high power laser system is up to half a meter, and numerous optical elements possess large surface curvatures. Given that the common interferometer or Hartmann–Shack sensor is incapable of measuring large phase slopes, the measurement of the complex reflectance or transmittance of these large elements is fairly challenging, but CDI method can be used to solve this problem.

Knowledge of the focus is important in conducting physical experiments. The focal spots of a high power laser system can be highly structured because of the complexity of the facility, which contains hundreds of optical surfaces. However, measurement of the focus field varies significantly[2, 40] because direct measurement of the focus is impossible at extreme intensities. Multiple focal-plane spatial phase retrieval for a chirped-pulse-amplification laser was demonstrated by Bahk in 2008[41, 42]. The wavefront measured by phase retrieval is used to predict focal spots at high energies via separate measurement of the differential wavefront changes. Comparison of the measured and the predicted focal spots is shown in Figure 12, where the directly measured focal spot is shown in Figure 12(a); this spot agrees well morphologically with the predicted focal spot shown in Figure 12(b). The application of phase retrieval helps to extend the capability of FSDs for high intensity lasers beyond conventional direct wavefront measurements.

In 2010, the CDI algorithm was used to form a FSD to predict the focus of an OMEGA EP laser at the University of Rochester's Laboratory for Laser Energetics[43]. The short-pulse diagnostic package contained a FSD which received a sampled beam downstream and a far-field CCD camera imaging the far-field intensity of the sample beam. The phase was initially retrieved from the sample beam to remove the differential piston uncertainty. Using a dense quasi-Newton downhill search algorithm to modify the optimization parameters for the subsequent iterations, the FSD prediction improved significantly compared with the initial FSD prediction, with the correlations for the diagnostic and target beams increased from 0.87 ± 0.04 and 0.82 ± 0.04 to 0.96 ± 0.01 and 0.87 ± 0.02, respectively. The error in the transfer wavefront measurement was then measured by CDI using the data recorded in the FSM. The correlation between the FSD prediction and the direct FSM measurement was improved to 0.93 ± 0.02 using the corrected transfer wavefront. Using the measured complex amplitude and transfer function, the focal spot of the main laser beam inside the chamber could be accurately predicted.

This diagnostic method was evaluated on the OMEGA EP laser beam over a population of 175 shots to illustrate its reliability and stability; the evaluation was conducted for approximately 18 months from 2010 to 2012[44]. Figure 13(a) shows the frequency of cross-correlation values between the FSD and the far-field CCD. A significant improvement in performance was obtained after the phase-retrieved corrections were applied; the mean cross-correlation increased from 0.83 to 0.96. Figure 13(b) shows the results of more stringent testing of the FSD measurement accuracy; the cross-correlation values (reliability >0.9 with $>95\%$ probability) between the measurements of FSD and FSM are also indicated. These diagnostics are used as a key tool for focal spot checking in the OMEGA EP laser.

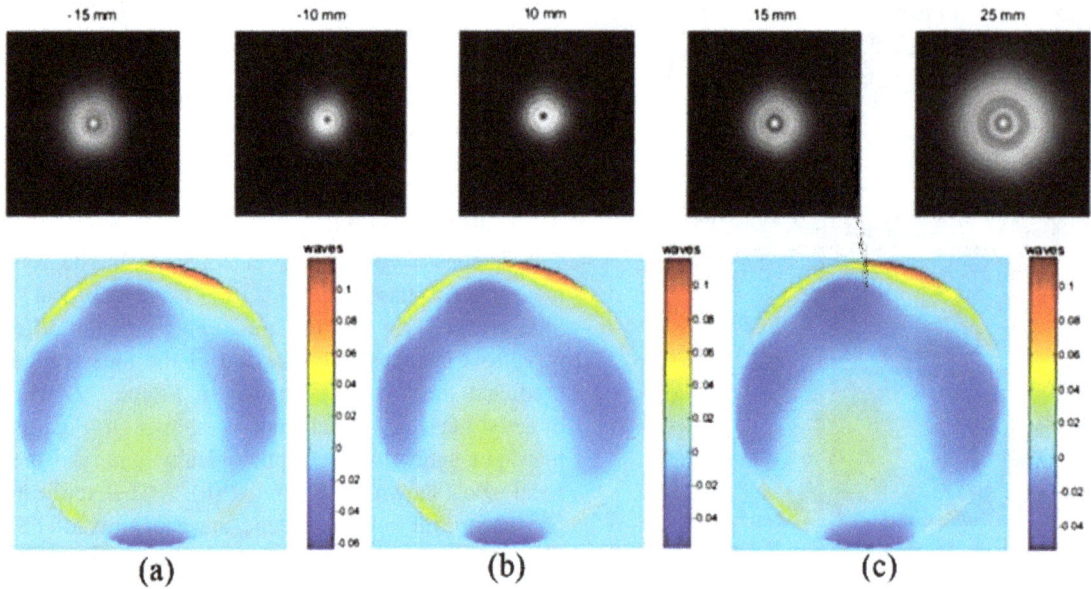

Figure 11. Recovered phases obtained by Brady and Fienup (from Ref. [39]).

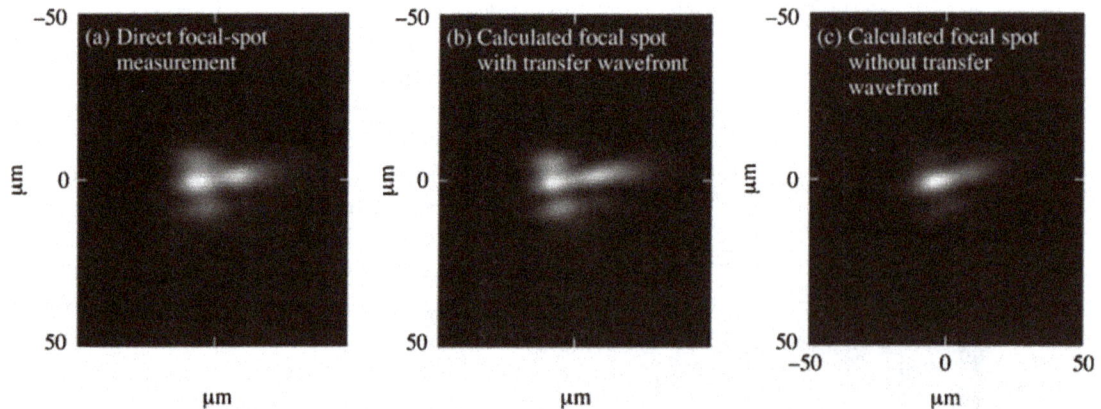

Figure 12. Linear scale comparison of the directly measured focal spot (a) in the presence of an aberrator with the focal spots calculated with and without the use ((b) and (c), respectively) of the transfer wavefront obtained from phase retrieval (from Ref. [42]).

4. PIE and its application

PIE is a newly developed CDI method proposed by Rodenburg in 2004 to overcome the disadvantages of traditional CDI[30]. Although PIE has been successfully used in electron and X-ray imaging, its high power laser applications have not been explored.

4.1. Basic principles

For clarity, the principles of PIE and ePIE are outlined in Figure 14; details can be found in the relevant literature[31, 45]. The object $O(x, y)$ under the illumination $P(x, y)$ can accurately shift across the optical axis; a set of diffraction patterns $I_n(x, y)$ is recorded by the CCD downstream of the object when the object is scanned to a series of positions

(Figure 15). Iterative calculations are performed between the object and the recording planes to retrieve the complex transmittance of the object and the complex amplitude of the illumination via the following procedure.

(1) The exit field at the current position R_i is calculated with two random guesses for $P_{n,g}(r)$ and $O_{n,g}(r - R_i)$

$$\Psi_{n,g}(r, R_i) = P_{n,g}(r) \cdot O_{n,g}(r - R_i), \quad (4)$$

where the subscript n represents the nth iteration.

(2) The wavefunction in the data recording plane is calculated from the Fourier transform of $\Psi_{n,g}(r, R_i)$:

$$\Psi_{g,n}(k, R_i) = \Im[\Psi_{g,n}(r, R_i)] = |\Psi_{g,n}(k, R_i)|e^{i\theta_n(k, R_i)}. \quad (5)$$

(3) The amplitude of $\Psi_{n,g}(r, R_i)$ is replaced by the square

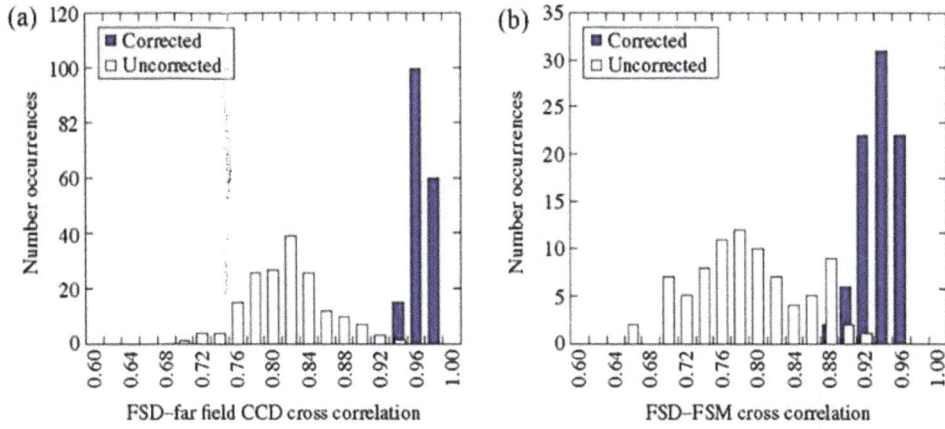

Figure 13. Histograms illustrating the effects of phase-retrieval improvements on a large population of measurements. (a) Sample beam focal-spot accuracy showing cross-correlation between the FSD prediction and the far-field CCD measurement. (b) Main-beam focal-spot accuracy showing cross-correlation between the FSD prediction and the FSM measurement (from Ref. [44]).

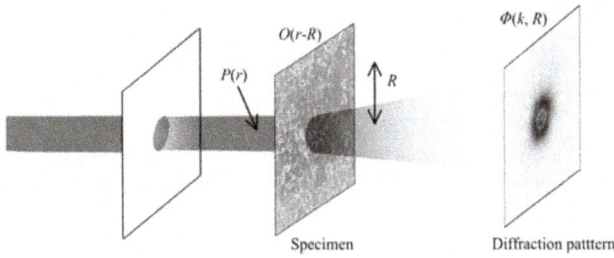

Figure 14. Principles of PIE and ePIE (from Ref. [46]).

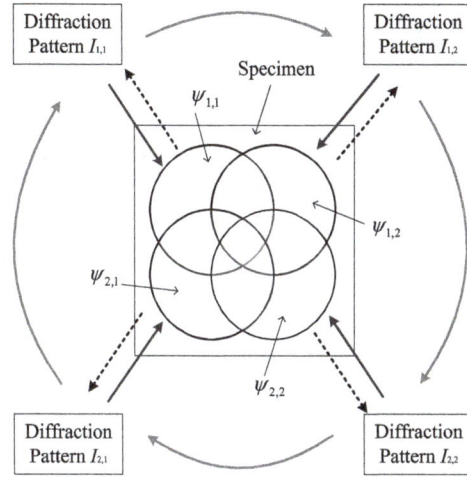

Figure 15. Diffraction patterns with the illumination beam in overlapping positions.

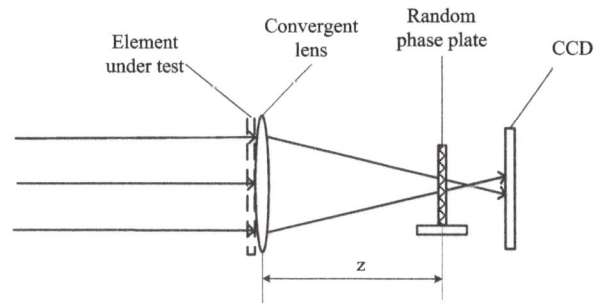

Figure 16. Experimental setup of the phase detection for large-aperture optical elements.

root of $I_n(x, y)$:

$$\Psi_{c,n}(k, R_i) = \sqrt{I_n}e^{i\theta_n(k, R_i)}. \tag{6}$$

(4) The guess is updated at the exit field by inverse Fourier transform:

$$\Psi_{c,n}(r, R_i) = \Im^{-1}[\Psi_{c,n}(k, R_i)]. \tag{7}$$

(5) The functions $P(r)$ and $O(r, R_i)$ are updated with the following formulas:

$$O_{\text{new}}(r, R_i) = O_{g,n}(r, R_i) + \frac{|P_n(r)|}{|P_n(r)|_{\max}} \frac{P_n^*(r)}{[|P_n(r)|^2 + \alpha]}$$
$$\times [\Psi_{c,n}(r, R_i) - P_n(r) \cdot O(r, R_i)], \tag{8}$$
$$P_{\text{new}}(r) = P(r) + \frac{|O(r, R_i)|}{|O(r, R_i)|_{\max}} \frac{O^*(r, R_i)}{[|O(r, R_i)|^2 + \alpha]}$$
$$\times [\Psi_{c,n}(r, R_i) - P_n(r) \cdot O(r, R_i)]. \tag{9}$$

Steps 1 to 5 are repeated until satisfactory images are generated.

4.2. Measurement of the transmittance of a large optical element with the use of ePIE

Illumination on the specimen in Figure 16 can be measured with the use of ePIE, and this property provides a new

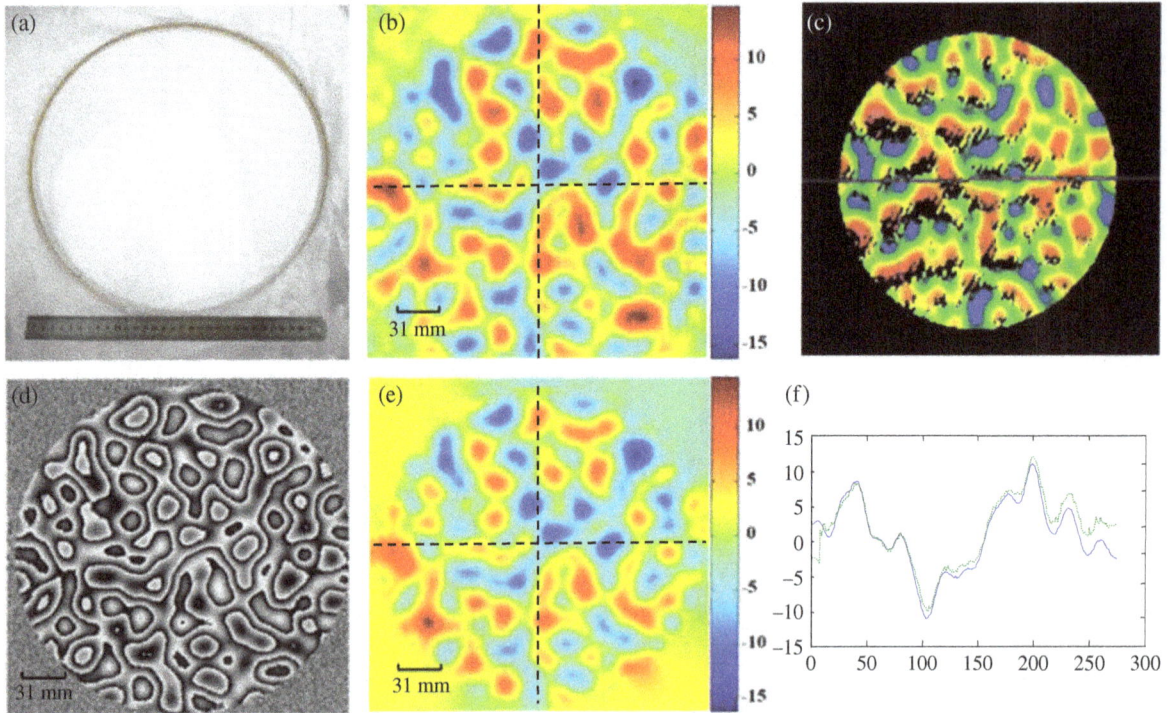

Figure 17. (a) Manufactured CPP, (b) CPP design value, (c) measurement result of a Zygo interferometer, (d) wrapped phase of the measured modulation function, (e) unwrapped phase of the measured modulation function, and (f) the measured result and designed value along the horizontal lines of (b) and (e) (from Ref. [46]).

method for measuring the complex transmittance of large optical elements used in high power laser applications[46]. Practical measurement was conducted to illustrate the feasibility of this method by CPP, which is a key element in ICF systems that smoothens the laser beam to ensure an ideal focal spot with a highly irregular surface profile[47, 48]. The optical setup used is shown in Figure 16. The wavefront $\varphi_0(x, y)$ of the light leaving the convergent lens can be measured using the ePIE algorithm when the CPP is removed (Figure 16); the wavefront $\varphi_1(x, y)$ of the light leaving the CPP can also be measured. By subtracting $\varphi_0(x, y)$ from $\varphi_1(x, y)$, the PM of the CPP is obtained.

Figure 17(a) shows a photograph of the manufactured CPP plate with approximately 31 cm diameter. Figure 17(b) shows the design value of the surface profile. Figure 17(c) shows the measurement result obtained using a Zygo interferometer, where the black areas indicate invalid measurements. The slope of the surface profile is very steep at these black areas and the interference fringes are too dense to resolve, resulting in the existence of these immeasurable areas. As such, the measurement result cannot indicate the real surface profile of the CPP. Figure 17(d) shows the wrapped phase of the measured PM function and Figure 17(e) shows the corresponding unwrapped phase distribution. The aperture of the measurement is 28 cm, which is decided by the parallel beam illuminating the CPP. The values along the vertical lines in Figures 17(b) and (e)

are plotted in Figure 17(f), and the maximum difference between the measured and designed values is approximately 2.1 rad. This research illustrates that ePIE is a promising technique for measuring large optical elements.

4.3. PM technique and its potential application

The data acquisition time of ePIE is several minutes; thus, it cannot be used to measure dynamic wavefronts. To overcome this disadvantage, Zhang proposed the PM method[49]. The basic principle and the experimental setup of the phase modulation technique are shown in Figures 18(a) and (b).

The modulator with a designed transmission function is located between the entrance and detector domains, while the entrance plane is the focal plane of the incident wave to be measured. As one frame of the diffraction pattern is recorded, the illumination on the modulator plane can be iteratively reconstructed at high accuracy. This method features a very simple structure and short data acquisition time; thus, it is suitable for FSDs and laser plasma imaging, where the laser beam is pulsed and most of the common techniques do not work well. Proof of concept is demonstrated by measuring a seriously distorted convergent He–Ne laser beam; Figures 19(a) and (b) show the reconstructed phase and modulus of the field incident on the phase modulator. Figures 19(c) and (d) show the predicted and measured focal spots; the difference between the two focal spots is difficult

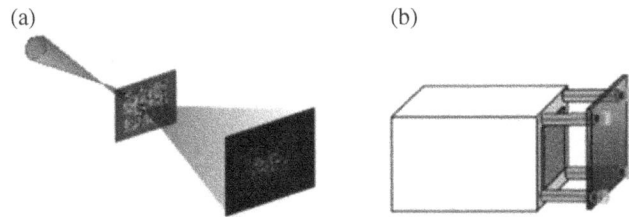

Figure 18. Schematic of the PM technique. (a) Basic principle and (b) experimental setup.

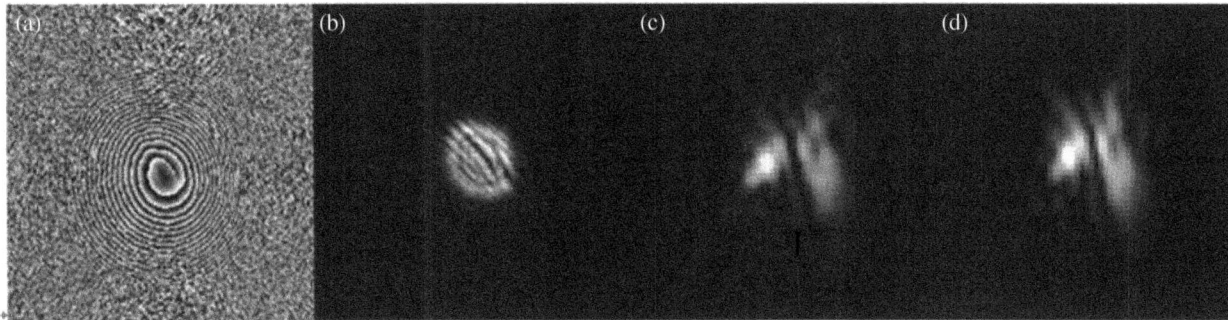

Figure 19. Experimental results of PM. (a) Reconstructed phase, (b) reconstructed modulus, (c) predicted focal spot with the PM technique, and (d) the measured focal spot.

to distinguish with the naked eye. The details of this research will be reported in a separate paper.

Summary

The use of several traditional phase measurement methods to measure the wavefront in high power laser applications was evaluated; their advantages, disadvantages, and applications were also discussed. As a classical technique, interferometry features the highest accuracy and resolution among the techniques studied. However, in most cases, interferometry is only used to measure the static properties of optical elements because of its complex structure and critical environmental requirements. The Hartmann–Shack wavefront sensor shows high measurement speed and may be used to realize feedback wavefront control. However, this sensor cannot detect high frequency components. CDI techniques have been increasingly adopted in high power laser applications and are to be considered as promising techniques in special cases. CDI techniques are based on principles that are entirely different from those of traditional techniques and were developed particularly for imaging using short-wavelength radiation.

Acknowledgement

This research is supported by the One Hundred Talents Project of the Chinese Academy of Sciences, China (Grant No. 1204341-XR0).

References

1. D. J. Trummer, R. J. Foley, and G. S. Shaw, *Third International Conference on Solid State Lasers for Application to Inertial Confinement Fusion*, Monterey, CA, USA, p. 363 (1999).
2. R. A. Zacharias, N. R. Beer, E. S. Bliss, S. C. Burkhart, S. J. Cohen, S. B. Sutton, R. L. Van Atta, S. E. Winters, J. T. Salmon, M. R. Latta, C. J. Stolz, D. C. Pigg, and T. J. Arnold, Opt. Eng. **43**, 2873 (2004).
3. J. Qiao, A. Kalb, M. J. Guardalben, G. King, D. Canning, and J. H. Kelly, Opt. Express **15**, 9562 (2007).
4. W. H. Williams, J. M. Auerbach, and M. A. Henesian, Proc. SPIE **3264**, 93 (1998).
5. W. H. Williams, J. M. Auerbach, M. A. Henesian, J. K. Lawson, P. A. Renard, and R. A. Sacks, Proc. SPIE **3492**, 22 (1998).
6. B. M. Van Wonterghem, J. T. Salmon, and R. W. Wilcox, Inertial Confinement Fusion Quarterly Report **5**, 4251 (1994).
7. V. I. Bespalov and V. I. Talanov, JETP Lett. **3**, 307 (1996).
8. T. J. Brennan and D. J. Wittich, Opt. Eng. **52**, 071416 (2013).
9. A. R. Saad and J. B. Martin, Appl. Opt. **52**, 5523 (2013).
10. J. Bromage, S.-W. Bahk, D. Irwin, J. Kwiatkowski, A. Pruyne, M. Millecchia, M. Moore, and J. D. Zuegel, Opt. Express **16**, 16561 (2008).
11. C. Ai, R. Knowlden, and J. Lamb, Proc. SPIE **3134**, 47 (1997).
12. T. Ling, D. Liu, L. Sun, Y. Yang, and Z. Cheng, Proc. SPIE **8838**, 88380J (2013).
13. D. Malacara, *Optical Shop Testing* (John Wiley & Sons, New York, 2007).
14. D. Liu, Y. Yang, J. Weng, X. Zhang, B. Chen, and X. Qin, Opt. Commun. **275**, 173 (2007).
15. C. R. Wolfe, J. D. Downie, and J. K. Lawson, Proc. SPIE **2870**, 553 (1996).
16. D. M. Aikens, C. R. Wolfe, and J. K. Lawson, Proc. SPIE **2576**, 281 (1995).
17. M. Bray, *Third International Conference on Solid State Lasers for Application to Inertial Confinement Fusion*, Monterey, CA, p. 946 (1998).

18. S. Wen, Q. Shi, H. Yan, Y. Zhang, C. Yang, and J. Wang, High Power Laser Particle Beams **24**, 2296 (2012).
19. J. Thunen and O. Kwon, Proc. SPIE **351**, 19 (1982).
20. W. Chow and G. Lawrence, Opt. Lett. **8**, 468 (1983).
21. M. Otsubo, K. Okada, and J. Tsujiuchi, Opt. Eng. **33**, 608 (1994).
22. S. Chen, S. Li, Y. Dai, and Z. Zheng, Appl. Opt. **46**, 3504 (2007).
23. J. Fleig, P. Dumas, P. E. Murphy, and G. W. Forbes, Proc. SPIE **5188**, 296 (2003).
24. D. E. Vandenberg, W. D. Humbel, and A. Wertheimer, Opt. Eng. **32**, 1951 (1993).
25. L. Yuan and Z. Wu, Proc. SPIE **7654**, 765402 (2010).
26. H. Song, G. Vdovin, R. Fraanje, G. Schitter, and M. Verhaegen, Opt. Lett. **34**, 61 (2009).
27. B. Xuan, J. Li, S. Song, and J. Xie, Proc. SPIE, 6723, 67230A. (2007).
28. J. R. Fienup, Opt. Lett. **3**, 27 (1978).
29. J. M. Zuo, I. Vartanyants, M. Gao, R. Zhang, and L. A. Nagahara, Science **300**, 1419 (2003).
30. G. J. Williams, H. M. Quiney, B. B. Dhal, C. Q. Tran, K. A. Nugent, A. G. Peele, D. Paterson, and M. D. de Jonge, Phys. Rev. Lett. **97**, 025506 (2006).
31. J. M. Rodenburg and H. M. L. Faulkner, Appl. Phys. Lett. **85**, 4795 (2004).
32. U. Weierstall, Q. Chen, J. C. H. Spence, M. R. Howells, M. Isaacson, and R. R. Panepucci, Ultramicroscopy **90**, 171 (2002).
33. H. N. Chapman, Nat. Mater. **8**, 299 (2009).
34. F. Hüe, J. M. Rodenburg, A. M. Maiden, F. Sweeney, and P. A. Midgley, Phys. Rev. B **82**, 121415 (2010).
35. M. J. Humphry, B. Kraus, A. C. Hurst, A. M. Maiden, and J. M. Rodenburg, Nat. Commun. **3**, 730 (2012).
36. E. Osherovich, Y. Shechtman, A. Szameit, P. Sidorenko, E. Bullkich, S. Gazit, S. Shoham, E. B. Kley, M. Zibulevsky, I. Yavneh, Y. C. Eldar, O. Cohen, and M. Segev, *Precision Imaging and Sensing (CF3C), San Jose, CA, USA* (2012).
37. A. Szameit, Y. Shechtman, E. Osherovich, E. Bullkich, P. Sidorenko, H. Dana, S. Steiner, E. B. Kley, S. Gazit, T. Cohen-Hyams, S. Shoham, M. Zibulevsky, I. Yavneh, Y. C. Eldar, O. Cohen, and M. Segev, Nat. Mater. **11**, 455 (2012).
38. S. Matsuoka and K. Yamakawa, J. Opt. Soc. Am. B **17**, 663 (2000).
39. G. R. Brady and J. R. Fienup, *Optical Fabrication and Testing, Rochester, NY, USA* (2006).
40. T. J. Kessler, J. Bunkenburg, H. Huang, A. Kozlov, and D. D. Meyerhofer, Opt. Lett. **29**, 635 (2004).
41. S. W. Bahk, J. Bromage, J. D. Zuegel, and J. R. Fienup, *Conference on Lasers and Electro-Optics, San Jose, CA, USA* (2008).
42. S. W. Bahk, J. Bromage, I. A. Begishev, C. Mileham, C. Stoeckl, M. Storm, and J. D. Zuegel, Appl. Opt. **47**, 4589 (2008).
43. B. E. Kruschwitz, S. W. Bahk, J. Bromage, D. Irwin, M. D. Moore, L. J. Waxer, J. D. Zuegel, and J. H. Kelly, *Conference on Lasers and Electro-Optics, San Jose, CA, USA* p. JThE113 (2010).
44. B. E. Kruschwitz, S. W. Bahk, J. Bromage, and D. Irwin, Opt. Express **20**, 20874 (2012).
45. H. M. L. Faulkner and J. M. Rodenburg, Ultramicroscopy **103**, 153 (2005).
46. H. Y. Wang, C. Liu, and S. P. Veetil, Opt. Express **22**, 2159 (2014).
47. C. Haynam, P. Wegner, J. Auerbach, M. Bowers, S. Dixit, G. Erbert, G. Heestand, M. Henesian, M. Hermann, and K. Jancaitis, Appl. Opt. **46**, 3276 (2007).
48. J. Neauport, X. Ribeyre, J. Daurios, D. Valla, M. Lavergne, V. Beau, and L. Videau, Appl. Opt. **42**, 2377 (2003).
49. F. Zhang and J. M. Rodenburg, Phys. Rev. B **82**, 121104 (2010).

7

Laser-induced damage tests based on a marker-based watershed algorithm with gray control

Yajing Guo[1,2], Shunxing Tang[1], Xiuqing Jiang[1,2], Yujie Peng[1,2], Baoqiang Zhu[1], and Zunqi Lin[1]

[1]*Joint Laboratory on High Power Laser and Physics, Shanghai Institute of Optics and Fine Mechanics, Chinese Academy of Sciences, Shanghai 201800, PR China*
[2]*University of Chinese Academy of Sciences, Beijing 100039, PR China*

Abstract
An effective damage test method based on a marker-based watershed algorithm with gray control (MWGC) is proposed to study the properties of damage induced by near-field laser irradiation for large-aperture laser facilities. Damage tests were performed on fused silica samples and information on the size of damage sites was obtained by this new algorithm, which can effectively suppress the issue of over-segmentation of images resulting from non-uniform illumination in dark-field imaging. Experimental analysis and results show that the lateral damage growth on the exit surface is exponential, and the number of damage sites decreases sharply with damage site size in the damage site distribution statistics. The average damage growth coefficients fitted according to the experimental results for Corning-7980 and Heraeus-Suprasil 312 samples at 351 nm are 1.10 ± 0.31 and 0.60 ± 0.09, respectively.

Keywords: damage growth; laser-induced damage; marker-based watershed algorithm with gray control

1. Introduction

Laser-induced damage is a major factor limiting the lifetime of optical components that can increase transmission losses and generate additional damage to optics downstream as a result of beam modulation[1–5]. The initial damage threshold and damage growth are typically used to estimate the properties of optical damage[6–9]. The morphology of damage sites is of great importance in investigating damage mechanisms[10, 11]. The lifetime of optics is mainly determined by the damage growth resulting from repeated pulses, including the growth in lateral size and number of damage sites[12]. The performance of the optics after subsequent shots might be considered acceptable in many applications if an initiated damage site with a typical diameter of tens of microns is very stable. However, if the laser-initiated site is not stable and increases in size then the performance of the optics will be degraded with further laser shots until it can no longer be used[13]. On the other hand, the modulation effects induced by multiple damage sites on the transmitted beam quality are more serious than those from a single site[14].

Damage detection is one of the main methods to directly evaluate the damage properties of optical components. In 1997, Lawrence Livermore National Laboratory (LLNL)[15] began research on damage detection in optics and further applied it to the National Ignition Facility (NIF). They adopted dark-field imaging technology to detect the damage as the damage image captured had a high contrast, which was beneficial to distinguish damage sites from the background. However, damage sites in the damage image have blurry edges and require an effective edge extraction method. Considering the significance of damage site information in the evaluation of damage properties, a method of information extraction with sufficient accuracy to determine the number and dimensions of damage sites is required. Typically, a binarization threshold technique is widely used to extract and measure damage areas in damage images. However, the extraction of information on damage sites by binarization depends greatly on the threshold selection, which is sensitive to the uniformity of illumination and the gray contrast between the object and the background. Furthermore, the binarization threshold extraction method focuses on the gray information and ignores the spatial information[16]. Han *et al.*[11] used a marker-controlling watershed algorithm to investigate the fine morphology of damage sites, which can segment adjacent objects with blurry edges. Limited by the object distance and the field of view of the microscope, the damage images captured by the CCD have disadvantages such as shadowing, non-uniform illumination and image contrast, burst noise and background gray variation, which can lead to over-segmentation in the marker-controlling watershed algorithm.

Correspondence to: Shunxing Tang. Email: leo@siom.ac.cn

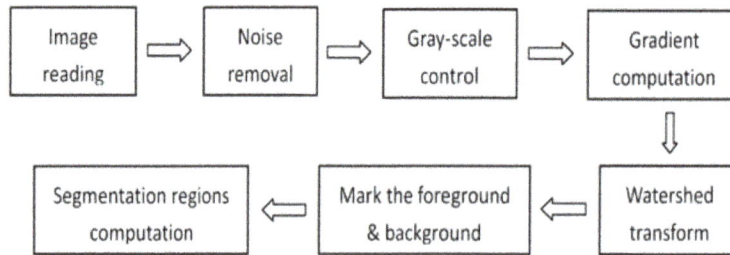

Figure 1. Flow chart of the imaging process using the MWGC.

Table 1. Comparison of Damage Site Sizes Obtained by Different Measurement Methods.

Diameter by MWGC (μm)	59.0	84.8	115.5	135.5	27.1	37.8	50.5	74.6
Diameter by OM (μm)	54.5	79.6	115.5	138.5	30.1	32.9	44.5	81.8
Absolute error (μm)	4.5	5.2	0	−3.0	−3.0	4.9	6.0	−7.2
Relative error (%)	8.3	6.5	0	−2.1	−1.0	14.9	13.4	−8.8
Average absolute error (μm)	0.93							
Average relative error (%)	2.8							

In this paper, gray control is included in the watershed algorithm to accurately extract information on damage in fused silica samples induced by large-aperture laser irradiation. Fused silica samples including Corning-7980 and Heraeus-Suprasil 312 were tested experimentally and the damage images were processed by the marker-based watershed algorithm with gray control (MWGC). The results show that the growth of damage on the exit surface is exponential and the number of damage sites decreased sharply with damage site size. The average damage growth coefficients at 351 nm are fitted to be 1.10 ± 0.31 and 0.60 ± 0.09 for Corning-7980 and Heraeus-Suprasil 312 samples, respectively.

2. Marker-based watershed algorithm with gray control

When considering multiple damage sites in one damage image, damage site extraction requires a clear edge contour to calculate the damage site area. To obtain the dimensions of damage sites and their distribution, the MWGC is applied to segment the damaged regions and accurately extract the lateral size information. Vincent et al.[17] proposed a immersion simulation-based watershed segmentation algorithm, which used regions growing from local minima in nature. Benefiting from the advantage of fast computation speed, closed contours, and accurate positioning, the watershed algorithm is applied in many fields. Moreover, this algorithm has a sensitive response to weak edges. Because of the non-uniform illumination field in dark-field imaging systems[15] and the presence of optical aberrations, both noise and dark textures appear in the damage image and give rise to fake local minima. If the watershed algorithm is applied to these fake minima then image over-segmentation will possibly occur. To overcome the problem of over-segmentation of

the image, gray control is implemented before the watershed calculation to remove noise and gradient non-uniformity.

The gray control is realized by setting a threshold that directly affects the image processing results. The threshold is determined by the gray level histogram of pixels in the damage image. For a damage image obtained by dark-field imaging technology, the image background has a low pixel gray value and maximum probability in the pixel gray histogram statistic. When the probability drops to the minimum from the peak, the gray value corresponding to the minimum probability is defined as the threshold of gray control. Figure 1 shows the flow chart of the improved algorithm.

Figure 2 shows the damage images extracted by different segmentation techniques. Figure 2(a) is the original damage image obtained by a microscope with CCD. The image binarization (Figure 2(b)) is merely an image denoising process for the gray information and cannot divide the damage sites. Since the selected threshold values in the threshold segmentation algorithm are generally determined by the gray scale for different regions, region segmentation based on this will be over-sensitive, which makes the damage pits inside one site divided into multiple sites unfavorably. This can be seen from damage regions with low gray contrast between the object and background in Figure 2(c). Over-segmentation can be observed in Figure 2(d) using marker-based watershed algorithms without gray control. In contrast, the MWGC clearly marks the different damage sites, which facilitates obtaining the damage growth and damage site size distribution. The different colors in Figures 2(c)–(e) represent different segmented damaged regions.

Table 1 shows a comparison between the diameters d_1 of damage site areas extracted using the MWGC and the diameters d_2 measured using an optical microscope (OM) with $200\times$ magnification. The morphology of damage sites obtained by the OM is shown in Figure 3. The average absolute error $\langle d_1 - d_2 \rangle$ is 0.93 μm, which is much less than the pixel size (as shown in Table 2). Also, the average relative error $\langle \frac{d_1}{d_2} - 1 \rangle$ is 2.8%, indicating that the sizes of damage sites extracted by the MWGC are believable. It is clear that the segmentation algorithm incorporating gray control is effective in processing the damage image produced by a large-aperture laser.

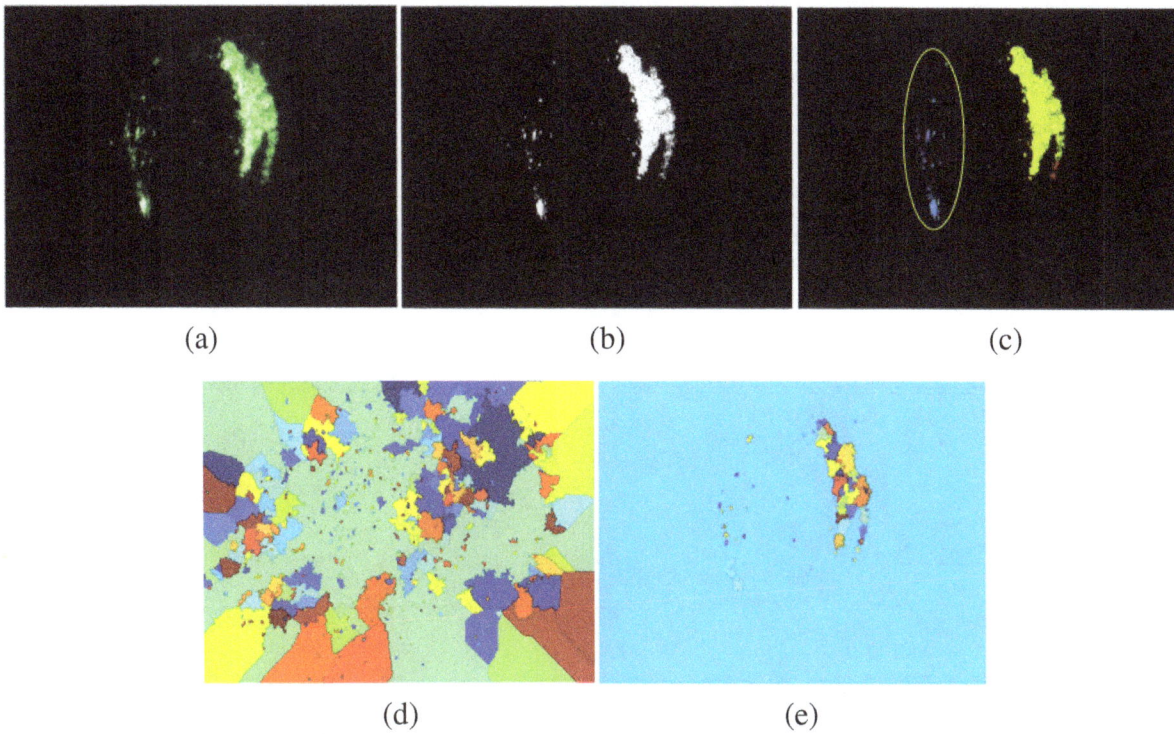

Figure 2. Results of image segmentation. (a) Original damage image; (b) by image binarization; (c) by threshold segmentation; (d) by marker-based watershed algorithm without gray control; (e) by MWGC.

Table 2. Pixel Size Calibration for Different Damage Images.

No.	Corresponding sample	Pixel size (μm/pixel)
1	Corning-7980	7.09
2	Heraeus-Suprasil 312	7.32

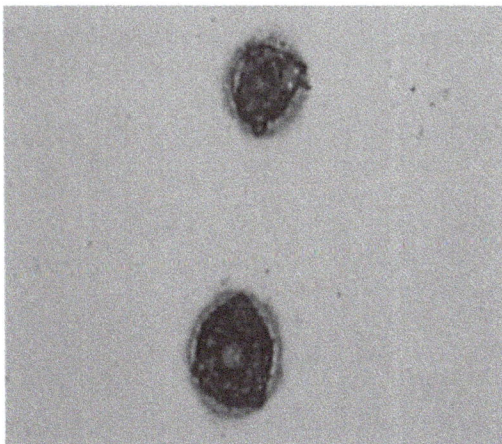

Figure 3. Damage site morphology captured by OM.

3. Experimental set-up

The experimental set-up is shown schematically in Figure 4, and consists of modules for third-harmonic generation, laser parameter measurements, damage initiation, and image capture. A 1053 nm laser with a beam aperture of $1.8 \times 1.8\,\mathrm{cm}^2$ exits from a four-pass amplifier, the pulse width of which is 3 ns, as shown in Figure 5. The 351 nm laser is generated through frequency conversion of 1053 nm laser using two KDP crystals. By sampling the laser output from KDP crystals, the laser parameters are measured. Next, fused silica samples are exposed to the laser transmitted through a focusing lens. The beam sizes on the damage samples are 0.56–0.86 cm and are achieved by adjusting the distance between the samples and the lens (1 and 1.5 m in focal lengths for Corning-7980 and Heraeus-Suprasil 312 samples, respectively). The laser energies and near-field energy density distribution are measured by the measurement set of laser parameters. Figure 6 shows the near-field energy density distribution at 351 nm with a beam contrast (peak to average) of 1.75. The damage images after each shot are observed by a microscope with a CCD. The microscope magnification is calibrated using Group 0-4 of a 1951USAF resolution test chart with 1.41 line pairs per millimeter. Subsequently, the pixel size of the damage image is calibrated as shown in Table 2.

In our experiments, Corning-7980 and Heraeus-Suprasil 312 (secondary cleanliness) samples with a size of $50 \times 50 \times 8\,\mathrm{mm}^3$ were prepared to test the damage behavior and the average damage growth and damage site distribution were calculated to illustrate the damage properties of the optical surface.

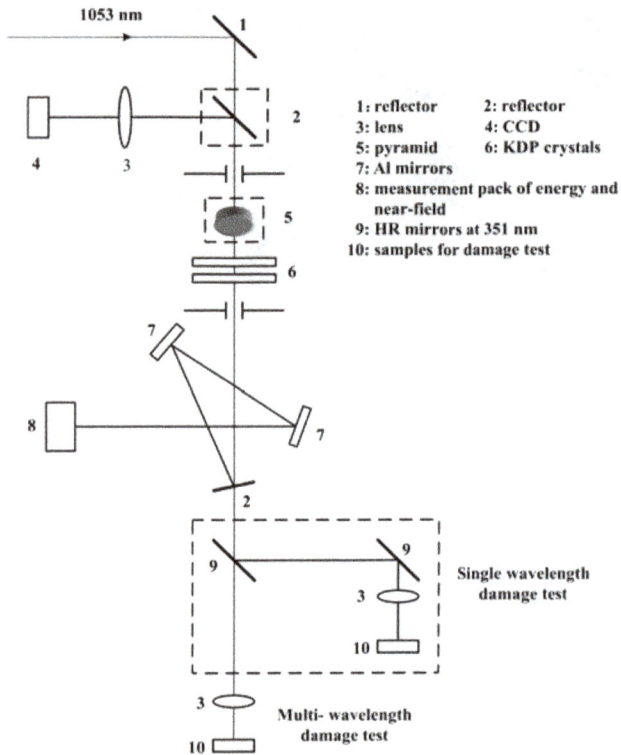

1: reflector   2: reflector
3: lens   4: CCD
5: pyramid   6: KDP crystals
7: Al mirrors
8: measurement pack of energy and
   near-field
9: HR mirrors at 351 nm
10: samples for damage test

Figure 4. Experimental set-up for laser-induced damage testing.

4. Results and discussion

4.1. Damage growth coefficients

Damage tests at 351 nm are carried out to analyze the damage growth and damage site distributions on Corning-7980 and Heraeus-Suprasil 312 (secondary cleanliness) samples. For successive laser shots, five damage sites are selected to calculate the damage growth coefficients for the two samples. The damage growths for different damage sites are exponential, as shown in Figure 7, which is in good agreement with the reported results[8, 12]. Table 3 shows

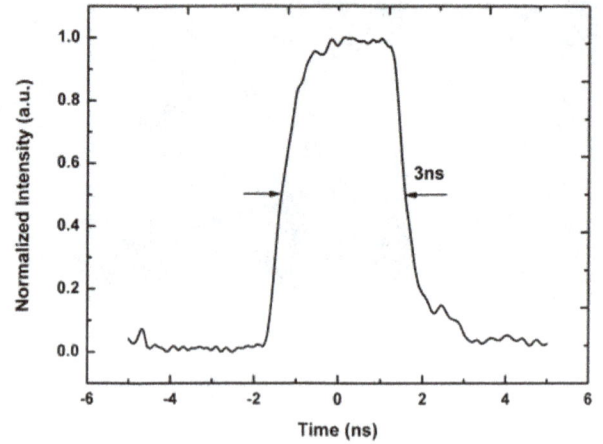

Figure 5. Temporal profile of a 3 ns pulse at 1053 nm.

Table 3. Damage growth for Corning-7980 and Heraeus-Suprasil 312 samples.

Sample	Average fluence (J cm^{-2})	Average growth coefficient	Standard deviation of growth coefficient	Fitting goodness R^2 (%)
Corning-7980	9.60	1.10	0.31	98.8
Heraeus-Suprasil 312	8.56	0.60	0.09	97.3

the damage growths for the two samples. The damage growth coefficients α are 1.10 ± 0.31 with an average fitting goodness R^2 of 98.8% and 0.60 ± 0.09 with an average fitting goodness R^2 of 97.3%, respectively.

As seen from Figure 7, the damage growth coefficients are distinctly different for the five tested sites in the same sample. The difference possibly results from the site morphologies tested and the local fluence. Numerous researchers[9, 13, 18, 19] have demonstrated that the complex damage growth process is affected not only by the various laser parameters but also the intrinsic structural features of the optical components. In general, there are various precursors on the optical surface that induce damage sites with different morphologies. Damage sites with different

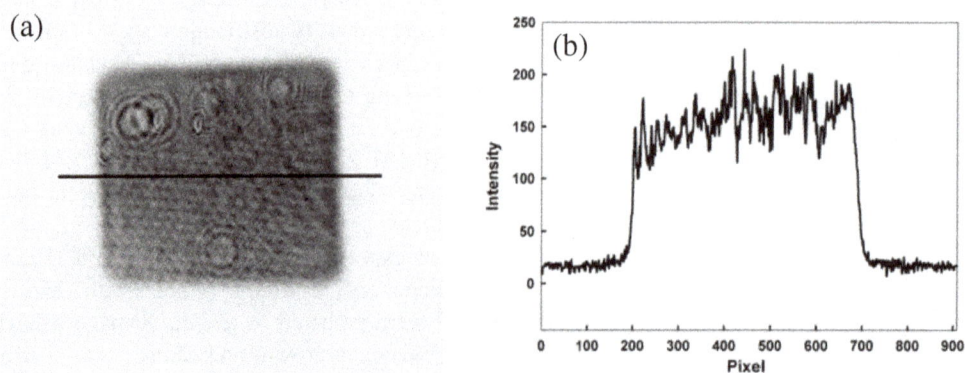

Figure 6. (a) Near-field energy density distribution at 351 nm. (b) Profile along the line in (a).

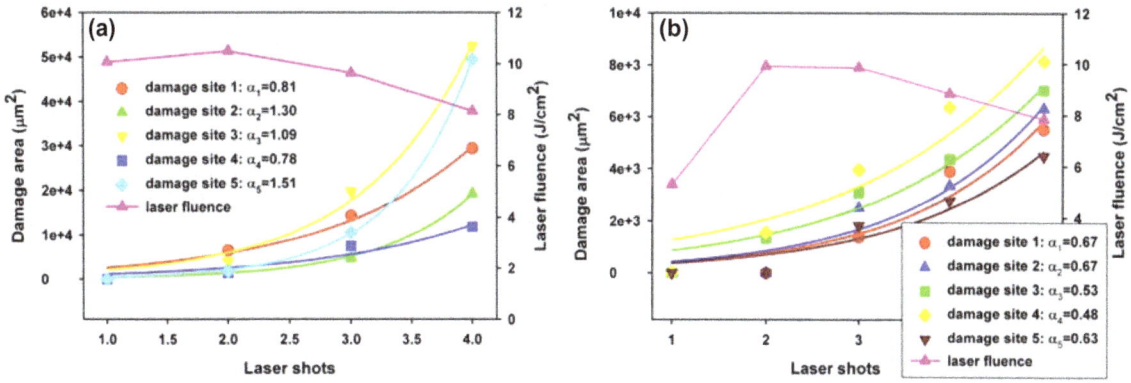

Figure 7. Damage growth of (a) Corning-7980 and (b) Heraeus-Suprasil 312 samples at 351 nm.

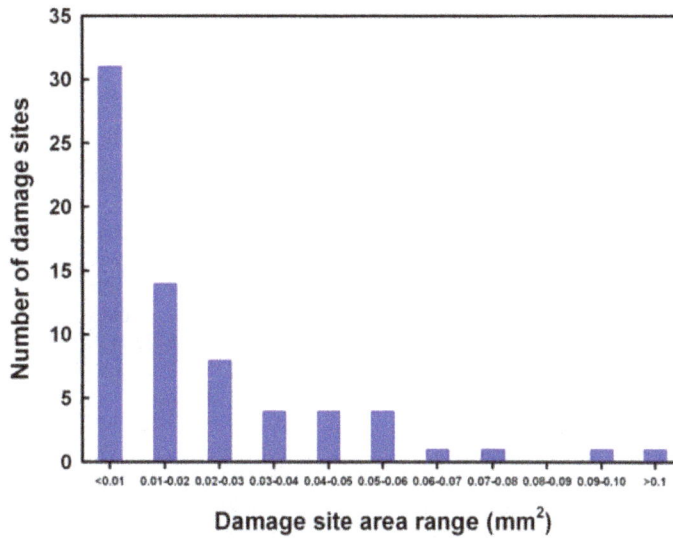

Figure 8. Damage site size distribution for the Corning-7980 sample at 351 nm.

morphologies grow differently. Therefore, damage sites grow with different coefficients after subsequent laser shots.

Note that the standard deviation of the growth coefficient for the Heraeus-Suprasil 312 sample is smaller than that for the Corning-7980 sample. This can be explained by the cleanliness of the optical surface. Most fragile precursors are removed from the Heraeus-Suprasil 312 sample surface with secondary cleanliness treatment. The damage sites induced by residual precursors are possibly similar. It can be inferred that the error bar of the growth coefficient (± 0.09) of Heraeus-Suprasil 312 is less than that of Corning-7980 (± 0.31).

4.2. Damage site distribution

Next, the damage site distributions are obtained after one laser shot, as shown in Figures 8 and 9. The number of damage sites decreased sharply with lateral damage site size. The average fluence of the 351 nm laser on Heraeus-Suprasil 312 sample is 22.60 J cm^{-2}, larger than that for the

Corning-7980 sample ($\sim$11 J cm^{-2}). Therefore, the number of large-size damage sites (area > 0.1 mm^2) for Corning-7980 sample is less than that for the Heraeus-Suprasil 312 sample.

5. Conclusions

Based on a near-field laser beam combined with the MWGC, the damage behaviors of Corning-7980 and Heraeus-Suprasil 312 samples are investigated in terms of the damage growth and damage site size distribution. The near-field laser provides the required beam size and fluence necessary to create damage. The damage image processing algorithm can effectively suppress the over-segmentation of the damage image and obtain accurate size information for the damage sites. Using the damage test method described, the average damage growth coefficients and damage site size distribution for fused silica samples are obtained to illustrate the damage behavior of optical component surfaces. It indicates the damage test method is effective and beneficial

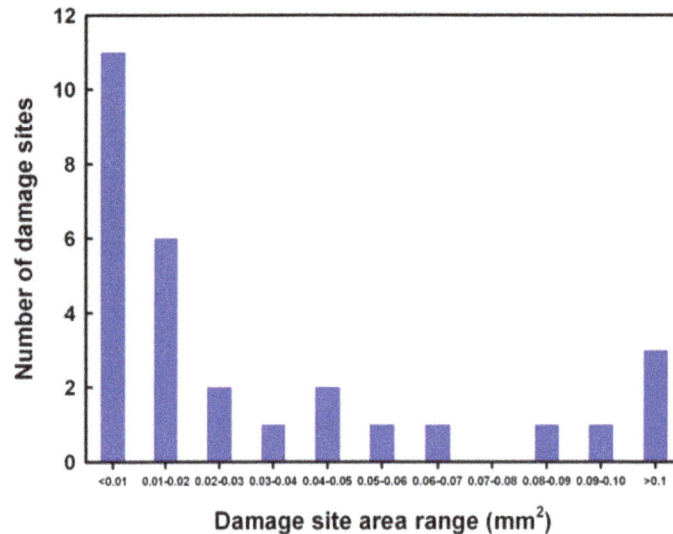

Figure 9. Damage site size distribution for the Heraeus-Suprasil 312 sample at 351 nm.

for further studies into large-aperture laser-induced damage characteristics, which play a key role in the assessment of the damage resistance of optical components.

References

1. W. Huang, W. Han, F. Wang, Y. Xiang, F. Li, B. Feng, F. Jing, X. Wei, W. Zheng, and X. Zhang, Proc. SPIE **7276**, 72760A (2008)
2. J. T. Hunt, *National ignition facility performance review 1998 USA: LLNL* (1999).
3. S. G. Demos, M. Staggs, K. Minoshima, and J. Fujimoto, Opt. Express **10**, 1444 (2002).
4. C. W. Carr, M. D. Feit, M. C. Nostrand, and J. J. Adam, Meas. Sci. Technol. **17**, 1985 (2006).
5. J. T. Hunt, K. R. Manes, and P. A. Renard, Appl. Optim. **32**, 5973 (1993).
6. W. Deng and C. Jin, Chin. Opt. Lett. **11**, S10702 (2013).
7. L. Lamaignere, G. Dupuy, and T. Donval, Proc. SPIE **8530**, 85301F (2012)
8. M. A. Norton, L. W. Hrubesh, Z. Wu, E. E. Donohue, M. D. Feit, M. R. Kozlowski, D. Milan, K. P. Neeb, W. A. Molander, A. M. Rubenchik, W. D. Sell, and P. Wegner, *Growth of laser initiated damage in fused silica at 351 nm, LLNL UCRL-JC-139624* (2001).
9. Z. M. Liao, G. M. Abdulla, R. A. Negres, D. A. Cross, and C. W. Carr, Opt. Express **20**, 15569 (2012).
10. J. Yoshiyama, F. Y. Genin, A. Salleo, I. Thomas, M. R. Kozlowski, L. M. Sheehan, I. D. Hutcheion, and D. W. Camp, Proc. SPIE **2744**, 220 (1997).
11. H. Han, Y. Li, T. Duan, C. He, G. Feng, and L. Yang, Optik **124**, 1940 (2012).
12. A. M. Rubenchik and M. D. Feit, Proc. SPIE **4679**, 79 (2002).
13. M. A. Norton, E. E. Donohue, W. G. Hollingsworth, M. D. Feit, A. M. Rubenchik, and R. P. Hackel, Proc. SPIE **5647**, 197 (2005).
14. G. Zhang, X. Lu, H. Cao, X. Yin, F. Lv, Z. Zhang, J. Li, R. Wang, W. Ma, and J. Zhu, Acta Phys. Sin. **61**, 024201 (2012).
15. F. Rainer, Proc. SPIE **3244**, 272 (1997).
16. J. He, H. Ge, and Y. Wang, Comput. Eng. Sci. **31**, 58 (2009).
17. L. Vincent and P. Soille, IEEE Trans. Pattern Anal. Mach. Intell. **13**, 583 (1991).
18. S. G. Demos, M. Staggs, and M. R. Kozlowski, Appl. Optim. **41**, 3628 (2002).
19. R. A. Negres, D. A. Cross, and C. W. Carr, Opt. Express **18**, 19966 (2010).

8

Overview of the HiLASE project: high average power pulsed DPSSL systems for research and industry

M. Divoky[1], M. Smrz[1], M. Chyla[1], P. Sikocinski[1], P. Severova[1], O. Novak[1], J. Huynh[1], S.S. Nagisetty[1], T. Miura[1], J. Pilař[1], O. Slezak[1], M. Sawicka[1], V. Jambunathan[1], J. Vanda[1], A. Endo[1], A. Lucianetti[1], D. Rostohar[1], P.D. Mason[2], P.J. Phillips[2], K. Ertel[2], S. Banerjee[2], C. Hernandez-Gomez[2], J.L. Collier[2], and T. Mocek[1]

[1]*HiLASE, Institute of Physics, AS CR, v.v.i., Na Slovance 2, 182 21 Prague, Czech Republic*
[2]*STFC Rutherford Appleton Laboratory, Didcot OX11 0QX, United Kingdom*

Abstract

An overview of the Czech national R&D project HiLASE (High average power pulsed laser) is presented. The project focuses on the development of advanced high repetition rate, diode pumped solid state laser (DPSSL) systems with energies in the range from mJ to 100 J and repetition rates in the range from 10 Hz to 100 kHz. Some applications of these lasers in research and hi-tech industry are also presented.

Keywords: DPSSL; Yb^{3+}:YAG; thin-disk; multi-slab; pulsed high average power laser

1. Introduction

Efficient diode pumping of solid-state lasers (DPSSL) has enabled lasers to reach CW output powers in the region of 100 kW[1, 2]. High energy pulsed DPSSLs are far behind the CW systems in respect of average power, but the development of new technology [thin disk, multi-slab] will enable operation at comparable average powers. Many laser projects are trying to reach the barrier of 1 kW average power in pulsed operation using either thin disks at high repetition rates[3–7] or a high energy and lower repetition rate with conventional designs[8], TRAM (Total Reflection Active Mirror)[9], thick disks[10] or multi-slabs[11, 12]. The HiLASE (High average power pulsed laser) project is aimed at the development of the next generation of pulsed DPSSL for hi-tech industrial applications, such as laser-induced damage testing, extreme ultra-violet (EUV) light generation, surface cleaning, precise manufacturing, laser peening, etc. HiLASE will be a user facility with several laser systems, with output parameters ranging from a few picosecond pulses with energies of 5 mJ–0.5 J and repetition rates of 1–100 kHz (thin disk technology) to systems with 100 J output energy in nanosecond pulses with a repetition rate of 10 Hz (multi-slab technology).

In this paper, an overview of the HiLASE activities, including laser development and laser applications, will be presented.

Correspondence to: Email: divoky@fzu.cz

2. Kilowatt-class thin disk laser system

For efficient generation of EUV and mid-IR light, a laser producing several mJ per pulse at a repetition rate of 1–100 kHz is required. For industrial applications, it is important to realize a robust, compact, and low-cost alternative to Ti:sapphire-based pulsed laser systems. Thin-disk lasers with their feature of a high pulse energy in the sub-picosecond region are one of the best devices suited for this application.

A thin-disk laser is based on an amplifier concept[13] that utilizes a very thin (~200 μm) laser active medium on a heatsink. The front face of the disk is AR coated, while the back face is HR coated and the disk works as an active mirror. Since the diameter of the disk is much larger than its thickness, the heat flux is mostly axial and the transverse temperature gradient is low. The path of the beam in the disk is short and thermal lens effects and mechanical deformation do not affect the beam quality much. Additionally, a low material path minimizes nonlinear phenomena such as self-focusing.

On the other hand, the thinness of the disk causes minimal pump light absorption and laser light gain. Therefore, the number of passes of both pump and laser light must be high. The pump light is sequentially reflected back to the laser disk by a parabolic reflector and roof prisms (Figure 1), so the absorption can exceed 90%. The number of laser light passes is usually increased in a regenerative cavity or in a multi-pass amplifier with many extraction passes.

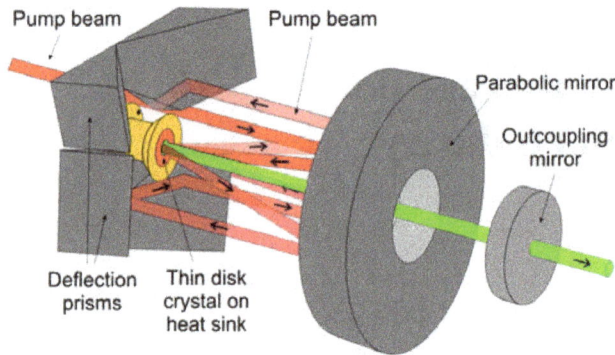

Figure 1. Schematic of the thin-disk cavity consisting of a parabolic mirror focusing the pump beam onto the thin-disk crystal. Multiple passes of the pump beam are made by means of deflection prisms. The cavity for laser beam extraction is formed by the thin-disk crystal and the outcoupling mirror [14].

Within the HiLASE project, three thin-disk-based kW-class laser beamlines are being developed, each delivering different output parameters. Beamline A will deliver a 750 mJ pulse energy at a 1.75 kHz repetition rate. This beamline is subcontracted to Dausinger and Giesen GmbH in order to reduce the overall project risk associated with the high demands. The HiLASE research group is developing Beamlines B and C with output parameters of 500 mJ at a 1 kHz repetition rate and 5 mJ at a 100 kHz repetition rate, respectively. All beamlines will provide a pulse duration of 1–3 ps. The output of Beamline B could be diverted into a 10 Hz repetition rate cryogenic amplifier that would later be upgraded to multi-joule output at repetition rates up to 120 Hz. Figure 2 shows the block diagrams of each beamline in the HiLASE project.

2.1. Beamline A

Beamline A consists of a fiber front-end that includes a pulse stretcher, pulse picker, and optical isolator. The front-end produces laser pulses with an energy of 1 μJ at a repetition

rate of 1.75 kHz. These pulses are further amplified in a regenerative amplifier to an energy around 150 mJ, then in a linear amplifier to an energy around 0.9 J. The amplified pulses are compressed in a grating pulse compressor to below 3 ps.

2.2. Beamline B

Beamline B[15] starts with an Yb-doped fiber oscillator operating at a 50 MHz repetition rate, delivering an output power of 2 W at a center wavelength of 1030 nm with a bandwidth of more than 20 nm. Pulses from the fiber laser are stretched by a Martinez-type stretcher up to 500 ps. A gold-coated grating with a groove density of 1740 l mm^{-1} was employed, providing a group delay dispersion of 1.41×10^8 fs^2/rad. After the pulse stretching, pulses are coupled into the regenerative amplifier cavity, which contains an Yb^{3+}:YAG thin disk, a thin film polarizer (TFP), a Pockels cell with a 10 mm × 10 mm × 22 mm BBO crystal, high-reflection mirrors, and a quarter-wave plate. Pulses from the stretcher are captured in the amplifier cavity when a quarter-wave voltage of more than 10 kV is applied to the Pockels cell, and they travel inside the cavity as long as the high voltage is applied. After amplification, the amplified pulse is ejected through the TFP and passes through an optical isolator consisting of a half-wave plate and a Faraday rotator. Afterwards, the amplified beam is diverted from the input beam by the polarizing beam splitter (PBS) and sent to the pulse compressor. A schematic of the current status of Beamline B is shown in Figure 3.

In order to reduce the thermally induced stress between the heatsink and the gain media, CuW is adopted as the heatsink material, because it has similar linear expansion coefficient to YAG. The Yb^{3+}:YAG crystal, doped to 7 at.%, has thickness of 0.22 mm and is soldered on the CuW heatsink using gold-tin solder. The mounted disk is pumped by a fiber-coupled diode laser module delivering an optical power

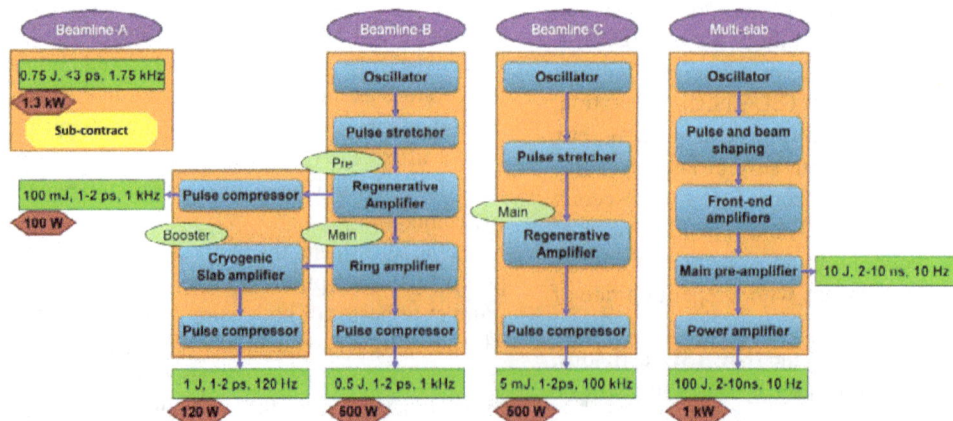

Figure 2. Overview of the HiLASE beamlines.

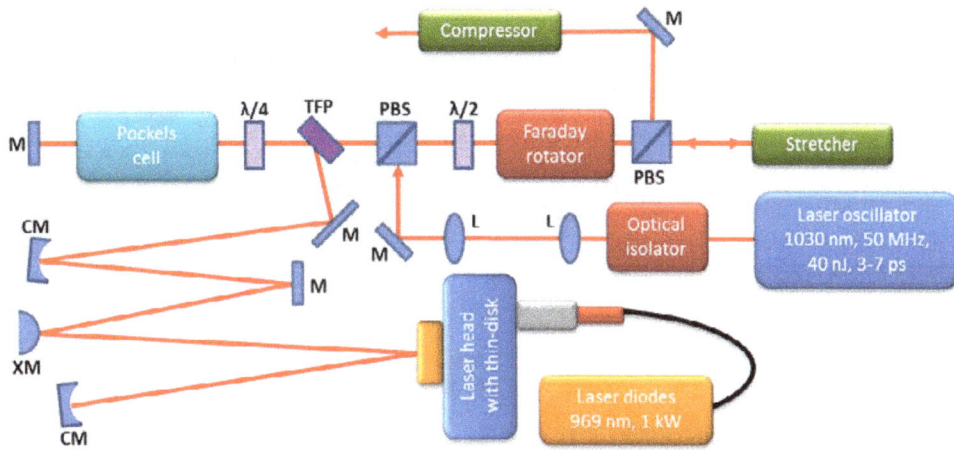

Figure 3. Schematic of the current status of Beamline B. Shown are flat mirrors (M), convex mirrors (XM), concave mirrors (CM), lenses (L), half and quarter waveplates (λ/2, λ/4), PBS, and a thin film polarization beam splitter (DP).

of up to 1 kW at a wavelength of 969 nm. The pump spot size on the disk was set to 4.8 mm to achieve an output of 45 mJ, and the amplifier cavity was designed so that the cavity mode was matched to the designed pump spot size. The optical-to-optical efficiency was close to 20%. The amplified laser pulses had a bandwidth of 1.5 nm, so they could be compressed down to 1 ps.

The laser cavity will later be upgraded with a second thin-disk head to reach an output energy of 100 mJ. Additionally, the Martinez-type stretcher will be replaced with a fiber-chirped Bragg grating stretcher that allows better control of dispersion and is more stable and compact. Finally, the amplified pulses will be directed to a second regenerative amplifier that will be constructed in 2014.

2.3. Beamline C

Beamline C[16] is aimed at achieving a pulse energy of 5 mJ at a 100 kHz repetition rate. In order to meet these requirements, an intense study has been conducted to develop a high repetition rate regenerative amplifier. The target specifications of Beamline C will be achieved after completing three milestones determined by the pulse energies, namely 0.5, 2, and 5 mJ. The experimental setup for reaching the first milestone (0.5 mJ) is shown in Figure 4. The regenerative amplifier is seeded by an Yb^{3+}-doped fiber oscillator, as in Beamline B, but the pulses are stretched by a chirped volume Bragg grating (CVBG) up to ~160 ps. The dispersion of the CVBG is 60 ps nm^{-1}. The FWHM spectral bandwidth and the clear aperture are 2.2 ± 0.5 nm and 8 mm × 8 mm, respectively. The dimensions of the BBO crystal for the Pockels cell are 5 mm × 5 mm × 25 mm and its quarter-wave voltage is 5.2 kV. An Yb^{3+}:YAG thin-disk with a free aperture of 8 mm and a thickness of 220 μm is installed in the laser head. The disk is pumped by a 1 kW fiber-coupled diode laser module at a wavelength of 969 nm

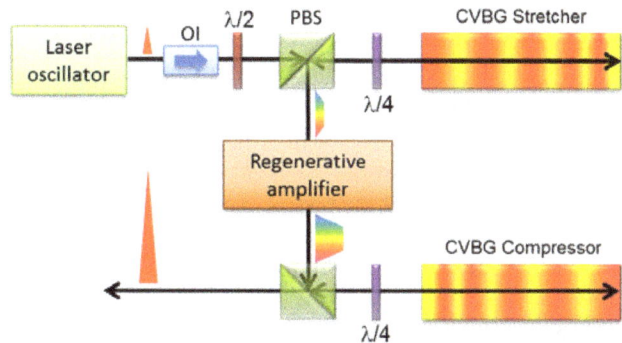

Figure 4. Schematic of the current status of Beamline C. Shown are an optical isolator (OI), λ/2, λ/4, PBS, and a CVBG stretcher and compressor.

with a pump spot diameter of 2.8 mm. A schematic of the beamline is shown in Figure 4.

The observed output energy was 830 μJ at a 100 kHz repetition rate. The low pulse energy enabled compression in a highly efficient CVBG compressor. The compressed pulse energy and the efficiency of the CVBG were 730 μJ and 88%, respectively. The output pulse had a spectral bandwidth of 1.2 nm and was compressed only to 4 ps pulse duration because the CVBG did not account for the dispersion of material in the path of the beam. By adding an additional diffraction grating compressor, the duration of the compressed pulses was decreased below 2 ps. The output energy will be increased by using more intense pump light and by modifying the thin-disk head and the cavity. Technical difficulties connected to further development of all the mentioned thin-disk beamlines are connected mostly to thermal management of the thin-disk modules and the availability of pump modules at a wavelength of 969 nm.

The current status of thin-disk beamlines is indicated in Table 1.

Table 1. Status of kW-class Thin-disk Beamlines.

Laser system	Beamline A	Beamline B	Beamline C	Cryogenic beamline
Completed	Front-end	Regenerative amplifier with one thin-disk head	All, except high power pump modules	None
Under development	Regenerative amplifier (May 2014)	Add second thin-disk head into regenerative amplifier	Add high power pump modules	10 Hz concept amplifier
Achieved energy	1 µJ	45 mJ	0.8 mJ	NA
Next milestone energy	150 mJ	100 mJ	2 mJ	1 J (10 Hz)
Final energy	750 mJ	500 mJ	5 mJ	1 J (100 Hz)
Operational	Q2 2015	Q2 2015	Q2 2015	2016

3. Kilowatt-class multi-slab laser system

To generate high energy pulses at low/moderate repetition rates, it is necessary to adopt an effective cooling mechanism and geometry. One of the solutions is to use an active medium in a slab geometry with active cooling of the slab faces, called a multi-slab, firstly adopted on the Mercury laser at Lawrence Livermore National Laboratory [17]. Room-temperature helium gas was used for cooling the slabs and 60 J at a 10 Hz repetition rate was obtained. The material of choice for next-generation high energy solid-state lasers is Yb^{3+}:YAG ceramic. However, Yb^{3+}:YAG is a quasi-three-level system that requires a high pump intensity for laser operation, thus increasing the number and cost of the pumping diodes. By cooling the crystal to low temperatures, the energy scheme changes to four levels, thus decreasing the threshold intensity by several orders of magnitude. Such a concept was introduced by DiPOLE[11], where an energy of 10 J was obtained. The cooling of the slabs was done using helium gas at a temperature of around 150 K. The next step is to demonstrate more than 100 J at a 10 Hz repetition rate, which will eventually lead to the delivery of 1 to 10 kJ pulses in a single beam at a repetition rate of about 10 Hz and a wall-plug efficiency of more than 10%[18, 19]. A 100 J-class laser is now under development at the Central Laser Facility (CLF) in collaboration with HiLASE.

The system incorporates a low-energy, fiber-based front end oscillator ($\sim$nJ), followed by a regenerative amplifier that increases the output energy to the mJ level and a thin-disk Yb^{3+}:YAG multi-pass booster amplifier to raise the output to 100 mJ. Two diode-pumped, helium-gas-cooled large-aperture power amplifiers then increase the output energy to between 7 and 10 J (Main Pre-amplifier) and finally to 100 J (Power Amplifier). The schematic of the system is shown in Figure 5.

3.1. Front end

The front end starts with a temperature-stabilized tunable CW fiber oscillator. The wavelength of the oscillator is matched to the peak of the gain curve of the cryogenically cooled amplifiers. The CW beam is then temporally shaped

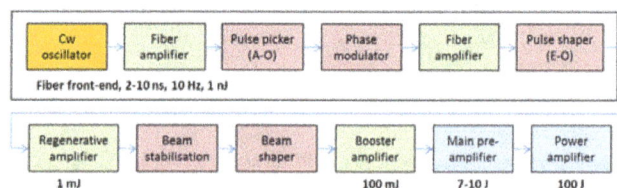

Figure 5. Schematic of the 100 J multi-slab laser system. The numbers represents the energy after the respective element.

in an acousto-optic (A-O) modulator to limit the repetition rate to 10 kHz and subsequently shaped by an electro-optic (E-O) modulator to produce 2–10 ns pulses with a semi-triangular shape. The temporal resolution of the shaper is below 200 ps. The pulses are further phase modulated by 2 and 4 GHz modulators to increase the bandwidth of the pulses and prevent stimulated Brillouin scattering (SBS) and stimulated Raman scattering (SRS) in the amplifier chain. Then the repetition of the pulses is decreased by a pulse picker and pulses are amplified in a thin regenerative disk amplifier to $\sim$1 mJ. The Gaussian beam coming from the regenerative amplifier is then spatially shaped to a square cross-section super-Gaussian profile in a beam shaper. Then it is further amplified to $\sim$100 mJ in a multi-pass booster amplifier. The booster amplifier preserves the square super-Gaussian beam profile that is injected into the 10 J main pre-amplifier.

3.2. 10 J main pre-amplifer

The 10 J main pre-amplifier is based on a multi-slab design. It consists of four circular Yb^{3+}:YAG slabs with two doping levels of Yb^{3+} (1.1, 2.0 at.%). The different doping levels are needed to uniformly divide the heat load among the slabs. The volume of each circular slab is diameter in 45 mm with a thickness of 5 mm and the pumped area is square 23 mm $\times$ 23 mm. The pump beam is homogenized light from diode stacks operating at 939 nm and producing 700 µs long laser pulses at a repetition rate of 10 Hz. The Yb^{3+}:YAG is clad with a 5 mm Cr^{4+}:YAG absorber (absorption coefficient 6 cm^{-1}) that prevents amplified spontaneous emission (ASE) and parasitic oscillations. The amplifier is cooled by forced helium gas flow and operates between 150 and 170 K.

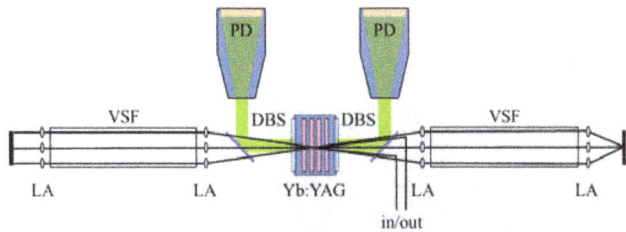

Figure 6. Schematic of the 10 J cryogenic multi-slab amplifier. It consists of Yb^{3+}:YAG ceramic slabs in the laser head (Yb:YAG), dichroic beam splitters (DBSs), lens arrays (LAs), vacuum spatial filters (VSFs), and homogenized pump diode laser modules (PDs).

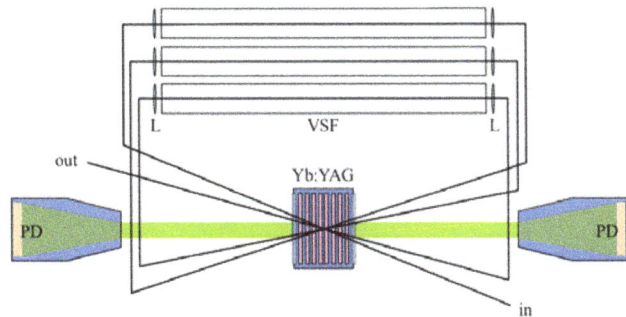

Figure 7. Schematic of the 100 J cryogenic multi-slab amplifier. It consists of Yb:YAG, lenses (L), VSF, and PD.

The extraction scheme of the multi-pass amplifier is shown in Figure 6. The beam is injected into the amplifier through a dichroic mirror and then image-relayed by a spatial filter ($f = 1$ m) to a back reflector and back to the amplifier head. There is one spatial filter on each side of the amplifier head. Each pass is propagated by a set of separate mirrors. A deformable mirror is placed in the amplifier after the third pass. After seven or eight passes, the beam is ejected from the amplifier with a pulse energy of 7 J in the beam and a size of 20 mm × 20 mm.

3.3. 100 J power amplifier

The 100 J power amplifier is also based on the multi-slab design. It consists of six square Yb^{3+}:YAG slabs with three doping levels of Yb^{3+} (0.4, 0.6, 1.0 at.%). The volume of each slab is 100 mm × 100 mm × 8.5 mm and the square pumped area is around 75 mm × 75 mm. The parameters of the pump light are similar to the 10 J amplifier. The Yb^{3+}:YAG is clad with a 10 mm wide Cr^{4+}:YAG absorber (absorption coefficient 3 cm^{-1}) that prevents ASE and parasitic oscillations.

The extraction scheme of the multi-pass amplifier is shown in Figure 7. The beam is injected into the amplifier at a range of angles around 5°, so any overlap with the pump beam outside the amplifier head is avoided and no dichroic mirrors are used. Each additional pass is image-relayed by a dedicated spatial filter ($f = 3$ m) back to the amplifier head.

Table 2. Status of kW-class Multi-slab Beamline.

Laser system	Beamline A
Completed	10 J main pre-amplifier
Under development	100 J power amplifier
Achieved energy	10 J
Next milestone energy	50 J
Final energy	100 J
Operational	Q3 2015

After the first pass, a deformable mirror is implemented to prevent degradation of the wavefront on subsequent passes. After four passes, the beam is ejected from the amplifier with a pulse energy of up to 120 J and a beam size around 75 mm × 75 mm. The average fluence in the amplified beam is 2 J cm^{-2}. The amplifier is cooled by forced helium gas flow and operates between 150 and 170 K.

The current status of the multi-slab laser system is indicated in Table 2.

3.4. Numerical modeling

The HiLASE team has undertaken extensive energetics, thermal and fluid-mechanical modeling in order to optimize various amplifier parameters.

For energetics modeling, we have developed a MATLAB code[20] for the evaluation of stored energy in the laser amplifier that includes ASE. This model calculates the pump energy absorbed by the gain medium, excited ion density, ASE and heat sources in the Yb^{3+}:YAG laser slabs by solving the rate equations in discrete time steps. During each step, the absorption of the pump radiation and spontaneous emission are calculated independently. In the absorption phase, the energy from a polychromatic pump source is absorbed in the medium in accordance with a probability proportional to the number of unexcited active ions and the absorption cross-section dependent on the pump wavelength. In the ASE phase, spontaneously emitted photons with random polarization are generated by the Monte Carlo method using the excited ion density inside the slab as the probability distribution. Their wavelength distribution is based on a probability density function derived from the emission cross-section. Rays containing thousands of photons are traced through the medium and amplified proportionally to the population inversion and the path length in each cell. All slabs of the amplifier are included in the model so the propagation of the rays among the slabs can be modeled. All surfaces can be treated as Fresnel reflecting, AR coated, or absorbing. Side faces can also include scattering. The slab model includes the active Yb^{3+}:YAG core and the absorbing Cr^{4+}:YAG cladding. A schematic of the slab and the calculated heat deposition are shown in Figure 8. Results were obtained for a slab size 100 mm × 100 mm, pump size 75 mm × 75 mm, Cr^{4+}:YAG cladding 10 mm,

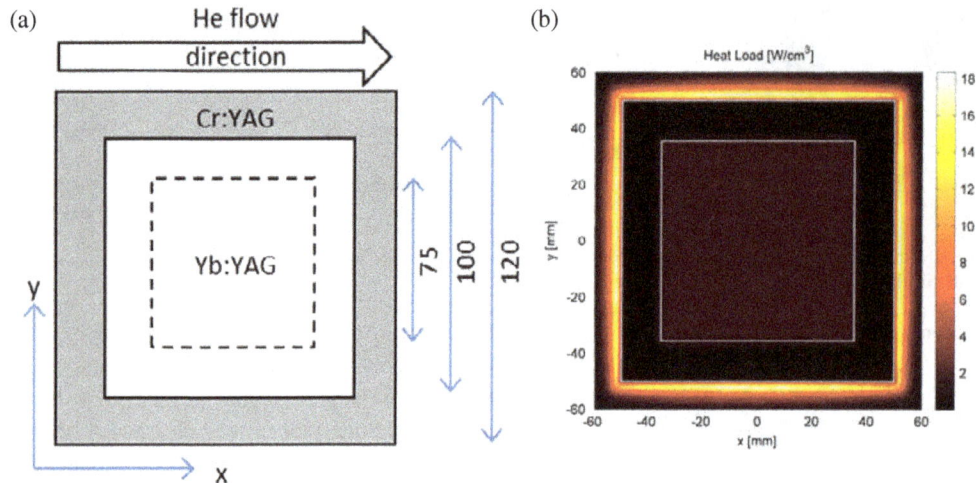

Figure 8. (a) Schematic of the laser slab with dimensions in mm, dashed line shows the spot of the pump beam in the Yb^{3+}:YAG part of the slab that is clad by Cr^{4+}:YAG. (b) Transverse heat load of the slab used for the calculations (assumed constant in the longitudinal direction).

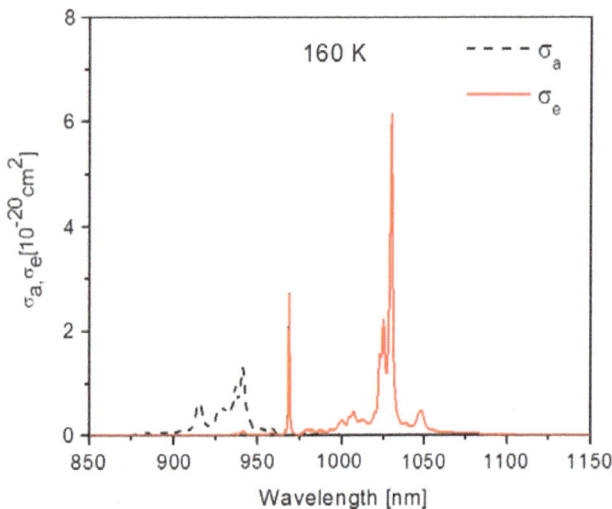

Figure 9. Measured absorption and emission cross-sections of the Yb^{3+}:YAG at a temperature of 160 K.

absorption coefficient 3 cm^{-1}, pump duration 1 ms, intensity 5 kW cm^{-2}, and temperature 160 K.

The wavelength-resolved absorption and emission cross-sections and lifetime on the upper laser level for a given temperature were obtained experimentally [21]. The cross-section and lifetime measurement station supports spectral resolution down to 15 pm and temporal resolution down to 100 ps. An example of the measurement of the absorption and emission cross-section of Yb^{3+}:YAG at a temperature of 160 K is shown in Figure 9.

A three-dimensional finite-element method (FEM) using Comsol Multiphysics software was chosen to model the thermal and stress effects in the amplifiers. The sources of heat were calculated in the ASE code. The lateral surfaces of the slabs are assumed to be cooled by flowing

helium gas at 160 K. The spatially resolved heat transfer coefficient was derived from a two-dimensional model of a turbulent flow of helium gas at 160 K using the standard $k\varepsilon$ model[22] together with Kays–Crawford heat exchange in the turbulent boundary layer [23, 24]. The slab was assumed to have no thermal contact with its 2 cm thick Invar holder; hence all the heat is removed by convection through the faces. From the temperature and stress maps of the slab (Figure 10), optical path difference (OPD) and birefringence depolarization losses (Figure 11) were calculated for a single slab according to a prior approach[25]. The gradual decrease of heat exchange efficiency in the direction of gas flow, caused by coolant heating, is the reason for the loss of left–right symmetry of the temperature, stress, depolarization, and OPD maps.

A beam propagation model of the 100 J power amplifier was created in MIRÓ using a Fresnel diffraction integral for propagation and the Frantz–Nodvik equation for amplification. The model was used to estimate beam aberrations, taking into consideration only the thermal OPD. The results of the beam intensity and OPD are shown in Figure 12. The resulting wavefront was corrected numerically. The reliability of the wavefront correction code was verified experimentally in a slab simulator[26].

The numerical model for wavefront correction calculates influence functions from a plate equation describing the bending of the thin facesheet for each individual actuator of the deformable mirror. The deformable mirror consists of a continuous gold facesheet (size of 100 mm × 100 mm, thickness of 1 mm) on which lateral forces are applied by piezoelectric stack actuators. The actuators form an equidistantly spaced rectangular array of 6 × 6 actuators and are capable of push/pull operation. The deformation of the mirror is computed as a superposition of the influence functions and the algorithm minimizes the rms OPD value.

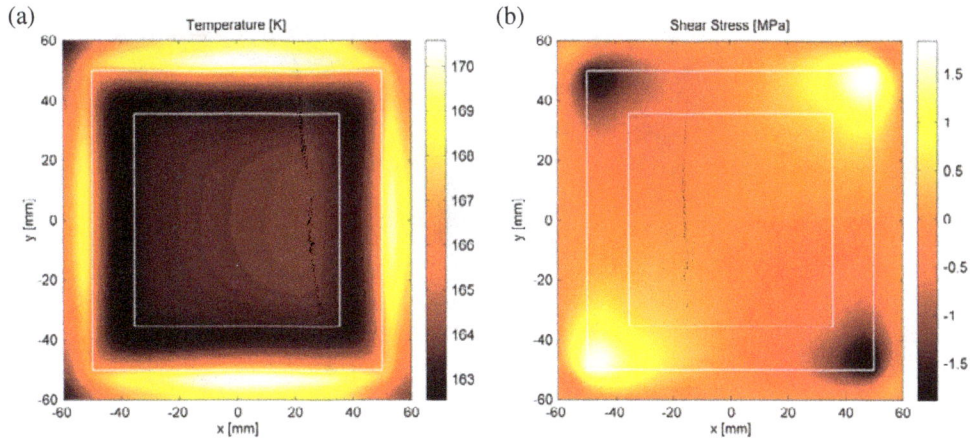

Figure 10. (a) Transverse distribution of temperature and (b) transverse distribution of the xy stress component in a longitudinal cut in the center of the laser slab.

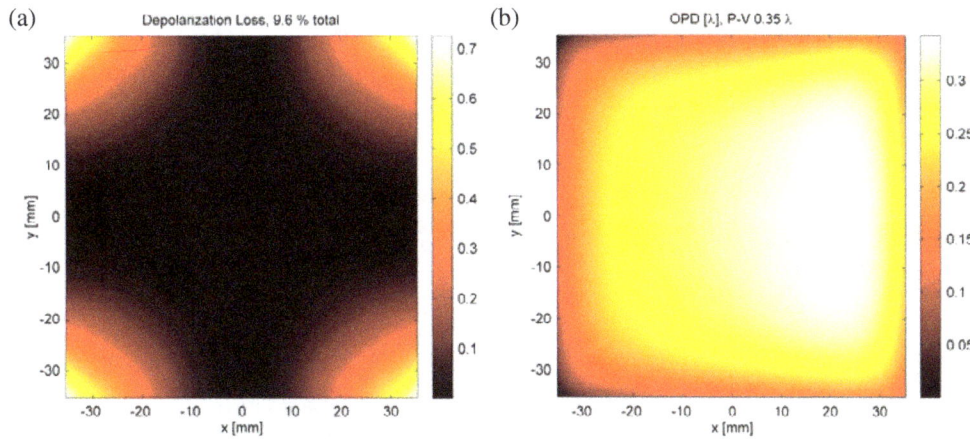

Figure 11. (a) Depolarization of the beam at the output of the amplifier (after four passes through six laser slabs) caused by stress-induced birefringence. (b) Stress- and temperature-induced OPD after a single pass through the laser head (after one pass through six laser slabs).

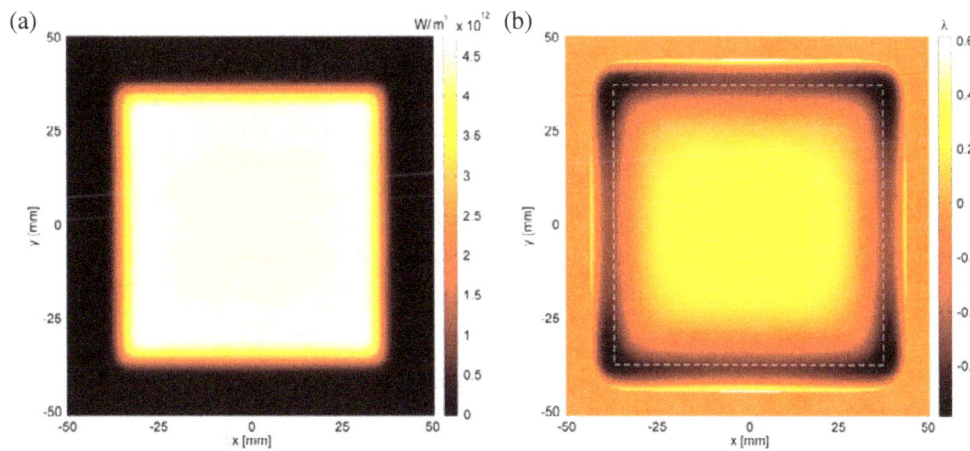

Figure 12. (a) Beam profile and (b) OPD of the beam at the output of the 100 J multi-slab system calculated in MIRÓ. Dashed square indicates the position of the laser beam.

The OPD after subtraction of defocus and tilt and the OPD corrected by the deformable mirror are shown in Figure 13.

The rms value of the OPD was reduced from 111 nm down to 31 nm.

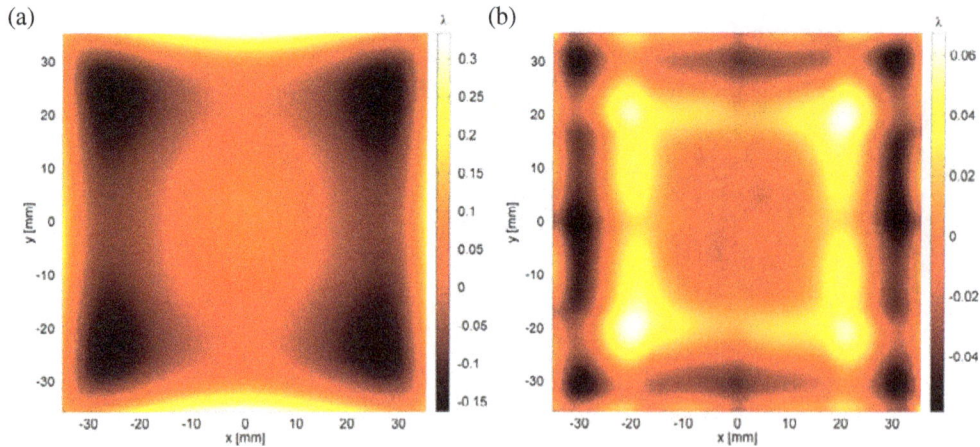

Figure 13. (a) Output OPD calculated in MIRÓ and shown in Figure 12(b) after subtraction of tilt and defocus. (b) Residual OPD after correction by the deformable mirror with 36 actuators.

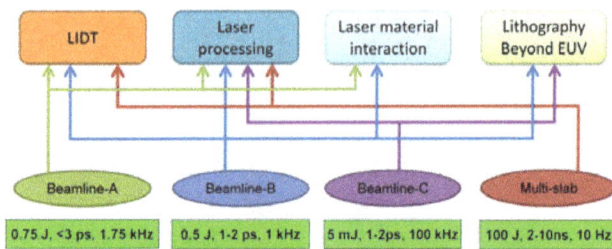

Figure 14. Schematic of the HiLASE application program.

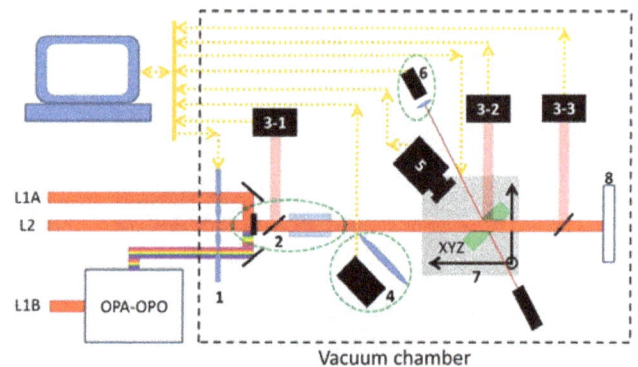

Figure 15. Schematic of the LIDT measurement station: (1) high-speed shutter, (2) beam positioning and focus, (3) beam diagnostics, (4) scattered light damage detection and fluorescence collector, (5) slow-motion camera, (6) interference damage detection; (7) XYZ tower, (8) beam dump. It uses laser pulses from Beamline A (L1A), Beamline B (L1B), and the Multi-slab (L2) laser system.

4. Applications

One of the long-term objectives of HiLASE is the identification of new and promising industrial applications and technologies using the DPSSL systems that were described above. Once commissioned in the HiLASE center these advanced DPSSL systems will enable, for example, research relevant to the testing of new dielectric optical components with high damage thresholds, prototyping new pump lasers for OPCPA (Optical Parametric Chirped Pulse Amplification) systems, driving high yield secondary photon and particle sources, and industrial applications related to efficient processing of materials (ablative removal of thin layers, cutting of optically transparent materials, laser peening, surface structuring and modifications, etc.). An overview of the HiLASE laser application program is shown in Figure 14.

4.1. Laser-induced damage threshold testing

First, the laser-induced damage threshold (LIDT) automated experimental station would be introduced. The station design allows one to measure the LIDT under a wide range of laser parameters: from the irradiation of small spots with 1–2 ps laser pulses at various wavelengths and a 1 kHz repetition rate to the irradiation of large spots with 2–10 ns laser pulses at 1030 nm with a 10 Hz repetition rate. The main advantage

of this station is real-time monitoring of laser damage with an acquisition frequency of up to 1 kHz. This station will allow the determination of the damage occurrence, as well as following the damage growth and damage threshold variation under repetitive irradiation. A schematic of the station is shown in Figure 15. The particular design of the LIDT station permits investigations of samples of up to 1 kg with surface sizes of 100 mm × 100 mm.

4.2. Mid-IR optical parametric generator

For the investigation of laser–material interactions and processing, as well as the thin-disk and multi-slab systems, a mid-IR pulse source with a high repetition rate and an average power of 10 W [27] is being developed. The proposed scheme of the wavelength conversion setup (Figure 16) uses the picosecond output from the thin-disk regenerative amplifier as the pump beam. The seed beam for the wave-

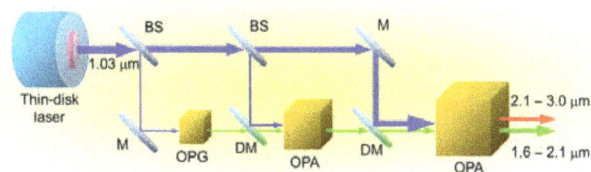

Figure 16. Schematic of the mid-IR parametric generator and amplifier. It consists of the thin-disk laser system, beam splitters (BS), mirrors (M), dichroic mirrors (DM), an OPG, and OPA.

length conversion setup comes from the optical parametric generator (OPG) and the signal is tunable between 1.6 and 2.1 μm. Only the signal beam is amplified by the optical parametric amplifier (OPA) chain. Idler beams having wavelengths from 2 to 3 μm can be extracted from the last parametric amplifier.

4.3. EUV light generation

For EUV generation, powerful CO_2 lasers are used to evaporate tin (Sn) droplets to generate a plasma that emits light at 13.5 nm. The CO_2 laser provides a much higher average power and higher conversion efficiency to UV light, but the laser footprint and plasma size are large[28], so it is suited for high-volume manufacturing. There is also a need for a small-scale EUV source with very high brightness for method and component testing. This compact source requires a stable beam with a higher beam quality than the beam quality provided by CO_2 lasers but at the same time the average power output required is only in the region of 1 kW. For this reason, we are constructing an EUV generation station using a thin-disk amplifier with a high average power. This source will be employed to study processes during EUV light generation.

Acknowledgements

This work benefited from the support of the Czech Republic's Ministry of Education, Youth and Sports to the HiLASE (CZ.1.05/2.1.00/01.0027), DPSSLasers (CZ.1.07/2.3.00/20.0143), and Postdok (CZ.1.07/2.3.00/30.0057) projects, co-financed by the European Regional Development Fund. This research has been partially supported by grant RVO 68407700.

References

1. E. Shcherbakov, V. Fomin, A. Abramov, A. Ferin, D. Mochalov, and V. P. Gapontsev, In *Advanced Solid-State Lasers Congress* (Optical Society of America, Washington, DC, 2013) ATh4A.2.

2. Joint High Power Solid-State Laser, http://www.northropgrumman.com/Capabilities/SolidStateHighEnergyLaserSystems/Pages/JointHighPowerSolidStateLaser.aspx (January 29, 2014).

3. T. Metzger, A. Schwarz, C. Teisset, D. Sutter, A. Killi, R. Kienberger, and F. Krausz, Opt. Lett. **34**, 2123 (2009).

4. C. Y. Teisset, M. Schultze, R. Bessing, M. Häfner, S. Prinz, D. Sutter, and T. Metzger, In *Advanced Solid-State Lasers Congress* (Optical Society of America, Washington, DC, 2013) JTh5A.1.

5. R. Jung, J. Tümmler, Th. Nubbemeyer, I. Will, W. Sandner, G. Erbert, and W. Pittroff, Disk lasers for powerful picosecond pulses with 100 Hz repetition rate. In *2nd Disk Laser Workshop* (Dausinger and Giessen, Stuttgart) (2012).

6. J.-P. Negel, A. Voss, M. A. Ahmed, D. Bauer, D. Sutter, A. Killi, and T. Graf, Opt. Lett. **38**, 5442 (2013).

7. C. J. Saraceno, F. Emaury, C. Schrieber, M. Hoffmann, M. Golling, T. Sudmeyer, and U. Keller, Opt. Lett. **39**, 9 (2014).

8. M. Hornung, S. Keppler, R. Bodefeld, A. Kessler, H. Liebetrau, J. Koerner, M. Hellwing, F. Schorcht, O. Jackel, A. Savert, J. Polz, A. K. Arunachalam, J. Hein, and M. C. Kaluza, Opt. Lett. **38**, 718 (2013).

9. M. Divoky, S. Tokita, H. Furuse, K. Matsumoto, Y. Nakamura, and J. Kawanaka, In *Advanced Solid-State Lasers Congress* (Optical Society of America, Washington, DC, 2013) AF2A.5.

10. T. Goncalves-Novo, D. Albach, B. Vincent, M. Arzakantsyan, and J. C. Chanteloup, Opt. Express **21**, 855 (2013).

11. K. Ertel, S. Banerjee, P. Mason, P. Phillips, R. Greenhalgh, C. Hernandez-Gomez, and J. Collier, Proc. SPIE **8780**, 87801W (2013).

12. A. Bayramian, S. Aceves, T. Anklam, K. Baker, E. Bliss, C. Boley, A. Bullington, J. Caird, D. Chen, R. Deri, M. Dunne, A. Erlandson, D. Flowers, M. Henesian, J. Latkowski, K. Manes, W. Molander, E. Moses, T. Piggott, S. Powers, S. Rana, S. Rodriguez, R. Sawicki, K. Schaffers, L. Seppala, M. Spaeth, S. Sutton, and S. Telford, Fusion Sci. Technol. **60**, 28 (2011).

13. A. Giesen, H. Hugel, A. Voss, K. Wittig, U. Brauch, and H. Opower, Appl. Phys. B **58**, 365 (1994).

14. C. Kränkel, R. Peters, K. Petermann, P. Loiseau, G. Aka, and G. Huber, J. Opt. Soc. Am. B **26**, 1310 (2009).

15. M. Chyla, T. Miura, M. Smrz, P. Severova, O. Novak, S.S. Nagisetty, A. Endo, and T. Mocek, Proc. SPIE **8780**, 87800A (2013).

16. M. Smrz, T. Miura, M. Chyla, A. Endo, and T. Mocek, In *IEEE Photonics Conference* (IEEE, Washington, 2013) MD1.4.

17. A. Bayramian, J. Armstrong, G. Beer, R. Campbell, B. Chai, R. Cross, A. Erlandson, Y. Fei, B. Freitas, R. Kent, J. Menapace, W. Molander, K. Schaffers, C. Siders, S. Sutton, J. Tassano, S. Telford, C. Ebbers, J. Caird, and C. Barty, J. Opt. Soc. Am. B **25**, B57 (2008).

18. A. C. Erlandson, S. M. Aceves, A. J. Bayramian, A. L. Bullington, R. J. Beach, C. D. Boley, J. A. Caird, R. J. Deri, A. M. Dunne, D. L. Flowers, M. A. Henesian, K. R. Manes, E. I. Moses, S. I. Rana, K. I. Schaffers, M. L. Spaeth, C. J. Stolz, and S. J. Telford, Opt. Mater. Express **1**, 1341 (2011).

19. P. D. Mason, K. Ertel, S. Banerjee, P. Phillips, C. Hernandez-Gomez, and J. Collier, Proc. SPIE **8780**, 87801X (2013).

20. M. Sawicka, M. Divoky, J. Novak, A. Lucianetti, B. Rus, and T. Mocek, J. Opt. Soc. Am. B **29**, 1270 (2012).

21. V. Jambunathan, J. Koerner, P. Sikocinski, M. Divoky, M. Sawicka, A. Lucianetti, J. Hein, and T. Mocek, Proc. SPIE **8780**, 87800G (2013).

22. B. E. Launder and D. B. Spalding, Comput. Meth. Appl. Mech. Eng. **3**, 269 (1974).

23. W. M. Kays, *Convective Heat and Mass Transfer* 3rd edition (McGraw-Hill Inc., 1993).

24. H. Schlichting and K. Gersten, *Boundary Layer Theory* 8th edition (Springer Verlag, Berlin, 2000).

25. O. Slezak, A. Lucianetti, M. Divoky, M. Sawicka, and T. Mocek, IEEE J. Quantum Electron. **49**, 960 (2013).

26. J. Pilar, M. Divoky, P. Sikocinski, O. Slezak, A. Lucianetti, V. Kmetik, S. Bonora, and T. Mocek, Proc. SPIE **8780**, 878011 (2013).

27. O. Novak, T. Miura, P. Severova, M. Smrž, A. Endo, and T. Mocek, In *Advanced Solid-State Lasers Congress* (Optical Society of America, Washington, DC, 2013) JTh2A.29.

28. A. Endo, In *Source: Lithography* M. Wang, (eds.) (INTECH, Croatia, 2010) chap. 9.

9

High harmonic generation and ionization effects in cluster targets

M. Aladi, I. Márton, P. Rácz, P. Dombi, and I.B. Földes

Wigner Research Centre for Physics of the Hungarian Academy of Sciences, Association EURATOM HAS, H-1121 Budapest, Konkoly-Thege u. 29-33, Hungary

Abstract

High harmonic generation in gas jets was investigated in different gases up to more than 14 bar backing pressure. The observation of increase of harmonic intensity with increasing pressure and laser intensity shows evidence of the presence of clusters in Xe with an increased efficiency compared with He, whereas Ar is an intermediate case for which clusters will start to dominate above a certain backing pressure. Spectral investigations give evidence for tunable harmonic generation in a broad spectral range. A spectral shift of opposite signature caused by the free electrons in the focal volume and the nanoplasmas inside the cluster was observed.

Keywords: clusters; high harmonic generation; nanoplasmas; ultrashort pulses

1. Introduction

The most flexible tool for generating coherent extreme ultraviolet (EUV) radiation by an intense laser pulse is the generation of its high harmonics. The harmonics of ultrashort laser pulses in gases cover practically the full spectral range from the visible to kilo-electronvolt x-ray. In the interaction of an intense laser pulse with a gas consisting of atoms, molecules or – in the present case – clusters, optical ionization may occur as a result of the distortion of the Coulomb potential by the intense electric field[1, 2]. The freed electron is driven by the laser electric field, and may recombine with its parent ion, emitting the excess energy in the form of a high energy photon. As this process is repeated every half-cycle, the temporal periodicity of the process leads to the appearance of discrete spectral lines at harmonics of the laser frequency. This half-cycle periodicity is the reason why in most cases odd harmonics are generated in gases up to a limit defined by the laser intensity and the ionization potential of the gas, giving an upper limit for the generated photon energies. A great deal of effort has been expended towards increase of the conversion efficiency of harmonics for better usability. High harmonic generation (HHG) from clusters seems to be a possible candidate for an efficient light source giving higher emission frequencies and higher conversion efficiency[3, 4].

Although the higher conversion efficiency and higher observable harmonics are advantageous, the investigation of HHG from clusters is not strongly prevalent because of two main reasons. The first reason is that although the conversion can be higher than in atomic gases, the ionization of clusters sets an upper limit for the obtainable intensity, similarly to that of atoms. Therefore, most of the recent efforts aim to use loose focusing[5–7] and consequently a long homogeneous target material for harmonic generation. Clearly, it is not easy to realize an elongated cluster source. The other obstacle for a broader application is that the mechanism of HHG from clusters is still a subject of intense debate[8]. Apart from the traditional three-step model which is based on the recombination to the same atom, models have been developed in which recombination is considered to the neighbouring atoms[9, 10], which may even produce incoherent radiation as there will be no phase locking between the two atomic wavefunctions[11]. Ruf et al.[8] suggests an alternative with tunnel ionization from a partly delocalized electron wavefunction and recombination to this wavefunction, i.e., to the cluster itself. In experiments the difficulty of separation of harmonics from the monomer atoms and from clusters makes it difficult to determine the actual mechanism of harmonic generation.

In the present work we aim to progress towards the ionization limit, i.e., we are investigating the limit where the ionization sets in with the signature of free electrons in the free space and inside the cluster as well. HHG is investigated in different gases, namely in He which is purely atomic, in Xe which generally forms clusters, and in transitory Ar where significant cluster generation sets in

Correspondence to: I. B. Földes, Wigner Research Centre for Physics, H-1525 Budapest, POB 49, Hungary. Email: foldes.istvan@wigner.mta.hu

above a certain backing pressure of the pulsed valve source. Investigation of the pressure dependence of the harmonic intensity shows the effects of cluster generation. Detailed analysis of the harmonic spectra shows the tunability of the harmonic wavelength by the appearance of free electrons in the interaction range, and also the effect of free electrons inside the clusters.

2. Experimental

Harmonics were generated in gas jet targets using commercial valves of Parker Hannifix, Series 9. The original conical nozzle had an orifice of 1mm diameter. In order to improve the gas jet target parameters, an additional nozzle with a cylindrical orifice of 0.7 mm diameter was used. The total gas density was measured using an x-ray shadowgraphic method for Ar and Xe, and thus the average density as well as the density distribution was determined in our earlier investigations[12] from 2–12 bar backing pressure. In the case of a typical 1 ms opening time of our experimental series the density at $\sim$1 mm from the nozzle tip was found to be as high as 10^{19} atoms cm^{-3} – depending slowly on the pressure – for both gases. Therefore, it seems to be reasonable to assume a similar density for He, as well.

The formation of clusters can be approximated by the semiempirical Hagena scaling parameter[13], $\Gamma^* = k(((d/\tan\alpha)^{0.85})/T_0^{2.29})p_0$, in which d is the diameter of the orifice in μm, p_0 is the backing pressure in mbar, T_0 is the temperature in Kelvin, and k is the condensation parameter[13], which is equal to $k = 5500$ for Xe, $k = 1650$ for Ar and $k \approx 4$ for He. We approximate the jet expansion half-angle by $\alpha \approx 45°$. Using this scaling parameter the average cluster size can be estimated[13] as $\bar{N} = 33(\Gamma^*/1000)^{2.35}$, which can be plotted in our range of interest, i.e., 1–20 bar backing pressure at room temperature, for the gases used in Figure 1. Our earlier experimental investigations[12] with the additional nozzle showed (Figure 3 therein) that the jet expansion half-angle was really $\alpha \approx 45°$ without the additional nozzle, and even smaller, $\alpha \approx 30°$, with the additional nozzle, giving even higher cluster sizes.

It can be seen that for He no clusters can be expected. The cluster sizes for Xe are significantly larger than for Ar, but with increasing pressure the size of the clusters is well above 1000 atoms even for Ar. These curves are estimated from the scaling law above[13], which will be used throughout the remainder of the paper. The dotted horizontal line gives the limit above which the cluster effect seems to play a significant role in our observations. As we shall see below, its intersection with the Xe and Ar data is in agreement with the harmonic results which suggest the effect of clusters above 6 bar backing pressure in Ar and even for the lowest Xe backing pressures.

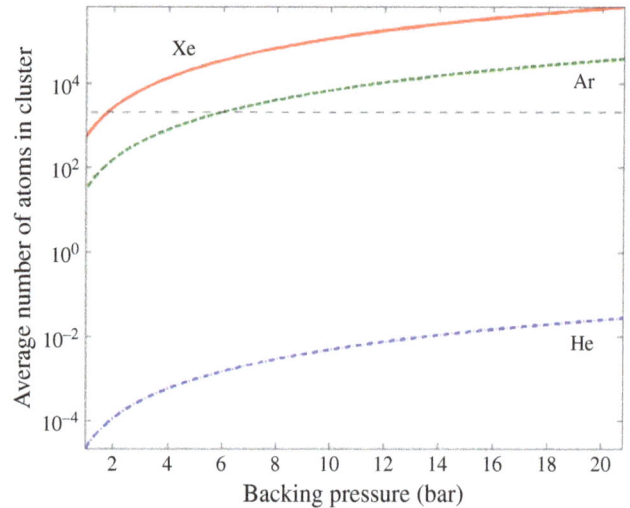

Figure 1. Pressure dependence of the average cluster size for He (dashed–dotted line), Ar (dashed line) and Xe (solid line) according to the Hagena scaling.

A Ti:sapphire laser beam was used in the experiments with 800 nm central wavelength and 1 kHz repetition rate. The pulse duration was 40 fs with 4 mJ pulse energy. The $\sim$1 cm diameter beam was focused into the vacuum chamber by an $f = 30$ cm lens. The position of the valve could be varied relative to the focal plane by moving the lens parallel to the beam. A toroidal holographic grating (Jobin-Yvon) of 550 lines mm^{-1} collected the emitted high harmonic in the vacuum ultraviolet (VUV). Due to the loose focusing there was a danger that in this imaging spectrometer in which the valve–grating distance was only $\sim$32 cm the grating might be damaged. Therefore the arrangement of Peatross *et al.*[14] was used, which is a combination of a beam block at the centre of the incident beam and an aperture before the grating which suppresses the fundamental beam. The detector was a microchannel plate (MCP) with a phosphor screen. The visible light of the screen was imaged onto a CCD detector. This single-shot spectrometer provided a spectrum between 20 and 50 nm with a resolution of $\sim$0.5 nm.

In the experiments, the intensity and spectral dependence of high harmonics were investigated for different gases in dependence on the pressure. For a comparison we chose the 30–50 nm wavelength range; thus, high enough harmonics could be observed with a single shot without moving the grating but in a range in which the reflectivity of the grating was still acceptable (note that it starts to decrease below 35 nm).

3. Results

HHG was observed in each gas from 1 to 20 bar backing pressure. Figure 2 illustrates the spectrum obtained from Ar at 14 bar backing pressure for a laser intensity of

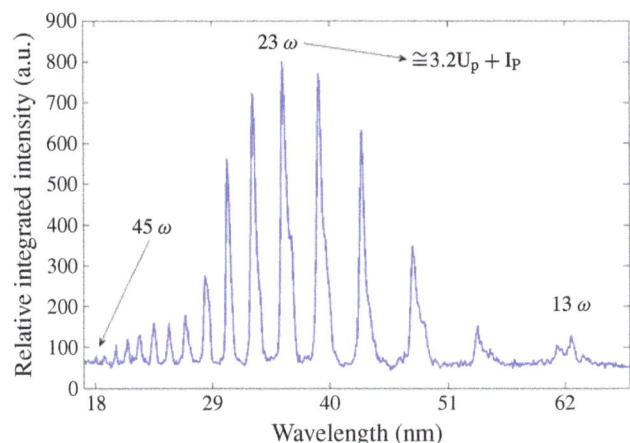

Figure 2. The HHG spectrum from Ar at 14 bar backing pressure and 10^{14} W cm^{-2} intensity.

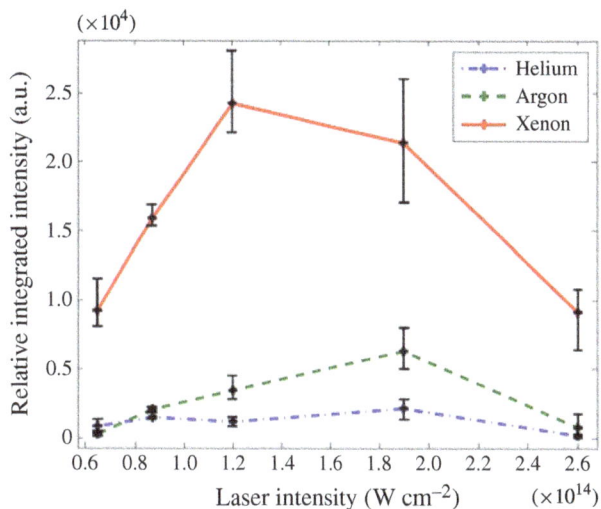

Figure 3. Intensity dependence of the 15th harmonic for different gases at 6 bar backing pressure.

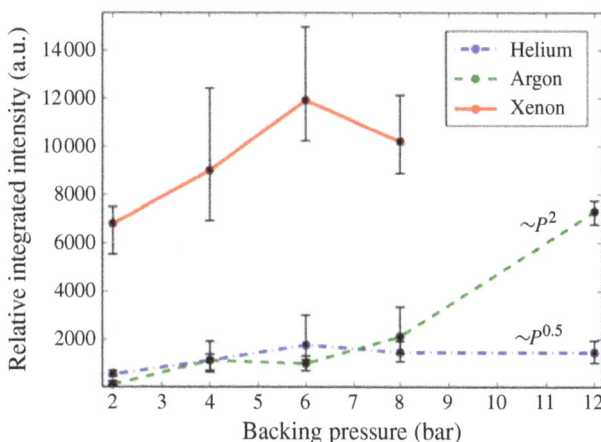

Figure 4. Comparison of the pressure dependence of the 25th harmonic at 6.5×10^{13} W cm^{-2} intensity in He, Ar, and Xe.

10^{14} W cm^{-2}, in which case harmonics up to the 45th order were observed, which is nearly a factor of 2 higher than the ponderomotive cutoff limit at $I = 3.2U_p + I_p$, in which I_p is the ionization potential and U_p is the ponderomotive energy. Clearly, the intensity of the harmonics starts to drop at this limit. We can mention here that according to the documentation of the grating as obtained from Francelab its efficiency drops below 30 nm wavelength; therefore, the real intensity of the high harmonic orders is relatively higher than in Figure 2.

In order to compare the intensity dependence of HHG for different gases we chose a given harmonic order. Although the dependences are similar for different harmonic orders, the selection of a given pressure and harmonic order reduces the uncertainties, and it is thus illustrative for showing the qualitative behaviour. Figure 3 compares the intensity of a given harmonic order (15th in this case) for different gases, and here we can see that the conversion efficiency is highest for Xe, and it is significantly lower in the other two gases. In each case, after an initial increase of conversion efficiency with increasing intensity, it turns to saturation and then to a decrease of conversion efficiency above $1–2 \times 10^{14}$ W cm^{-2}. While the highest observed conversion efficiency in Xe cannot be fully attributed to the presence of clusters (even the atomic conversion efficiency of xenon is the highest one) the earlier saturation refers to the lower ionization potential.

The existence of clusters can be confirmed when investigating the pressure dependence of HHG, especially for Ar where the change is abrupt. Figure 4 compares the efficiency of the 25th harmonic for the three gases as a function of backing pressure. At low argon pressures and for He the increase of intensity with pressure is slow, even less than linear. In the case of argon a steep increase of efficiency starts above 6 bar backing pressure. This steeper increase can be attributed to the effect of clusters that appear with increasing pressure. Although the range of steeper increase is not sufficient for fitting, the steeper increase is

approximately quadratically dependent on the pressure. This steeper dependence is expected for clusters[3], especially above 6 bar in Ar where the size of the clusters starts to become significant, i.e., more than 1000 atoms. It should be noted that this cluster size is the same as for Xe already at 1 bar; therefore, we can assume that for Xe cluster effects dominate for the whole pressure range of our investigations. In Xe the pressure dependence starts steeply – similarly to Ar above 6 bar – then it shows a saturation, probably because of propagation effects in the high density material.

It should be noted that our observations differ slightly from those of Donnelly et al.[3], as they claimed an even stronger, cubic pressure dependence in the case of cluster targets. This difference can be partly caused by the different pulse duration of the laser, which was significantly longer in their work, 150 fs as compared with our 40 fs duration. Another possibility is that due to the different shapes of the valves

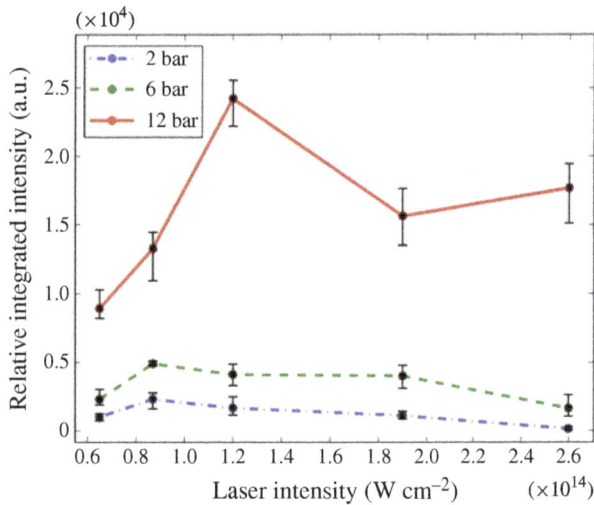

Figure 5. Intensity dependence of 25ω generation for Ar at 2, 6 and 12 bar backing pressure.

the densities were different in the different experiments[12]. Last but not least, propagation effects may play an important role in the not too steep pressure dependence in our case, as this explanation is supported by the observation of intensity saturation in Xe for high pressure, as illustrated in Figure 4.

The observation that argon seems to display a sharply increased conversion efficiency with the appearance of clusters shows that we can investigate the intensity dependence of harmonic conversion for different backing pressures when clusters are generated, i.e., above 6 bar, and when not. Figure 5 illustrates the intensity dependence for 2 bar backing pressure where cluster formation is negligible, for 12 bar

with strong clusterization, and for 6 bar which is intermediate. Clearly, harmonic conversion is significantly higher for the largest clusters (12 bar), and at the lowest intensities, the conversion efficiency increase is steeper than for the cases with lower pressures and modest cluster formation. On the other hand, the saturation due to ionization of the gas is similar for clusters to that for atomic gases. Therefore, we can see that although clusters may really increase the conversion efficiency of harmonics, the ionization limit is not changed significantly.

Special emphasis was directed towards detailed spectral analysis of the harmonics. Here, the results were collected for the case in which the gas jet was behind the focus of the laser, i.e., in the diverging beam; thus, throughout the experiments the short trajectories of electrons in the HHG process were probably dominant[15]. As was expected, the spectral width of the harmonics was significantly broader in Xe, where the beam interacted with clusters, than in He, in which no clusters were present. While the typical spectral width of harmonics in the wavelength range of 30–50 nm did not show significant pressure dependence and its full width varied between 1.4 and 2 nm, the spectral width of the harmonics for Xe showed a near linear increase with increasing pressure. As an example, in the case of the 15ω radiation the spectral width was 1.9 nm for 2 bar backing pressure, and it reached 4.5 nm width in the case of 12 bar. These observations agree with earlier results, e.g., with fullerene targets[16].

It was discovered in the 1990s that in the case of moving the gas target relative to the focal plane a spectral blue shift can be observed due to the varying free electron density, thus providing even a tunability of high harmonics[17]. Figure 6 clearly shows this effect, namely that a significant blue

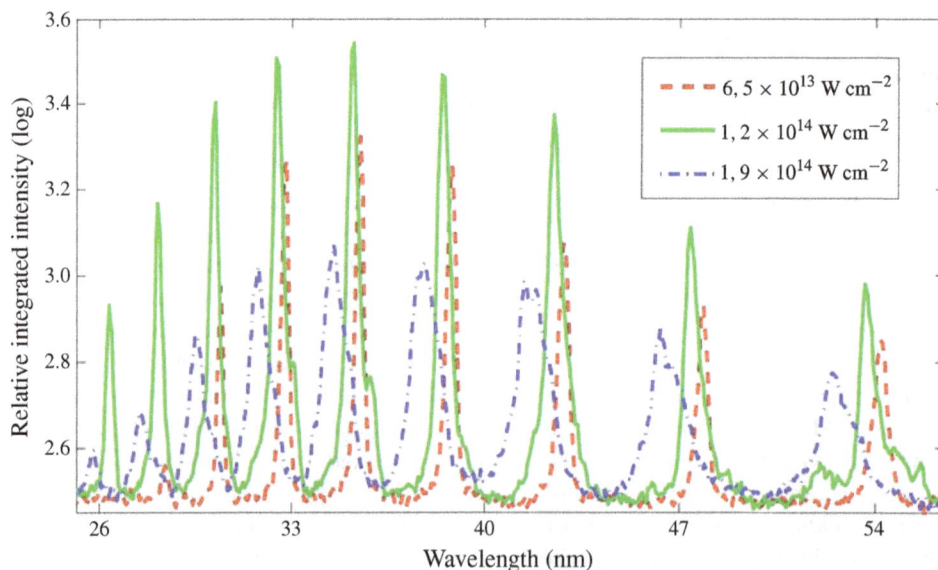

Figure 6. Increasing blue shift of high harmonics with increasing intensity for 12 bar Ar backing pressure.

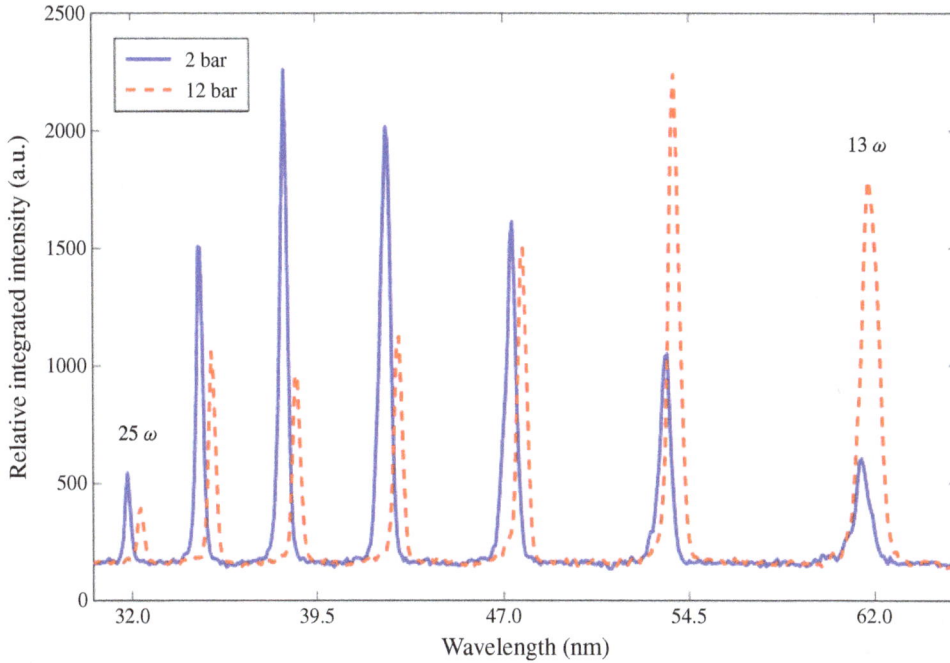

Figure 7. Harmonic spectra from Xe at 6.5×10^{13} W cm^{-2} intensity for 2 and 12 bar (dotted line) backing pressure.

shift of the harmonics can be observed which increases with increasing intensity. A special case of 12 bar backing pressure, i.e., high density of Ar, is illustrated here. It must be added that the spectral shift is dependent, in addition to the intensity, on the density of material and on the material as well, the results being different for different gases for the same backing pressure. Due to the strong density dependence, the blue shift is very sensitively dependent on the exact distance from the valve. Although the large spectral shift opens the possibility of tuning the harmonic wavelengths on the full observable spectral range in the VUV, application requires further studies with accurate measurement of the gas density.

An interesting phenomenon can be observed when at relatively low intensity we increase the backing pressure and thus the size of the clusters. Figure 7 shows the observed harmonic spectra from Xe in the case of 6.5×10^{13} W cm^{-2} intensity for low (2 bar) and high (12 bar) backing pressure. The spectral shift towards longer wavelengths is clearly visible.

In order to give a more quantitative insight we illustrate in Figure 8 the 21st harmonic in the case of xenon gas for the lowest intensity applied, i.e., 6.5×10^{13} W cm^{-2}. We start with 2 bar backing pressure – when the clusters consist of a maximum of 1000 particles according to Figure 1 – as the one with zero spectral shift. On increasing the backing pressure a red shift, i.e., a spectral shift with the opposite signature to the one caused by the free electrons, can be observed. Although this type of spectral shift is always nearly an order of magnitude lower than the contribution

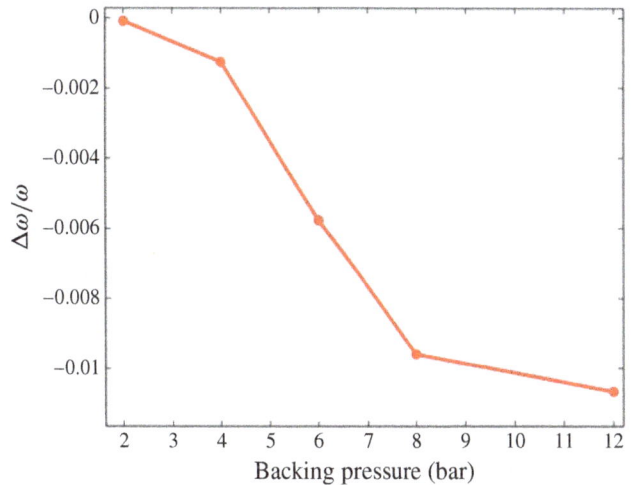

Figure 8. Red shift of the 21st harmonic for Xe at 6.5×10^{13} W cm^{-2} intensity with increasing pressure.

from free electrons it can be as high as $\Delta\omega/\omega \approx 10^{-2}$ for the highest applied pressure of 12 bar, in which case clusters with sizes of $\sim 10^5$ particles are expected. This contribution is expected to that caused by the nanoplasmas inside the clusters, as was suggested by Tisch[18] and which we shall discuss below. It must be emphasized that as it is significantly lower than the blue shift caused by the free electrons it is only a relative red shift, often suppressed by the larger free-electron contribution, especially for higher intensities when the free electrons will dominate in the propagation effects.

4. Discussion

On the one hand, the experimental results confirm the earlier observations of intense HHG from cluster targets; on the other hand, the steep intensity increase of HHG with increasing pressure provides proof of the existence of clusters. The observation of a power law dependence, namely that the exponent is different from some earlier observations[3], is probably only partially caused by the difference of the pulse duration of the applied lasers. As is evident from the spectral observations, the results are very sensitively dependent on the actual experimental parameters, especially the gas density and the cluster size.

Indeed, the most interesting result is the observed spectral structure of the generated harmonics. As is illustrated in Figure 6, the spectral shift caused by intensity variation is comparable with the distances between the subsequent harmonics. It should be borne in mind that by changing the gas and its pressure, this shift can be further increased, i.e., a quasi-continuously tunable coherent VUV and EUV light source can be generated. This may even serve as a seed pulse for an x-ray laser amplifier[19].

The blue shift of the harmonics due to the free electrons, which increases both by increasing the pressure and by increasing the intensity, was explained by the phase matching condition, which for the laser and harmonics of order q can be given as

$$\Delta k = k_{q\omega} - q k_\omega = \Delta k_{disp} + \Delta k_{geom} + \Delta k_{electron}, \quad (1)$$

where $k_{q\omega}$ and k_ω are the wavevectors for the harmonics and the laser radiation, respectively. The subscript *disp* is the dephasing contribution from atomic dispersion, *geom* is the geometric contribution determined by the focusing geometry, and *electron* is the contribution from free electrons in the interaction range. Our main interest here is the free-electron contribution, which can be written as

$$\Delta k_{electron} \approx \frac{6\pi}{\lambda_\omega} \left(n_{q\omega} - n_\omega \right)$$
$$= \frac{6\pi}{\lambda_\omega} \left(\sqrt{1 - \frac{\omega_p^2}{q^2\omega^2}} - \sqrt{1 - \frac{\omega_p^2}{\omega^2}} \right) \geqslant 0. \quad (2)$$

According to usual notation, ω_p is the plasma frequency. This term, as seen here, is always positive; thus, it gives a blue shift in frequency which can be approximated[20] by

$$\Delta \omega_{electron} = \frac{\omega}{2 n_e c} \frac{\partial \langle n_e \rangle}{\partial t} l, \quad (3)$$

in which the averaged density $\langle n_e \rangle$ along the pathlength l is given, which is a sort of self-phase modulation.

The effect of nanoplasmas inside the clusters on the dephasing was estimated by Tisch[18], based on the simple Drude model for the dielectric function, i.e.,

$$\varepsilon = 1 - \frac{\omega_p^2}{\omega(\omega + i\nu)}, \quad (4)$$

which, in general, uses the electron–ion collision frequency ν. The clusters are assumed to be dielectric spheres of radius r with a dipole moment of

$$p = \left(\frac{\varepsilon - 1}{\varepsilon - 2} \right) r^3 E_0, \quad (5)$$

for the field strength E_0. Thus, the linear susceptibility can be estimated for the cluster density n_{cl} by

$$\chi_c = \frac{n_{cl} p}{E_0} = n_{cl} \left(\frac{\varepsilon - 1}{\varepsilon + 2} \right) r^3. \quad (6)$$

The refractive index can now be estimated for the collisionless case by

$$n(\omega) = \sqrt{1 + 4\pi\chi} \approx 1 + 2\pi\chi = 1 - \frac{2\pi n_e r^3 n_{cl}}{3 n_{crit} - n_e}, \quad (7)$$

where n_e is the electron density inside the cluster and n_{cr} is the critical electron density. This means that $n(\omega) > 1$ for $n_e > 3 n_{crit}$ and, on the other hand, $n(\omega) < 1$ for $n_e < 3 n_{crit}$, where n_e is the electron density in the cluster. Consequently, we can estimate the dephasing as

$$\Delta k_{nanoplasma} = \frac{6\pi}{\lambda_\omega} \left(1 - \frac{2\pi n_e r^3 n_{cl}}{3 n_{qcrit} - n_e} \right)$$
$$- \frac{6\pi}{\lambda_\omega} \left(1 - \frac{2\pi n_e r^3 n_{cl}}{3 n_{crit} - n_e} \right). \quad (8)$$

The first term in Equation (8) is smaller than 1 due to the high frequency of the harmonics, and the second term is >1. Therefore, it has the opposite signature to Equation (2); consequently the nanoplasma can give a negative contribution with opposite sign to the free electrons in the interaction range. Thus, we can confirm that the spectral contribution of the nanoplasmas in the clusters can give a spectral shift of opposite signature to the free electrons; therefore, the observation of this red shift may also serve as evidence of cluster generation.

However, it must be noted that the above-mentioned blue and red shifts rarely appear separately in a clean form; they are strongly dependent on the parameters of the clusters and the lasers. Parametrization of the full range of observations can be carried out by using independent diagnostics of the cluster size. It must also be mentioned that in our estimations we used the simple analytical estimations of Tisch[18]. Clearly, a full computer modelling of phase matching effects for different propagation geometries and different cluster sizes would be a great step forward.

5. Conclusion

We can conclude that the HHG in different gases gives a clear signature of the existence of clusters. Clusters can be used as a possible method to increase the conversion efficiency of HHG, but in this case the ionization threshold gives an upper limit, similarly to atomic gases.

Spectral investigation of high harmonics gives evidence of a possible tunable coherent radiation source in the whole VUV and EUV spectral range. The opposite signatures of the spectral shifts caused by the free electrons in the focal volume and the nanoplasmas inside the cluster can be applied as a further signature of cluster diagnostics, for which a comparison with the cluster size is under progress. As a further remark, it can be mentioned that a recent idea using dual-gas multijet arrays[21] combined with cluster generation could become an even more efficient source of high harmonics.

Acknowledgements

This work, supported by the European Communities under the contract of the Association between EURATOM and the Hungarian Academy of Sciences, was carried out within the framework of the European Fusion Development Agreement. The views and opinions expressed herein do not necessarily reflect those of the European Commission. It was also supported by the Hungarian contract No. ELI 09-1-2010-0010 hELIos-ELI, project 109257 of the Hungarian Scientific Research Fund and the COST MP1208 and MP1203 activities. P.R. was supported by a postdoctoral fellowship of the Hungarian Academy of Sciences.

References

1. K. J. Schafer, B. Yang, L. F. DiMauro, and K. C. Kulander, Phys. Rev. Lett. **70**, 1599 (1993).
2. P. Corkum, Phys. Rev. Lett. **71**, 1994 (1993).
3. T. D. Donnelly, T. Ditmire, K. Neumann, M. D. Perry, and R. W. Falcone, Phys. Rev. Lett. **76**, 2472 (1996).
4. C. Vozzi, M. Nisoli, J.-P. Caumes, G. Sansone, S. Stagira, S. De Silvestri, M. Vecchiocattivi, D. Bassi, M. Pascolini, L. Poletto, P. Villoresi, and G. Tondello, Appl. Phys. Lett. **86**, 111121 (2005).
5. Y. Wu, E. Cunningham, H. Zang, J. Li, M. Chini, X. Wang, Y. Wang, K. Zhao, and Z. Chang, Appl. Phys. Lett. **102**, 201104 (2013).
6. P. Rudawski, C. M. Heyl, F. Brizuela, J. Schwenke, A. Persson, E. Mansten, R. Rakowski, L. Rading, F. Campi, B. Kim, P. Johnsson, and A. L'Huillier, Rev. Sci. Instrum. **84**, 073103 (2013).
7. E. J. Takahashi, P. Lan, O. D. Mücke, Y. Nabekawa, and K. Midorikawa, Nature Commun. **4**, 3691 (2013).
8. H. Ruf, C. Handschin, R. Cireasa, N. Thiré, A. Ferré, S. Petit, D. Descamps, E. Mével, E. Constant, V. Blanchet, B. Fabre, and Y. Mairesse, Phys. Rev. Lett. **110**, 083902 (2013).
9. P. Moreno, L. Plaja, and L. Roso, Europhys. Lett. **28**, 629 (1994).
10. V. Véniard, R. Taieb, and A. Maquet, Phys. Rev. A **65**, 013202 (2001).
11. D. F. Zaretsky, Ph. Korneev, and W. Becker, J. Phys. B **43**, 105402 (2010).
12. R. Rakowski, A. Bartnik, H. Fiedorowicz, R. Jarocki, J. Kostecki, J. Mikolajczyk, A. Szczurek, M. Szczurek, I. B. Földes, and Zs. Tóth, Nucl. Instrum. Methods A **551**, 139 (2005).
13. O. F. Hagena, Rev. Sci. Instrum. **63**, 2374 (1992).
14. J. Peatross, J. L. Chaloupka, and D. D. Meyerhofer, Opt. Lett. **19**, 942 (1994).
15. F. Krausz and M. Ivanov, Rev. Mod. Phys. **81**, 163 (2009).
16. R. A. Ganeev, L. B. Elouga Bom, M. C. H. Wong, J.-P. Brichta, V. R. Bhardwaj, P. V. Redkin, and T. Ozaki, Phys. Rev. A **80**, 043808 (2009).
17. C. Altucci, R. Bruzzese, C. de Lisio, M. Nisoli, S. Stagira, S. de Silvestri, O. Svelto, A. Boscolo, P. Ceccherini, L. Poletto, G. Tondello, and P. Villoresi, Phys. Rev. A **61**, 021801(R) (1999).
18. J. W. G. Tisch, Phys. Rev. A **62**, 041802 (R) (2000).
19. Ph. Zeitoun, G. Faivre, S. Sebban, T. Mocek, A. Hallou, M. Fajardo, D. Aubert, Ph. Balcou, F. Burgy, D. Douillet, S. Kazamias, G. de Lacheze-Murel, T. Lefrou, S. le Pape, P. Mercere, H. Merdji, A. S. Morlens, J. P. Rousseau, and C. Valentin, Nature **431**, 426 (2004).
20. H. J. Shin, D. G. Lee, Y. H. Cha, J.-H. Kim, K. H. Hong, and C. H. Nam, Phys. Rev. A **63**, 053407 (2001).
21. A. Willner, F. Tavella, M. Yeung, T. Dzelzainis, C. Kamperidis, M. Bakarezos, D. Adams, M. Schulz, R. Riedel, M. C. Hoffmann, W. Hu, J. Rossbach, M. Drescher, N. A. Papadogiannis, M. Tatarakis, B. Dromey, and M. Zepf, Phys. Rev. Lett. **107**, 175002 (2011).

10

Progress of the 10 J water-cooled Yb:YAG laser system in RCLF

Jian-Gang Zheng, Xin-Ying Jiang, Xiong-Wei Yan, Jun Zhang, Zhen-Guo Wang, Deng-Sheng Wu, Xiao-Lin Tian, Xiong-Jun Zhang, Ming-Zhong Li, Qi-Hua Zhu, Jing-Qin Su, Feng Jing, and Wan-Guo Zheng

Research Center of Laser Fusion (RCLF), CAEP, P.O. Box 919-988, Mianyang, Sichuan 621900, China

Abstract

The high repetition rate 10 J/10 ns Yb:YAG laser system and its key techniques are reported. The amplifiers in this system have a multi-pass V-shape structure and the heat in the amplifiers is removed by means of laminar water flow. In the main amplifier, the laser is four-pass, and an approximately 8.5 J/1 Hz/10 ns output is achieved in the primary test. The far-field of the output beam is approximately 10 times the diffraction limit. Because of the higher levels of amplified spontaneous emission (ASE) in the main amplifier, the output energy is lower than expected. At the end we discuss some measures that can improve the properties of the laser system.

Keywords: Yb:YAG; V-shape amplifier; water-cooled; 10 J laser system

1. Introduction

It is becoming a reality that the energy crisis may be resolved by means of laser fusion, with significant on-going international research on inertial confinement fusion (ICF)[1] and the National Ignition Facility (NIF)[2, 3] coming closer to ignition. However, the laser driver of the NIF is based on Nd-doped glass[4, 5], which can operate only at low repetition rates (typically one shot every few hours)[6]. This kind of laser system cannot meet the requirements of inertial fusion energy (IFE). Therefore, it is necessary to develop a new type of laser driver for fusion that can work at a high repetition rate (typically 10–20 Hz)[7–11].

The following issues must be considered to obtain a high repetition rate for the IFE laser driver. Firstly, the laser system, especially the main amplifier system of the laser driver, must be actively thermal managed. Secondly, the gain medium of the amplifier must be able to operate at a high repetition rate – namely, when the laser system works repeatedly, the gain medium must dissipate heat to the outside to maintain itself in good order. Thirdly, the laser system must have a high energy conversion efficiency, so that electricity can be supplied from the power station.

Yb:YAG is an excellent gain medium for IFE laser systems[12–14] because of the long radiative lifetime of the upper energy level, which implies greater energy storage.

Yb:YAG has a high thermal conductivity, thereby enabling the Yb:YAG laser system to operate repeatedly. The medium is free from several loss mechanisms (e.g. excited-state absorption, concentration quenching and upconversion) even at high dopant concentrations. In addition, Yb:YAG exhibits low quantum defects between the pump photons and the laser photons, which results in low thermal loading. The use of this material has increased following the emergence of Yb:YAG ceramics that meet the requirements of high energy in a single beam. Given these advantages, Yb:YAG (crystal or ceramics) may be widely used in laser systems with high repetition rates and high energy conversion efficiencies[15–22].

Unfortunately, Yb:YAG is a quasi-three-level system, and its properties are strongly affected by temperature[23, 24]. Hence, the thermal management for the amplifier must be carefully designed to avoid a high rise in temperature[25]. The thermal conduction, absorption and emission cross-section of Yb:YAG are high at low temperatures[26]. Low-temperature operation requires a large amount of additional energy, which reduces the overall energy conversion efficiency of the laser system. A low-temperature cooling setup also increases the complexity of the laser system.

Water is a raw material that exhibits a high thermal capacity and fluidity. As it is readily available and inexpensive, water is a suitable cooling medium for IFE laser drivers. The laser system, especially the main amplifier, must be properly designed to utilize water as a coolant.

Correspondence to: Zhen-Guo Wang, Mianshan Road No.64, Mianyang, Sichuan province, China, ZIP code: 621900. Email: zjg8861@163.com

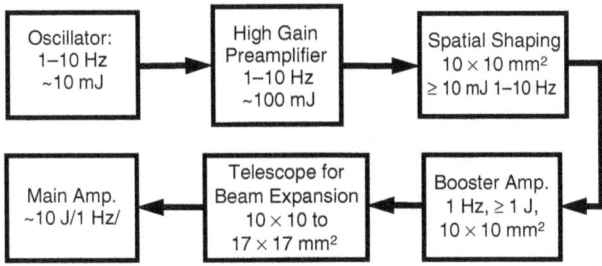

Figure 1. Structure of the proposed laser system.

In this paper, a room-temperature water-cooled 10 J Yb:YAG laser system is introduced. We present the structure of the laser system and describe the performance of its subsystem. Experimental results are discussed and solutions to increase the performance of the laser system are proposed.

2. Laser system design

2.1. Schematic of the laser system

The laser system (Figure 1) comprises a pulse generator, high-gain preamplifier, spatial shaping unit, booster am-

plifier, telescope for beam expansion, and main amplifier. The generator can produce laser pulses of approximately 10 mJ/10 ns, which are used as seed pulses for the laser system. The preamplifier provided an approximately 10 times gain and amplified the seed pulses to 100 mJ. The laser beam was expanded and clipped to a 10 mm × 10 mm square profile in the spatial shaping unit, reducing the energy of the laser pulse to approximately 10 mJ. The profile of the laser beam was configured to have a flat top, so that the beam could be properly amplified. The laser pulse was amplified to 1 J in the booster amplifier to achieve efficient energy conversion in the main amplifier. Finally, the main amplifier raised the energy of the laser pulses to approximately 10 J.

Figure 2 shows the layout of the laser system. To increase the efficiency of energy extraction, multi-pass amplification was used in the amplifiers. This process included non-image-relaying multi-pass V-shape amplification in the preamplifier, close-to-image-relaying multi-pass V-shape amplification in the booster amplifier, and image-relaying multi-pass V-shape amplification in the main amplifier. Image-relaying amplification enabled the uniform flat-top beam to be propagated to the gain medium for uniform amplification and avoided damage in the gain medium caused by near-field unevenness.

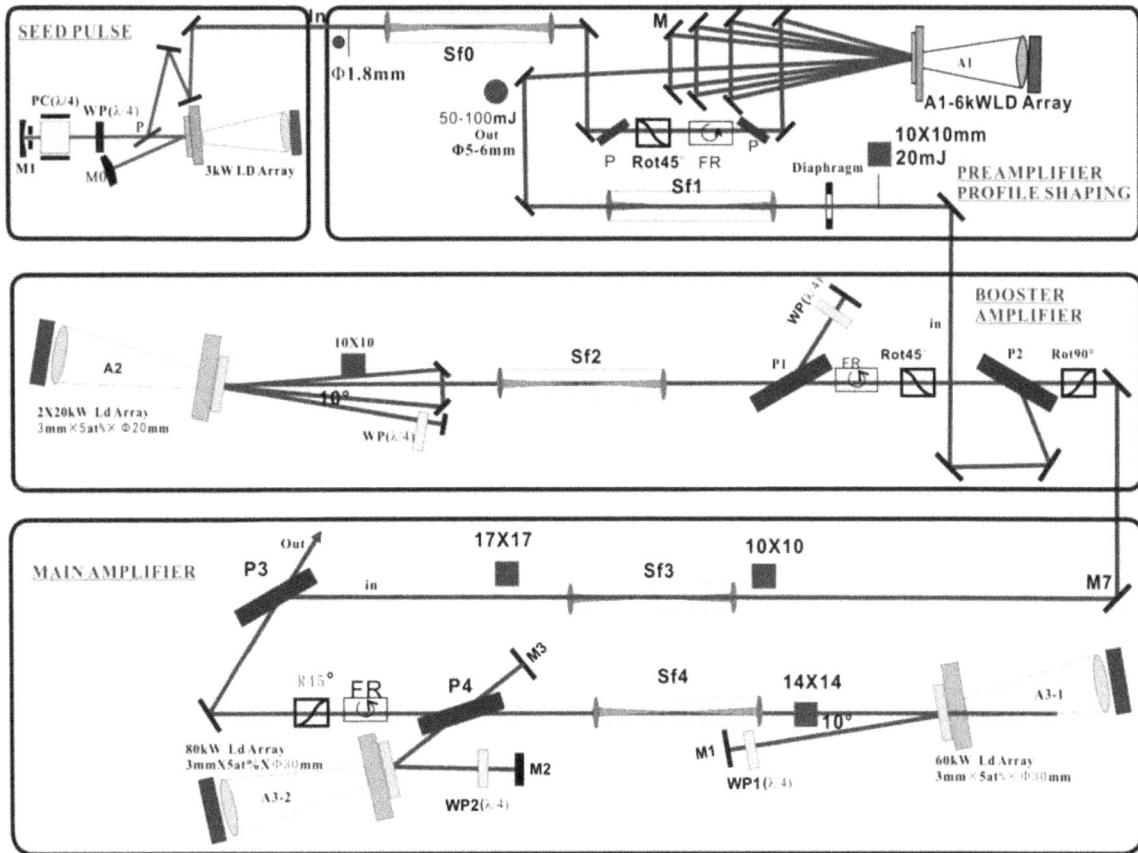

Figure 2. Layout of the laser system.

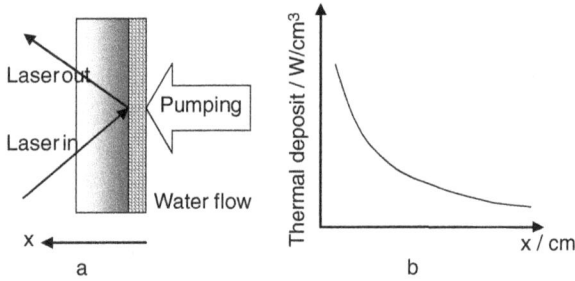

Figure 3. (a) Amplifier configuration and (b) thermal deposition in the gain medium as a function of x (in cm).

2.2. Configuration and thermal management of the amplifier

Figure 3(a) shows the amplifier configuration. The laser and pump light were located on opposite sides of the Yb:YAG crystal, with water flowing on the pump light side to cool the gain medium. Different dual-wavelength coatings were applied to the crystal: a coating that could reflect 1030 nm laser light and transmit 940 nm pump light was applied on the water-cooling side, while a coating that could transmit 1030 nm laser light and reflect 940 nm pump light was applied on the opposite side. This coating system ensured that the pump light and laser propagated through the crystal twice so that the pump light was adequately absorbed by the gain medium and the laser was amplified with high efficiency. The amplifier used was of the back-water-cooled type. One advantage of this type of amplifier lies in the ease of collection of water and cost efficiency; another advantage is the increase in thermal deposition on the pumping side (see Figure 3(b)). The amplifier configuration allows large amounts of heat to be extracted from the gain medium.

2.3. Pumping module for main amplifier

Laser diode (LD) arrays were used as a pumping source to improve the efficiency of energy conversion. The LD was modularized (Figure 4(a)) so that the diode could be produced in batches and easily used in large laser systems. In the 10 J laser system, the power of a single LD pump module is 10 kW (also called a 10 kW LD array). This system includes two planar arrays, each of which comprises about 24 bars; the power of the LD bar is about 200 W for a 200 A pump current. If the number of bars is less than 20 then the number of LD arrays and modules in the laser system increases and difficulties of system integration increase. The LD arrays were cooled by water flow through micro-channels. The LD bars were collimated by micro-lenses to ensure good propagation of the pump light.

High pump power densities are necessary for the Yb:YAG amplifier because some ion population is observed at lower lasing levels even at room temperature. An approximately 15–20 kW cm^{-2} power density is needed to achieve adequate energy storage in the gain medium. However, since

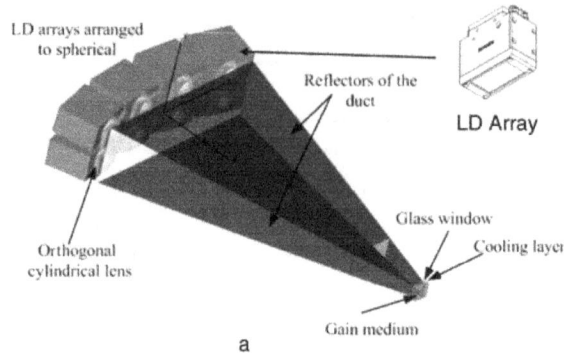

Figure 4. (a) Arrangement of LD modules with the duct and (b) output distribution.

the power density of the LD array is only 2–3 kW cm^{-2}, the requirements of a highly efficient laser amplifier cannot be met. Hence, the pump light must be focused and shaped to improve the power density and uniformity on the gain medium.

In our system, the LD arrays were arranged on a spherical segment (Figure 4(a)). The pump light was concentrated by lenses and ducts, which comprised two reflective plates covered with a silver coating. This LD array arrangement was selected to use planar LD arrays and test their properties. Moreover, in this arrangement, several planar LD arrays may be used, so the same optical components could focus the pump light emitted by the LD array. Finally, the reflection times of pump light on the ducts could be determined by this arrangement. In our pump module, over 23 kW cm^{-2} of pump power is achieved in a gain medium with an approximately 1.1 kW cm^{-2} LD array emission; the pumping efficiency reaches approximately 84%. Figure 4(b) shows the output of the pump module. The pumping is uniform. The maximum, minimum, and average values in the available area are approximately 111, 70, and 90, respectively, and the corresponding modulation in the

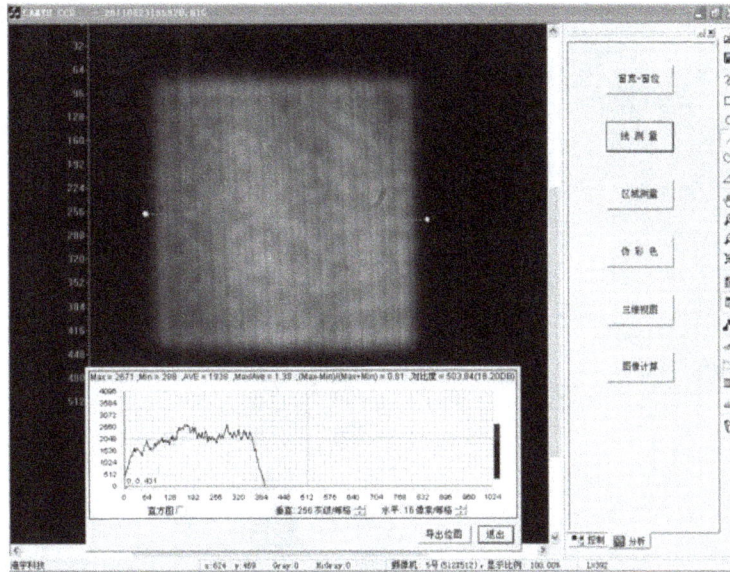

Figure 5. Laser beam near-field from the preamplifier.

available area is about 1.4:1. Pumping modulation may be attributed to the LD bars assembled in the LD arrays.

The total pumping power in the main amplifier was about 140 kW, resulting from fourteen 10 kW LD arrays; these LD arrays were arranged in two amplifier modules. One amplifier comprised eight 10 kW LD arrays, and the other comprised six 10 kW LD arrays.

3. Primary experimental results

The generator produces a 10 mJ/10 ns laser pulse with a Gaussian near-field and a wavelength of approximately 1030 nm. Moreover, the generator can work at 10 Hz. The seed pulse is eight-pass V-shape amplified to approximately 100 mJ in the preamplifier. At this stage, the beam is circular and exhibits a Gaussian profile, which is unfavourable for amplification; thus, the beam must be shaped into a square flat-top form. The beam energy must be at least 10 mJ so that enough energy can be obtained from the amplifiers. In our system, the laser energy is about 10 mJ after the shaping stage. The beam size is 10 mm × 10 mm, and ratio of the maximum and minimum modulation is about 1.38 (Figure 5), it basically meets the requirement of main amplifier.

The square flat-top beam is eight-pass V-shape amplified in the booster amplifier, after the booster amplifier, the fraction of maximum and minimum modulation is about 1.39 (Figure 6). The output energy of laser pulse is adjusted by controlling the pumping power. The maximum output energy reaches 1.2 J (Figure 7). The output energy of the beam is less than that of the initial design because of its small aperture and the low overlap between the laser beam and pumping area. However, for the 10 J laser system, a 1 J output energy is adequate for the subsequent amplification.

The laser energy results mainly from the main amplifier. We assessed the amplified spontaneous emission (ASE) in the main amplifier. The optimized product of concentration and thickness is about 15–20 at.% mm from the absorption efficiency. In our system, the doped ion concentration and thickness in the gain medium are 5 at.% and 3 mm, respectively. We evaluated the ASE using the spontaneous radiation signal from the amplifier. The top and bottom panels in Figure 8 show the spontaneous radiation signals from the 60 and 80 kW laser heads, respectively. The top panel (Figure 8) shows the spontaneous radiation signal increases rapidly until 600 μs; this increase in signal slows down thereafter, implying that energy storage increases within the duration of pumping. However, in the 80 kW laser head, the spontaneous radiation signal increases during pumping until 700 μs, then slows down, and remains constant thereafter. The energy storage fails to increase beyond 700 μs. Hence, the pump durations are set to 1 ms and 700 μs for the 60 and 80 kW laser heads, respectively.

A V-shaped multi-pass amplification was applied in the main amplifier (Figure 2) to improve the extraction efficiency of energy storage. A laser pulse was injected into the main amplifier through a polarizer (P4) and then passed through a spatial filter (SF4) to decrease the beam aperture from 17 mm × 17 mm to 14 mm × 14 mm. This pulse was subsequently double passed through the 60 kW laser head and V-shape amplified twice. In this section, the laser beam passed twice through the quarter-wave plate (WP1) and the direction of polarization was rotated by 90°. Hence, the laser pulse was reflected into the 80 kW laser head when it returned to the polarizer (P4). In the 80 kW amplifier, the laser was initially V-shape amplified twice, double passing through another quarter-wave plate (WP2);

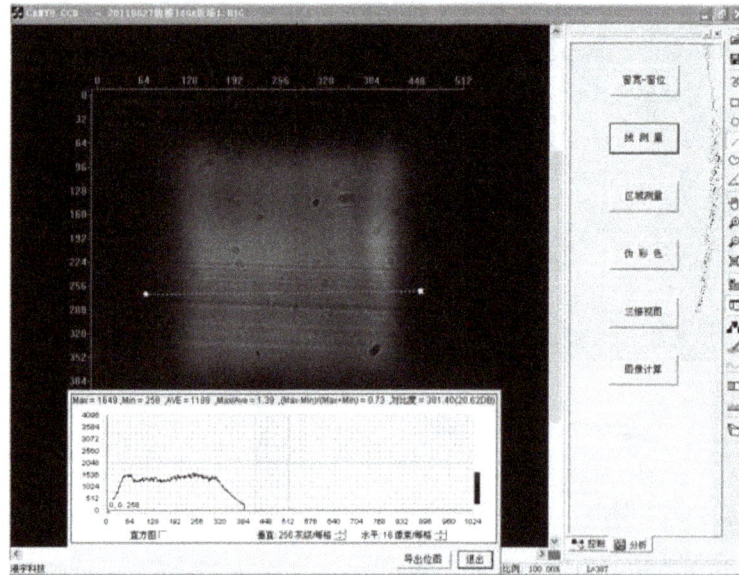

Figure 6. Laser beam near-field from the booster amplifier.

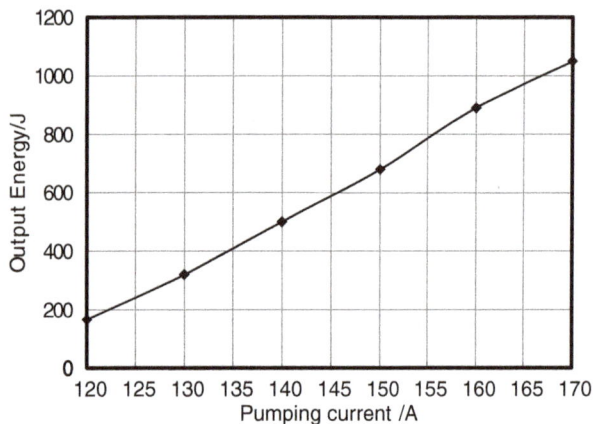

Figure 7. Output energy from the booster amplifier with different pumping currents.

its polarization was again rotated by 90°. When the laser pulse returned from the 80 kW laser head, it passed through the polarizer (P4), was directly reflected back by a mirror (M3), and then passed through the polarizer (P4) again back to the 80 kW laser head to perform a second series of double amplifications. Following the second series of amplifications the polarization of laser beam was rotated by 90° again. When the laser pulse returned from 80 kW laser head to the polarizer, it was reflected to the 60 kW laser head for yet another cycle of amplification. Following all the amplifications (four times each in the 80 and 60 kW amplifiers), the polarization of the laser beam was rotated 180°. From here, the laser pulse passed through the polarizer (P4) from the main amplifier, then through the Faraday rotator, and was exported from the laser system. The

maximum energy obtained in the primary experiment is approximately 8.5 J (Figure 9); the near-field modulation is approximately 1.5 and the far-field is approximately 10 times the diffraction limit (Figure 10).

Near-field modulation is attributed to the pumping. Strip-shaped modulation is observed in the pumping output – in agreement with the arrangement of the LD bars. Hence, the best way to reduce modulation is to eliminate pumping modulation. The far-field of the system output is approximately 10 times the diffraction limit because of thermal aberrations in the gain medium. Even low thermal aberrations causes serious effects on the beam quality because of the small aperture of the pump area and the concentration of thermal deposition. Therefore, an enhanced procedure for thermal management must be considered; some techniques could include heating the edge of the gain medium and using adaptive mirrors.

4. Brief discussion of the experimental results

4.1. Output energy of the laser system

The output energy of the laser system in the integration experiment is less than expected in the initial design.

This result is attributed to the marked ASE in the amplifier, especially in the 80 kW laser head. The spontaneous radiation signal no longer increases after 700 μs pumping in this laser head, which implies that the ASE and parasitic oscillations increase; ASE especially decreases energy storage. Figure 11 shows the small signal gain (SSG) for different thicknesses of gain medium with the same product of concentration and thickness (15 at.% mm) under a 200 A pump current. The SSG improves markedly on

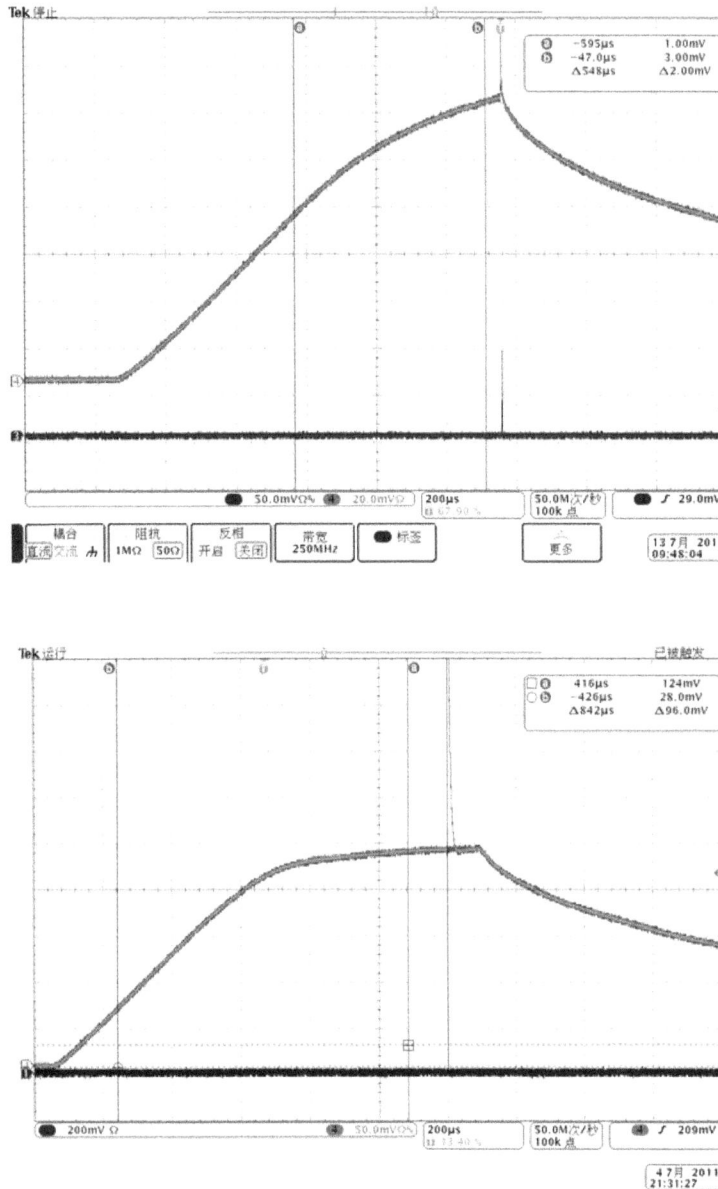

Figure 8. Amplifier fluorescence obtained at the heads with 60 kW (top) and 80 kW (bottom) pump powers.

increasing the medium thickness and decreasing the dopant concentration.

The low output energy may also be explained by the low overlap between the laser beam and the pumping area. A slight shift between the laser beam and active area may seriously affect this overlap because the apertures of the pump light and laser beam are small. Because the pump light exhibits a high angle of divergence in the proposed lens + duct LD pumping system, the active area increases with increasing distance from the incidence plane of the pump light, the laser exhibits good directivity, which results in low overlap. Hence, overlap must be considered during optimization of the concentration and thickness of the gain medium to yield high SSG.

The LD array comprises several LD bars – the directivity of these bars and the power uniformity of the LD arrays markedly affect the distribution of pump light on the gain medium. Figure 4 reveals that the profile of pump light on the gain medium exhibits modulation in the direction of the fast axes of the bars, which may cause modulation of the gain and energy storage in the amplifier. Similar modulation on the near-field of the output laser beam also results in damage to optical components. Hence, uniformity of the pump light must be improved to improve the output energy of the laser pulse.

The gain medium in our system is not surrounded by absorptive cladding, which could also absorb ASE from the gain area. Hence, the ASE light is reflected into the gain

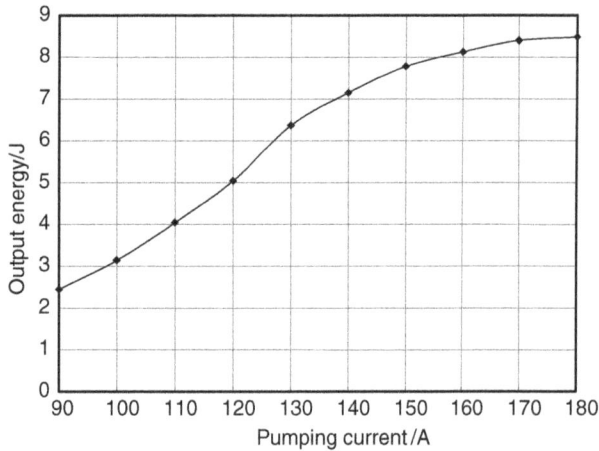

Figure 9. Output energy of the laser system at different currents.

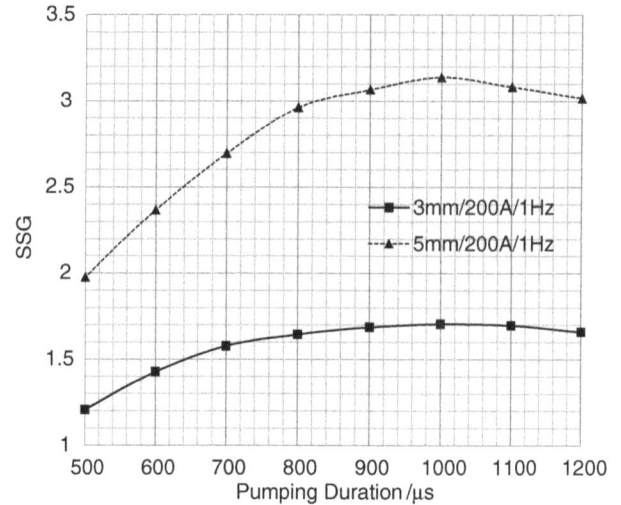

Figure 11. SSG at different pumping currents for 5 mm@3 at.% and 3 mm@5 at.%.

area from the edge of the gain medium and energy storage in the amplifier is reduced. Even if the thickness of the gain medium is increased and the concentration of doped ions is decreased, re-amplification of the ASE cannot be eliminated. Therefore, the gain medium must be surrounded by absorptive cladding to avoid re-amplification of the ASE.

4.2. Effects of repetition rate

Thermal deposition and management differ respectively under various doping rates and thicknesses of the gain media. The laser performance of Yb:YAG crystal or ceramics is affected by temperature. Hence, we tested two types of Yb:YAG crystal with the same product of concentration and thickness (5 at.%@3 mm and 10 at.%@1.5 mm). The repetition rate ranges from 1 to 10 Hz and the pump current is 200 A. Figure 12 reveals that the SSG initially decreases with the repetition rate and then reaches an optimum value. This result is attributed to the low energy storage in the gain medium at the beginning of pumping – hence, the ASE and parasitic oscillations are low. A maximum SSG is observed with increasing pumping duration because of the effects of

ASE, parasitic oscillations and temperature on the energy storage in the amplifier. In our system we set the repetition rate to 1 Hz during the primary integration.

5. Summary and conclusion

We have integrated a 10 J Yb:YAG laser system, in which the output energy of laser pulse is approximately 8.5 J under a 1 Hz repetition rate. The output energy of laser is slightly lower than that of the initial design because ASE and parasitic oscillations decrease the energy storage. Given the marked effects of temperature on the laser performance of Yb:YAG, the laser system works only at 1 Hz. For high repetition rates, the working point of the amplifier must be optimized. We aim to optimize the working point of the amplifier as well as the parameters of the gain medium in future research. During further development, we will heat the edge of the gain medium to balance the thermal aberrations in the pump area and clad the gain medium with an absorber to avoid re-amplification of the reflected ASE.

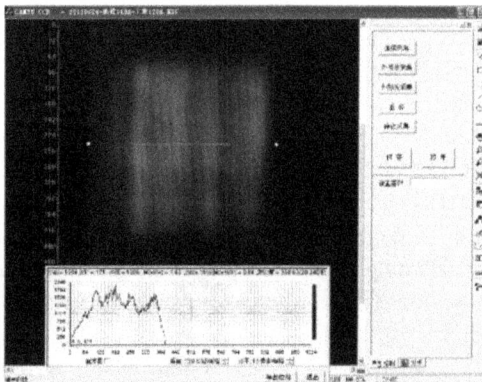

Figure 10. Near-field (left) and far-field (right) image of the laser beam.

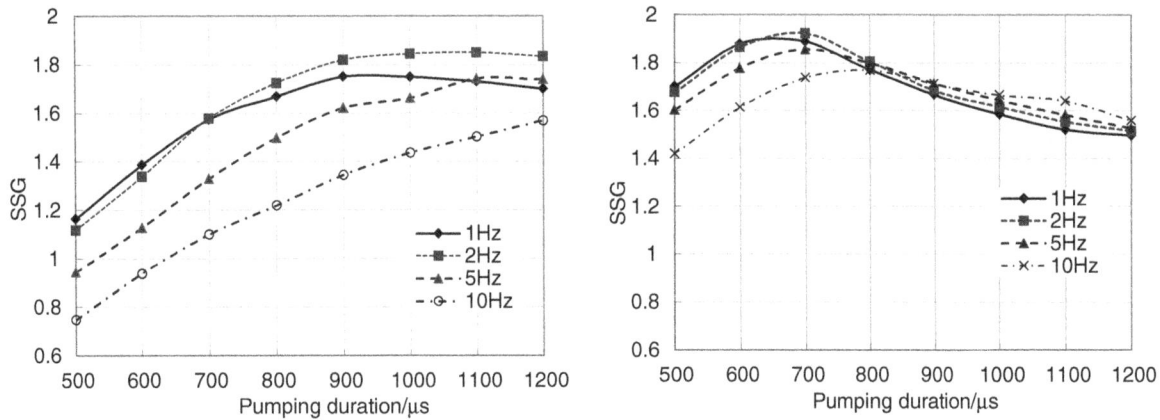

Figure 12. SSG in the amplifier with 5 at.%@3 mm (left) and 10 at.%@1.5 mm (right).

References

1. T. Land, *International Laser Operations Workshop, Livermore, CA, United States* (2013).
2. NIF closes in on alpha heating milestone (2014); Available from:
3. O. A. Hurricane, D. A. Callahan, D. T. Casey, P. M. Celliers, C. Cerjan, E. L. Dewald, T. R. Dittrich, T. Döppner, D. E. Hinkel, L. F. Berzak Hopkins, J. L. Kline, S. Le Pape, T. Ma, A. G. MacPhee, J. L. Milovich, A. Pak, H.-S. Park, P. K. Patel, B. A. Remington, J. D. Salmonson, P. T. Springer, and R. Tommasini, Nature **506**, 343 (2014).
4. NIF Conceptual Design Team. National Ignition Facility Conceptual Design Report Volume 3: Conceptual (1994).
5. J. Horvath, UCRL-JC-124520 (1996).
6. N. Team, *National Ignition Facility User Guide* (2012).
7. C. Marshall, C. Bibeau, A. Bayramian, R. Beach, C. Ebbers, M. Emanuel, B. Freitas, S. Fulkerson, E. Honea, B. Krupke, J. Lawson, C. Orth, S. Payne, C. Petty, H. Powell, K. Schaffers, J. Skidmore, L. Smith, S. Sutton, and S. Telford, UCRL-JC-128084-REV-2 (1998).
8. M. Dunne, HiPER: a laser fusion facility for Europe (2007).
9. J. A. Caird, V. Agrawal, A. Bayramian, R. Beach, J. Britten, D. Chen, R. Cross, C. Ebbers, A. Erlandson, M. Feit, B. Freitas, C. Ghosh, C. Haefner, D. Homoelle, T. Ladran, J. Latkowski, W. Molander, J. Murray, S. Rubenchik, K. Schaffers, C. W. Siders, E. Stappaerts, S. Sutton, S. Telford, J. Trenholme, and C. P. J. Barty, in *Proceedings of 18th TOFE Conference* (2008).
10. F. Romanelli, Fusion Electricity A roadmap to the realisation of fusion energy, Specific Terms to prepare a technical roadmap to fusion electricity by 2050 (2012).
11. K. Yoshida, M. Yamanaka, M. Nakatsuka, T. Sasaki, and S. Nakai, Proc. SPIE **2966**, 2 (1997).
12. X. Wang, X. Xu, Z. Zhao, B. Jiang, J. Xu, G. Zhao, P. Deng, G. Bourdet, and J.-C. Chanteloup, Opt. Mater. **29**, 1662 (2007).
13. A. Bayramian, *7th International HEC-DPSSL Workshop 2012: Lake Tahoe, California* (2012).
14. S. M. A. A. C. Erlandson, A. J. Bayramian, A. L. Bullington, R. J. Beach, J. A. C. C. D. Boley, R. J. Deri, A. M. Dunne, D. L. Flowers, M. A. Henesian, E. I. M. K. R. Manes, S. I. Rana, K. I. Schaffers, M. L. Spaeth, C. J. Stolz, and S. J. Telford, Opt. Mater. Express **1**, 1341 (2011).
15. K. Ertel, S. Banerjee, P. D. Mason, P. J. Phillips, C. Hernandez-Gomez, and J. L. Collier, in *Proceedings of High Intensity Lasers and High Field Phenomena* (2012).
16. P. D. Mason, S. B. K. Ertel, P. J. Phillips, C. Hernandez-Gomez, and J. L. Collier, in *Proceedings of Advanced Solid-State Photonics* (2012).
17. W. Huang, J. Wang, X. Lu, and X. Li, Laser Phys. **23**, 035804 (2013).
18. D. Albach, M. Arzakantsyan, T. Novo, B. Vincent, and J.-C. Chanteloup, *7th International HEC-DPSSL Workshop 2012: Lake Tahoe, California* (2012).
19. P. D. Mason, K. Ertel, S. Banerjee, P. J. Phillips, C. Hernandez-Gomez, and J. L. Collier, in *Proceedings of CLEO 2012* CM3D.1 (2012).
20. I. B. Mukhin, O. V. Palashov, E. A. Khazanov, A. Ikesue, and Y. L. Aung, Opt. Express **13**, 5983 (2005).
21. J. Kawanaka, N. Miyanaga, T. Kawashima, K. Tsubakimoto, Y. Fujimoto, H. Kubomura, S. Matsuoka, T. Ikegawa, Y. Suzuki, N. Tsuchiya, T. Jitsuno, H. Furukawa, T. Kanabe, H. Fujita, K. Yoshida, H. Nakano, J. Nishimae, M. Nakatsuka, K. Ueda, and K. Tomabechi, J. Phys. Conf. Ser. **112**, 032058 (2008).
22. H. Yoshioka, S. Nakamura, T. Ogawa, and S. Wada, Opt. Express **17**, 8919 (2009).
23. J. Dong and P. Deng, J. Phys. Chem. Solids **64**, 1163 (2003).
24. A. Yoshida, S. Tokita, J. Kawanaka, T. Yanagitani, H. Yagi, F. Yamamura, and T. Kawashima, J. Phys. Conf. Ser. **112**, 032062 (2008).
25. C. Honninger, I. Johannsen, M. Moser, G. Zhang, A. Giesen, and U. Keller, Appl. Phys. B **65**, 423 (1997).
26. X. Li, X. Shi, P. Shi, M. Guo, G. Zhang, Y. Lu, and Q. Hu, Acta Opt. Sin. **21**, 1268 (2001).

11

Theory of light sail acceleration by intense lasers: an overview

Andrea Macchi

National Institute of Optics, National Research Council (CNR/INO), Research Unit 'Adriano Gozzini',
Department of Physics 'Enrico Fermi', University of Pisa, largo Bruno Pontecorvo 3, I-56127 Pisa, Italy

Abstract

A short overview of the theory of acceleration of thin foils driven by the radiation pressure of superintense lasers is presented. A simple criterion for radiation pressure dominance at intensities around 5×10^{20} W cm^{-2} is given, and the possibility for fast energy gain in the relativistic regime is discussed.

Keywords: light sail; radiation pressure; laser–plasma acceleration of electrons and ions; laser-driven acceleration

1. Introduction

It has been known since the discovery of Maxwell's equations that light, i.e., electromagnetic (EM) radiation, exerts a pressure on a reflecting object, and thus may accelerate it. In 1925, Zander[1] suggested exploiting the radiation pressure of the Sun for space travel using light sails, i.e., mirrors of large area and small thickness.

The scattering of an EM wave by a particle also leads to momentum absorption and acceleration. In 1957, Veksler[2] suggested that Thomson scattering by a small cluster containing N electrons may accelerate the cluster to high velocities. The fundamental point of Veksler's proposal was that the radiation force on the cluster scaled as N^2, providing an example of his new principle of coherent acceleration, i.e., the use of collective effects to accelerate large amounts of particles to high energies.

After the invention of the laser, Forward in 1962[3, 4] and Marx in 1966[5] proposed using an Earth-based laser system to accelerate a rocket up to relativistic velocities. Marx's paper included a relativistic analysis of the motion of a sail, i.e., a plane perfect mirror, accelerated by radiation pressure, based on the equations

$$\frac{\mathrm{d}(\gamma V)}{\mathrm{d}t} = \frac{2}{\sigma_0 c} I(t - X/c) \frac{1 - V/c}{1 + V/c}, \quad \frac{\mathrm{d}X}{\mathrm{d}t} = V, \quad (1)$$

where $I = I(t)$ is the intensity of the laser pulse, σ_0 is the surface mass density of the sail, and $\gamma = (1 - V^2/c^2)^{-1/2}$. The concept is sketched in Figure 1. The most interesting

Correspondence to: A. Macchi, Dipartimento di Fisica 'Enrico Fermi', largo Bruno Pontecorvo 3, I-56127 Pisa, Italy. Email:andrea.macchi@ino.it. Web: www.df.unipi.it/~macchi.

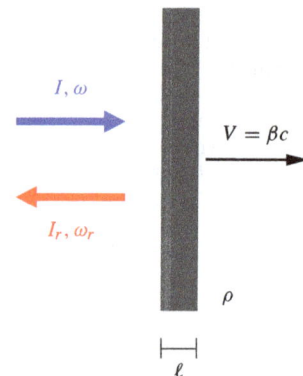

Figure 1. The light sail concept. The sail is modeled as a perfect mirror of surface density $\sigma = \rho\ell$, with ρ the mass density and ℓ the thickness. The sail is pushed by a plane wave of intensity I and frequency ω. Notice that the equations of motion for the sail given in (1) and the expression for the mechanical efficiency may be simply obtained by considering the Doppler shift of the reflected radiation [$\omega_r = \omega(1 - \beta)/(1 + \beta)$] and the conservation of the 'number of photons'; see, for example, Ref. [6].

result (but also the subject of a long-lasting controversy[7]) was the expression for the mechanicalefficiency $\eta = 2\beta/(1 + \beta)$ (with $\beta = V/c$), which reaches 100% in the relativistic limit $\beta \to 1$. Equations (1), hereafter referred to as the light sail (LS) equations, have the same form as for the motion of the Thomson scattering particle[8], evidencing the connection with Veksler's proposed mechanism.

In 2004, using particle-in-cell (PIC) simulations of the acceleration of a thin plasma foil by a laser pulse with intensity $I > 10^{23}$ W cm^{-2}, Esirkepov *et al.*[9] showed that the motion of the foil was also well fitted by the above-mentioned equation, giving evidence that the foil was driven from radiation pressure. The scaling of the LS equations to foreseeable laser and target parameters showed the possibility of reaching the relativistic velocity of the foil, corresponding to an energy per nucleon above the GeV barrier. The coherent motion of the foil also implied an inherent mono-energetic spectrum, which would be crucial for most applications. Such features have then stimulated a strong interest in LS acceleration.

In this paper, we give a brief overview of the research on LS acceleration in the past decade, mostly focusing on theoretical aspects and open issues. A simple criterion for radiation pressure dominance at intensities around 5×10^{20} W cm^{-2} is given, and the possibility for fast energy gain in the relativistic regime is pointed out. A more comprehensive presentation of experimental and simulation results may be found in recent review papers on laser-driven ion acceleration[10–13].

2. One-dimensional dynamics

For an arbitrary pulse profile $I(t)$, the final value of γ is obtained from Equations (1) as

$$\gamma_\infty \equiv \gamma(t = \infty) = 1 + \frac{\mathcal{F}^2}{2(\mathcal{F}+1)}, \quad \mathcal{F} = \frac{2}{\sigma c^2} \int_0^\infty I(t') \mathrm{d}t'. \tag{2}$$

For a flat-top intensity profile, i.e., a constant value of I between $t = 0$ and τ_L, Equations (1) can be solved exactly. Here, we just give the limiting cases of $\beta \ll 1$ and $\beta \to 1$, for which the integration is straightforward (notice that, for $\beta \simeq 1$, $(1+\beta)/(1-\beta) \simeq 4\gamma^2$):

$$\gamma(t) = \begin{cases} 1 + [1 - \exp(-2\Omega t)]^2/8 & (\Omega t \ll 1) \\ (3\Omega t/4)^{1/3} & (\Omega t \gg 1), \end{cases} \tag{3}$$

where $\Omega = 2I/\sigma_0 c^2$. Equations (2) and (3) may be used to obtain the acceleration time and length in the laboratory for a given value of the final energy per nucleon $\mathcal{E}_{max} = m_p c^2 (\gamma_\infty - 1)$ (notice that it would be incorrect to plug the pulse duration τ_L in Equation (3) to obtain $\mathcal{E}_{max}$).

It is evident that the energy gain is quite fast for $\beta \ll 1$ but becomes much slower in the relativistic regime as $\beta \to 1$. In a realistic multi-dimensional scenario, this is a possible issue, because of laser pulse diffraction on distances larger than the Rayleigh length. Fortunately, as discussed below, the energy gain may be faster in three-dimensional (3D) geometry thanks to the target rarefaction.

Obviously, the lighter the sail the higher the energy for a given laser pulse. However, if the foil target is too thin, then it becomes transparent to the laser pulse, and the radiation pressure boost drops down. Based on the simple model of a delta-like foil and purely transverse electron motion[14, 15], the threshold for transparency due to relativistic effects is given by

$$a_0 \simeq \zeta, \tag{4}$$

where $a_0 = (I/m_e n_c c^3)^{1/2}$, $\zeta = \pi\sigma_0/(Z m_i n_c \lambda)$, $n_c = \pi m_e c^2/(e^2\lambda^2) = \pi/(r_c\lambda^2)$ is the cut-off density, and λ is the laser wavelength. Despite the very simplified underlying model, Equation (4) describes fairly well the onset of transparency and the breakdown of LS acceleration observed in 1D simulations[16]. Actually, Equation (4) may be considered as slightly pessimistic, because, as the foil moves, the reflectivity increases, due to the decrease of the pulse frequency in the moving frame[17]. The situation is more complex for finite-width pulses in multi-dimensional geometry, because the transverse expansion of the foil leads to a decrease of the surface density σ in time.

The above modeling considers the sail as a neutral rigid body with electrons comoving with ions. Indeed, charge separation effects are crucial in the 'inner' dynamics of LS acceleration. Figure 2 shows the initial stages of ion acceleration. Electrons are pushed into the target by the secular ponderomotive force per unit volume $\mathbf{f}_p = \langle \mathbf{J} \times \mathbf{B} \rangle$, where the brackets denote a cycle average. The ponderomotive force sweeps and piles up the electrons, creating a charge depletion layer until f_p is exactly balanced by an electrostatic field E_x (this corresponds to the balance between P_{rad} and the electrostatic pressure on ions; see, for example, Refs. [16, 17]). In turn, E_x accelerates the ions as shown in Figure 1. In a first stage, the ions in the layer where the EM field penetrates are accelerated up to a velocity v_i within a time t_c, given by[18]

$$\frac{v_i}{c} \simeq \left(\frac{I}{\rho c^3}\right)^{1/2} = \left(\frac{Z m_e n_c}{A m_p n_e}\right)^{1/2} a_0,$$

$$t_c \simeq \frac{1}{\omega a_0} \left(\frac{A m_p}{Z m_e}\right)^{1/2}, \tag{5}$$

where ρ is the mass density (for simplicity we assume non-relativistic motion; see [19] for relativistically corrected expressions). At $t = t_c$, the accelerated ions have piled up at the position $x = x_s \simeq v_i t_c$. If this position coincides with the rear surface of the foil, the acceleration cycle may be repeated, and eventually the sequence of acceleration stages converges to the motion described by (1)[20].

The correct balance of electrostatic and radiation pressure shows that only a fraction $F \simeq 1 - a_0/\zeta$ of the ions is accelerated coherently as a sail, even if the motion of the latter is still described by Equation (1) with σ_0 including the *total* mass of the foil[16, 17]. During the motion, as long as the geometry is one dimensional, the electrostatic pressure on the sail depends only on the total charge behind the sail, while the radiation pressure decreases by a factor $(1 - \beta)/(1 + \beta)$. Thus, to maintain the pressure balance, additional ions are progressively trapped in the sail[17, 21]. Applying the pressure balance as in Ref. [16] with the velocity correction yields the final fraction of accelerated ions:

$$F \simeq 1 - \frac{a_0}{\zeta} \left(\frac{1 - \beta_\infty}{1 + \beta_\infty}\right)^{1/2}, \tag{6}$$

where $\beta_\infty = [(1+\mathcal{F})^2 - 1]/[(1+\mathcal{F})^2 + 1]$. This shows that all ions are eventually accelerated in the relativistic limit.

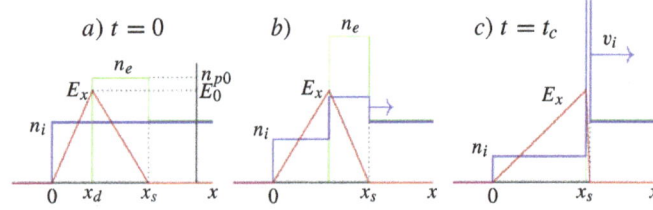

Figure 2. The first stage of ion acceleration driven by radiation pressure[18]. The densities of ions (n_i) and electrons (n_e) are approximated by step-like functions. Ions initially in the $x_d < x < x_s$ layer are accelerated by the charge separation field E_x up to velocity v_i at time $t = t_c$.

3. Radiation pressure dominance

Since a thin plasma foil is not a perfect mirror, it is not trivial that irradiation by intense light should result in LS acceleration. In most accessible laser–plasma interaction conditions, strong heating of electrons occurs, and the resulting kinetic pressure exceeds the radiation pressure; in such a situation, the plasma foil expands and the resulting ion energy spectrum is very different from the LS case. The situation is somewhat reminiscent of the Crookes radiometer or light mill, where the vanes are white (reflecting) on one side and black (absorbing) on the other side: the mill rotates in the direction *opposite* to what would be expected from radiation pressure being higher on the white side than on the black side, because the effects of heating and thermal pressure dominate.

To find the conditions in which the radiation pressure $P_{\rm rad}$ will dominate the acceleration, let us briefly recall the heating dynamics of electrons. At normal incidence, electrons are driven in the direction perpendicular to the target surface by the $\mathbf{v} \times \mathbf{B}$ force, which for linear polarization (LP) has an oscillating term at 2ω (where ω is the laser frequency) in addition to the secular ponderomotive force. Heating of electrons occurs via oscillations across the laser–plasma interface driven by the oscillating term, which vanishes for circular polarization (CP)[18]. The use of CP pulses has then been proposed by several authors[22–24] to obtain an efficient LS regime at 'any' intensity. Detailed 3D simulations in the relativistic regime[25] also showed that for CP pulses higher energies and better collimation of the ion beam are obtained with respect to LP pulses. Experiments performed so far, however, have shown a limited impact of the use of CP[26–29] and non-LS effects such as species separation in the spectrum[28–30] (in the ideal LS regime, all species move at the same velocity; thus the energy per nucleon is independent on the mass number). These data suggest that tight focusing of the laser pulse and, possibly, imperfect conversion to CP may prevent efficient LS operation, at least in the intensity regime investigated so far, i.e., $I \simeq (2 \times 10^{18} \div 2 \times 10^{21})$ W cm^{-2}.

In view of future experiments at higher intensities and of possible technical difficulties for producing ultraintense CP pulses, it appears important to discuss possible conditions for radiation pressure dominance also for LP, when electron heating is important. Heuristically, the transfer of energy to ions via $P_{\rm rad}$ can be efficient if it is 'faster' than the heating of electrons, which occurs on a laser halfcycle. Esirkepov et al.[9] suggested that ions should become promptly relativistic, i.e., reaching a velocity close to c within one cycle, so that they would 'stick' to electrons. To estimate the corresponding laser intensity for such a regime, let us assume $v_i \simeq c/2$ in Equation (5): this gives

$$a_0 \simeq 30 \left(\frac{n_e}{n_c} \right)^{1/2}, \qquad (7)$$

which for $n_e/n_c \simeq 100$ gives $I\lambda^2 > 10^{23}$ W cm^{-2} μm^2, which is the typical intensity of the simulations in Ref. [9]. These values are not currently available, although they may be reached in the laboratory within the next decade.

Here, we propose a different condition, which leads to a more accessible intensity threshold. The above-defined ion acceleration time t_c may be taken as the relevant temporal scale for energy transfer to ions. For electrons, acceleration occurs on a laser halfcycle being driven by the oscillating force at 2ω. Thus we suggest $t_c < \pi/\omega$ as the condition for energy transfer to ions being more efficient than to electrons. This leads to the threshold for the laser amplitude

$$a_0 > \frac{1}{\pi} \left(\frac{A m_p}{Z m_e} \right)^{1/2} \simeq 19, \qquad (8)$$

which is equivalent to $I\lambda^2 > 5 \times 10^{20}$ W cm^{-2} μm^2, independently of the plasma density. This estimate is in qualitative agreement with LS signatures being observed in current experiments at similar intensities[28]. A slightly greater intensity threshold of 10^{21} W cm^{-2} μm^2 has been suggested by Qiao et al.[31] on a different basis, i.e., by comparing the ion energy gain due to the radiation pressure push with that in the fast electron sheath.

4. Fast gain regimes: 'unlimited' acceleration

In a realistic situation, the laser pulse has a finite width, and it drives a cocoon deformation and transverse expansion of the target. This unavoidable effect may lead to early breakthrough of the laser pulse and termination of the LS stage; thus the use of a smooth transverse profile to keep a nearly plane geometry was suggested by several simulation studies. In contrast, Bulanov et al.[32] suggested that the decrease of target density due to transverse expansion may lead, in proper conditions, to acceleration up to higher energies than in the planar case, at the expense of the number of accelerated ions. This has been named the 'unlimited' acceleration regime.

In the following, we give a brief and simplified account of the detailed theory developed by Bulanov et al.[33]. The basic modification of Equation (1) for the longitudinal motion of the sail is that the surface density now depends on time due to the transverse expansion,

$$\sigma = \sigma(t) = \frac{\sigma(0)}{\Lambda^{D-1}(t)}, \tag{9}$$

where $\Lambda(t)$ describes the dilatation of the transverse position of a fluid element of the sail, i.e., $r_\perp(t) = \Lambda(t) r_\perp(0)$, and D is the dimensionality of the system; $D = 1$ corresponds to planar geometry (constant σ), $D = 2$ to two-dimensional Cartesian geometry, and $D = 3$ to three-dimensional geometry with cylindrical symmetry. Now, it is assumed that for a given element the motion is ballistic after an initial kick by the laser pulse delivering a transverse momentum $p_\perp = m_i \varpi_0 r_\perp(0)$, i.e., proportional to the initial position. This relation might be justified by observing that such kick comes from the transverse ponderomotive force, which is proportional to the gradient of the intensity and thus would be a linear function of position for a parabolic profile. It is further assumed that $p_\perp \ll p_\parallel$, with $p_\parallel$ the longitudinal momentum. The transverse velocity thus decreases as a result of the increasing longitudinal momentum. This leads to the equation for Λ,

$$\frac{d\Lambda}{dt} = \frac{\dot{r}_\perp(t)}{r_\perp(0)} = \frac{\varpi_0}{\gamma(t)}, \quad \gamma(t) \simeq (p_\parallel^2 + m_i^2 c^2)^{1/2}, \tag{10}$$

with the coupled equation for $p_\parallel = \gamma\beta_\parallel$:

$$\frac{d(\gamma\beta_\parallel)}{dt} = \frac{2I}{\sigma_0 c^2} \Lambda^{D-1}(t) \frac{1 - \beta_\parallel}{1 + \beta_\parallel}. \tag{11}$$

For further simplification, we consider the asymptotic ultra-relativistic limit, in which $\beta_\parallel \to 1$ and $(1 - \beta_\parallel)/(1 + \beta_\parallel) \simeq (2\gamma)^{-2}$. In this limit, we find a solution

$$\gamma = \left(\frac{t}{\tau_k}\right)^k, \quad k = \frac{D}{D + 2}, \tag{12}$$

with the time constants given for $D = 1, 2, 3$ by

$$\tau_{1/3} = \left(\frac{3}{4\Omega}\right), \quad \tau_{2/4} = \left(\frac{1}{\Omega\varpi_0}\right)^{1/2}, \tag{13}$$

$$\tau_{3/5} = \left(\frac{48}{125\Omega\varpi_0^2}\right)^{1/3}.$$

Fast energy gain in this regime thus depends on the initial conditions via the parameter ϖ_0. Assuming the initial transverse kick to be of the same order as the longitudinal one, $\varpi_0 \simeq \Omega$ may be assumed for a quick estimate. For a given temporal profile $I(t)$, the final energy and surface density along the axis may be obtained by integrating Equations (10) and (11) with respect to the proper time $t' = t - X/c$, with $dt' = (1 - \beta_\parallel)dt$. Bulanov et al.[33] also discuss 'optimal' pulse profiles to maximize the acceleration; heuristically, since the 'unlimited' mechanism is actually limited by the onset of transparency, one argues that the decrease of the density may be matched with the decrease of the pulse frequency in the sail frame to keep a constant reflectivity.

5. 3D simulations

The above-outlined theory shows that, differently from other acceleration mechanisms[34, 35], the energy gain may be higher in a realistic 3D geometry that with respect to a 1D plane case. Confirmation of this theory in numerical experiments thus requires fully 3D large-scale simulations, which are feasible on the most powerful present-day parallel supercomputers.

Simulations by Tamburini et al.[25] (Figure 3(a)) have given a first evidence of the energy enhancement in fully 3D simulations. These simulations also indicated a baseline for LS operation in the relativistic regime, showing that the use of CP leads to higher energies and a more collimated beam with respect to the LP case, and also to negligible radiation friction effects. To evaluate the energy gain at the end of the acceleration stage, larger computational resources have been necessary to extend the simulation time by four times. In such simulations, the temporal dependence of maximum ion energy is in good agreement with the $\sim t^{3/5}$ scaling given by Equation (12), as shown in Figure 3(b))[12]. Ultimately, the acceleration is stopped by the onset of transparency. These simulations have been performed on the FERMI supercomputer at CINECA, Italy.

6. Conclusions and perspectives

The laser-driven light sail concept, which was first studied as a visionary approach to interstellar travel, currently represents an implementation of Veksler's coherent acceleration paradigm and a possible route towards a laser–plasma accelerator. Experiments are just entering the regime of intensities exceeding 5×10^{20} W cm^{-2} where, according to our discussion, the radiation pressure push is expected to be the dominant acceleration mechanism. Recent progress in both achieving extremely high-contrast pulses and manufacturing ultrathin targets has been crucial for light sail experiments[26–30], and results such as the observation of the fast scaling of ion energy in the non-relativistic limit[28]

a)

b)

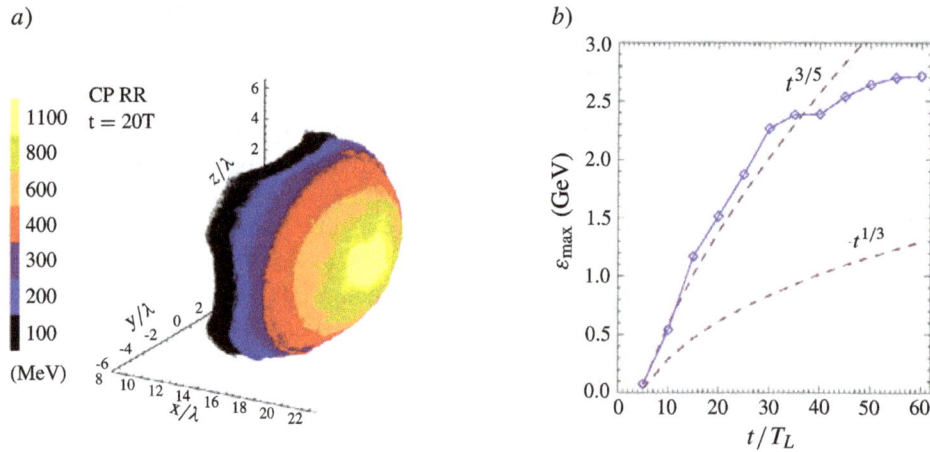

Figure 3. 3D particle-in-cell simulations of thin foil acceleration. (a) Space and energy distribution of ions[25] (reproduced by permission of APS) at $t = 20T$ from the acceleration start ($T = 2\pi/\omega$ laser period). (b) Maximum ion energy versus time[12] (reproduced by permission of IOP Publishing). Both simulations have been performed for a $9\lambda \times (10\lambda)^2$ pulse (FWHM values) with peak amplitude $a_0 = 198$ and circular polarization, and a hydrogen plasma foil with surface density $\sigma = 64 m_p n_c \lambda$, so that $a_0 \simeq \zeta$. See the references for details.

are promising. However, several open issues are apparent, such as achieving mono-energetic spectra, and the effect of parameters such as the laser pulse focusing and duration still needs to be completely understood and optimized.

With the availability of next-generation lasers at extreme intensities, success of the light sail approach in producing relativistic ions will depend on the possibility of achieving and controlling the so-called 'unlimited' regime based on a suitable (and possibly self-regulated) transverse expansion of the target. On this route one expects technical challenges, such as clean circular polarization for ultraintense pulses, as well as other possible issues not considered in this paper, such as the target stability.

Acknowledgements

It is a pleasure to thank S. V. Bulanov and F. Pegoraro for scientific inspiration and enlightening discussions, and T. V. Liseykina, A. Sgattoni, S. Sinigardi, and M. Tamburini for hard simulation work. Support from the Italian Ministry of University and Research via the FIR project 'SULDIS' is acknowledged.

References

1. F. A. Zander, Technika i Zhizn **13**, 15 (1924) (in Russian).
2. V. I. Veksler, At. Energy **2**, 525 (1957).
3. R. L. Forward, J. Spacecraft **21**, 187 (1984).
4. P. Gilster, *Centauri Dreams: Imagining and Planning Interstellar Exploration* chapter 6 (Springer Science + Business Media, 2004).
5. G. Marx, Nature **211**, 22 (1966).
6. A. Macchi, *A Superintense Laser–Plasma Interaction Theory Primer*, chapter 5, SpringerBriefs in Physics, (Springer, 2013).
7. J. F. L. Simmons and C. R. McInnes, Am. J. Phys. **61**, 205 (1993).
8. L. D. Landau and E. M. Lifshitz, *The Classical Theory of Fields* 2nd edn chapter 78 pp. 250 (Elsevier, Oxford, 1962).
9. T. Esirkepov, M. Borghesi, S. V. Bulanov, G. Mourou, and T. Tajima, Phys. Rev. Lett. **92**, 175003 (2004).
10. H. Daido, M. Nishiuchi, and A. S. Pirozhkov, Rep. Prog. Phys. **75**, 056401 (2012).
11. A. Macchi, M. Borghesi, and M. Passoni, Rev. Mod. Phys. **85**, 751 (2013).
12. A. Macchi, A. Sgattoni, S. Sinigardi, M. Borghesi, and M. Passoni, Plasma Phys. Control. Fusion **55**, 124020 (2013).
13. J. C. Fernandez, B. J. Albright, F. N. Beg, M. E. Foord, B. M. Hegelich, J. J. Honrubia, M. Roth, R. B. Stephens, and L. Yin, Nucl. Fusion **54**, (2014).
14. V. A. Vshivkov, N. M. Naumova, F. Pegoraro, and S. V. Bulanov, Phys. Plasmas **5**, 2727 (1998).
15. A. Macchi, *A Superintense Laser–Plasma Interaction Theory Primer*, chapter 3, SpringerBriefs in Physics, pp. 52–53. (Springer, 2013).
16. A. Macchi, S. Veghini, and F. Pegoraro, Phys. Rev. Lett. **103**, 085003 (2009).
17. A. Macchi, S. Veghini, T. V. Liseykina, and F. Pegoraro, New J. Phys. **12**, 045013 (2010).
18. A. Macchi, F. Cattani, T. V. Liseykina, and F. Cornolti, Phys. Rev. Lett. **94**, 165003 (2005).
19. A. P. L. Robinson, P. Gibbon, M. Zepf, S. Kar, R. G. Evans, and C. Bellei, Plasma Phys. Control. Fusion **51**, 024004 (2009).
20. M. Grech, S. Skupin, A. Diaw, T. Schlegel, and V. T. Tikhonchuk, New J. Phys. **13**, 123003 (2011).
21. B. Eliasson, C. S. Liu, X. Shao, R. Z. Sagdeev, and P. K. Shukla, New J. Phys. **11**, 073006 (2009).
22. X. Zhang, B. Shen, X. Li, Z. Jin, and F. Wang, Phys. Plasmas **14**, 073101 (2007).
23. O. Klimo, J. Psikal, J. Limpouch, and V. T. Tikhonchuk, Phys. Rev. ST Accel. Beams **11**, 031301 (2008).
24. A. P. L. Robinson, M. Zepf, S. Kar, R. G. Evans, and C. Bellei, New J. Phys. **10**, 013021 (2008).
25. M. Tamburini, T. V. Liseykina, F. Pegoraro, and A. Macchi, Phys. Rev. E **85**, 016407 (2012).

26. A. Henig, S. Steinke, M. Schnürer, T. Sokollik, R. Hörlein, D. Kiefer, D. Jung, J. Schreiber, B. M. Hegelich, X. Q. Yan, J. Meyer ter Vehn, T. Tajima, P. V. Nickles, W. Sandner, and D. Habs, Phys. Rev. Lett. **103**, 245003 (2009).

27. F. Dollar, C. Zulick, A. G. R. Thomas, V. Chvykov, J. Davis, G. Kalinchenko, T. Matsuoka, C. McGuffey, G. M. Petrov, L. Willingale, V. Yanovsky, A. Maksimchuk, and K. Krushelnick, Phys. Rev. Lett. **108**, 175005 (2012).

28. S. Kar, K. F. Kakolee, B. Qiao, A. Macchi, M. Cerchez, D. Doria, M. Geissler, P. McKenna, D. Neely, J. Osterholz, R. Prasad, K. Quinn, B. Ramakrishna, G. Sarri, O. Willi, X. Y. Yuan, M. Zepf, and M. Borghesi, Phys. Rev. Lett. **109**, 185006 (2012).

29. B. Aurand, S. Kuschel, O. Jaeckel, C. Roedel, H. Y. Zhao, S. Herzer, A. E. Paz, J. Bierbach, J. Polz, B. Elkin, G. G. Paulus, A. Karmakar, P. Gibbon, T. Kuehl, and M. C. Kaluza, New J. Phys. **15**, 033031 (2013).

30. S. Steinke, P. Hilz, M. Schnürer, G. Priebe, J. Bränzel, F. Abicht, D. Kiefer, C. Kreuzer, T. Ostermayr, J. Schreiber, A. A. Andreev, T. P. Yu, A. Pukhov, and W. Sandner, Phys. Rev. ST Accel. Beams **16**, 011303 (2013).

31. B. Qiao, S. Kar, M. Geissler, P. Gibbon, M. Zepf, and M. Borghesi, Phys. Rev. Lett. **108**, 115002 (2012).

32. S. V. Bulanov, E. Yu. Echkina, T. Zh. Esirkepov, I. N. Inovenkov, M. Kando, F. Pegoraro, and G. Korn, Phys. Rev. Lett. **104**, 135003 (2010).

33. S. V. Bulanov, E. Yu. Echkina, T. Zh. Esirkepov, I. N. Inovenkov, M. Kando, F. Pegoraro, and G. Korn, Phys. Plasmas **17**, 063102 (2010).

34. A. Sgattoni, P. Londrillo, A. Macchi, and M. Passoni, Phys. Rev. E **85**, 036405 (2012).

35. E. d'Humières, A. Brantov, V. Yu. Bychenkov, and V. T. Tikhonchuk, Phys. Plasmas **20**, 023103 (2013).

12

High energy density physics research at IMP, Lanzhou, China

Yongtao Zhao, Rui Cheng, Yuyu Wang, Xianming Zhou, Yu Lei, Yuanbo Sun, Ge Xu, Jieru Ren, Lina Sheng, Zimin Zhang, and Guoqing Xiao

Institute of Modern Physics, Chinese Academy of Sciences, Lanzhou 730000, China

Abstract
Recent research activities relevant to high energy density physics (HEDP) driven by the heavy ion beam at the Institute of Modern Physics, Chinese Academy of Sciences are presented. Radiography of static objects with the fast extracted high energy carbon ion beam from the Cooling Storage Ring is discussed. Investigation of the low energy heavy ion beam and plasma interaction is reported. With HEDP research as one of the main goals, the project HIAF (High Intensity heavy-ion Accelerator Facility), proposed by the Institute of Modern Physics as the 12th five-year-plan of China, is introduced.

Keywords: heavy ion beam; high energy density physics; ion beam and plasma interaction; radiography

1. Introduction

High energy density (HED) is generally defined as a state with energy content above 10^{11} J m^{-3}, or equivalently with pressure above 1 Mbar. HED states widely exist in the universe, for example the hydrogen in the core of the Sun or in the core of the Jupiter, the iron in the core of the Earth, the water in the core of Uranus, and so on. The creation of such an extreme state in the laboratory and the study of its properties is very challenging and of key importance in the areas of astrophysics, planetary sciences, geophysics, inertial fusion sciences and applications, and so on. In recent years, the primary emphasis in HED research has been given to the physical properties of warm dense matter (WDM), which is a special HED state typically with temperature of the order of eV and density of the order of solid density, ± 2 magnitudes. In the WDM region, standard theoretical techniques break down, and experiments are badly needed.

As a potential driver for HED matter or inertial fusion energy (IFE), the heavy ion beam from an accelerator is unique, with the advantages of high repetition rate, large size of sample, homogeneous physical condition, good reproducibility and isometric heating of any target at high density, in addition to the ability to compress matter with a front/side shock at very low entropy. There are also some other important applications of ion accelerators in HED physics (HEDP) research, for example, in studying the atomic processes in plasma, in diagnostics of HED samples

by methods of high energy proton/ion radiography and in fast ignition of a compressed fuel. Associated investigations have been pursued with increasing intensity by major accelerator laboratories and institutions in Europe, the USA, Russia, and Japan, where significant progress has been made during the last few decades[1–14].

The Institute of Modern Physics, Chinese Academy of Sciences (IMP) also addresses key issues of HEDP research with the heavy ion beam at the Heavy Ion Research Facility in Lanzhou (HIRFL). In this paper, investigations of radiography by the fast extracted high energy carbon ion beam from the CSR (the Cooling Storage Ring) are introduced, studies on the interaction of a low energy heavy ion beam with plasma are discussed, the project HIAF (High Intensity heavy-ion Accelerator Facility), proposed by the IMP as the 12th five-year-plan of China, and the related parameters and proposals for HEDP research at HIAF are introduced.

2. Radiography of static objects by the fast extracted high energy carbon beam from the CSR

In a typical dynamic experiment, diagnostics of the spatial, density and element distribution of a bulk target and their evolution are of key importance. Apart from imaging with self-radiation such as x-rays or neutrons from the target, radiography with a separate bright source is also commonly utilized. Compared with conventional x-radiography, high energy proton or heavy ion radiography is very promising, in particular as it ensures long penetration

Correspondence to: Y. Zhao, Nanchang Road 509, Lanzhou 730000, China. Email: zhaoyt@impcas.ac.cn

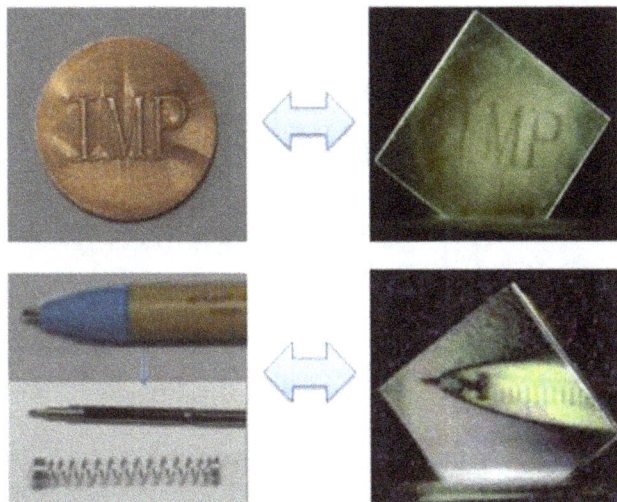

Figure 1. Typical radiographic images (right) of static objects (left) by the method of marginal range radiography.

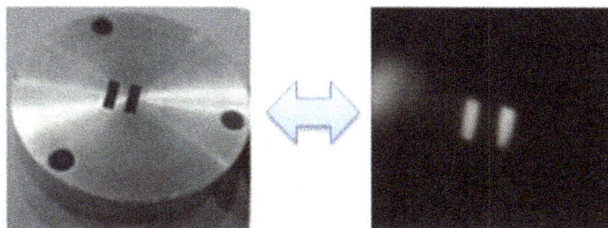

Figure 2. A radiographic image (right) of a static object (left) by the method of magnetic imaging radiography.

distance, high spatial resolution, large dynamic range and high sensitivity to the material density. Many successful experiments on high energy proton radiography have been performed in recent decades[15–18]. In this section, recent results on high energy carbon beam radiography of static objects are introduced. Methods of both marginal range radiography and magnetic imaging lens radiography have been utilized. The difference of heavy ion radiography from proton radiography will be discussed.

The experiments were performed at the HIRFL, where beams of protons with a maximum energy of 2.8 GeV or carbon with a maximum energy of 1 GeV/u can be provided. Detailed introduction of the facilities has been reported elsewhere[10].

Figure 1 shows typical radiographic images of an IMP sign and a ballpoint pen, where the marginal range radiography method was used. Approximately 10^8 carbon ions with an energy of 240 MeV/u were extracted from the CSR at HIRFL in a fast extraction mode, with a pulse duration of about 650 ns. Very good spatial resolution was achieved in the experiment; for example, the diameter of the steel wire inside the ballpoint pen was 400 μm; the measurement from the radiographic experiment was 389 ± 21 μm.

Since the Bragg peak for heavy ions in matter is much sharper, and its transverse distribution at the range margin is much narrower than that for protons, better spatial and density resolution can be expected for heavy ion marginal radiography than for proton marginal radiography (for more details, see our previous report[18]).

Radiography with magnetic imaging lenses has been tested at the CSR as well. Figure 2 shows the first radiographic image with a 600 MeV/u (7.2 GeV in total) carbon beam by utilizing a magnetic imaging lens system with magnifier of around 1:1. The beam intensity is

approximately 5×10^9 with a pulse duration of about 500 ns. The static object is a piece of 10 mm thick aluminum plate with two parallel slits; each slit is 10 mm in length and 3 mm in width. Since the imaging lenses include two dipole magnets, a bright spot on the left-hand side of the image of the slits was formed by the ions passing through the thick plate. It is calculated that the spatial resolution of the radiographic image was about 65 μm[19].

As we know that the principle of magnetic imaging radiography is very similar to transmission electron microscopy (TEM), magnetic imaging lenses in the system can overcome the scattering blur, while a collimator in between the magnets can optimize the contrast. The total momentum and its dispersion, together with the magnifier of the imaging system, may influence the spatial resolution as well. In general, the spatial resolution can be described as following

$$\Delta x \propto L_c \cdot \phi \cdot \frac{\Delta p}{p}, \qquad (1)$$

where L_c, ϕ, and $\Delta p/p$ are the chromatic length, the scattering angle, and the momentum dispersion of the beam after transmitting the object, respectively. It is always emphasized that a better resolution could be expected by increasing the momentum (or energy) of the particles, since a higher particle energy may cause less scattering blur, lower momentum dispersion and lower chromatic aberration. By increasing the particle mass (with the same projectile velocity), the same effects (less scattering blur, and so on) can be expected as well; in other words, high energy heavy ion radiography may have a better resolution than proton radiography, if the fragmentation influence and other effects are not taken into account. Moreover, heavy ions always have a much higher energy deposition, so that less beam intensity is required to produce a similarly bright image on the screen, compared with protons.

3. Interaction of a low energy heavy ion beam with plasma

Investigation of the interaction processes of ion beams with plasma has attracted a lot of attention during recent decades. The motivations are mainly as follows: (1) the energy deposition process of heavy ions in ionized matter is one of

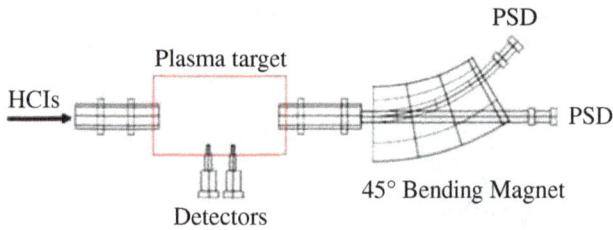

Figure 3. The experimental terminal for studies of low energy ion and plasma interaction at the IMP.

the most important processes in heavy-ion-driven HED and in the burning of inertial confinement fusion (ICF) fuel; (2) plasma devices could serve as important accelerator equipment to focus an ion beam (so-called plasma lens) and/or to strip an ion beam (so-called plasma stripper)[20–26]. Apart from these applications, such research is also an important fundamental topic in understanding the atomic processes in plasma, such as the di-electron recombination process, the free electron capture process, the effective charge in the Coulomb interaction process, and so on.

As has been shown in previous experiments, the stopping power of ionized matter is increased compared with that of cold, non-ionized matter. Enhancement factors of the order of 2–3 have been observed at high projectile energies ($E \sim$ MeV/u), depending on the projectile ion species and the free electron density of the plasma[21, 22]. This effect is especially pronounced at lower ion energies ($E \sim$ keV/u), where an enhancement factor of up to 35 has been observed[23]. Due to the strong nonlinear effects and their special importance in ICF research, more and more emphasis has been given to investigations of ion beams in the low energy range and/or of plasma with high intensity[18, 24–26]. In this section, the recent progress in research on low energy ion interaction with plasma is briefly introduced.

As shown in Figure 3, an experimental terminal for studies of low energy ion and plasma interaction has been set up on the High Voltage Electronic Cyclotron Resonance ion source (HV-ECR) platform at the IMP. The platform can provide both proton and heavy ion beams with intensity of the order of tens to several thousands of enA, and with energy up to $320*q$ keV (q is the charge state of the ions). The energy resolution of the 45° bending magnet is around 1%, depending on the beam size and the beam divergence.

A gas discharging plasma device, as shown in Figure 4, can generate plasma by igniting an electric discharge in two collinear quartz tubes, and then produce fully ionized hydrogen plasma with a free electron density of 10^{16}–10^{17} cm^{-3} and with a temperature of 1–2 eV. The capacitor bank of about 3 µF, discharging at a voltage of up to 5 kV, produces the electrical current going in two opposite directions in either of the two quartz tubes. Such a design of the plasma target is able to suppress the accelerating effect caused by the electric field between the anode and cathodes. Detailed

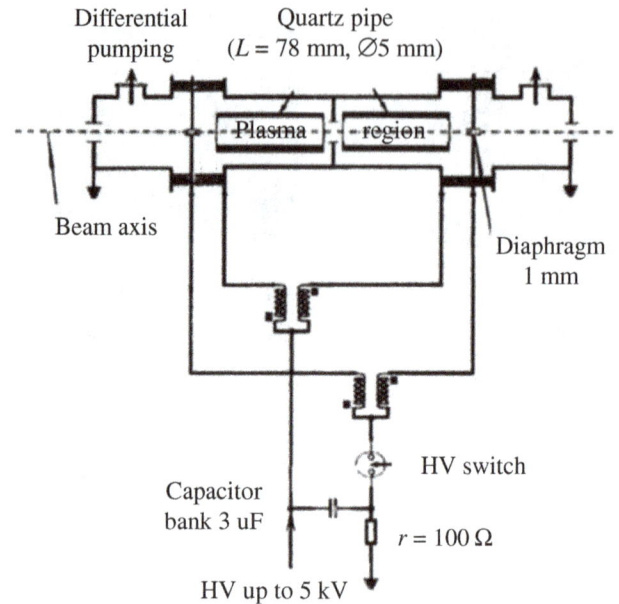

Figure 4. The scheme of the plasma device.

information about the experimental setup and the test results, as well as some simulation results on beam transportation, can be found in previous reports[18, 20].

4. High Intensity heavy-ion Accelerator Facility (HIAF)

After successful construction of the CSR at HIRFL, a large scale scientific research platform, named the HIAF, was proposed in light of the trend and development in nuclear physics and the associated high energy heavy ion research fields. The proposed platform will be one of the projects for basic sciences and technologies as the 12th five-year-plan in China; it will be a laboratory open to the outside world, similar to CSR which was built as the 9th five-year-plan in China. The main goals of this platform are the following: (1) exploration of the effective interactions inside nuclei and the formation of elements heavier than iron in the universe, and other fields related to nuclear physics and nuclear astrophysics; (2) investigation of HEDP and the basic techniques for ICF driven by an intense heavy ion beam; (3) development of the biology and material sciences related to particle irradiation, and so on.

The HIAF complex, as shown in Figure 5, includes a superconductive electronic cyclotron resonance (SECR) ion source, an ion linear accelerator (i-Linac), a booster ring (BRing), a spectrometer ring (SRing), a compression ring (CRing) and about eight experimental terminals; the beam lines or terminals marked in gray in the figure are currently not included in the first stage of the HIAF project. The HIAF will be capable of accelerating ions ranging from proton to uranium, for example, 9.3 GeV protons with an intensity of 1×10^{12} ppp (particles per pulse) or 800 MeV/u U^{34+} with

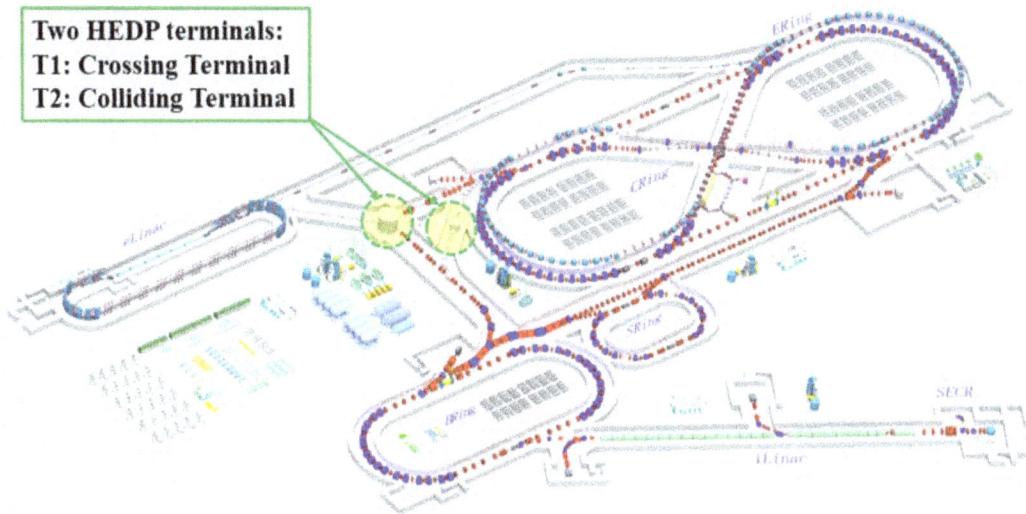

Figure 5. The layout of the HIAF complex and the two terminals for HEDP research.

an intensity of 5×10^{11} can be extracted from the BRing, while 12 GeV protons with an intensity of 2×10^{12} ppp or 1.1 GeV/u U^{34+} with an intensity of 1×10^{12} can be extracted from the CRing.

Since both the BRing and the CRing can produce high energy and high intensity ion beams, two terminals for HEDP research were proposed at the HIAF, one for crossing (T1) and the other for colliding (T2) of the beams from the BRing and the CRing in the target area. Due to budget limitations and technical challenges, only T1 is included in the first stage of the HIAF project, where HED matter will be produced by the beam from the CRing and diagnosed by the proton radiographic beam from the BRing. High energy electron radiography was proposed to be utilized in the HEDP experiments at the HIAF[27, 28].

Compared with FAIR (Facility for Anti-proton and Ion Research), the power of the final beam for driving a HED sample at the HIAF will be improved due to the advanced design, in particular in the following aspects: (1) the higher energy of the beam from the Linac and the booster ring (serving as injectors) will ensure a higher space charge limit; (2) the larger acceptance of the booster ring will ensure a higher intensity of the injection beam; (3) the powerful electron cooler in the CRing will ensure better focusing and better compression. The key parameters related to HEDP research at the HIAF and other advanced heavy ion drivers are listed in Table 1, where E_0, N, E_{total}, S_f, t, and E_ρ are the particle energy, the beam intensity (in units of ppp), the total beam energy per pulse, the FWHM of the beam spot, the pulse duration, and the energy density in a lead target, respectively[2–4].

The main topics of HEDP@HIAF will include, for example, the properties of WDM and the related hydrodynamic instabilities, plenary sciences, beam–plasma interaction, fast flyers driven by an intense ion beam, target physics associated with ICF and the related accelerator physics and

Table 1. Key parameters related to HEDP research at the HIAF and other advanced heavy ion drivers (for a uranium beam).

	SIS-18	FAIR (Ph-I)	HIAF (Ph-1)[a]
E_0	0.4 GeV/u	1 GeV/u	1.1 GeV/u
N	4×10^9	4×10^{11}	1×10^{12}
E_{total}	0.06 kJ	15 kJ	41 kJ
S_f	~1 mm	~1 mm	1–0.5 mm
t	130 ns	50 ns	130–30 ns
E_ρ	2×10^{10} J m^{-3}	2.4×10^{12} J m^{-3}	6–24 $\times 10^{12}$ J m^{-3}

[a] The upper limit may rely on the budget or upgrading of the HIAF.

technologies. The HIAF is a laboratory open to the outside world; worldwide proposals are welcome.

Acknowledgements

This work was supported by the Major State Basic Research Development Program of China ('973' Program, grant number 2010CB832902) and the National Natural Science Foundation of China (grant numbers 11105192, 11075192, 11275241, 11375034, and 11275238). We thank D. Hoffmann, R. Piriz, A. Golubev and his colleagues from ITEP, N. Tahir, J. Jocoby, G. Logan, O. Rosmej, Th. Stöhlker, W. L. Zhang and his colleagues from CAEP, X. A. Zhang, Z. Xu, Y. N. Wang, G. Q. Wang and the *HIAF committee at IMP* for their contributions in the past which have made this work possible.

References

1. J. Meyer-ter-Vehn, Nucl. Fusion **22**, 561 (1986).
2. D. H. H. Hoffmann, V. E. Fortov, M. Kuster, V. Mintsev, B. Y. Sharkov, N. A. Tahir, S. Udrea, D. Varentsov, and K. Weyrich, Astrophys. Space Sci. **322**, 167 (2009).
3. N. A. Tahir, Th. Stöhlker, A. Shutov, I. V. Lomonosov, V. E. Fortov, M. French, N. Nettelmann, R. Redmer, A. R. Piriz, C. Deutsch, Y. Zhao, P. Zhang, H. Xu, G. Xiao, and W. Zhan, New J. Phys. **12**, 073022 (2010).

4. N. A. Tahir, C. Deutsch, V. E. Fortov, V. Gryaznov, D. H. H. Hoffmann, M. Kulish, I. V. Lomonosov, V. Mintsev, P. Ni, D. Nikolaev, A. R. Piriz, N. Shilkin, P. Spiller, A. Shutov, M. Temporal, V. Ternovoi, S. Udrea, and D. Varentsov, Phys. Rev. Lett. **95**, 035001 (2005).

5. A. A. Golubev and V. B. Mintsev, At. Energy **112**, 147 (2012).

6. B. Sharkov, Plasma Phys. Control. Fusion **43**, A229 (2001).

7. G. Logan, F. Bieniosek, F. Bieniosek, E. Henestroza, J. Kwan, E. P. Lee, M. Leitner, L. Prost, P. Roy, P. A. Seidl, S. Eylon, J.-L. Vay, W. Waldron, S. Yu, J. Barnard, D. Callahan, R. Cohen, A. Friedman, D. Grote, M. K. Covo, W. R. Meier, A. Molvik, S. Lund, R. Davidson, P. Efthimion, E. Gilson, L. Grisham, I. Kaganovich, H. Qin, E. Startsev, D. Rose, D. Welch, C. Olson, R. Kishek, P. O'Shea, and I. Haber, Nucl. Instrum. Methods A **544**, 1 (2005).

8. A. R. Piriz, Y. B. Sun, and N. A. Tahir, Phys. Rev. E **88**, 023026 (2013).

9. S. Kawata, Phys. Plasmas **19**, 024503 (2013).

10. Y. Zhao, G. Xiao, H. Xu, H. Zhao, J. Xia, G. Jin, X. Ma, Y. Liu, Z. Yang, P. Zhang, Y. Wang, D. Li, H. Zhao, W. Zhan, Z. Xu, D. Zhao, F. Li, and X. Chen, Nucl. Instrum. Methods B **267**, 163 (2009).

11. F. Genco and A. Hassanein, Laser Part. Beams **32**, 217 (2014).

12. F. Genco and A. Hassanein, Laser Part. Beams **32**, 305 (2014).

13. M. Roth, T. E. Cowan, and M. H. Key, Phys. Rev. Lett. **86**, 436 (2001).

14. H. Qin, R. C. Davidson, and B. G. Logan, Phys. Rev. Lett. **104**, 254801 (2010).

15. A. M. Koehler and V. W. Steward, Nature **245**, 38 (1973).

16. P. A. Rigg, C. L. Schwartz, R. S. Hixson, G. E. Hogan, K. K. Kwiatkowski, F. G. Mariam, M. Marr-Lyon, F. E. Merrill, C. L. Morris, P. Rightly, A. Saunders, and D. Tupa, Phys. Rev. B **77**, 220101 (2008).

17. A. A. Golubev, V. S. Demidov, E. V. Demidova, S. V. Dudin, A. V. Kantsyrev, S. A. Kolesnikov, V. B. Mintsev, G. N. Smirnov, V. I. Turtikov, A. V. Utkin, V. E. Fortov, and B. Yu. Sharkov, Tech. Phys. Lett. **36**, 177 (2010).

18. Y. Zhao, Z. Hu, R. Cheng, Y. Wang, H. Peng, A. Golubev, X. Zhang, X. Lu, D. Zhang, X. Zhou, X. Wang, G. Xu, J. Ren, Y. Li, Y. Lei, Y. Sun, J. Zhao, T. Wang, Y. Wang, and G. Xiao, Laser Part. Beams **30**, 679 (2012).

19. L. Sheng, Y. Zhao, G. Yang, T. Wei, X. Jiang, X. Zhou, R. Cheng, Y. Yan, P. Li, J. Yang, Y. Yuan, J. Xia, and G. Xiao, Laser Part. Beams (2014).

20. R. Cheng, Y. Zhao, X. Zhou, Y. Li, Y. Wang, Y. Lei, Y. Sun, X. Wang, Y. Yu, J. Ren, S. Liu, G. Xiao, and D. H. H. Hoffmann, Phys. Scr. T **156**, 014074 (2013).

21. K. G. Dietrich, D. H. H. Hoffmann, E. Boggasch, J. Jacoby, H. Wahl, M. Elfers, C. R. Haas, V. P. Dubenkov, and A. A. Golubev, Phys. Rev. Lett. **69**, 3623 (1992).

22. D. H. H. Hoffmann, K. Weyrich, H. Wahl, D. Gardés, R. Bimbot, and C. Fleurier, Phys. Rev. A **42**, 2313 (1990).

23. J. Jacoby, D. H. H. Hoffmann, W. Laux, R. W. Müller, H. Wahl, K. Weyrich, E. Boggasch, B. Heimrich, C. Stöckl, H. Wetzler, and S. Miyamoto, Phys. Rev. Lett. **74**, 1550 (1995).

24. A. Golubev, M. Basko, A. Fertman, A. Kozodaev, N. Mesheryakov, B. Sharkov, A. Vishnevskiy, V. Fortov, M. Kulish, V. Gryaznov, V. Mintsev, E. Golubev, A. Pukhov, V. Smirnov, U. Funk, S. Stoewe, M. Stetter, H.-P. Flierl, D. H. H. Hoffmann, J. Jacoby, and I. Iosilevski, Phys. Rev. E **57**, 3363 (1998).

25. A. Frank, A. Blažević, V. Bagnoud, M. M. Basko, M. Börner, W. Cayzac, D. Kraus, T. Heßling, D. H. H. Hoffmann, A. Ortner, A. Otten, A. Pelka, D. Pepler, D. Schumacher, An. Tauschwitz, and M. Roth, Phys. Rev. Lett. **110**, 115001 (2013).

26. C. Teske, B. Lee, J. Jacoby, W. Schweizer, and J. C. Sun, Rev. Sci. Instrum. **81**, 046101 (2010).

27. F. Merrill, A. Hunt, F. Mariam, K. Morley, C. Morris, A. Saunders, and C. Schwartz, Nucl. Instrum. Methods B **261**, 382 (2005).

28. Y. T. Zhao, Z. M. Zhang, and W. Gai, *et al.* 2013 A high resolution spatial–temporal imaging diagnostic for high energy density physics experiments (submitted).

13

The all-diode-pumped laser system POLARIS – an experimentalist's tool generating ultra-high contrast pulses with high energy

Marco Hornung[1,2], Hartmut Liebetrau[2], Andreas Seidel[2], Sebastian Keppler[2], Alexander Kessler[1], Jörg Körner[2], Marco Hellwing[2], Frank Schorcht[1], Diethard Klöpfel[2], Ajay K. Arunachalam[1], Georg A. Becker[2], Alexander Sävert[1,2], Jens Polz[2], Joachim Hein[1,2], and Malte C. Kaluza[1,2]

[1] *Helmholtz-Institute Jena, Germany*

[2] *Institute of Optics and Quantum Electronics, Jena, Germany*

Abstract

The development, the underlying technology and the current status of the fully diode-pumped solid-state laser system POLARIS is reviewed. Currently, the POLARIS system delivers 4 J energy, 144 fs long laser pulses with an ultra-high temporal contrast of 5×10^{12} for the ASE, which is achieved using a so-called double chirped-pulse amplification scheme and cross-polarized wave generation pulse cleaning. By tightly focusing, the peak intensity exceeds 3.5×10^{20} W cm^{-2}. These parameters predestine POLARIS as a scientific tool well suited for sophisticated experiments, as exemplified by presenting measurements of accelerated proton energies. Recently, an additional amplifier has been added to the laser chain. In the ramp-up phase, pulses from this amplifier are not yet compressed and have not yet reached the anticipated energy. Nevertheless, an output energy of 16.6 J has been achieved so far.

Keywords: design; high power laser; laser amplifiers; laser plasmas interaction; laser systems; modelling; optimization; ultra-intense; ultra-short pulse laser interaction with matter

1. Introduction

Chirped-pulse amplification (CPA[1]) laser systems with output powers of several terawatts or even petawatts, which can be focused to intensities in excess of 10^{22} W cm^{-2} are widely used to study laser–matter interactions. During the past three decades this field of science has been growing rapidly and it has been shown that the laser performance in terms of pulse duration, pulse energy, and temporal intensity contrast strongly affects the experimental results. Experiments of particular interest are electron[2, 3] or ion acceleration[4], the laser-based generation of X-rays[5], high-energy physics[6] or laser-based proton radiography[4, 7]. The community interested in these high-intensity phenomena is spread worldwide, operating several dozens of high-intensity lasers, the majority of which are based on direct (e.g., for Nd:Glass-systems) or indirect (e.g., for Ti:Sapphire-systems, which are pumped by flashlamp-pumped frequency-doubled Nd:YAG Lasers) pumping of the active material by flash lamps.

However, for more than one decade strong efforts have been made to establish diode-pumped solid-sate laser (DPSSL) technology for generating high-energy femtosecond or picosecond laser pulses[8]. The most commonly used Yb^{3+}-doped amplification media have already been used for the amplification of ns-laser pulses to energies in excess of 10 J, as recently shown in the projects LUCIA (Yb:YAG, 14 J[9]), DIPOLE (Yb:YAG, 10 J[10]), and MERCURY (Yb:S-FAP, 55 J[11]). Furthermore, a number of projects have started to investigate and to develop high-energy class DPSSLs (HECDPSSL) all over the world[12].

At the Helmholtz-Institute and the Institute of Optics and Quantum Electronics in Jena, Germany, the POLARIS laser system[13] has been developed and commissioned during the past decade. It has commenced its daily operation. Within its experimental program, more than 16,000 shots have been delivered on target during the past two years. The POLARIS project was started in 1999 in order to develop a high-intensity HECDPSSL which could be used in laser–matter interaction experiments. The current key parameters of POLARIS are: 1030 nm centre wavelength, up to 6.5 J pulse energy (4 J on target), 144 fs pulse duration, 7.1 μm^2 focal-spot size, and a temporal contrast for the

Correspondence to: Dr. Marco Hornung, Helmholtz-Institute Jena, Fröbelstieg 3, 07743 Jena, Germany. Email: Marco.Hornung@uni-jena.de

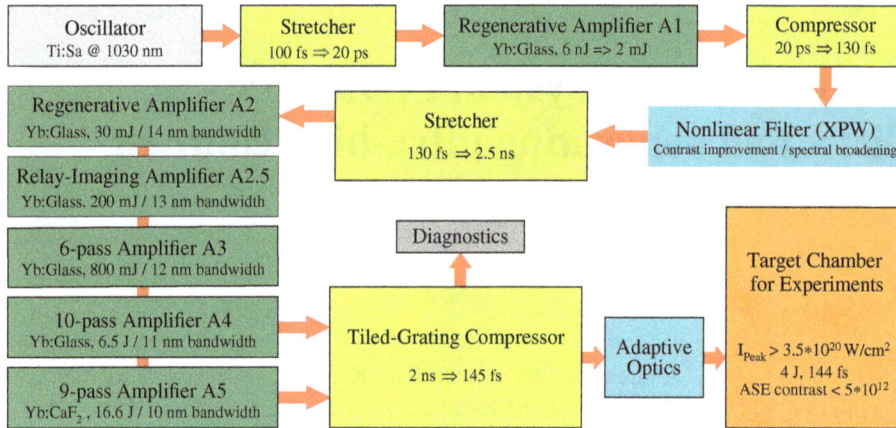

Figure 1. Schematic overview of the POLARIS laser system. An oscillator and two stretcher–compressor stages are used together with six amplifiers (green boxes). A nonlinear filter based on XPW broadens the spectrum and enhances the temporal contrast. An adaptive optics system is used to flatten the wavefront before the pulses enter the target chamber for focusing.

amplified spontaneous emission (ASE) of 5×10^{12}. With these parameters, a peak intensity of 3.5×10^{20} W cm^{-2} is available for experiments. To the best of our knowledge POLARIS is currently the most powerful and intense diode-pumped laser system. Nevertheless, the laser is still under continued development in order to further increase the pulse energy, decrease the pulse duration, and to better meet the requirements of experiments.

In this paper we describe the architecture of POLARIS, including the recently commissioned amplifier A5[14] and a newly installed stretcher–compressor system (double-CPA[15]). After the first CPA stage, the pulses are used to generate a cross-polarized wave (XPW[16]), thereby significantly improving the temporal intensity contrast. Furthermore, we present a detailed characterization of the amplified, compressed and focused pulses with respect to their temporal and spatial properties. The performance of the laser system is finally quantified for application in high-intensity experiments in terms of peak intensity, temporal contrast and shot-to-shot stability.

2. Architecture of the POLARIS laser

In Figure 1 the layout of the POLARIS laser is shown. The system utilizes two subsequent CPA units and six amplification stages to amplify the laser pulses.

After pulse compression a radiation-shielded bunker with a target chamber is available for experiments. The seed pulses for the laser chain are generated in a commercial mode-locked Ti:sapphire oscillator (Coherent MIRA 900) running at a central wavelength of 1030 nm with a pulse energy of 7 nJ and a spectral bandwidth of 20 nm (FWHM). Before entering the first regenerative amplifier the pulses are temporally stretched to 20 ps. The amplifier A1 increases the pulse energy to 2 mJ. Afterwards the pulses are re-

compressed to 130 fs before they enter the nonlinear XPW-filter realized by a BaF$_2$-crystal.

The contrast-cleaned pulses are then stretched once more by the second stretcher (cf. [13, 17]) to a pulse duration of 2.5 ns and furtheramplified by the second regenerative amplifier A2 to a pulse energy of 30 mJ. The amplification to the Joule-level is accomplished with a relay-imaging amplifier (A2.5: $E_{out} = 200$ mJ) and two multipass non-imaging amplifiers (A3: $E_{out} = 800$ mJ[18], and A4: $E_{out} = 6.5$ J). In all of these amplifiers Yb^{3+}-doped fluoride-phosphate glass[19–22] is used as the active material.

The amplified pulses can then either be sent to the main compressor or used as a seed for the final amplifier A5. This amplifier uses Yb:CaF$_2$ in a nine-pass configuration as the active material[23–25] and is currently able to deliver a maximum output energy of 16.6 J (with 2.7 J seed energy). The active material of this amplifier is pumped by 120 laser diode stacks in a 2.5 ms long pump pulse at 940 nm with a 300 kW pump power. A detailed technical description of this amplifier is given in[14]. The beam line for the compression and focusing of the A5-amplified pulses is currently under development and will be finished soon.

For all the experiments shown here, only pulses amplified up to A4 were used. They were compressed with a tiled-grating compressor[26] followed by an adaptive optics system to improve the focusability of the beam. By focusing the beam in the target chamber with a $f/2$ parabola, a peak intensity in excess of 3.5×10^{20} W cm^{-2} is available.

3. Double-CPA and XPW for temporal contrast improvement and spectral broadening

Since the temporal intensity contrast has been shown to be one of the most important parameters for the laser's successful application in high-intensity experiments, we have continuously optimized the contrast of POLARIS. To apply a nonlinear filtering via XPW generation for contrast improvement, the front end was extended by the above-mentioned picosecond-CPA stage.

The two-pass Öffner-type stretcher consists of a 50×50 mm^2 gold grating with a line density of 1200 lines per mm, a 50.8 mm concave mirror with a focal length of 200 mm, and a convex mirror with a focal length of -100 mm. Back reflection from a hollow-roof mirror accomplishes the second pass. Due to the small stretching factor the footprint of the whole setup is only 40×20 cm^2.

In the subsequent regenerative amplifier A1 the stretched 20 ps pulses are amplified up to energies of 3 mJ without any spectral distortion due to self-phase modulation. The output spectrum of the amplifier has a bandwidth (FWHM) of 12 nm, supporting a re-compressed pulse duration of 130 fs if bandwidth-limited Gaussian pulses are assumed. To avoid the generation of post- and pre-pulses the round-trip time of the amplifier was matched to the pulse repetition time of the oscillator[27]. The beam profile of the output pulse has a smooth Gaussian shape (FWHM-diameter of 1 mm) without any hotspots.

Because the nonlinear filter requires an unstretched bandwidth-limited input, the pulse has to be re-compressed after amplification in A1. The compression is achieved by a grating compressor, which consists of two gold gratings similar to the one in the stretcher. The gratings are separated by 95 mm and again a hollow-roof mirror is used for the second pass. The compressor can be operated in air. Pulses with a duration of 130 fs, which is the bandwidth limit of the amplified spectral intensity profile, are achieved.

After this compression the pulse is sent to the XPW-stage, where a 2 mm thick holographic cut BaF$_2$-crystal is used as the nonlinear element. The beam is focused and recollimated with two lenses, both of them having a focal length of 1 m. The BaF$_2$-crystal can be rotated around its surface normal for the XPW optimization.

In order to adjust the required intensity the crystal is placed slightly behind the focal plane. Placing the whole focused beam between the two lenses, including the crystal and its mount, in a vacuum chamber avoids nonlinear effects in air. To separate the XPW and the input signal the setup is placed between two crossed polarizers with a extinction ratio better than 2×10^{-6}. Thus, considering a 6% conversion efficiency, an increase of the intensity contrast of more than four orders of magnitude is achievable.

The generated cross-polarized signal beam exhibits a TEM$_{00}$-mode with a maximum pulse energy of 140 µJ, and a spectral bandwidth of 21 nm (FWHM). It is used in the following as the seed for the remaining amplifier chain after passing the main ns-stretcher (Figure 1). Since due to the XPW process the spectrum of the seed pulses was broader than the bandwidth achieved with the conventional amplification-only setup (cf. [13]), a shorter pulse duration after the final compression could be achieved.

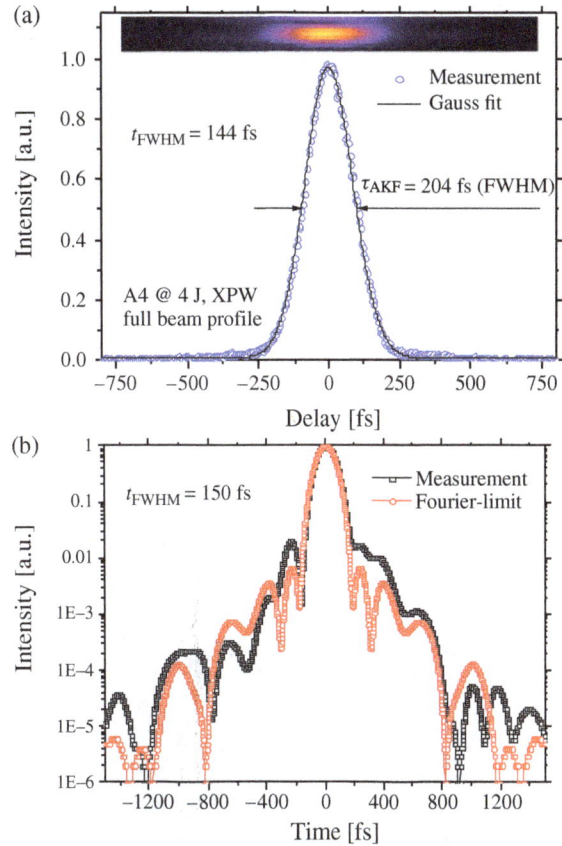

Figure 2. Pulse duration measurements of the compressed pulses. (a) Second-order autocorrelation of a 4 J pulse energy laser pulse which was amplified with A4. A Gaussian distribution fits well to the measured data. Blue: measurement, black: Gaussian fit. (b) High-dynamic Wizzler measurement of the A2 pulses (10 mJ pulse energy). Black: measurement, red: Fourier-limit.

4. Pulse duration measurements

In this section we present the temporal properties of the final compressed laser pulses.

In Figure 2(a) a measurement of the temporal profile of a laser pulse with 4 J energy is shown. This measurement was performed with a home-built second-order autocorrelator, where the pulse is focused in one dimension into a BBO-crystal using a cylindrical lens. The upper inset displays the spatially resolved autocorrelation signal. An FWHM-pulse duration of 144 fs is measured over the full beam profile. It fits well to a Gaussian pulse shape (black line). The broader spectrum of the XPW (cf. [13]) allows the pulse duration including amplification up to the amplifier A4 to be reduced from formerly 164 fs to currently 144 fs. In the former setup only a conventional, single-stage CPA layout was implemented.

In Figure 2(b) a high-dynamic measurement of the pulse duration of the compressed laser pulses (from the front end, i.e., up to amplifier A2) with 10 mJ pulse energy is shown. For this measurement a self-referenced spectral

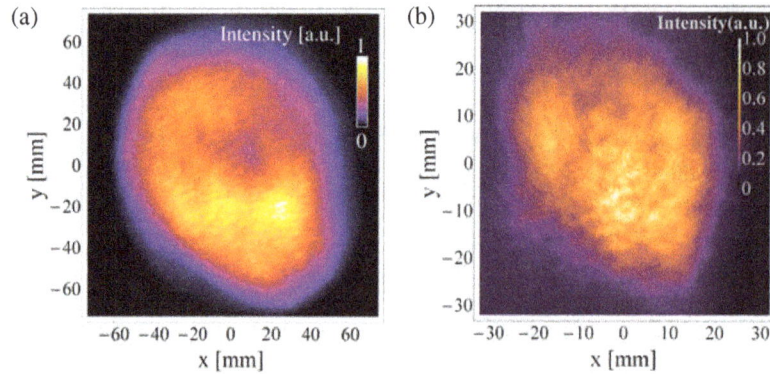

Figure 3. Measurements of the near-field intensity distribution of the amplified laser pulses. (a) Beam profile of the fourth amplifier (A4) measured in front of the focusing parabola after pulse compression. (b) Beam profile of the fifth amplifier (A5) measured directly after amplification with a pulse energy of 16.6 J.

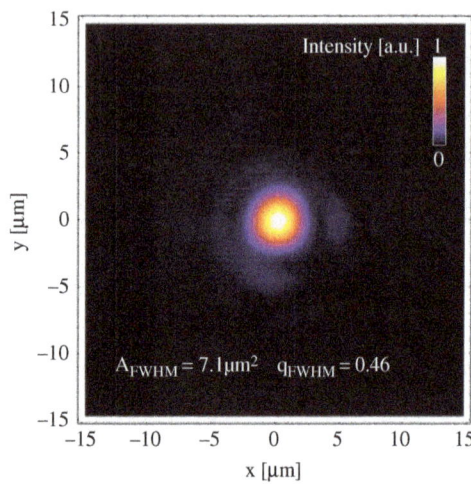

Figure 4. Measured transverse far-field profile of the A4-amplified and compressed laser pulses. The area within which the intensity is larger than $I_{max}/2$ is 7.1 μm^2 and contains 46% of the pulse energy ($q = 0.46$).

interferometer (Wizzler) was used and the spectral phase was flattened using a Dazzler (both Fastlite Techn.) positioned directly behind the oscillator without using the XPW-stage. This measurement highlights the capability of generating nearly transform-limited laser pulses with high temporal contrast. However, in order to use the Dazzler's spectral shaping capability together with the XPW-cleaned pulses the device needs to be inserted behind the XPW-stage.

5. Spatial pulse characterization

A homogeneous super-Gaussian-like near-field profile of the laser pulses is desirable for an efficient energy extraction and a maximum compressor throughput. The maximum extractable pulse energy of the main amplifiers A4 and A5 is limited by the fluence threshold at which laser-induced damage occurs on the surface or in the volume of the active material. The Yb:glass, which is used in amplifier A4,

withstands a fluence of 3 J cm^{-2} in long-term operation, whereas the fluence on the Yb:CaF$_2$-crystal used in amplifier A5 is currently limited by the damage threshold of the AR coatings to 2 J cm^{-2}. In Figure 3 the near-field intensity distributions are shown for pulses that are amplified in A4 and A5, respectively.

The measurement for the A4-amplified pulse in Figure 3(a) was done in front of the focusing parabola. Between the exit of the last amplifier and the compressor entrance the laser beam diameter is expanded by a 5.6× magnification telescope. The enlarged beam diameter ensures a safe operation in terms of laser-induced damage of the compressor gratings ($F_{damage} = 200$ mJ cm^{-2}).

The beam profile of the A5-amplified pulses, shown in Figure 3(b), was recorded after the last passage through the active material. In order to realize a sufficient beam diameter for the pulse compressor a 2.5× magnification telescope is placed between the amplifier A5 and the pulse compressor.

To generate a high peak intensity for laser–matter experiments the pulses are focused in the target chamber with an off-axis parabola ($f = 300$ mm, $f/2$, cf. Figure 1). Since pulses from the amplifier A5 have not been compressed yet, we have only been using the A4-amplified pulses for experiments so far. The compression and focusing of the A5 pulses is currently under investigation.

For the improvement of the focusability of the pulses from A4 we installed an adaptive optics system. The combination of a 160-mm-diameter adaptive mirror and a wavefront sensor helps to flatten the wavefront and to reduce the focal-spot size as well as to increase the energy content within the focal-spot area (for details see [13]).

The focal spot of the amplified laser pulses is shown in Figure 4. For this measurement the amplified pulses were strongly attenuated in the laser chain. The area within which the intensity is larger than $I_{max}/2$ (A_{FWHM}) is 7.1 μm^2 and contains 46% of the total laser pulse energy ($q_{FWHM} = 0.46$). The Strehl ratio, defined as the ratio of the achieved peak intensity to the calculated peak intensity assuming a flat wavefront, is 0.45 and likely limited by residual chromatic and wavefront aberrations.

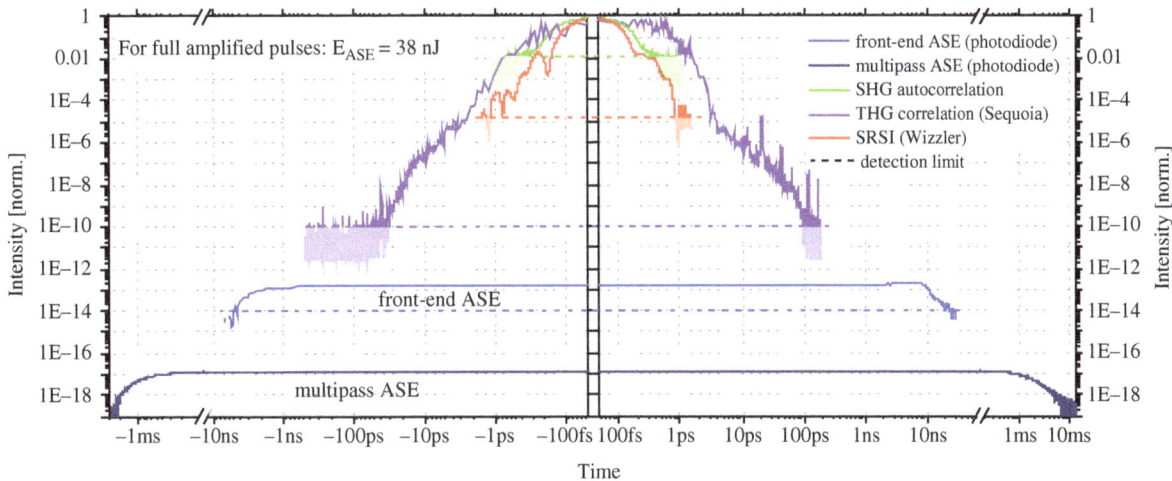

Figure 5. High-dynamic temporal characterization of the amplified and compressed laser pulses. The intensity is given as the relative on-target intensity using $f/3$-focusing and negative times are defined as the times before the laser pulse. The laser pulse is characterized using different measurements: red: Self-Referenced Spectral Interferometry (Wizzler), green: SHG-correlator, pink: THG-correlator, blue and dark blue: photodiode. The detection limit of each measurement is marked as a dashed line.

6. High-dynamic temporal pulse characterization

As mentioned in the introduction, the temporal intensity contrast is one of the most important parameters for the application of high-intensity laser pulses in experiments. In this section we will quantify the temporal contrast by giving an overview of the temporal behaviour of the compressed pulses on different timescales with a high-dynamic range.

In Figure 5 the temporal characterization of the POLARIS pulses is shown on a log–log scale. The graph combines different types of measurements for the main laser pulse, the front-end ASE and the multipass-amplifier ASE. The relative on-target intensity has been recorded using the above-mentioned $f/2$-focusing parabola currently installed in our target chamber. Due to the different methods used, the characterization spans both a range of 17 orders of magnitude for the relative intensity and a time-window from femtoseconds to milliseconds. The individual detection limits for each individual measurement are indicated by dashed lines in the corresponding colours.

The red (Wizzler) and green (SHG-autocorrelation) curves in Figure 5 are the pulse duration measurements which are shown above. They are able to measure the temporal intensity profile of the laser pulse with a dynamic range of 1.5 (SHG-autocorrelator) and 5 (Wizzler) orders of magnitude.

To further resolve the temporal structure of the compressed laser pulses a commercial third-order cross-correlator (Sequoia, Amplitude Technologies) has been used. It is able to resolve 10 orders of magnitude relative intensity for pulses centred at 1030 nm. Here we mention that the measurement with the Sequoia is taken in the near-field of the compressed pulse and since we are using a tiled-grating compressor some spectral components of the pulses are missing depending on the lateral position in the laser pulse[17]. This may lead to a

smaller relative intensity contrast for the laser pulse (and its leading edge) as compared to the contrast in the focal spot where the pulses are applied in experiments.

However, even the dynamic range of the third-order cross-correlator is not sufficient to resolve the relative intensity contrast of our pulses when using the XPW front end. Apart from some residual pre-pulses, the relative contrast ratio of the pulses can no longer be resolved for $t < -40$ ps before the main pulse due to the detection limit of 10^{-10} for the relative intensity contrast. The residual pre-pulses are likely generated by reflections inside the amplifier chain[28]. Their elimination is the subject of ongoing work.

The ASE of the amplifiers was measured by operating the laser with fully pumped amplifiers, but blocking pulses from the oscillator. Due to the non-saturated amplification within all amplifiers of POLARIS the generated ASE without seed is identical to the ASE generated during the amplification. Using a high-sensitivity photodiode and calibrated ND filters we are able to measure the temporally dependent relative intensity of the ASE. Furthermore, with our focal-spot diagnostic we could record the spatial far-field distribution of the ASE contributions from the different amplifiers, finally leading to the relative ASE intensities on target. This method is described in detail in[29].

Due to its architecture the POLARIS laser emits two different types of ASE. The first is the so-called front-end ASE which is generated in the first amplifier (A1), reduced by the XPW-stage but further amplified by the second amplifier, with a pulse duration of 13 ns, which corresponds to the round-trip times in the regenerative amplifiers. The blue curve in Figure 5 displays a photodiode measurement of the front-end ASE having a intensity of 2×10^{-13} relative to the main pulse.

The second part is the so-called multipass-ASE, which is generated in the multipass amplifiers A2.5, A3, and A4. Due to their multipass architecture, which does not implement a cavity, the pulse duration is on the order of a few milliseconds. This corresponds to the pump duration and the fluorescence lifetime of the active material.

Note that the front-end ASE is also amplified in the multipass amplifiers since this front-end ASE is seeding – together with the main laser pulse – all subsequent amplifiers. Due to the long pulse duration of 2 ms, the multipass-ASE energy and intensity can be significantly decreased by temporal gating of the main laser pulse during the amplification. This is accomplished by two Pockels cells, which are installed before and after the amplifier A4. Their gate duration is set to be as short as 10 ns.

The relative intensity of the multipass-ASE with respect to the main pulse is 1.5×10^{-17} and displayed in Figure 5 as the dark blue line. This measurement was taken with a photodiode where the amplifier chain was blocked in front of amplifier A2.5. Due to the missing cavities in the multipass amplifiers the focusability of the multipass-ASE is strongly degraded and leads to a one order of magnitude larger focal spot as compared to the main pulse. Since the front-end ASE is generated by two regenerative amplifiers, which define the spatial profile and divergence of the main pulse, they both have a nearly identical focal-spot size.

Additionally, we measured the energy of the ASE to be 38 nJ, which is the sum of all contributions. Compared to an ASE energy of 130 µJ and a relative ASE intensity of 10^{-9} of our conventional and formerly used front end (without XPW, cf. [13]) we have significantly improved the experimental performance in terms of temporal intensity contrast with the double-CPA and XPW front end. As a consequence, up to 2 ns before the main laser pulse a contrast of 5×10^{12} (relative intensity of 2×10^{-13}), and up to 1 ms a temporal intensity contrast of 7×10^{16} (relative intensity of 1.5×10^{-17}) has been achieved.

7. Experimental performance

With the currently installed $f/2$-focusing parabola the POLARIS laser is capable of delivering a peak intensity of 3.5×10^{20} W cm^{-2} with a repetition rate of 1/40 Hz. The pulses are as short as 144 fs, with a pulse energy of 4 J on target.

With the final commissioning of amplifier A5, the on-target pulse energy will likely exceed 10 J[14]. To reproducibly operate the laser on a daily basis we have optimized and installed enclosures to minimize air fluctuations and reduce dust contamination on the optics. Furthermore, most of the optics which were used for online alignment (e.g., thermal drift compensation) are motorized for remote control. In Figure 6(a), a measurement of the pulse duration of 300 consecutive (taken over 3.3 h during an experiment)

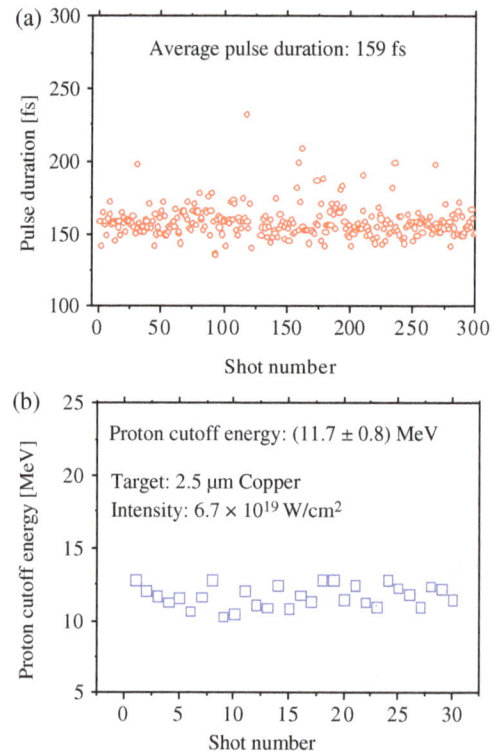

Figure 6. Stability measurements. (a) Pulse duration measurement with a single-shot SHG-autocorrelator during 300 consecutive full energy shots (taken over 3.3 h). (b) Cutoff energy of TNSA-accelerated protons from a 2.5-µm-thick copper foil versus shot number (30 consecutive shots).

full energy shots is shown. The sometimes increased pulse duration is likely induced by mechanical vibrations on the very sensitive tiled-grating arrangement in the pulse compressor.

Up to now we have performed several experimental campaigns over the past two years in order to accelerate protons or electrons. As an example, protons have been accelerated from thin foils by the so-called target-normal sheath acceleration (TNSA[4]). To investigate the stability of the entire laser system we have used such an experiment to record 30 consecutive shots over a time of 20 min while measuring the cutoff energy of the accelerated protons. In Figure 6(b) the data is displayed. In this particular experiment a 2.5-µm-thick copper foil was used as target and an average proton energy of 11.7 MeV was achieved. While keeping all parameters constant during these 30 shots the standard deviation for the measured proton energy was as small as 7%. Note that for our parameters the proton cutoff energy only weakly depends on the pulse duration[30].

8. Conclusion and outlook

In conclusion we present an overview of the fully diode-pumped solid-state laser POLARIS. With a peak intensity of 3.5×10^{20} W cm^{-2} it is currently, to the best of our knowledge, the most intense DPSSL worldwide. Moreover, the pulses are generated with an ultra-high temporal contrast of 5×10^{12} for the ASE of the laser system.

Furthermore, we have shown that the pulse energy can be increased with a diode-pumped $Yb:CaF_2$-crystal to 16.6 J. In the near future, the pulses from this amplification stage will be compressed and focused to be available for high-intensity laser–matter experiments.

Finally, we investigate the operation performance of PO-LARIS in laser–matter interaction experiments, where a stable generation of protons has been achieved with a standard deviation of 7% for the cutoff energy.

Acknowledgements

The research leading to these results has received funding from the European Commission's (EC) 7th Framework Programme (LASERLAB-EUROPE, grant no. 228334) and from the Bundesministerium für Bildung und Forschung (BMBF) (03ZIK445 and 03Z1H531).

References

1. D. Strickland and G. Mourou, Opt. Comm. **56**, 219 (1985).
2. E. Esarey, C. B. Schroeder, and W. P. Leemans, Rev. Mod. Phys. **81**, 1229 (2009).
3. H. T. Kim, K. H. Pae, H. J. Cha, I. J. Kim, T. J. Yu, J. H. Sung, S. K. Lee, T. M. Jeong, and J. Lee, Phys. Rev. Lett. **111**, 165002 (2013).
4. A. Macchi, M. Borghesi, and M. Passoni, Rev. Mod. Phys. **85**, 751 (2013).
5. S. Corde, K. T. Phuoc, A. Beck, G. Lambert, R. Fitour, E. Lefebvre, V. Malka, and A. Rousse, Rev. Mod. Phys. **85**, 1 (2013).
6. Y. Izawa, N. Miyanaga, J. Kawanaka, and K. Yamakawa, J. Opt. Soc. Korea **12**, 178 (2008).
7. H. Johnston, Physicsworld **26**, 50 (2013).
8. M. Siebold, J. Hein, M. Hornung, S. Podleska, M. C. Kaluza, S. Bock, and R. Sauerbrey, Appl. Phys. B **90**, 431 (2008).
9. T. Goncalves-Novo, D. Albach, B. Vincent, M. Arzakantsyan, and J. C. Chanteloup, Opt. Exp. **21**, 855 (2013).
10. S. Banerjee, K. Ertel, P. D. Mason, P. J. Phillips, M. Siebold, M. Löser, C. Hernandez-Gomez, and J. L. Collier, Opt. Lett. **37**, 2175 (2012).
11. A. Bayramian, P. Armstrong, E. Ault, R. Beach, C. Bibeau, J. Caird, R. Campbell, B. Chai, J. Dawson, C. Ebbers, A. Erlandson, Y. Fei, B. Freitas, R. Kent, Z. Liao, T. Ladran, J. Menapace, B. Molander, S. Payne, N. Peterson, M. Randles, K. Schaffers, S. Sutton, J. Tassano, S. Telford, and E. Utterback, J. Fusion Sci. Technol. **52**, 383 (2007).
12. T. Töpfer, J. Neukum, J. Hein, and M. Siebold, Laser Focus World **46**, 64 (2010).
13. M. Hornung, S. Keppler, R. Bödefeld, A. Kessler, H. Liebetrau, J. Körner, M. Hellwing, F. Schorcht, O. Jäckel, A. Sävert, J. Polz, A. K. Arunachalam, J. Hein, and M. C. Kaluza, Opt. Lett. **38**, 718 (2013).
14. A. Kessler, M. Hornung, S. Keppler, F. Schorcht, M. Hellwing, H. Liebetrau, J. Körner, A. Sävert, M. Siebold, M. Schnepp, J. Hein, and M. C. Kaluza, Opt. Lett. **39**, 1333 (2014).
15. M. P. Kalashnikov, E. Risse, H. Schönnagel, A. Husakou, J. Herrmann, and W. Sandner, Opt. Exp. **12**, 5088 (2004).
16. A. Jullien, O. Albert, G. Chériaux, J. Etchepare, S. Kourtev, N. Minkovski, and S. M. Saltiel, J. Opt. Soc. Am. B **22**, 2635 (2005).
17. M. Hornung, R. Bödefeld, M. Siebold, A. Kessler, M. Schnepp, R. Wachs, A. Sävert, S. Podleska, S. Keppler, J. Hein, and M. C. Kaluza, Appl. Phys. B **101**, 93 (2010).
18. J. Hein, S. Podleska, M. Siebold, M. Hellwing, R. Bödefeld, R. Sauerbrey, D. Ehrt, and W. Wintzer, Appl. Phys. B **79**, 419 (2004).
19. T. Töpfer, J. Hein, J. Philipps, D. Ehrt, and R. Sauerbrey, Appl. Phys. B **71**, 203 (2000).
20. T. Töpfer, J. Hein, W. Wintzer, D. Ehrt, and R. Sauerbrey, Glass Sci. Technol. **75**, 223 (2002).
21. D. Ehrt, Curr. Opin. Solid State Mater. Sci. **7**, 135 (2003).
22. S. Paoloni, J. Hein, T. Töpfer, H. G. Walther, R. Sauerbrey, D. Ehrt, and W. Wintzer, Appl. Phys. B **78**, 415 (2004).
23. M. Siebold, M. Hornung, R. Bödefeld, S. Podleska, S. Klingebiel, C. Wandt, F. Krausz, S. Karsch, R. Uecker, A. Jochmann, J. Hein, and M. C. Kaluza, Opt. Lett. **33**, 2770 (2008).
24. M. Siebold, S. Bock, U. Schramm, B. Xu, J. L. Doualan, P. Camy, and R. Moncorge, Appl. Phys. B **97**, 327 (2009).
25. J. Körner, C. Vorholt, H. Liebetrau, M. Kahle, D. Klöpfel, R. Seifert, J. Hein, and M. C. Kaluza, J. Opt. Soc. Am. B **29**, 2493 (2012).
26. M. Hornung, R. Bödefeld, M. Siebold, M. Schnepp, J. Hein, R. Sauerbrey, and M. C. Kaluza, Appl. Opt. **46**, 7432 (2007).
27. S. Keppler, R. Bödefeld, M. Hornung, A. Sävert, J. Hein, and M. C. Kaluza, Appl. Phys. B **104**, 11 (2011).
28. S. Keppler, M. Hornung, R. Bödefeld, M. Kahle, J. Hein, and M. C. Kaluza, Opt. Exp. **20**, 20742 (2012).
29. S. Keppler, M. Hornung, R. Bödefeld, A. Sävert, H. Liebetrau, J. Hein, and M. C. Kaluza, Opt. Express **22** (9), (2014) (accepted for publication).
30. J. Schreiber, F. Bell, F. Grüner, U. Schramm, M. Geissler, M. Schnürer, S. Ter-Avetisyan, B. M. Hegelich, J. Cobble, E. Brahmbrink, J. Fuchs, P. Audebert, and D. Habs, Phys. Rev. Lett. **97**, 045005 (2006).

14

Independent and continuous third-order dispersion compensation using a pair of prisms

Qingwei Yang, Xinglong Xie, Jun Kang, Haidong Zhu, Ailin Guo, and Qi Gao

National Laboratory on High Power Laser and Physics, Shanghai Institute of Optics and Fine Mechanics, Chinese Academy of Sciences, No. 390, Qinghe Road, Jiading District, Shanghai 201800, China

Abstract

The dispersion of a pair of prisms is analyzed by means of a ray-tracing method operating at other than tip-to-tip propagation of the prisms, taking into consideration the limited spectral bandwidth. The variations of the group delay dispersion and the third-order dispersion for a pair of prisms are calculated with respect to the incident position and the separation between the prisms. The pair of prisms can provide a wide range of independent and continuous third-order dispersion compensation. The effect of residual third-order dispersion on the pulse contrast ratio and pulse duration is also calculated. The residual third-order dispersion not only worsens the pulse contrast ratio, but also increases the pulse duration to the hundreds of femtosecond range for a tens of femtosecond pulse, even when the residual third-order dispersion is small. These phenomena are helpful in compensating for the residual high-order dispersion and in understanding its effect on pulse contrast ratios and pulse durations in ultrashort laser systems.

Keywords: chirped pulse amplification; dispersion; prisms; ultrashort pulses

1. Introduction

The chirped pulse amplification (CPA) technique[1, 2] is commonly used to build high-power laser systems for various experiments and applications, such as laser-driven plasma accelerators and fast ignition in inertial confinement fusion. Peak laser powers of up to 1 PW (1 PW = 10^{15} W) are now achievable, and are expected to be boosted to the EW level (1 EW = 10^{18} W) in the near future[3].

In CPA ultrashort laser systems, the dispersion elements include stretcher, compressor and amplifier material[4–10]. The compressor can completely compensate for group delay dispersion (GDD) and partly compensate for third-order dispersion (TOD), which are generated by the stretcher and amplifier material[6, 7]. However, compensating for the residual high-order dispersion is extremely important in the entire system to obtain the shortest pulse duration and highest pulse contrast, particularly for ultrashort pulses (<50 fs)[3, 5]. Thus, setting up an independent dispersion compensation element is necessary to compensate for the residual high-order dispersion of the system.

The prism pair is an important element in dispersion compensation. In 1984, Fork *et al.* indicated that a pair of prisms

could provide negative GDD and TOD, which are suitable for the dispersion compensation element[11–14]. However, the equations used by Fork to evaluate the GDD and TOD of a pair of prisms are only for tip-to-tip propagation in the prisms and do not consider the effect of the spectral bandwidth. Another approach to evaluate the dispersion of a pair of prisms was presented by Arissian *et al.*[15, 16]; the technique is based on accurate calculation of the optical path of the pair of prisms. However, the equations employed by Arissian *et al.* to evaluate GDD and TOD of the pair of prisms do not consider limited spectral bandwidth and continuous TOD compensation, which are very important for dispersion compensation. For a particular ultrashort laser system, dispersion compensation can be considered to be quantitatively meaningful only under the premise of a certain spectral bandwidth; otherwise, this method is qualitative. For example, for a 30 fs ultrashort laser system, the spectral bandwidth of dispersion compensation should remain greater than 140 nm in order to obtain a short compressed pulse and a high pulse contrast ratio. But, in practical applications, the possible spectral bandwidth is only 30 nm when the spectral bandwidth is not considered in the dispersion compensation. In that time, a short compressed pulse (e.g., 30 fs) cannot be obtained. Therefore, it is very important to consider the

Correspondence to: Q. Yang, No. 390, Qinghe Road, Jiading District, Shanghai 201800, China. Email: yqwphy@siom.ac.cn

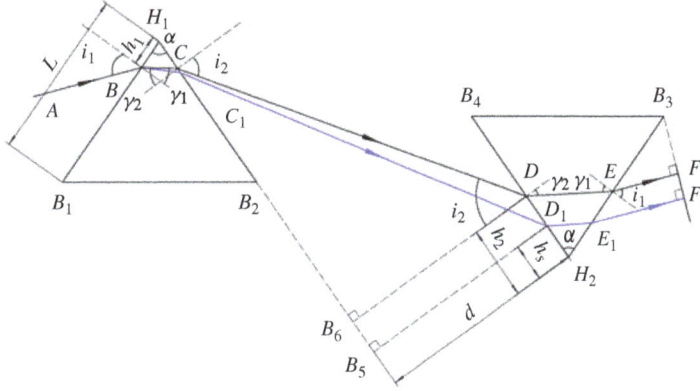

Figure 1. Ray-tracing sketch for a pair of identical isosceles prisms in a parallel face configuration.

spectral bandwidth in dispersion compensation, particularly in practical applications.

In this study, we focus on independent and continuous TOD compensation using a pair of prisms. The dispersion of the pair of prisms is analyzed in detail using a ray-tracing method operating at other than tip-to-tip propagation of the prisms. The variations of GDD and TOD for the pair of prisms are calculated with respect to the incident position and the separation between the prisms. A pair of prisms can provide a wide range of independent and continuous TOD compensation. The effect of residual TOD (RTOD) on the pulse contrast ratio and pulse duration is calculated. The RTOD not only worsens the pulse contrast ratio, but also greatly increases the pulse duration to the hundreds of femtosecond range for a tens of femtosecond pulse, even when small RTOD is employed. These phenomena are helpful in compensating for residual high-order dispersion and in understanding its effect on the pulse contrast ratio in ultrashort laser systems.

2. Model

Figure 1 shows the ray-tracing sketch for a pair of identical isosceles prisms in a parallel face configuration. The blue curve represents the shortest wavelength trajectory and the black curve represents any single wavelength trajectory after dispersion through the prism pairs.

Based on Figure 1, the optical path of any one wavelength is given by

$$p = 2[n(BC + DE) + CD + EF]$$
$$= 2\left\{ n\left[\frac{h_1 \sin \alpha}{\cos \gamma_2(\lambda)} + \frac{h_2 \sin \alpha}{\cos \gamma_1(\lambda)} \right] + \frac{d}{\cos i_2(\lambda)} \right.$$
$$\left. + \left[L - \frac{h_2 \cos \gamma_2(\lambda)}{\cos \gamma_1(\lambda)} \right] \sin i_1 \right\} \quad (1)$$

$$h_2 = h_s + d \tan i_2(\lambda_s) - d \tan i_2(\lambda) + \frac{h_1 \sin \alpha}{\cos \gamma_2(\lambda_s)} - \frac{h_1 \sin \alpha}{\cos \gamma_2(\lambda)}$$
$$(2)$$

where λ is the wavelength under consideration, λ_s is the shortest wavelength, $n(\lambda)$ is the refractive index of the prism, h_1 is the distance of the beam input trajectory to the apex of the prism, h_2 is the distance of the trajectory at a particular wavelength to the apex of the second prism, h_s is the distance of the trajectory of the shortest wavelength to the apex of the second prism, L is the waist length of the isosceles prisms, α is the apex angle of the prism, i_1 is the incidence angle, $i_2(\lambda)$ is the exit angle of the prism, $\gamma_1(\lambda)$ is the refraction angle in the incidence plane, and $\gamma_2(\lambda)$ is the incidence angle in the exit plane of the first prism at a particular wavelength.

To ensure high transmission efficiency of the prism pairs, Brewster angle incidence is adopted. The values of these angles can be calculated using $i_1 = \arctan[n(\lambda_0)]$, $\gamma_1(\lambda) = \arcsin[\frac{\sin i_1}{n(\lambda)}]$, $\gamma_2(\lambda) = \alpha - \gamma_1(\lambda)$, $i_2(\lambda) = \arcsin[n(\lambda) \sin \gamma_2(\lambda)]$, where λ_0 is the central wavelength.

The total phase for the pair of prisms yields

$$\phi(\omega) = \frac{\omega p}{c} \quad (3)$$

where ω is the angular frequency and c is the speed of light in vacuum.

According to Equation (3), the dispersions can be divided into various orders, such as the second-, third- and fourth-order dispersions (i.e., GDD, TOD and FOD, respectively).

$$\text{GDD}|_{\omega_0} = \frac{\lambda^3}{2\pi c^2} \frac{d^2 P}{d\lambda^2} \bigg|_{\lambda_0} \quad (4)$$

$$\text{TOD}|_{\omega_0} = -\frac{\lambda^4}{4\pi^2 c^3} \left(3\frac{d^2 P}{d\lambda^2} + \lambda \frac{d^3 P}{d\lambda^3} \right) \bigg|_{\lambda_0} \quad (5)$$

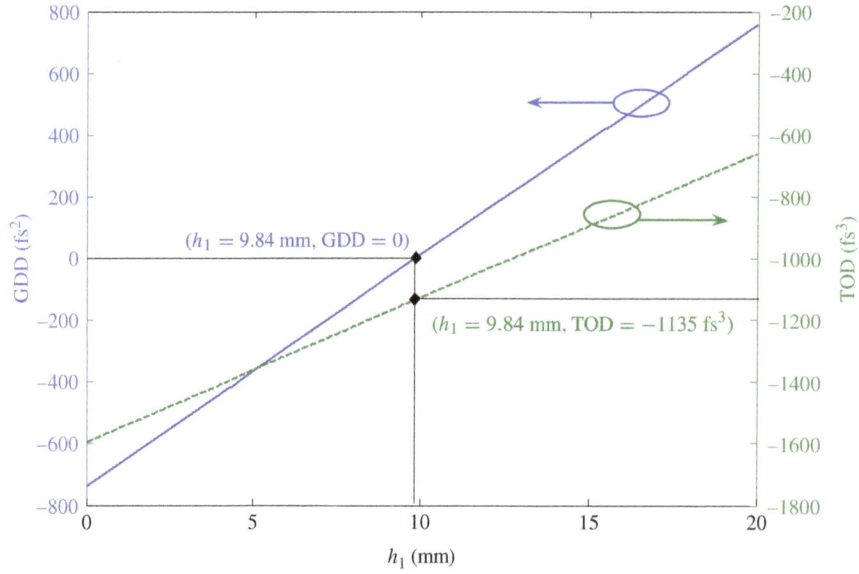

Figure 2. At the central wavelength, GDD and TOD change with h_1. Material: CaF$_2$; simulation parameters: $i_1 = 69.9°$, $\lambda_0 = 808$ nm, $\Delta\lambda = 140$ nm, $\lambda_s = 738$ nm, $d = 800$ mm, and $h_s = 1$ mm.

$$\text{FOD}|_{\omega_0} = \frac{\lambda^5}{8\pi^3 c^4}\left(12\frac{d^2 P}{d\lambda^2} + 8\lambda\frac{d^3 P}{d\lambda^3} + \lambda^2\frac{d^4 P}{d\lambda^4}\right)\Bigg|_{\lambda_0} \quad (6)$$

where ω_0 is the central angular frequency.

3. Numerical results

According to the model constructed in Section 2, we can calculate the GDD and TOD of the prisms. The influences of the distances h_1, h_s, and d on GDD and TOD are determined. We then discuss independent and continuous TOD compensation by changing the distances h_1 and d (or h_2 and d) simultaneously. The effect of RTOD on the pulse contrast ratio and pulse duration is also discussed.

During the simulation, we adopted a typical Sellmeier series equation to describe glass material dispersion[2]. The Sellmeier series equation is written as

$$n^2 = 1 + \frac{B_1\lambda^2}{\lambda^2 - C_1} + \frac{B_2\lambda^2}{\lambda^2 - C_2} + \frac{B_3\lambda^2}{\lambda^2 - C_3} \quad (7)$$

where the wavelength λ is expressed in μm. The Sellmeier coefficient values for $B_1, B_2, B_3, C_1, C_2, C_3$ are shown in Table 1.

3.1. Influence of distances h_1 and h_s on dispersion

At the central wavelength, GDD and TOD change with distance h_1 or h_s (Figures 2 and 3, respectively). GDD changes from negative to positive and TOD reduces rapidly when h_1 or h_s increases. For instance, when $h_1 < 9.84$ mm, the value of GDD is reduced rapidly (Figure 2). However,

when $h_1 > 9.84$ mm, the value of GDD increases rapidly. Thus, $h_1 = 9.84$ mm is the critical value; at $h_1 = 9.84$ mm, the value of GDD is zero and the value of TOD is -1135 fs^3. When $h_s < 11.7$ mm, the value of GDD reduces rapidly. However, when $h_s > 11.7$ mm, the value of GDD increases rapidly. Thus, $h_s = 11.7$ mm is the critical value; at $h_s = 11.7$ mm, the value of GDD is zero and the value of TOD is -1134 fs^3.

Figures 2 and 3 show that the simultaneous occurrence of GDD and TOD with the same or different signs is possible when h_1 or h_s changes. Therefore, the prisms can compensate for GDD and TOD with the same signs, as well as GDD and TOD with different signs, by changing h_1 or h_s.

In the simulation shown in Figure 2, the following parameters are employed: $i_1 = 69.9°$, $\lambda_0 = 808$ nm, $\Delta\lambda = 140$ nm, $\lambda_s = 738$ nm, $d = 800$ mm, and $h_s = 1$ mm; material: CaF$_2$.

In the simulation shown in Figure 3, the following parameters are employed: $i_1 = 69.9°$, $\lambda_0 = 808$ nm, $\Delta\lambda = 140$ nm, $\lambda_s = 738$ nm, $d = 800$ mm, and $h_1 = 1$ mm; material: CaF$_2$.

3.2. Influence of distance d on dispersion

At the central wavelength, GDD and TOD change with d (Figure 4). GDD and TOD change simultaneously from positive to negative with increasing d. For instance, when $d < 498.2$ mm, the value of GDD reduces rapidly. However, $d > 498.2$ mm, the value of GDD increases rapidly. Thus $d = 498.2$ mm is the critical value; at $d = 498.2$ mm, GDD is zero and TOD is -753 fs^3.

During the simulation, the following parameters are employed: $i_1 = 69.9°$, $\lambda_0 = 808$ nm, $\Delta\lambda = 140$ nm, $\lambda_s = 738$ nm, $h_1 = 6$ mm, and $h_s = 1$ mm; material: CaF$_2$.

Table 1. Sellmeier coefficients for Equation (7) obtained from the Thorlab catalog[17].

Material	B_1	B_2	B_3	C_1	C_2	C_3
CaF$_2$	5.67588800E−001	4.71091400E−001	3.84847230E+000	2.52643000E−003	1.00783330E−002	1.20055600E+003
SF10	1.61625977E+000	2.59229334E−001	1.07762317E+000	1.27534559E−002	5.81983954E−002	1.16607680E+002

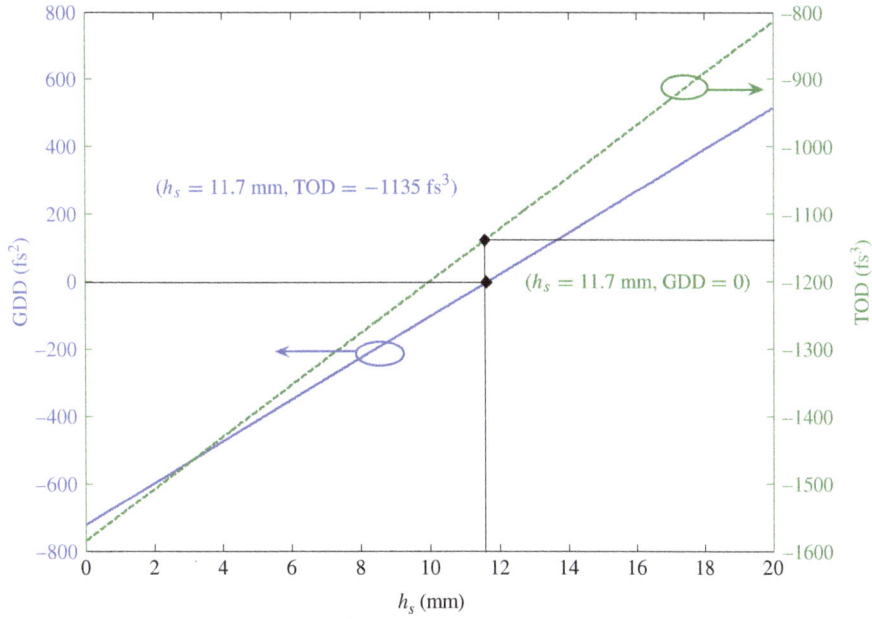

Figure 3. At the central wavelength, GDD and TOD change with h_s. Material: CaF$_2$; simulation parameters: $i_1 = 69.9°$, $\lambda_0 = 808$ nm, $\Delta\lambda = 140$ nm, $\lambda_s = 738$ nm, $d = 800$ mm, and $h_1 = 1$ mm.

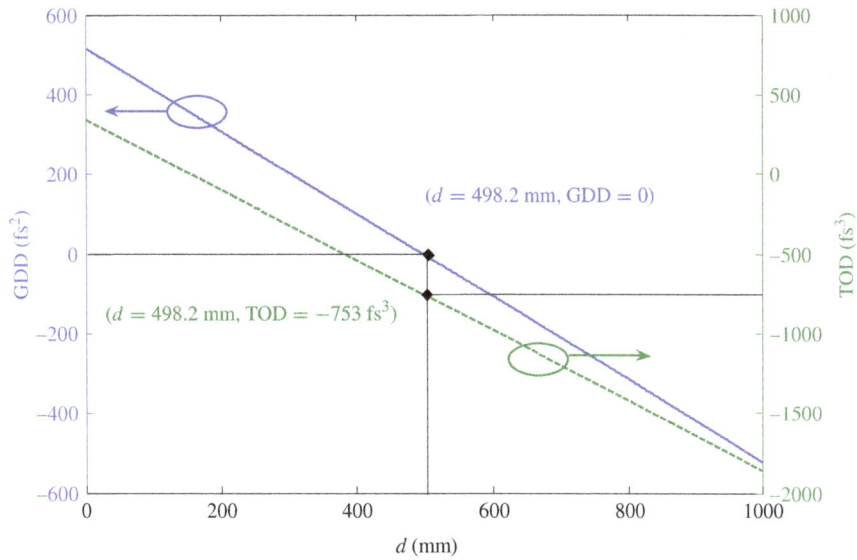

Figure 4. At the central wavelength, GDD and TOD change with d. Material: CaF$_2$; simulation parameters: $i_1 = 69.9°$, $\lambda_0 = 808$ nm, $\Delta\lambda = 140$ nm, $\lambda_s = 738$ nm, $h_1 = 6$ mm, and $h_s = 1$ mm.

3.3. TOD independent and continuous compensation

Based on the analyses in Sections 3.1 and 3.2, we can conclude that the prisms can independently compensate for the TOD of the system by selecting the appropriate values of h_1 and d or h_s and d. At the central wavelength, TOD changes with h_1 and d when GDD = 0 (Figures 5 and 6). The figures indicate group-specific values (h_1, d)

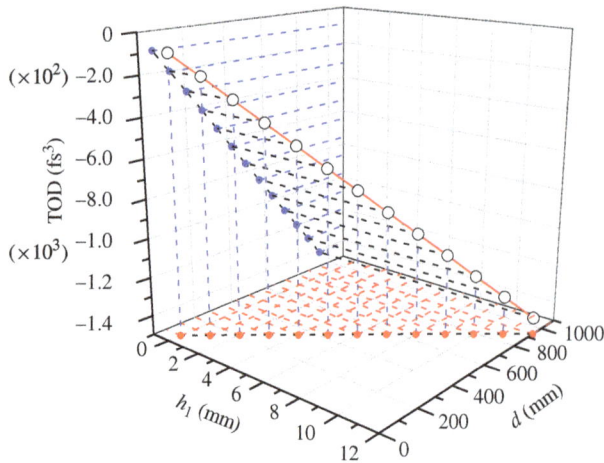

Figure 5. At the central wavelength, TOD changes with h_1 and d when GDD $= 0$. Simulation parameters: $i_1 = 69.9°$, $\lambda_0 = 808$ nm, $\Delta\lambda = 140$ nm, $\lambda_s = 738$ nm, and $h_s = 1$ mm; material: CaF_2.

Figure 6. At the central wavelength, TOD changes with h_1 and d when GDD $= 0$. Simulation parameters: $i_1 = 60.6°$, $\lambda_0 = 808$ nm, $\Delta\lambda = 140$ nm, $\lambda_s = 738$ nm, and $h_s = 1$ mm; material: SF10.

that correspond to a particular TOD compensation value. Thus, the RTOD of the laser system is compensated by adjusting distances h_1 and d simultaneously.

In the simulation shown in Figure 5, the following parameters are employed: $i_1 = 69.9°$, $\lambda_0 = 808$ nm, $\Delta\lambda = 140$ nm, $\lambda_s = 738$ nm, and $h_s = 1$ mm; material: CaF_2. In the simulation shown in Figure 6, the following parameters are employed: $i_1 = 60.6°$, $\lambda_0 = 808$ nm, $\Delta\lambda = 140$ nm, $\lambda_s = 738$ nm, and $h_s = 1$ mm; material: SF10. By comparison, we find that the CaF_2 prism pair can provide a smaller TOD, which is suitable for small RTOD compensation (Figure 5). By contrast, the SF10 prism pair can provide a larger TOD, which is suitable for large RTOD compensation (Figure 6) in the CPA laser system. The CaF_2 prisms are suitable for compensating RTOD under 10^3 fs^3, whereas the SF10 prisms are suitable for compensating RTOD under 10^4 fs^3.

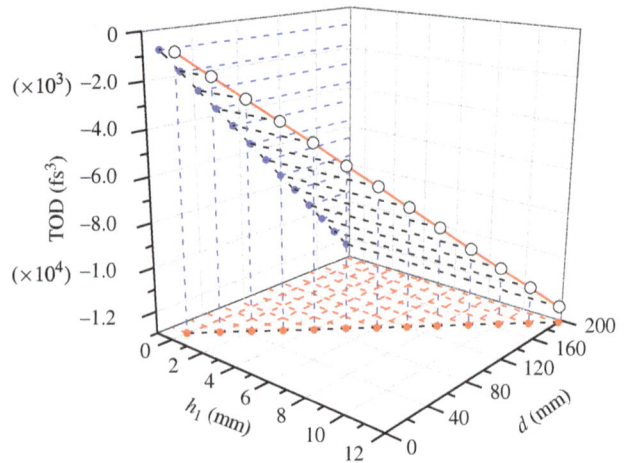

3.4. Effect of RTOD on pulse contrast ratio and pulse duration

To demonstrate the effect of RTOD on the pulse contrast in CPA lasers, we employed a 30 fs compressed pulse laser system as a example. The central wavelength is $\lambda_0 = 808$ nm and the spectral width is 140 nm. The spectral functions of the pulse exiting the compressor are assumed to have a Gaussian shape. The final output pulse contrast is calculated using a model in Ref. [2].

Figure 7 shows the effect of RTOD on the pulse contrast ratio. Even when RTOD is up to 10^4 fs^3, the output pulse contrast ratio in the 10 ps point is almost equal to the case when RTOD is zero (Figure 7(a)). However, the output pulse contrast ratios change in the range -300 to 300 fs (Figure 7(b)). For an ultrashort pulse (<50 fs), the RTOD not only affects the output contrast ratio, but also the pulse

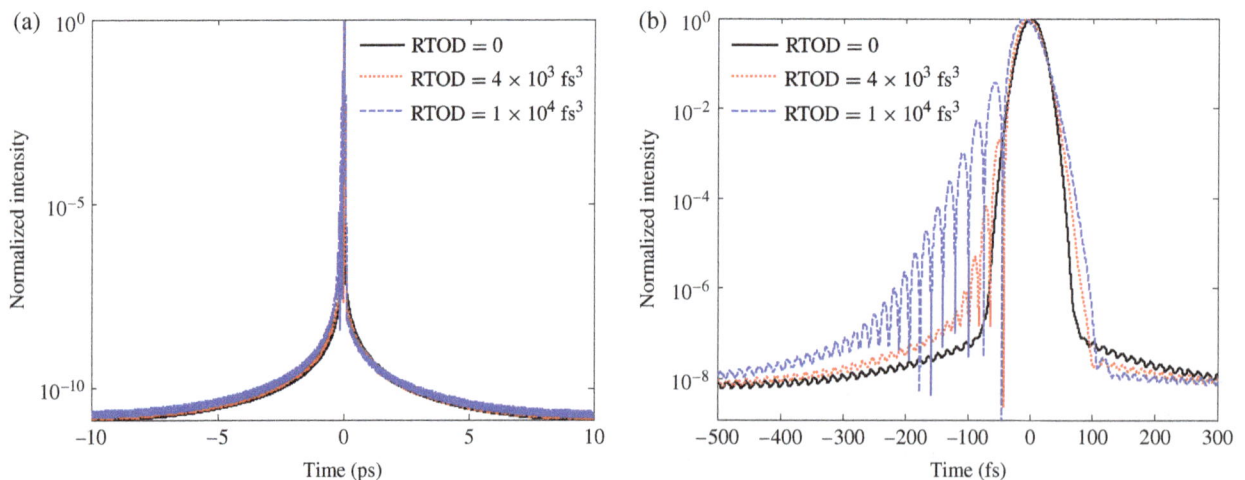

Figure 7. Effect of RTOD on the pulse contrast ratio. (b) is a magnified section of (a).

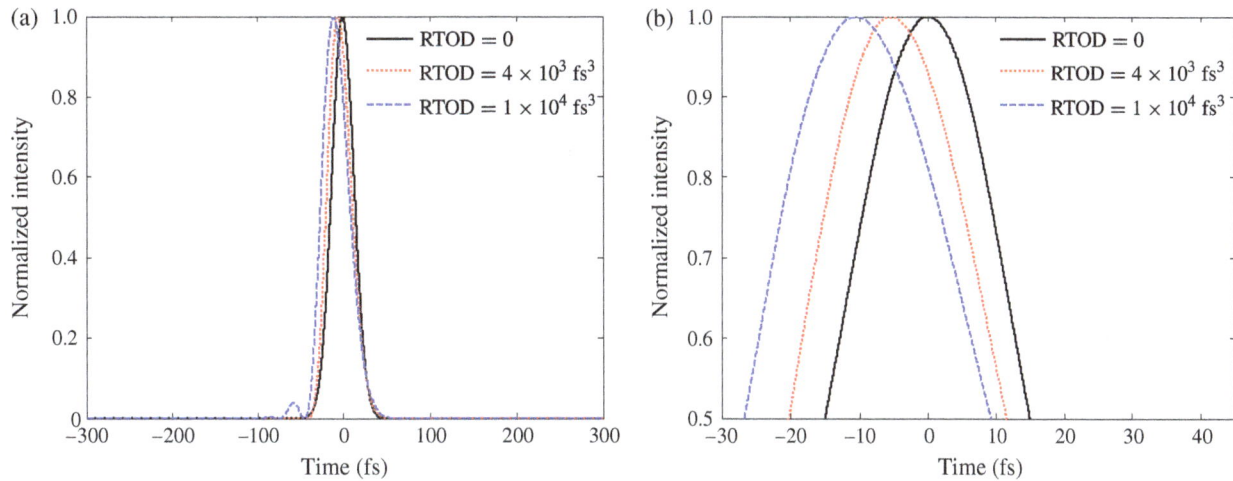

Figure 8. Effect of RTOD on pulse duration. The figure on the right is a magnified section of the figure on the left.

duration from -300 to 300 fs. When RTOD is 1×10^4 fs^3, the output pulse duration is increased from 30 to 36 fs (Figure 8).

4. Conclusion

In summary, a ray-tracing model is presented to calculate the dispersion of a pair of prisms operating at other than tip-to-tip propagation of the prisms. The pair of prisms can provide a wide range of independent and continuous TOD compensation by employing appropriate values of h_1 and d or h_s and d simultaneously, which is helpful in compensating the residual high-order dispersion of the CPA laser system. RTOD not only worsens the pulse contrast ratio, but also increases the pulse duration to the hundreds of femtosecond range for a tens of femtosecond pulse, even at small RTOD. These phenomena are helpful in understanding the effect of residual high-order dispersion on the pulse contrast ratio in ultrashort pulse laser systems.

Acknowledgements

This work is supported by the National High-Tech Committee of China and the National Nature Science Foundation of China.

References

1. D. Strickland and G. Mourou, Opt. Commun. **56**, 219 (1985).
2. G. A. Mourou, T. Tajima, and S. V. Bulanov, Rev. Mod. Phys. **78**, 309 (2006).
3. http://www.extreme-light-infrastructure.eu.
4. G. Cheriaux, P. Rousseau, F. Salin, and J. P. Chambaret, Opt. Lett. **21**, 414 (1996).
5. S. Backus, C. G. Durfee III, M. M. Murnane, and H. C. Kapteyn, Rev. Sci. Instrum. **69**, 1207 (1998).
6. E. B. Treacy, IEEE J. Quantum Electron. **QE-5**, 454 (1969).
7. Q. Yang, A. Guo, X. Xie, F. Zhang, M. Sun, Q. Gao, M. Li, and Z. Lin, Rev. Laser Eng. **36 (APLS)**, 1053 (2008).
8. Q. Yang, M. Liu, Y. Wang, X. Xie, A. Guo, and Z. Lin, Optik **123**, 1704 (2012).
9. Q. Yang, A. Guo, X. Xie, F. Zhang, M. Sun, Q. Gao, M. Li, and Z. Lin, Optik **121**, 696 (2010).
10. Q. Yang, M. Liu, Y. Wang, X. Xie, M. Sun, T. Xu, Q. Gao, A. Guo, and Z. Lin, Chin. Opt. Lett. **10**, S21401 (2012).
11. R. L. Fork, O. E. Martinez, and J. P. Gordon, Opt. Lett. **9**, 150 (1984).
12. R. E. Sherriff, J. Opt. Soc. Am. B **15**, 1224 (1998).
13. R. Zhang, D. Pang, J. Sun, Q. Wang, S. Zhang, and G. Wen, Opt. Laser Technol. **31**, 373 (1999).
14. E. Cojocaru, Appl. Opt. **42**, 6910 (2003).
15. L. Arissian and J. C. Diels, Phy. Rev. A **75**, 013814.1 (2007).
16. C. Y. R. Corral, M. R. Aguilar, and J. G. Mejia, J. Mod. Opt. **56**, 1659 (2009).
17. http://www.thorlabschina.cn/newgrouppage9.cfm?objectgroup_id=3242.

Tape casting of a YAG/Yb:YAG/YAG transparent ceramic for a broadband tunable laser

Chao Wang[1], Wenxue Li[1], Xianghui Yang[1], Dongbi Bai[1], Kangwen Yang[1], Xuewei Ba[2], Jiang Li[2], Yubai Pan[2], and Heping Zeng[1]

[1] State Key Laboratory of Precision Spectroscopy, East China Normal University, Shanghai 200062, China

[2] Key Laboratory of Transparent Opt-functional Inorganic Materials, Shanghai Institute of Ceramics, Chinese Academy of Sciences, Shanghai 200050, China

Abstract

A composite transparent YAG/Yb:YAG/YAG ceramic was prepared by a non-aqueous tape-casting method. An optical transmittance of 82% was obtained at visible wavelength and around 1100 nm. A low-threshold, broadband tunable continuous-wave (CW) laser at 1031 nm was further demonstrated from the ceramic sample, which was pumped by a 974 nm fiber-pigtailed laser diode. The threshold pump power was 0.45 W and the maximum output power was 3.2 W, corresponding to a slope efficiency of 20.4%. By inserting an SF57 prism in the laser cavity, the output wavelength could be tuned continuously from 1021 to 1058 nm.

Keywords: composite transparent ceramic; solid-state laser; broadband tunable wavelength

1. Introduction

Transparent ceramic materials have been widely used in building high-performance solid-state lasers due to their significant advantages over single-crystal materials such as large-size fabrication, reduced production cycle, high doping concentration and stable chemical properties[1–4]. Early research into potential substitutions of high-power laser gain media was focused on neodymium-doped YAG ceramics. In 1995, Ikesue demonstrated a 1.1 at.% Nd:YAG ceramic laser for the first time and observed that the continuous-wave (CW) laser had a slope efficiency of 28% at 1060 nm[5]. In 2001, Lu demonstrated laser oscillation with 1.0 at.% Nd:YAG ceramic and the output power reached 72 W[6]. One year later, a group led by Ueda increased the output power of a Nd:YAG ceramic laser to 1.46 kW, corresponding to a slope efficiency of 42%[7]. However, the gain bandwidth of the Nd^{3+} ion is merely 1 nm, limiting the output pulse width to picosecond magnitude in mode-locked operation for the purpose of generating ultrafast lasers[8–10].

Compared with Nd^{3+}, Yb^{3+} has been recognized as a more attractive alternative for high-power or ultrashort-pulse laser operation due to its simple two-manifold energy levels

and large emission cross-section, which is beneficial for the elimination of undesired effects such as up-conversion, cross relaxation and excited-state absorption[11–15]. In particular, the relatively broadband absorption cross-section of Yb ions enables high pump absorption efficiency from 941 to 970 nm, which is within the emission wavelength range of high-power, low-cost InGaAs laser diodes, thus making diode-pumped solid-state Yb-doped lasers attractive for practical high-power and high-energy applications[16–18]. Moreover, the large emission cross-section of Yb material enables high-efficiency and widely tunable laser operation[19–22].

However, with the substantial increase of output power in solid-state lasers, the heavy thermal load of the quasi-three-level structure has become a severe problem that restricts the power scaling of Yb-doped solid-state lasers. Therefore, the exploration of advanced materials with superior heat transfer performance has become a hot scientific focus. In 2006, Ikesue and Aung demonstrated the fabrication of a laser ceramic with a complicated structure without precise polishing and diffusion bonding; the advanced ceramic processing technique they presented paved the way to the design of laser gain media with flexibility and convenience[23]. Soon after, they carried out in-depth research into the advantages of composite-structure laser ceramic, which was reported in 2008. Through a comparison of the thermal distribution between a conventional ceramic

Correspondence to: W. Li, East China Normal University, State Key Laboratory of Precision Spectroscopy Room B113, Science Building, 3663 N. Zhongshan Road, Shanghai, CN 200062, China. Email: wxli@phy.ecnu.edu.cn

with uniform doping concentration and a novel ceramic with smooth gradient doping concentration in an edge-pumped laser, the experimental results showed that thermal lensing effects were efficiently suppressed in the composite ceramic due to the excellent gradient distribution of doping ions[24]. Moreover, the composite ceramic structure also allows for effective beam mode and pattern control of Q-switched or mode-locked lasers, which will promote the integration and downsizing of solid-state laser oscillators[25–27]. Recently, a diode corner-pumped high-power slab laser with an output power of more than 520 W and a slope efficiency of 32% was introduced due to the easy synthesis of ceramic active media and a simple pumping system[28].

In this paper, we report the fabrication of a multilayer composite YAG/5 at.% Yb:YAG/YAG laser ceramic by a nonaqueous tape-casting and vacuum-sintering process. The transmittance characteristics and laser performance of this composite ceramic are demonstrated. The experimental results indicate that low-threshold and continually tunable properties are achieved under high-brightness diode laser pumping.

2. Experimental setup and results

The non-aqueous tape-casting technique was adopted in our study to fabricate multilayer ceramics. To begin with, we prepared two groups of high-purity commercial powders as starting materials. An Al_2O_3 and Y_2O_3 mixture was used to produce undoped YAG slurry, while an Yb_2O_3, Al_2O_3 and Y_2O_3 mixture was used to produce 5 at.% Yb:YAG slurry. The weight of each powder was precisely controlled with a stoichiometric ratio of YAG and 5 at.% Yb:YAG, respectively. We used polyvinyl butyral, PEG-400 and butyl benzyl phthalate as the binder and the plasticizers. The binder and the plasticizers were added into the powders and the mixture was then ball-milled for 10 h. A degassing process was subsequently carried out in a vacuum device for 30 min. After that, the tape-casting process was implemented with a height between the blade and the substrate of 350 μm and a casting velocity of 100 mm min^{-1}. The eight stacked tapes obtained had a thickness of approximately 1 mm and were clipped into wafers with a diameter of 20 mm. We chose 15 undoped tapes as the bottom layer, stacked 10 doped tapes as the interlayer, and placed another 15 undoped tapes onto them as the top layer. These sandwich structure multi-tapes were laminated at 85 °C for 30 min under a pressure of 40 MPa to obtain the green samples. The samples were sintered at 1760 °C for 50 h under vacuum conditions at a pressure of less than 1×10^{-3} Pa and annealed at 1450 °C for 10 h in a stove to improve the optical quality of the ceramic samples.

Figure 1 shows photographs of YAG/Yb:YAG/YAG ceramic synthesized by the non-aqueous tape-casting method. The top image is the unannealed green body and the bottom image is the annealed laser ceramic. The color of the unannealed ceramic was green because of light absorption

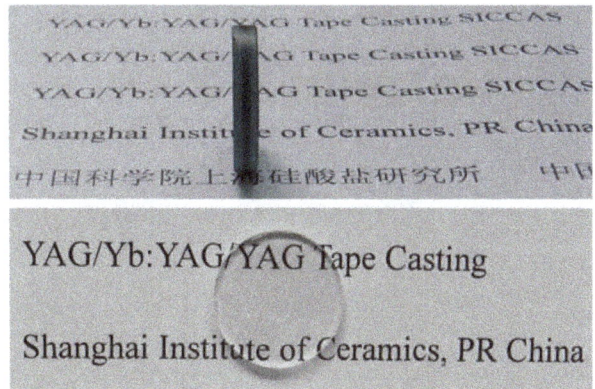

Figure 1. An image of the YAG/Yb:YAG/YAG (top: unannealed; bottom: annealed).

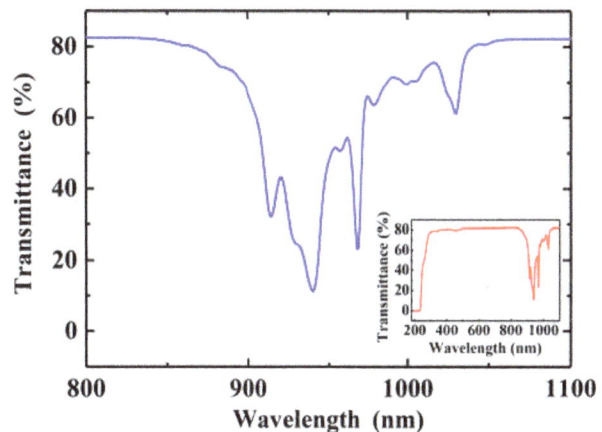

Figure 2. The transmittance curve of the YAG/Yb:YAG/YAG ceramic from 800 to 1100 nm. The inset is the transmittance curve from 200 to 1100 nm.

in the visible range; both of the two undoped layers and the doped interlayer in the unannealed ceramic are affected by lattice defects caused by oxygen vacancies. The doped interlayer exhibits a darker shade of green due to the effects of lattice defects as well as 'F' color centers caused by ion transformations from Yb^{3+} to Yb^{2+}[29, 30]. The annealed ceramic was completely transparent, which can be seen from the image. The optical transmittance of the ceramic from 200 to 1100 nm is shown in Figure 2. It can be seen that the transmittance is as high as 83% from 400 to 800 nm and attains a higher value around 1100 nm. From 800 to 1100 nm, there are three strong absorption peaks around 915, 940 and 970 nm, which correspond to the conventional broadband absorption spectrum of Yb^{3+} ions. The strongest absorption rate reaches 90% at 940 nm, which corresponds to the energy transition between two manifolds of Yb^{3+} ions.

In order to test the performance of the ceramic laser, a folded laser cavity was built, as schematically shown in Figure 3. The total cavity length was 450 mm, which was pumped by a high-brightness fiber-pigtailed diode laser with a maximum output power of 30 W at 974 nm. The core diameter of the fiber was 100 μm, and the numerical

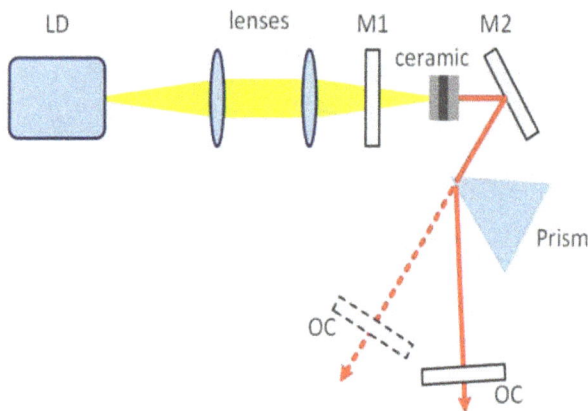

Figure 3. The experimental setup of the laser cavity.

aperture was 0.22. Two lenses were employed to build a 1:1 imaging system for efficient focusing of the pump beam into the surface of the ceramic sample with a spot size of about 100 μm. The size of the ceramic used in our experiment was 5 mm × 5 mm × 4 mm ($X \times Y \times Z$) and the doping concentration of Yb^{3+} ions was 5 at.%. The ceramic was wrapped with indium foil and fixed tightly into a three-dimensional adjustment stage. The temperature of the gain medium was controlled at 14 °C by a water-cooled copper heat sink. We adopted a stable three-mirror folded resonator. The resonator consisted of one dichroic input coupler (M1), one folding mirror (M2) with a radius of curvature of 300 mm and one output coupler (OC). The dichroic mirror M1 and folding mirror M2 were anti-reflection coated at 976 nm and high-reflection coated in a wavelength range from 1020 to 1120 nm. Three OCs with different transmissions of $T = 2\%$, $T = 5\%$, and $T = 10\%$ from 1020 to 1120 nm were used to measure the laser performance. The output CW laser was monitored by a fiber spectrometer and a power meter.

In the process of adjusting the laser, we found that the output power hardly changed when we moved the focus spot of the pumping laser diode on the ceramic by using the three-dimensional adjustment frame ($X \times Y$), which implied uniform distribution of the ceramic. Figure 4(a) shows

the CW laser spectrum of the YAG/5 at.% Yb:YAG/YAG ceramic. The output wavelength was about 1031 nm with a full width at half maximum (FWHM) of over 3 nm. The CW laser oscillation could be observed by the fiber spectrometer with a threshold incident pump power of 0.45 W when using the $T = 2\%$ OC. In order to optimize the laser performance of the ceramic laser, we monitored the output power obtained using OCs with $T = 2\%$, $T = 5\%$, and $T = 10\%$, as shown in Figure 4(b). The maximum output power of 3.2 W was obtained when using an OC with $T = 10\%$ with an incident pump power of 26.6 W, corresponding to a slope efficiency and optical conversion efficiency of 20.4% and 12.0%, respectively. Once the incident pump power exceeded 20 W, the output power tended to be unchanged with increase of pump power. This was mainly caused by the wavelength drift of the incident pump laser away from the narrow absorption bandwidth of 976 nm for the gain media due to the temperature change of the diode laser. By employing a pump laser with a stable wavelength and improving the cavity structure, a higher output power could be expected.

To further characterize the wavelength tunability of this YAG/Yb:YAG/YAG ceramic laser, we tuned the laser wavelength by inserting an SF57 Brewster prism into the resonator. It was located between the folded mirror and the OC at the Brewster angle. Under 5 W incident pump power, the Yb:YAG laser could be tuned from 1021 to 1058 nm, with a maximum output power of 178 mW at 1031 nm, as shown in Figure 5. The tuning curve was continuous and smooth, and the broadband tunability indicated the possibility of ultrashort femtosecond pulse generation by such a composite ceramic. The coating of all cavity mirrors had a reflection range starting from 1020 nm; by employing a reflection coating with a lower cut-off wavelength, laser oscillation at wavelengths shorter than 1020 nm would be expected.

3. Conclusion

In summary, a continuously tunable YAG/Yb:YAG/YAG composite ceramic laser was demonstrated in a three-mirror

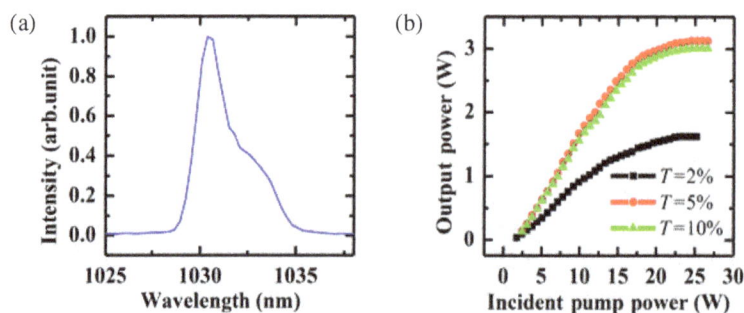

Figure 4. (a) The typical output laser spectrum and (b) the output power of the YAG/Yb:YAG/YAG multilayered ceramic laser using OCs with $T = 2\%$, $T = 5\%$, and $T = 10\%$.

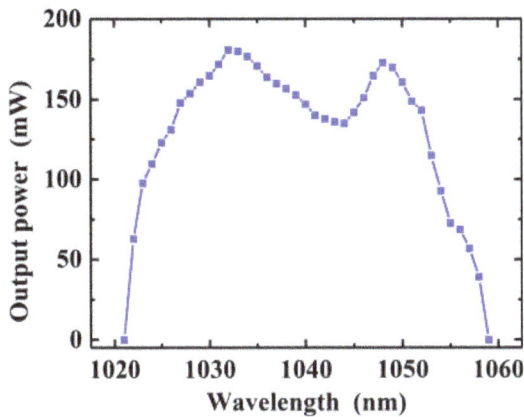

Figure 5. The tuning curve obtained with an intracavity dispersive prism for the YAG/5 at.% Yb:YAG/YAG ceramic laser.

folded cavity. The multilayer composite ceramic with a sandwich structure was fabricated by the non-aqueous tape-casting method. At 26.6 W incident pump power, a 3.2 W maximum average output power of the CW laser was achieved at 1031 nm with a slope efficiency of 20.4%. The minimum threshold incident pump power for laser oscillation was 0.45 W. By inserting an SF57 prism in the laser cavity, a smooth tunable curve from 1021 to 1058 nm was obtained. Such multilayer composite ceramics support different doping concentrations, with high transmission efficiency from visible to infrared and continuous broadband wavelength tunability. Our results demonstrated that composite ceramics fabricated by the tape-casting technique could be tailored with flexible structures and improved heat dissipation, which are desirable qualities for the next generation of high-power ultrashort solid-state lasers.

Acknowledgements

This work was supported by the Major Program of the National Natural Science Foundation of China (Grant No. 50990301) and the Project for Young Scientists Fund of the National Natural Science Foundation of China (Grant Nos. 51002172 and 51302298).

References

1. J. Sanghera, W. Kim, G. Villalobos, B. Shaw, C. Baker, J. Frantz, B. Sadowski, and I. Aggarwal, Opt. Mater. **35**, 693 (2013).
2. W. B. Liu, Y. P. Zeng, L. Wang, Y. Q. Shen, B. X. Jiang, J. Li, D. Zhang, and Y. B. Pan, Laser Phys. **22**, 1622 (2012).
3. J. Akiyama, Y. Sato, and T. Taira, Opt. Lett. **35**, 3598 (2010).
4. W. Liu, Y. Zeng, J. Li, Y. Shen, Y. Bo, N. Zong, P. Wang, Y. Xu, J. Xu, D. Cui, Q. Peng, Z. Xu, D. Zhang, and Y. Pan, J. Alloys Compd. **527**, 66 (2012).
5. A. Ikesue, T. Kinoshita, K. Kamata, and K. Yoshida, J. Am. Ceram. Soc. **78**, 1033 (1995).
6. J. Lu, J. Lu, T. Murai, K. Takaichi, T. Uematsu, K. Ueda, H. Yagi, T. Yanagitani, and A. A. Kaminskii, Jpn. J. Appl. Phys. **40**, L1277 (2001).
7. J. Lu, K. Takaichi, T. Uematsu, A. Shirakawa, M. Musha, K. Ueda, H. Yagi, T. Yanagitani, and A. A. Kaminskii, Appl. Phys. Lett. **81**, 4324 (2002).
8. T. Taira, A. Mukai, Y. Nozawa, and T. Kobayashi, Opt. Lett. **16**, 1955 (1991).
9. W. Li, Q. Hao, H. Zhai, H. Zeng, W. Lu, G. Zhao, L. Zheng, L. Su, and J. Xu, Opt. Express **15**, 2354 (2007).
10. H. Yoshioka, S. Nakamura, T. Ogawa, and S. Wada, Opt. Express **17**, 8919 (2009).
11. A. Shirakawa, K. Takaichi, H. Yagi, J. Bisson, J. Lu, M. Musha, K. Ueda, T. Yanaqitani, T. Petrov, and A. Kaminskii, Opt. Express **11**, 2911 (2003).
12. A. Shirakawa, K. Takaichi, H. Yagi, M. Tanisho, J. F. Bisson, J. Lu, K. Ueda, T. Yanagitani, and A. A. Kaminskii, Laser Phys. **14**, 1375 (2004).
13. X. J. Cheng, B. X. Jiang, L. Li, J. L. Wang, Z. G. Yang, W. B. Cheng, X. C. Shi, and Y. B. Pan, Laser Phys. **22**, 652 (2012).
14. H. Y. Zhu, C. W. Xu, J. Zhang, D. Y. Tang, D. W. Luo, and Y. M. Duan, Laser Phys. Lett. **10**, 75802 (2013).
15. C. W. Xu, D. Y. Tang, H. Y. Zhu, and J. Zhang, Laser Phys. Lett. **10**, 95702 (2013).
16. M. Tsunekane and T. Taira, Appl. Phys. Lett. **90**, 121101 (2007).
17. H. Zhou, W. Li, K. Yang, N. Lin, B. Jiang, Y. Pan, and H. Zeng, Opt. Express **20**, A489 (2012).
18. A. A. Kaminskii, S. N. Bagayev, K. Ueda, K. Takaichi, H. Yagi, and T. Yanagitani, Laser Phys. Lett. **3**, 124 (2006).
19. Q. Hao, W. Li, H. Zeng, Q. Yang, C. Dou, H. Zhou, and W. Lu, Appl. Phys. Lett. **92**, 211106 (2008).
20. Q. Hao, W. Li, H. Pan, X. Zhang, B. Jiang, Y. Pan, and H. Zeng, Opt. Express **17**, 17734 (2009).
21. N. Lin, W. Li, Y. Zhou, Y. Shi, M. Yan, K. Yang, J. Zhao, X. Yang, and H. Zeng, Laser Phys. Lett. **10**, 15103 (2013).
22. A. Pirri, M. Vannini, V. Babin, M. Niki, and G. Toci, Laser Phys. **23**, 95002 (2013).
23. A. Ikesue, Y. L. Aung, T. Taira, T. Kamimura, K. Yoshida, and G. L. Messing, Ann. Rev. Mater. Res. **36**, 397 (2006).
24. A. Ikesue and Y. L. Aung, Nat. Photon. **2**, 721 (2008).
25. J. Li, Y. Pan, Y. Zeng, W. Liu, B. Jiang, and J. Guo, Int. J. Refract. Met. Hard Mater. **39**, 44 (2013).
26. F. Tang, Y. Cao, J. Huang, H. Liu, W. Guo, and W. Wang, J. Am. Ceram. Soc. **95**, 56 (2012).
27. X. Ba, J. Li, Y. Pan, L. Jing, B. Jiang, W. Liu, H. Kou, and J. Guo, J. Rare Earths **31**, 507 (2013).
28. Q. Liu, M. Gong, F. Lu, W. Gong, and C. Li, Opt. Lett. **30**, 726 (2005).
29. P. Yang, P. Deng, and Z. Yin, J. Lumin. **97**, 51 (2002).
30. X. Xu, Z. Zhao, P. Song, G. Zhou, J. Xu, and P. Deng, J. Opt. Soc. Am. B **21**, 543 (2004).

16

Off-axis Fresnel numbers in laser systems

Yudong Yao, Junyong Zhang, Yanli Zhang, Qunyu Bi, and Jianqiang Zhu

Shanghai Institute of Optics and Fine Mechanics, Chinese Academy of Science, Shanghai 201800, China

Abstract

The physical meaning and essence of Fresnel numbers are discussed, and two definitions of these numbers for off-axis optical systems are proposed. The universal Fresnel number is found to be $N = (a^2/\lambda z) * C_1 + C_2$. The Rayleigh–Sommerfeld nonparaxial diffraction formula states that a simple analytical formula for the nonparaxial intensity distribution after a circular aperture can be obtained. Theoretical derivations and numerical calculations reveal that the first correction factor C_1 is equal to $\cos\theta$ and the second factor C_2 is a function of the incident wavefront and the shape of the diffractive aperture. Finally, some diffraction phenomena in off-axis optical systems are explained by the off-axis Fresnel number.

Keywords: correction factor; off-axis Fresnel number; off-axis optical system; Rayleigh–Sommerfeld diffraction integral

1. Introduction

Diffraction fields can be exactly solved by the Fresnel diffraction integral, but the calculation is highly complicated. The Fresnel number N allows a qualitative or semi-quantitative analysis of the diffraction field, thereby providing a clear physical picture and intuitive method. Some examples are the connections between the Fresnel number and the focal shift[1], the confirmation and application of the π phase jump in the boundary diffraction wave[2] and the Fresnel patterns in a system with a lens[3]; these examples are analyzed quantitatively by the Fresnel number.

If a plane wave is normally incident upon a circular aperture, the standard Fresnel number is defined as[4]

$$N_{st} = \frac{a^2}{\lambda z},$$

where a is the radius of the circular aperture, λ is the wavelength of the incident light, and z denotes the distance from the diffractive aperture to the axial point under the conditions $a \gg \lambda$, $z \gg a$.

The essence of the Fresnel number is the variation of the optical path in the propagation; the physical meaning is the number of Fresnel half-wave zones included in the diffractive aperture[4]. The diffracted field is analyzed by the nature of the central point using the Fresnel number. For axially diffracted fields, the diffraction pattern changes with the distance in the Fresnel diffraction region; maximal

Correspondence to: Yudong Yao, Shanghai Institute of Optics and Fine Mechanics, Chinese Academy of Science, No. 390, Qinghe Rd., Jiading, Shanghai 201800, China. Email: yaoyud1990@hotmail.com

and minimal on-axis intensities are observed with a parity change in N. In the Fraunhofer diffraction region ($N < 1$) the rays produced by diffractive apertures are superimposed at the central point with almost the same phase; hence, the diffraction pattern remains stable[3].

The Fresnel number can be expanded to the off-axis point; the number of Fresnel half-wave zones for this point can be calculated to elucidate the properties of radially diffracted fields. For example, the locations of the minimal and secondary maximal intensities in the radially diffracted field of a grating can be obtained by the half-wave zone method[5].

Light beams fall under normal or oblique incidences in optical systems, and their Fresnel number of normal incidence is given in Refs. [6–8]. The application of the Fresnel number in high power laser systems has been discussed in Ref. [9]. The physical meaning and application of complex Fresnel numbers in Gaussian beams diffracted by hard apertures have also been studied[10]. However, oblique incidence[11] and tilted optical elements[12] are often used in practice; hence, a universal Fresnel number to explain the diffraction phenomena in an off-axis optical system is necessary. In this paper an expression for the off-axis Fresnel number is provided through theoretical derivations and numerical calculations.

2. Definition of the off-axis Fresnel number

The Fresnel number is defined in two ways. First, the number physically represents the optical path difference between a wavelet from the aperture edge to the observation point and

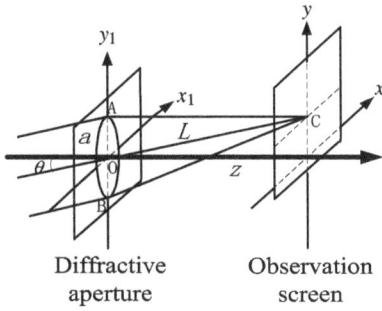

Figure 1. Diagram representing off-axis beams.

a wavelet from the aperture center to the observation point; the path difference is then divided by $\lambda/2$[3]. Second, the positions of extreme axial intensities correspond to integral Fresnel numbers[7]. Odd and even values of N yield the maximum and minimum axial intensities respectively.

By taking the direction of the center wavelet as the auxiliary axis L, the focal point C is defined as the intersection of the auxiliary axis and the observation screen (Figure 1); θ is the angle between the incident beam and the optical axis. The diffractive aperture is on the (x_1, y_1) plane and the calculated diffraction field is on the (x, y) plane, which is parallel to the (x_1, y_1) plane and has a normal distance z from it. Based on this point, the off-axis Fresnel number can be obtained using two methods. One is through the basic definition of the Fresnel number: the off-axis Fresnel number can be expressed as $N = 2\Delta/\lambda$, where Δ is the optical path difference between the wavelet of the aperture edge to the focal point and the wavelet of the aperture center to the focal point. The other is by defining an equivalent off-axis Fresnel number: the integral Fresnel numbers correspond to locations of intensity extrema on the auxiliary axis. The axial position z of the intensity extrema is obtained through the I–z curve (I is the normalized intensity on the auxiliary axis) derived from numerical calculations; z is sorted in descending order, the values of which correspond to Fresnel numbers $1, 2, \ldots, N$ (N is the number of intensity extrema) which are obtained by fitting the N–z curve. The off-axis Fresnel number is then corrected as

$$N = N_{st}C_1 + C_2, \qquad (1)$$

where C_1 is the correction factor determined by the incident angle and C_2 is the correction factor determined by the incident wavefront and the shape of the diffractive aperture.

3. Resolution of the expression for the off-axis Fresnel number

In the case of a plane wave, the off-axis Fresnel number is obtained through theoretical derivations and numerical calculations. The scope of application is discussed by comparison with the result obtained through numerical calculations.

3.1. Theoretical derivation

The coordinate system is chosen so that the incident plane wave is perpendicular to x_1, subtends an angle of $\pi + \theta/2$ with the y_1 axis in the counterclockwise direction and forms an angle of θ with the optical axis z. Taylor series expansion is carried out around L ($L = z/\cos\theta$) if L is taken as the auxiliary axis. The meridian plane comprises axes L and z; the sagittal surface consists of axes L and x_1. The optical path differences are calculated separately in these two surfaces (Figure 2).

3.1.1. Meridian plane
As shown in Figure 2(a), a plane wave is obliquely incident on the aperture, and the line DEB is the intersection line of the equiphase surface and the meridian plane. Optical paths before line DEB are equal, so only optical paths after it have to be calculated.

(1) The optical path of the upper wavelet (DAC).

$$L_{1m} = DA + AC = 2a\sin\theta + [z^2 + (z\tan\theta - a)^2]^{\frac{1}{2}}$$
$$= 2a\sin\theta + \frac{z}{\cos\theta}\left[1 + \left(\frac{a}{z/\cos\theta}\right)^2 - \left(\frac{a}{z/\sin 2\theta}\right)\right]^{\frac{1}{2}}$$
$$\approx a\sin\theta + \frac{z}{\cos\theta} + \frac{a^2\cos\theta}{2z}.$$

When $a \ll z$ the last two items in the bracket are much less than 1, so Taylor series expansion is performed around L ($L = z/\cos\theta$).

(2) The optical path of the lower wavelet (BC).

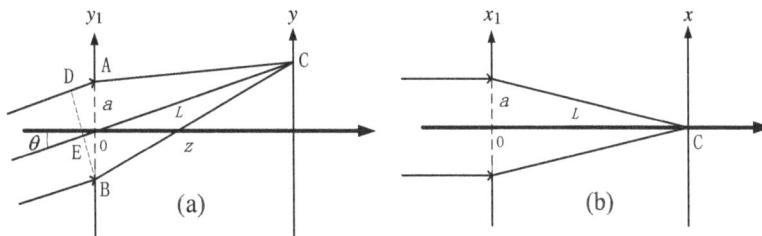

Figure 2. Beam propagation in the (a) meridian plane and (b) the sagittal surface.

$$L_{2m} = BC = [z^2 + (z\tan\theta + a)^2]^{\frac{1}{2}}$$

$$= \frac{z}{\cos\theta}\left[1 + \left(\frac{a}{z/\cos\theta}\right)^2 + \left(\frac{a}{z/\sin 2\theta}\right)\right]^{\frac{1}{2}}$$

$$\approx a\sin\theta + \frac{z}{\cos\theta} + \frac{a^2\cos\theta}{2z} \quad (a \ll z).$$

(3) The optical path of the center wavelet (EC).

$$L_{0m} = EC = a\sin\theta + \frac{z}{\cos\theta}.$$

It can be seen that the optical paths on the upper and lower edges are equal ($L_m = L_{1m} = L_{2m}$); the difference between optical paths on the edge and on the center is given as

$$\Delta_m = L_m - L_{0m} = \frac{a^2\cos\theta}{2z}.$$

3.1.2. Sagittal surface
(1) The optical paths of the wavelets from the upper and lower edges are equal (as shown in Figure 2(b)).

$$L_s = \left[\left(\frac{z}{\cos\theta}\right)^2 + a^2\right]^{\frac{1}{2}} \approx \frac{z}{\cos\theta} + \frac{a^2\cos\theta}{2z}.$$

(2) The optical path of the wavelet from the center.

$$L_{0s} = \frac{z}{\cos\theta}.$$

(3) The optical path difference.

$$\Delta_s = L_s - L_{0s} = \frac{a^2\cos\theta}{2z}.$$

The derivation implies that the optical path differences in the meridian plane and the sagittal surface are equal and are given as $\Delta = a^2\cos\theta/2z$; the off-axis Fresnel number can be expressed as

$$N = N_{st} * \cos\theta.$$

The equivalence of the optical path differences suggests that the diffraction patterns are the same in the horizontal and vertical directions.

3.2. Numerical calculation

The Rayleigh–Sommerfeld (R–S) nonparaxial diffraction integral is adopted to handle beams under oblique incidence; some approximations are introduced to avoid complicated calculations. A simple analytical formula for the nonparaxial intensity distribution is derived to reduce the computational complexity.

3.2.1. Nonparaxial intensity diffraction behind a circular aperture
The obliquely incident beam is no longer paraxial; hence, the R–S formula is used to calculate the distribution of the diffracted field[13, 14]. The coordinate system is selected as shown in Figure 1.

$$E(x, y, z) = \frac{1}{j\lambda}\int\int E_1(x_1, y_1, 0)\frac{\exp(jkR)}{R}$$

$$\times \left(1 + \frac{j}{kR}\right)\frac{z}{R}dx_1dy_1,$$

$$R = [(x - x_1)^2 + (y - y_1)^2 + z^2]^{\frac{1}{2}}, \qquad (2)$$

where a is the radius of the circular aperture, z is the distance from the diffraction screen to the observation screen, R denotes the distance from the source point $(x_1, y_1, 0)$ to the field point (x, y, z), and $E_1(x_1, y_1, 0) = \exp[jk(a + y_1)\sin\theta]$ is the amplitude distribution of the incident field.

Although the R–S formula can accurately calculate the nonparaxial scalar diffraction field and yield an accurate off-axis Fresnel number, it is difficult to obtain a universal expression for this number because of the complexity of the mathematical and numerical calculations. Thus, it is necessary to adopt an effective approximation to obtain a concise expression for the off-axis Fresnel number. R is expanded into a Taylor series around L[15]:

$$R = [(x - x_1)^2 + (y - y_1)^2 + z^2]^{\frac{1}{2}}$$

$$\approx L + \frac{x_1^2 - 2xx_1}{2L} + \frac{y_1^2 - 2yy_1}{2L}, \qquad (3)$$

where $L = \sqrt{x^2 + y^2 + z^2}$ is the distance along the auxiliary axis and $L^2 \gg (x_1^2 + y_1^2 - 2xx_1 - 2yy_1)_{\max}$.

The circular function is expressed as a series expansion with a complex Gaussian function[16]

$$circ\left(\frac{\sqrt{x_1^2 + y_1^2}}{a^2}\right) = \sum_{N=1}^{10} A_N \exp\left(-B_N\frac{x_1^2 + y_1^2}{a^2}\right),$$

where A_N and B_N are the expansion coefficients.

Using the expression into Equation (2) yields

$$E(x, y, z) = \frac{\pi\cos^2\theta}{j\lambda z}\exp\left[jk\left(\frac{z}{\cos\theta} + a\sin\theta\right)\right]$$

$$\times \sum_{N=1}^{10}\frac{A_N}{B_N/a^2 - jk\cos\theta/2z}$$

$$\times \exp\left\{\frac{(jk\cos\theta/z)^2[x^2 + (y - z\tan\theta)^2]}{4(B_N/a^2 - jk\cos\theta/2z)}\right\}. \qquad (4)$$

The coordinates of the focal point C are expressed as $x_0 = 0$, $y_0 = z\tan\theta$; substitution this expression into Equation (4) gives the complex amplitude distribution of the focal point C,

$$E(x_0, y_0, z) = \frac{\pi\cos^2\theta}{j\lambda z}\exp\left[jk\left(\frac{z}{\cos\theta} + a\sin\theta\right)\right]$$

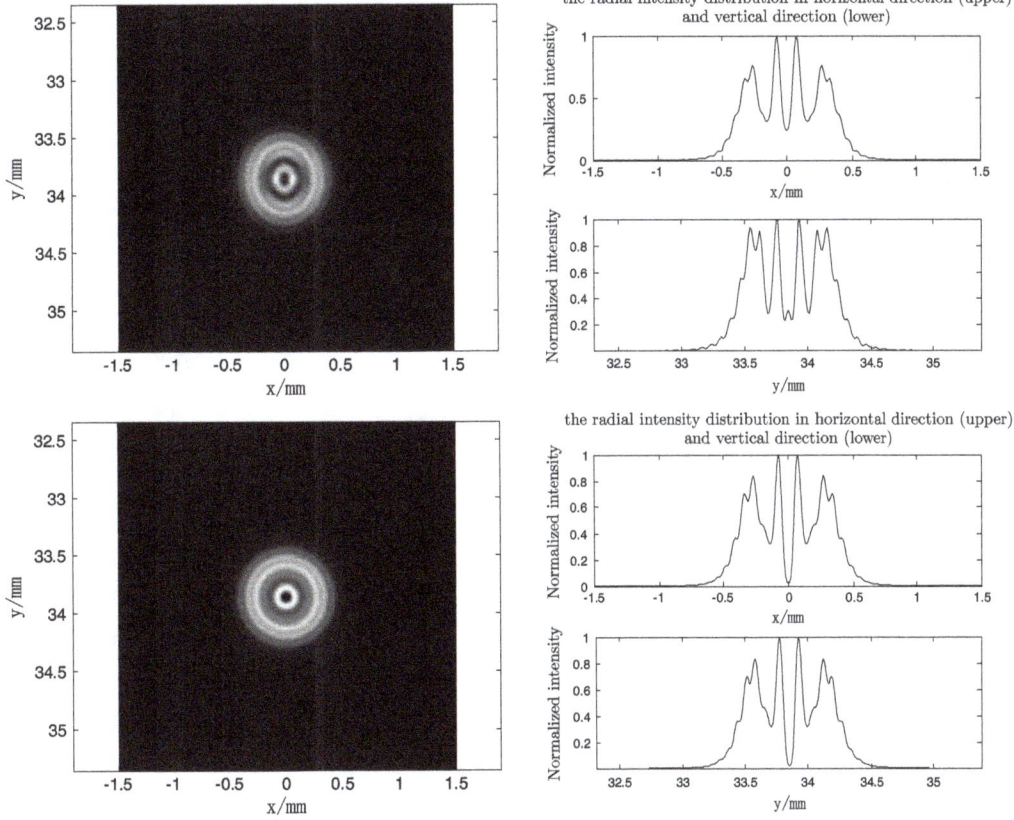

Figure 3. The diffraction patterns obtained by the R–S (upper panel) and analytical (lower panel) formulas.

$$\times \sum_{N=1}^{10} \frac{A_N}{B_N/a^2 - jk\cos\theta/2z}. \qquad (5)$$

The intensity of the focal point on the auxiliary axis is given as

$$I(x_0, y_0, z) = E(x_0, y_0, z) \cdot E^*(x_0, y_0, z). \qquad (6)$$

3.2.2. In silico simulations

(1) Comparison with the diffraction pattern.

Taking $a = 0.5$ mm, $\lambda = 632.8$ nm, $z = 93$ mm, $\theta = 20°$, the diffracted patterns (Figure 3) are simulated by using Equations (2) and (4); the analytical formula significantly reduces the numerical complexity effort because of the analytical treatment.

Both patterns are roughly similar; however, a difference still exists because the analytical formula is derived under approximation conditions. The diffraction pattern obtained by the R–S formula is not circular; the radial intensity distributions are different in the horizontal and vertical directions. The diffraction pattern obtained using the analytical formula exhibits the same intensity distribution in both directions; the patterns are coincident with those obtained in Section 3.1, in which the optical path differences are approximately equal in both directions. However, the focal point extrema using the two methods are consistent, which indicates the validity of the analytical formula in calculating the off-axis Fresnel number.

(2) Off-axis Fresnel number derived through numerical calculations.

From the definition of the equivalent off-axis Fresnel number, I–z and N–z curves are acquired from analytical calculations. Figure 4 compares the curves under normal and oblique incidence; the result implies that the position z of the intensity extrema under oblique incidence exhibits an offset to the left relative to the normally incident one. This offset is attributed to L, which is taken as the auxiliary axis by employing an axis transformation in the off-axis case (Figure 1). If the axial distance between the extreme values is defined as the diffractive period, the period becomes shorter when a beam is obliquely incident (Figure 4(a)).

The simple analytical formula is used to calculate the off-axis Fresnel number. Based on the standard Fresnel number, the off-axis Fresnel number is expressed as $N = N_{st} * C_1$, where C_1 is the correction factor. Through curve fitting, the correction factors under different incident angles are obtained from the simulation. The solid line in Figure 5 represents the curve of C_1, and the dots represent the value of $\cos\theta$; C_1 is substantially equal to $\cos\theta$.

The off-axis Fresnel number can be expressed as

$$N = N_{st} * \cos\theta. \qquad (7)$$

(a)

(b)

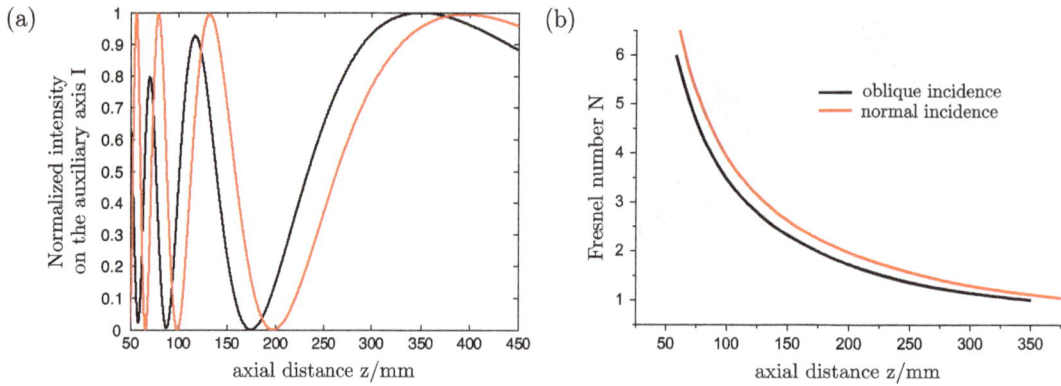

Figure 4. I–z (a) and N–z (b) curves.

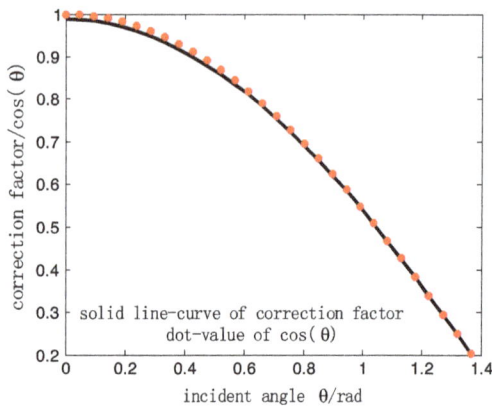

Figure 5. Curve of C_1.

Figure 6. N–z curves under different incident angles.

The same expression is obtained in Section 3.1. Comparison of the expression with the standard Fresnel number yields an equivalent propagation distance $z_{eff} = z/\cos\theta$ under oblique incidence. This parameter is the propagation distance along the auxiliary axis L. When θ approaches zero, Equation (7) becomes $N = a^2/\lambda z$, the normally incident Fresnel number.

3.3. Scope of application

The off-axis Fresnel number is established under certain approximations. Hence, the ratio of the propagation distance and radius has to meet specific conditions. The accuracy of the Fresnel number obtained by the R–S formula allows comparison of the N–z curves derived from the R–S and analytic formulas, which gives the scope of application of the expression. Figure 6 illustrates the N–z curves under different incident angles and upon setting $a = 0.5$ mm, $\lambda = 632.8$ nm.

The curves agree with each other well for $N \leqslant 8$; the expression derived in this paper is established for a relatively fine beam. When the propagation distance z is small with respect to the radius a, the R–S formula and numerical

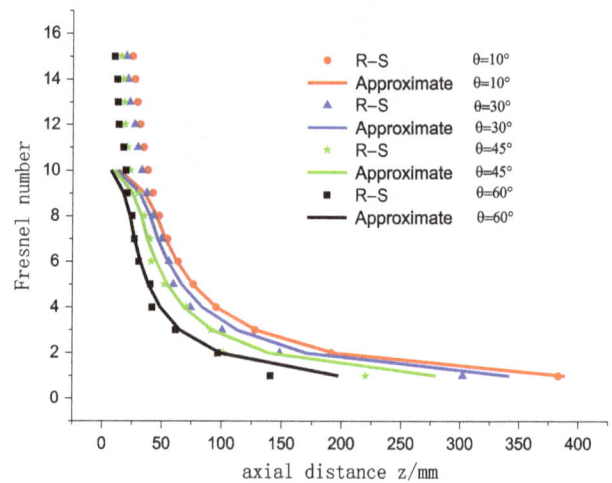

calculation are used to determine the Fresnel number. In laser systems, z is usually much larger than a; therefore the expression is generally applicable.

4. Further improvement

The correction factors C_1 and C_2 are introduced into the definition of the Fresnel number in Equation (1). The expression for C_1 is acquired as $C_1 = \cos\theta$, which implies consistency between normal and oblique incidences in the diffractive process; the two incidences can be unified by axis transformation.

To further explore the correction factor C_2 under oblique incidence, an obliquely incident spherical wave is selected with a curvature center at point S and a curvature radius of R_c (Figure 7). The optical path difference is calculated between the wavelet from the edge and the wavelet from the center to the focal point C, in both the meridian and sagittal planes; the difference remains equal:

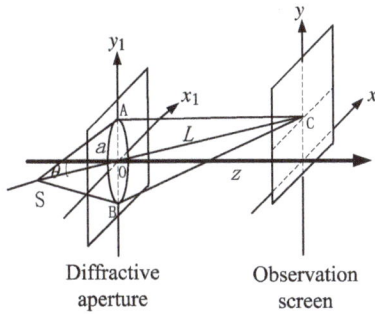

Figure 7. Diagram of an obliquely incident spherical wave.

$$\Delta = \frac{a^2 \cos\theta}{2\lambda} + \frac{a^2}{2R_c}.$$

The equivalence of the optical path differences yields the off-axis Fresnel number for an incident spherical wave:

$$N = \frac{a^2 \cos\theta}{\lambda z} + \frac{a^2}{\lambda R_c}. \qquad (8)$$

The correction factor $C_2 = a^2/\lambda R_c$ is identical to the factor under normal incidence, indicating that the factor is not related to the incident angle, but is determined by the curvature radius of the incident wavefront; this result proves the correctness of the off-axis Fresnel number. In the propagation process, the effect of the wavefront on the optical path is not related to the incident angle.

Equation (8) reveals that the Fresnel number N is related to a, z, R_c and θ, which explains the function of the off-axis optical system in the design of laser systems. Adjustment of θ by oblique incidence or tilted optical elements improves the radial intensity distribution, which reduces the harm to optical elements caused by vibration generated by diffraction.

The off-axis Fresnel number is also applicable for non-circular apertures and complex optical systems. For a non-circular aperture, C_2 is independent of the incident angle; this value is a function of the aperture shape. Hence, C_2 is similar to the correction factor under normal incidence[3]. C_2 is equal to 0.23 from our results for a square aperture (which is almost the same as the factor in Ref. [7]). In complex optical systems, the optical path difference can be solved by adopting matrix optics along the auxiliary axis L. The essence is the axis transformation in off-axis optical systems.

5. Conclusions

The off-axis Fresnel number is defined in this paper: $N = (a^2/\lambda z) * C_1 + C_2$. Theoretical analysis and numerical calculation yield $C_1 = \cos\theta$; C_2 is independent of the incident angle, but dependent on the incident wavefront and the aperture shape. In summary, the following conclusions are reached. (1) During propagation, normal and oblique incidences are consistent and can be unified by the axis transformation factor $C_1 = \cos\theta$. (2) The Fresnel number can be utilized to prove that the effects of the incident wavefront and the aperture shape are not related to the incident angle θ. (3) The Fresnel number has a significant function in the design of laser systems, such as the effect of off-axis optical systems. Change of θ by oblique incidence or tilted optical elements improves the radial intensity distribution. This improvement reduces harm to optical elements, which is caused by diffraction-generated vibrations.

The Fresnel number is no longer confined to normal incidence; the number can be used to explain more diffractive optical phenomena, such as the design of off-axis laser systems. The expression is also suitable for incident waves of arbitrary shape, non-circular diffractive apertures and complex optical systems.

Acknowledgements

This research is supported by the National Natural Science Foundation of China (Grant Nos 61205212, 11104296 and 61205210).

References

1. Y. Yu and H. Zappe, Opt. Lett. **37**, 1592 (2012).
2. S. Wang, Q. Lin, Q. Chen, and D. Zhao, Chin. J. Lasers **A26**, 347 (1999).
3. R.G. Wenzel, J.M. Telle, and J.L. Carlsten, J. Opt. Soc. Am. A **3**, 838 (1986).
4. M. Born and E. Wolf, Opt. Principles (1978).
5. L. Peng and J. Lu, Phys. Bull. **9**, 5 (1999).
6. S. Wang, E. Bernabeu, and J. Alda, Opt. Quantum Electron. **24**, 1351 (1992).
7. L. Yu, X. Xu, A. Lin, and S. Wang, Acta Opt. Sin. **20**, 1131 (2000).
8. D. Fan, Acta Opt. Sin. **3**, 319 (1983).
9. A.J. Campillo, J.E. Pearson, S.L. Shapiro, and N.J. Terrell, Appl. Phys. Lett. **23**, 85 (1973).
10. S. Wang, X. Jiang, and Q. Lin, Chin. J. Lasers **27**, 140 (2000).
11. A.M. Herkommer, H. Münz, and R. Reichle, Int. Soc. Opt. Proc. SPIE **8170**, 81700B (2011).
12. D.F. Gardner, B. Zhang, and M.D. Seaberg, Opt. Express **20**, 19050 (2012).
13. C.J.R. Sheppard and M. Hrynevych, J. Opt. Soc. Am. A **9**, 274 (1992).
14. J.E. Harvey, Am. J. Phys. **47**, 974 (1979).
15. D. Xiaojiu, G. Feng, L. Caixia, W. Fei, and H. Jigang, Acta Photon. Sin. **35**, 898 (2006).
16. Y. Qing and B. Lv, Laser Technol. **26**, 174 (2002).

17

Review of the current status of fast ignition research at the IAPCM

Hong-bo Cai[1,2], Si-zhong Wu[1], Jun-feng Wu[1], Mo Chen[1], Hua Zhang[1], Min-qing He[1], Li-hua Cao[1,2], Cang-tao Zhou[1,2], Shao-ping Zhu[1], and Xian-tu He[1,2]

[1]Institute of Applied Physics and Computational Mathematics, Beijing 100094, People's Republic of China
[2]Center for Applied Physics and Technology, Peking University, Beijing 100871, People's Republic of China

Abstract

We review the present status and future prospects of fast ignition (FI) research of the theoretical group at the IAPCM (Institute of Applied Physics and Computational Mathematics, Beijing) as a part of the inertial confinement fusion project. Since the approval of the FI project at the IAPCM, we have devoted our efforts to improving the integrated codes for FI and designing advanced targets together with the experimental group. Recent FI experiments [K. U. Akli et al., Phys. Rev. E 86, 065402 (2012)] showed that the petawatt laser beam energy was not efficiently converted into the compressed core because of the beam divergence of relativistic electron beams. The coupling efficiency can be improved in three ways: (1) using a cone–wire-in-shell advanced target to enhance the transport efficiency, (2) using external magnetic fields to collimate fast electrons, and (3) reducing the prepulse level of the petawatt laser beam. The integrated codes for FI, named ICFI, including a radiation hydrodynamic code, a particle-in-cell (PIC) simulation code, and a hybrid fluid–PIC code, have been developed to design this advanced target at the IAPCM. The Shenguang-II upgraded laser facility has been constructed for FI research; it consists of eight beams (in total 24 kJ/3ω, 3 ns) for implosion compression, and a heating laser beam (0.5–1 kJ, 3–5 ps) for generating the relativistic electron beam. A fully integrated FI experiment is scheduled for the 2014 project.

Keywords: fast ignition; integrated simulation codes

1. Introduction

Fast ignition (FI)[1–3] is an ignition scheme for inertial confinement fusion (ICF). In this two-step ICF scheme, the fuel pellet is first compressed to a high density $\sim300\,\mathrm{g\,cm^{-3}}$, and then this highly compressed deuterium–tritium pellet (10–20 µm at its core) is ignited by a ~10 ps, 10 kJ intense flux of MeV electrons (or ions). These high-energy particles are generated by the absorption of an ultra-intense petawatt laser, at the edge of the pellet, which is usually ~50 µm away from the dense core. Previous studies[4] showed that electrons with 1–3 MeV energy are optimal for FI. In order to accelerate electrons to these energies, the ultra-intense petawatt laser needs an intensity of $>10^{19}\,\mathrm{W\,cm^{-2}}$, a pulse duration of <20 ps, a spot size of <20 µm, and a laser contrast of $>10^8$. This two-step process, which separates fuel assembly and ignition, could relax the driver requirements and promise high gains[2].

Correspondence to: Hong-bo Cai, Institute of Applied Physics and Computational Mathematics, Beijing 100094, People's Republic of China. Email: cai_hongbo@iapcm.ac.cn

Investigations of the FI scheme are challenging and involve extremely high-energy-density physics, including, for example, ultra-intense lasers with intensities larger than 10^{19} W cm^{-2}, pressures in excess of 1 Gbar, magnetic fields in excess of 100 MG, and electric fields in excess of 10^{12} V m^{-1}. Addressing this complexity and the scale of the physical issue inherently requires high-energy and high-power laser facilities that are now becoming available, as well as the most advanced theory and computer simulation capability available[5]. Nowadays, a number of laser facilities are currently performing experiments for FI, including, for example, the OMEGA facility at the laboratory for laser energetic (LLE) at the University of Rochester, the FIREX-I at the Institute of Laser Engineering (ILE) in Osaka University, the Shenguang-II upgraded laser facility in Shanghai, the Trident laser facility at the Los Alamos National Laboratory, the Petal facility in France, and the Vulcan facility at the Rutherford Appleton Laboratory in the UK. New facilities are in different stages of design or construction, such as the NIF-ARC at the Lawrence Livermore National Laboratory (LLNL) and the FI Realization Experiment in Japan. Many experiments have been carried out to study the generation of high-energy particles using ultra-intense laser pulses and the feasibility of FI with these high-energy particles[6–8].

Research has shown that most potential sources of offset in fuel compression are also potential problems in standard conventional hot-spot shells and are well enough controlled that they are not a problem[9]. Now most FI research works is focused on the generation, transport, and deposition of high-energy particles[10–19]. There are usually three important stages in the interaction of an ultra-intense petawatt laser pulse with a solid target. In the first stage, the solid target is heated by the laser prepulse, resulting in the formation of the preplasma. Usually, the scale of this preplasma can be very large (~ 100 μm) depending on the prepulse level. After the formation of the preplasma, the subsequent main laser pulse propagates inside the underdense plasma, and the nonlinear interaction of the laser pulse and the relativistic plasma plays an important role in the laser propagation and relativistic electron beam (REB) generation. In this stage, the crucial point is how to produce a sufficient number of well-collimated relativistic electrons with an appropriate energy range. In the second stage, the relativistic electrons escape from the laser fields when they enter into the overdense plasma. The relativistic electrons transport from the critical density to the core densities ($n_e \sim 10^{26}$ cm^{-3}). During the transport, the electrons lose their energy before reaching the core due to the collisional effect and collective stopping. The crucial point in this stage is how to control the beam filamentation and collimate the REBs. In the third stage, the REBs reach the compressed fuel and deposit their energy in a small volume by collisions, and finally ignite the fuel.

We have been developing our integrated simulation codes to investigate FI sciences. The integrated codes include a radiation hydrodynamic code, a particle-in-cell (PIC) simulation code, and a hybrid PIC code. With the help of these codes, we designed an advanced cone–wire-in-shell target for FI, and undertook the systematic research for the generation, transport, and deposition of the fast electrons.

2. Integrated fast ignition simulation codes

While the basic idea for FI is straightforward, the realization of this technique, particularly for the injection of the petawatt ultra-intense laser beam, relies upon a systematic understanding of many complex physical issues, ranging from nonlinear laser-plasma interactions to REB transport through dense plasmas[20]. In order to deal adequately with the complexities of the FI physics, simulations should play an important role in understanding the physics and exciting new ideas for FI. However, the physical processes in FI are very complex. Since they have different time scales, large spatial ranges and multi-physics, it is impossible to simulate all the FI processes within one code. Recently, we have developed an 'Integrated Code for Fast Ignition'

at the IAPCM (ICFI), which consists of a 2D3V/3D3V (two/three dimensions in space and three dimensions in velocity) relativistic collisional PIC code ASCENT, two-dimensional radiation hydrodynamic codes LARED-S[21] and XB2D, hybrid fluid–PIC codes EBT2D&3D[22, 23] and HFPIC[24].

The typical scenario in the ICFI is summarized as follows. LARED-S, which is multi-dimensional, massively parallel and Eulerian-mesh-based radiation-hydrodynamic code, computes the implosion dynamics of a cone-in-shell FI target. ASCENT, which obtains the density profile at the maximum compression time from LARED-S, simulates the interaction of an ultra-intense petawatt laser pulse with this obtained-profile plasma target. The transport modeling is done with the hybrid fluid–PIC codes EBT2D&3D and HFPIC. The hybrid codes receive the distribution functions of the REBs produced by the ultra-intense laser pulse from ASCENT, and then models the PIC-simulated electron transport through a collisional plasma. Finally, the fuel heating and thermonuclear burn are simulated again with the radiation-hydrodynamic code LARED-S. A schematic diagram of the ICFI is presented in Figure 1. Besides the ICFI codes, we also developed a relativistic Fokker–Planck code which helps to model the energy deposition together with the hybrid–PIC codes. In the following sections, we will introduce these codes and the physical research.

3. 2D radiation hydrodynamics simulation

The implosion dynamics of a cone-in-shell FI target is simulated by LARED-S[21], which is a two-dimensional radiation hydrodynamics code including laser ray tracing, multi-group radiation diffusion, electron and ion thermal conduction, atomic physics, nuclear reaction, alpha particle transport, and the quotidian equations of state.

In order to increase the fast electron transport efficiency, a variant target design containing a reentrant cone coupled wire is introduced to guide the fast electrons. The wire attached to the tip of the gold cone is surrounded by low-Z materials, providing a high-resistivity-core–low-cladding target structure. Megagauss (MG) magnetic fields can be produced around the wire, which can collimate fast electrons and enhance the transport efficiency. But during the implosion, the nonthermal Au M-band lines in the emissions from the Au hohlraum are absorbed by the high-Z cone and wire. A dense vapor layer is generated on their surface, which is Raleigh–Taylor unstable against the pressures exerted by the lower-density gas escaping the collapsing shell and mixes with it[9]. In order to avoid the expansion of the Au cone, we use a plastic (CH) coating on the gold cone and wire surfaces. The implosions of the cone–wire-in-shell with and without the CH coating are investigated by LARED-S. As shown in Figure 2, the cone–wire structure survives at the maximal compression time.

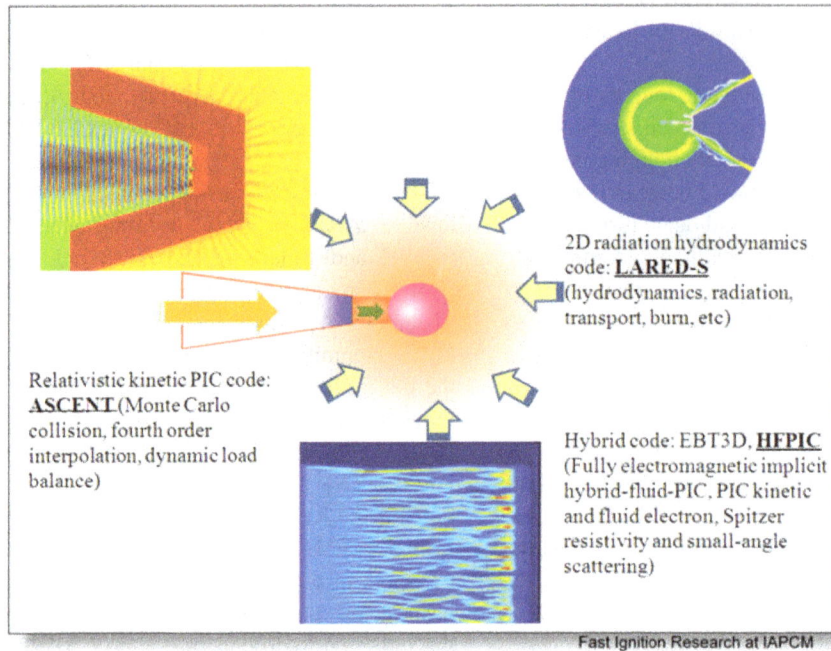

Figure 1. (Color online) A schematic diagram of the ICFI at the IAPCM.

However, the high-Z material is mixed and diffused into the core of the compressed core in the case without the CH coating. In contrast, in the case with the CH coating, the interface of the high-Z and low-Z material is very clear and the high-Z material does not enter into the compressed core. This means the CH coating works well in preventing the expansion of the high-Z plasma.

Another issue is that an ultraintense laser pulse is usually accompanied with a long prepulse of ASE (amplified spontaneous emission). The ASE pulses have a typical pulse of 0.5–8 ns and a contrast of 10^{-9}–10^{-6}. This ASE pulse usually creates a plasma corona. Here, the initial PIC simulation parameters of the preplasma of the gold cone target after the irradiation of the ASE pulse is also studied with the radiation hydrodynamic code. We will discuss it in the following section.

4. Laser–plasma interaction (LPI) modeling with the PIC code

Our LPI simulations are performed with the relativistic PIC code ASCENT. In many settings in plasma physics, a first-principle PIC simulation is still the most suitable tool for understanding the physics. However, by their nature, large-scale explicit PIC simulations require huge computing resources. Here, our code has recently been reconstructed on an infrastructure named JASMIN, which is aimed at structured or adaptive block-structured meshes for numerical simulations of complex systems on parallel computers[25]. Using JASMIN, our code can efficiently use thousands and tens of thousands of CPU processors.

In the previous section, we have shown that the cone–wire structure survives at the maximal compression time during the implosion. The collimation effect of quasistatic magnetic fields on the transport of REBs in these specially engineered structure targets was investigated by our PIC simulations[26–29]. The PIC simulations showed that in the high-resistivity-core–low-resistivity-cladding structure targets, the magnetic fields on the interfaces are generated by the gradients of the resistivity and the REB current, while in the low-density-core–high-density-cladding structure targets, the magnetic fields are generated by the nonparallel density gradients and the fast-electron current near the interface[27]. The generated quasistatic magnetic fields are as high as several or several tens of MG, which are strong enough to collimate fast electrons.

Such a strong imposed magnetic field has also been demonstrated with a capacitor–coil target and an ns–kJ laser without compression[30]. It seems that the imposed magnetic field raises great hopes by controlling the electron divergence and lowering the laser energy required to obtain the fuel burning for FI. In our work[31], the interaction of an ultraintense laser pulse with overdense plasma with different imposed magnetic fields is investigated by ASCENT. It has been shown that, in the case without imposed magnetic field, the fast electrons, only carrying about 7% of the total input laser energy, can transport 42 microns from the interaction region to the boundary. In comparison, in the case with $B_0 > 3$ MG, the fast electrons, carrying about 20% of the total input laser energy, can transport to the boundary. However, it was also found that the divergence angle increases with increase of the laser intensity. Therefore, an imposed

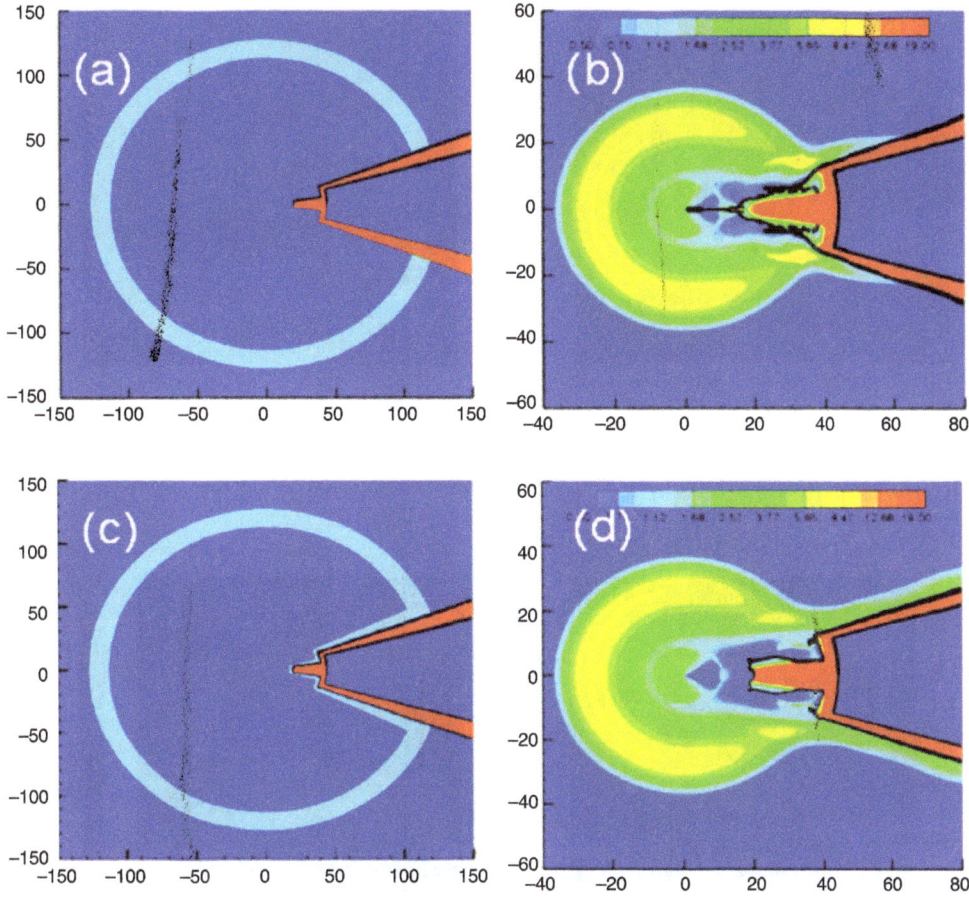

Figure 2. (Color online) Mass density contours of the cone–wire-in-shell target in LASER-S simulations. (a) and (b) are without the CH coating, (c) and (d) are with the CH coating. Black lines represent the contact surfaces between gold and the low-Z materials.

magnetic field with $B_0 = 3$–30 MG may be a suitable value for collimating the fast electrons. Furthermore, we also studied the effects of the background plasma temperature on the current filamentation instability and the growth of the fine-structure magnetic fields. We found a new way to suppress the current filamentation instability and its detrimental effects on the fast-electron beam divergence.

Studies have shown that the nanosecond ASE prepulse usually produces a large-scale preplasma with size 30–100 μm. The large-scale preplasma plays an important role in the generation of the fast electrons. Studies[4] have also shown that fast electrons with energy 1–3 MeV are optimal for heating the dense core. Our simulations[32, 35] show that, in the large-scale preplasma, stochastic heating can accelerate electrons to very high energies, carrying a significant fraction of input laser energy and decreasing the laser coupling efficiency to the compressed core. However, few simulations have been performed on how the preplasma inside the cone can affect the generation of the fast electrons for FI. MacPhee[36] and Akli[37] have simulated a cone with preplasma inside, and have shown that the preplasma significantly reduces the forward-going component of 2–4 MeV electrons. Both of the simulation parameters are fitted

to the Titan laser facility, which delivers (150 ± 10) J in 0.7 ± 0.2 ps. However, in the FI scheme, the heating laser beam has to deliver >200 kJ in 20 ps. On the other hand, we find that a specially designed target, such as a double cone, can reduce the detrimental effect of the prepulse (as shown as in Figure 3). Therefore, how the prepulse of the heating laser beam affects the forward-going fast electrons in the FI scenario still needs further research[31].

In our simulations, this physics is modeled in two parts: (a) the radiation-hydrodynamic code XB2D is used to calculate the distribution of the preformed plasma created by the laser prepulse inside the inner cone, and the obtained electron and ion density profiles are exported to the PIC code ASCENT; (b) ASCENT is used to study the interaction of an ultra-intense ultra-short laser pulse with the preformed plasma from the XB2D code with a high fidelity. Figure 3(a) is a sketch of the geometry of the simulations. In our double-cone target the inner cone wall is isolated from the background plasma (corona plasma) by a vacuum gap. The width of the inner cone wing is $5\lambda_0$ and the width of the gap is $3\lambda_0$. Here, λ_0 is the wavelength of the incident laser pulse. The plasma consists of three species: electrons, protons with $m_p/m_e = 1836$ outside the cone, and heavy ions (the Au

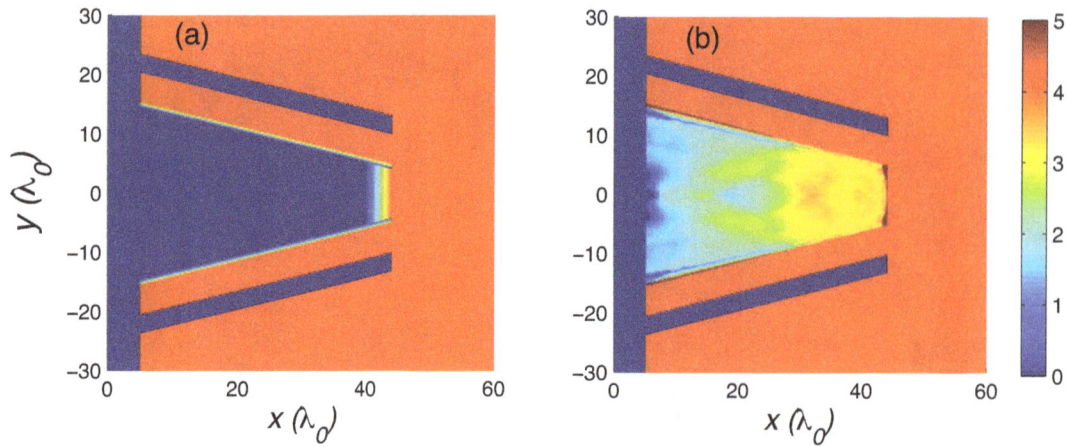

Figure 3. (Color online) Initial density profile of the double-cone target (a) without and (b) with large-scale preplasma.

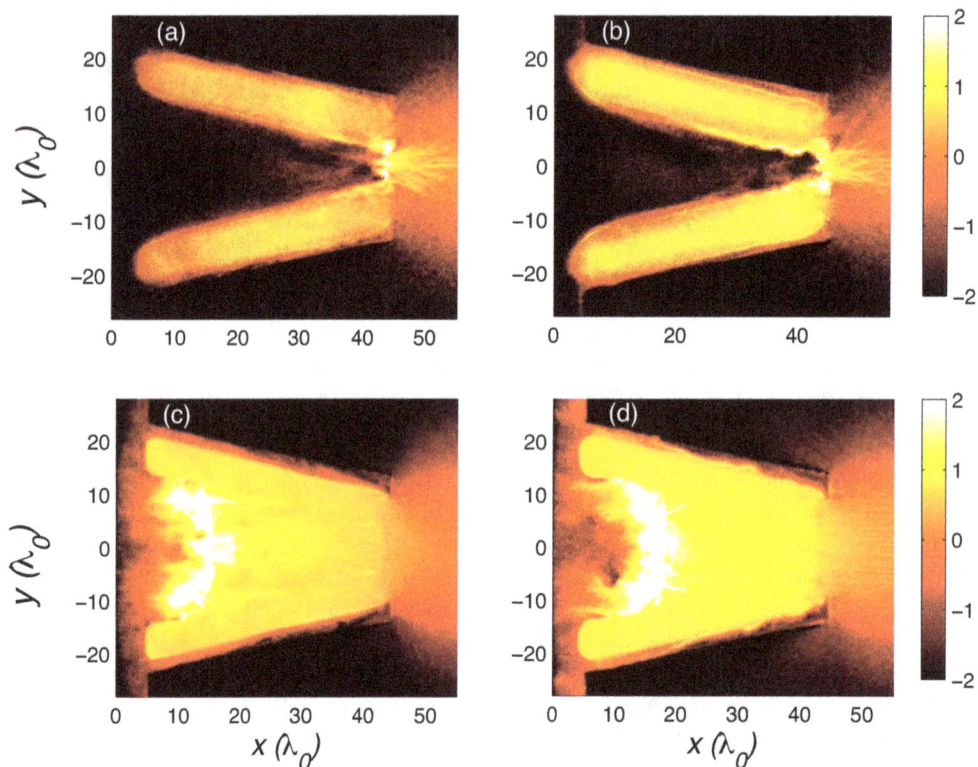

Figure 4. (Color online) The natural logarithm of the electron energy density for (a), (b) without and (c), (d) with large-scale preplasma at time (a), (c) $t = 500$ fs, and (b), (d) $t = 1000$ fs.

ion with an assumed charge state $Z_i = 40$). Both the plasma density of the gold cone and that of the hydrogen plasma are $40n_c$. The p-polarized laser pulse at a wavelength of $\lambda_0 = 1.06$ µm and with a intensity of 1.2×10^{19} W cm^{-2} irradiates the target from the left boundary.

In Figure 4, the energy density distributions of electrons with energies between $0.5 \leqslant E[MeV] \leqslant 2$ are plotted. It is clearly seen that the fast electrons are produced at different positions. In the small-scale preplasma case, the

fast electrons are mainly produced at the cone tip, as shown in Figure 4(a). Alternatively, in the large-scale preplasma case, the fast electrons are mainly generated at the critical surface which is tens of microns away from the cone tip, as shown in Figure 4(c). In the large-scale preplasma case, the fast electrons that propagate inside the preplasma and cone target experience instabilities and collisions, resulting in energy loss and decreasing the coupling efficiency. Furthermore, since the fast electrons are divergent, more electrons will escape from the side wall in the large-scale preplasma case. We can use a double-cone target to avoid this loss[38]. As a consequence, most of the fast electrons are still collimated and transport forward to the cone tip, as shown in Figure 4(d). We also check the forward fast-electron flux, and find that, with the help of the double-cone target, the forward electron flux in the large-scale preplasma case only decreases 2% in comparison with the small-scale preplasma case. This is not so serious compared with the simulations in Refs. [36] and [37].

5. Electron transport modeling with the hybrid PIC code

The generation and transport of fast electrons, which are important issues in FI, have been studied with various simulation methods[39, 40]. The simulations are usually broken up into two steps because of the disparate spatial and temporal scales involved in the LPI and fast-electron transport. The LPI and generation of fast electrons are simulated with our PIC code ASCENT, while the electron transport is modeled with a hybrid PIC method which permits more coarse temporal and spatial resolutions. The study of fast-electron transport in overdense matter is essential to the success of FI. Here, we developed two different hybrid PIC codes: HFPIC[41] and EBT2D&3D[22, 23].

The numerical algorithms of the hybrid code HFPIC are similar to those developed by Welch et al.[39], in which both PIC kinetic and fluid electrons are included. The scheme uses a direct implicit particle push and an implicit electro-magnetic solver, which relaxes the usual PIC restrictions on the temporal and spatial resolutions and still maintains a good energy conservation. Moreover, the Spitzer resistivity and small-angle scattering are both included, thus allowing us to investigate energetic electron transport in very-high-density plasmas. Such hybrid simulation algorithms have been very successful in modeling the REB transport in a solid or highly compressed high-temperature plasma target[41–45].

In order to study the field generation and fast-electron collimation at the material interface, a solid target consisting of two materials (Cu inside and Al outside) is considered. For the injected beam with currents greater than the Alfven limit, a return current moves in the opposite direction to establish approximately current-neutral equilibrium, as shown in Figure 5(a). A huge magnetic field of the order of tens of MG is produced as the current filamentation instability develops. The magnetic field at the interface, which is mainly due to the resistivity gradient, can collimate the fast electrons, as shown in Figure 5(b) and 5(d). The plasma is heated due to the Ohmic heating because of the return current (see Figure 5(c)), and the resistivity as well as the space-charge field is thus reduced, allowing the fast electrons to propagate further into the overdense plasmas. In order to enhance the material effects on the transport of

the fast electrons, different materials (Au, Cu, Al, and C) are considered. It is found that the use of a low-Z target material is more efficient for collimating beam electrons as well as generating higher-energy ions[45].

The numerical algorithms of the hybrid code EBT2D&3D is different[22, 23]. The background target is treated as a cold and stationary fluid, while the fast electrons are described by the Fokker–Planck equation and solved by the PIC method[40]. Collisions between fast electrons are ignored and the background particles give only drag and random angular scattering terms. The electric field is calculated by Ohms law and the magnetic field is calculated from Faraday's law. Thus, the effects of the self-generated electric and magnetic fields are considered. Recently, the BPS model for the Coulomb logarithm including the quantum and coupling effects for a wide range of plasma conditions has been supplemented in the code[46–48]. These algorithms are much simpler and enable us to do larger-scale simulations. The detailed physics will be published in the near future.

6. Energy deposition modeling with the relativistic Fokker–Planck code

Besides the ICFI codes, we developed a relativistic Fokker–Planck code to model the collisions between fast electrons and core plasma particles[49, 50]. Energy deposition of fast electrons into the core DT plasmas is one of the crucial processes in the FI scenario which will directly influence the overall coupling efficiency and the thermal property of the hot spot formed at the core edge. In the context of energy deposition, the core plasma density is extremely high, reaching up to nearly 10^{26} cm^{-3} (for mass density 300 g cm^{-3}), and the energy loss mechanisms dominated by plasma collective behavior such as micro-instabilities during beam–plasma interaction, self-field generation and heating, etc., will be greatly or totally suppressed. Therefore, the fast electrons (ignitor) lose their energy mainly due to collisional process. Since the collision time between a fast electron and a core plasma particle is very fast, typically on the order of 10^{-16} s, this means that a kinetic approach will be of great necessity in order to gain more knowledge during energy deposition.

Two kinds of collision type are involved in our Fokker–Planck code, namely, short-range and long-rang collisions. The former refers to the binary Coulomb collisions with impact parameter less than the Debye length; they are well described by the Fokker–Planck collision operator. The latter is associated with the collective response of core plasmas, which is often treated with the Balescu–Lenard collision term. These collision operators were simplified in an exact way. It should be noted that the authors of Ref. [51] had also tried to study energy deposition with a relativistic Fokker–Planck model, in which a simple and linearized collision term is obtained under a cold plasma approximation. In

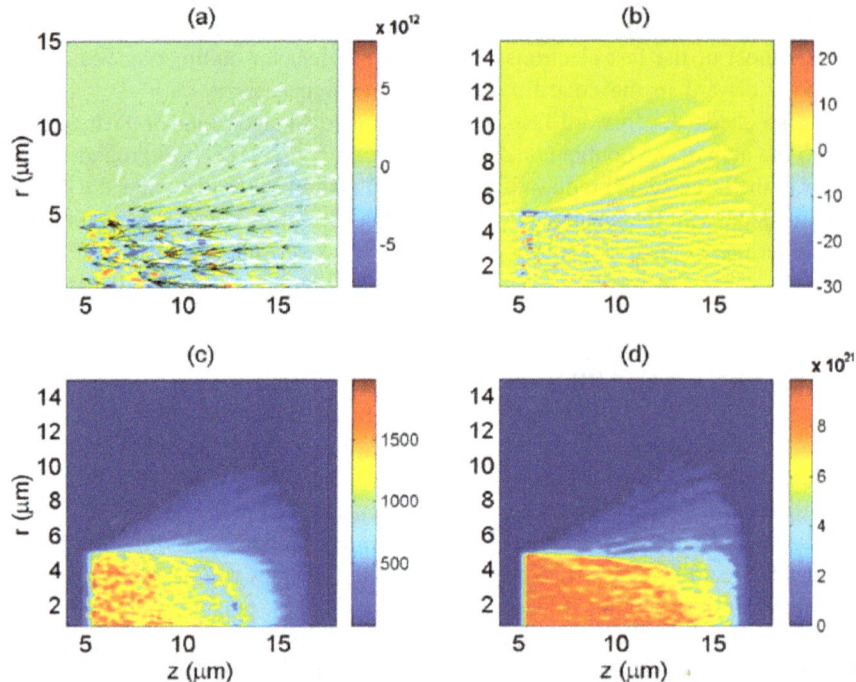

Figure 5. (Color online) Snapshots of (a) the longitudinal current density Jz(z, r) (in A cm^{-2}), (b) the azimuthal magnetic B_θ (in MG), (c) the temperature T_e (in eV) of the target plasma electrons, and (d) the beam density nb (in cm^{-3}) (from Zhou *et al.* 2008[42]).

our model, the exact collision term was converted from an initial complex integrodifferential form to a differential expression with analogous Rosenbluth potential functions. The explicit expansion of potential functions in terms of spherical harmonics allowed us to rewrite the collision operator in a compact form which contains only simple integrations and differentiations, readily suitable for rapid numerical evaluation as well as analytic work. The corresponding Fokker–Planck code is developed and benchmarked physically. Several numerical techniques are developed to deal with the involved complexity of the relativistic Fokker–Planck model. The collisional coefficients is calculated with a multi-dimensional integration algorithm with high resolution. An implicit finite-volume method is adopted to recover the conservative properties of the kinetic equation and a nonlinear flux limiter is introduced to guarantee a non-negative distribution function.

Preliminary numerical studies on fast-electron energy deposition is made with our relativistic Fokker–Planck code. In the context of energy deposition, a collision generally has two effects on the incident fast electrons. On the one hand, it will cause a slowing down, leading to energy loss. On the other hand, it will scatter electrons to deviate from their initial incident direction, resulting in beam blooming. Two quantities are very important in evaluating the energy deposition profile, and they have been extensively studied by many other authors[52]: the continuous path length before they are stopped (continuous range) and the depth it can reach along the initial direction (penetration depth). They

can be obtained straightforwardly with Fokker–Planck code. Details can be found in Ref. [50].

7. Discussion and summary

FI, which is considered to provide an alternative way of achieving ICF ignition with a considerably smaller laser energy, is being studied by many groups worldwide using ultra-intense short-pulse lasers. In the last ten years, it has opened new and very promising perspectives for relativistic laser–plasma interaction, and has made significant progress in understanding and improving ignition physics. However, there are still many issues associated with FI, such as inefficient laser–target energy coupling. It should be noted that the present low laser coupling efficiency is obtained from experiments on petawatt laser facilities with output energy smaller than 1 kJ and a pulse duration smaller than 1 ps. Actually, in FI, the laser beams have to deliver > 200 kJ in 20 ps. Since the laser energy and pulse duration play an important role in the generation of fast electrons, it is still hard to extend the physics from these hundreds-of-Joule laser facilities to a thousand times bigger one (200 kJ). Further efforts associated with this issue will be carried on in the future.

In the IAPCM, significant progress in code developing and FI physics understanding has been achieved since the FI project was formally started in 2009. Reliable PIC code, hybrid PIC code, and radiation hydrodynamic code have been developed and checked. The physics of pellet compression, and the production, transportation, and deposition of fast electrons in overdense plasmas have been studied with our codes. In particular, a number of ideas to control the beam divergence of the fast electrons have been described with our PIC code and hybrid PIC codes. Furthermore, the FI target design has been carried out with our hydrodynamic code and demonstrated in the experiments on Shenguang series laser facilities. As the next step, the coupling efficiency of FI will be investigated with our integrated simulation codes and the Shenguang-II upgraded laser facility.

Acknowledgements

The authors gratefully acknowledge C. Y. Zheng, Z. J. Liu for fruitful discussions. This work was supported by the National Natural Science Foundation of China (Grant Nos. 11275028, 11105016 and 91230205).

References

1. M. Tabak, J. Hammer, M. E. Glinsky, W. L. Kruer, S. C. Wilks, J. Woodworth, E. M. Campbell, M. D. Perry, and R. J. Mason, Phys. Plasmas **1**, 1626 (1994).

2. S. Atzeni, A. Schiavi, J. J. Honrubia, X. Ribeyre, G. Schurtz, Ph. Nicolai, M. Olazabal-Loume, C. Bellei, R. G. Evans, and J. R. Davies, Phys. Plasmas **15**, 056311 (2008).

3. J. J. Honrubia and J. Meyer-ter-Vehn, Plasma Phys. Control. Fusion **51**, 014008 (2009).

4. M. H. Key, J. C. Adam, K. U. Akli, M. Borghesi, M. H. Chen, R. G. Evans, R. R. Freeman, H. Habara, S. P. Hatchett, J. M. Hill, A. Heron, J. A. King, R. Kodama, K. L. Lancaster, A. J. Mackinnon, P. Patel, T. Phillips, L. Romagnani, R. A. Snavely, R. Stephens, C. Stoeckl, R. Town, Y. Toyama, B. Zhang, M. Zepf, and P. A. Norreys, Phys. Plasmas **15**, 022701 (2008).

5. An assessment of the prospects for inertial fusion energy, The National Academies Press, Washington, DC (2013).

6. R. Kodama, H. Shiraga, K. Shigemori, Y. Toyama, S. Fujioka, H. Azechi, H. Fujita, H. Habara, T. Hall, Y. Izawa, T. Jitsuno, Y. Kitagawa, K. M. Krushelnick, K. L. Lancaster, K. Mima, K. Nagai, M. Nakai, H. Nishimura, T. Norimatsu, P. A. Norreys, S. Sakabe, K. A. Tanaka, A. Youssef, M. Zepf, and T. Yamanaka, Nature **418**, 933 (2002); R. Kodama, P. A. Norreys, K. Mima, A. E. Dangor, R. G. Evans, H. Fujita, Y. Kitagawa, K. Krushelnick, T. Miyakoshi, N. Miyanaga, T. Norimatsu, S. J. Rose, T. Shozaki, K. Shigemori, A. Sunahara, M. Tampo, K. A. Tanakaka, Y. Toyama, T. Yamanaka and M. Zepf Nature **412**, 798 (2001).

7. W. Theobald, A. A. Solodov, C. Stoeckl, K. S. Anderson, R. Betti, T. R. Boehly, R. S. Craxton, J. A. Delettrez, C. Dorrer, J. A. Frenje, V. Yu. Glebov, H. Habara, K. A. Tanaka, J. P. Knauer, R. Lauck, F. J. Marshall, K. L. Marshall, D. D. Meyerhofer, P. M. Nilson, P. K. Patel, H. Chen, T. C. Sangster, W. Seka, N. Sinenian, T. Ma, F. N. Beg, E. Giraldez, and R. B. Stephens, Phys. Plasmas **18**, 056305 (2011).

8. T. Ma, H. Sawada, P. K. Patel, C. D. Chen, L. Divol, D. P. Higginson, A. J. Kemp, M. H. Key, D. J. Larson, S. Le Pape, A. Link, A. G. MacPhee, H. S. McLean, Y. Ping, R. B. Stephens, S. C. Wilks, and F. N. Beg, Phys. Rev. Lett. **108**, 115004 (2012).

9. R. B. Stephens, S. P. Hatchett, M. Tabak, C. Stoeckl, H. Shiraga, S. Fujioka, M. Bonino, A. Nikroo, R. Petrasso, T. C. Sangster, J. Smith, and K. A. Tanaka, Phys. Plasmas **12**, 056312 (2005).

10. K. L. Lancaster, J. S. Green, D. S. Hey, K. U. Akli, J. R. Davies, R. J. Clarke, R. R. Freeman, H. Habara, M. H. Key, R. Kodama, K. Krushelnick, C. D. Murphy, M. Nakatsutsumi, P. Simpson, R. Stephens, C. Stoeckl, T. Yabuuchi, M. Zepf, and P. A. Norreys, Phys. Rev. Lett. **98**, 125002 (2007).

11. J. S. Green, V. M. Ovchinnikov, R. G. Evans, K. U. Akli, H. Azechi, F. N. Beg, C. Bellei, R. R. Freeman, H. Habara, R. Heathcote, M. H. Key, J. A. King, K. L. Lancaster, N. C. Lopes, T. Ma, A. J. Mackinnon, K. Markey, A. McPhee, Z. Najmudin, P. Nilson, R. Onofrei, R. Stephens, K. Takeda, K. A. Tanaka, W. Theobald, T. Tanimoto, J. Waugh, L. Van Woerkom, N. C. Woolsey, M. Zepf, J. R. Davies, and P. A. Norreys, Phys. Rev. Lett. **100**, 015003 (2008).

12. S. Kar, A. P. L. Robinson, D. C. Carroll, O. Lundh, K. Markey, P. McKenna, P. Norreys, and M. Zepf, Phys. Rev. Lett. **102**, 055001 (2009).

13. A. P. L. Robinson, M. Sherlock, and P. A. Norreys, Phys. Rev. Lett. **100**, 025002 (2008); A. P. L. Robinson and M. Sherlock Phys. Plasmas **14**, 083105 (2007).

14. B. Ramakrishna, S. Kar, A. P. L. Robinson, D. J. Adams, K. Markey, M. N. Quinn, X. H. Yuan, P. McKenna, K. L. Lancaster, J. S. Green, R. H. H. Scott, P. A. Norreys, J. Schreiber, and M. Zepf, Phys. Rev. Lett. **105**, 135001 (2010).

15. A. P. L. Robinson, M. H. Key, and M. Tabak, Phys. Rev. Lett. 125004 (2012).

16. R. H. H. Scott, C. Beaucourt, H.-P. Schlenvoigt, K. Markey, K. L. Lancaster, C. P. Ridgers, C. M. Brenner, J. Pasley, R. J. Gray, I. O. Musgrave, A. P. L. Robinson, K. Li, M. M. Notley, J. R. Davies, S. D. Baton, J. J. Santos, J.-L. Feugeas, Ph. Nicolai, G. Malka, V. T. Tikhonchuk, P. McKenna, D. Neely, S. J. Rose, and P. A. Norreys, Phys. Rev. Lett. **109**, 015001 (2012).

17. Y. Sentoku, E. D'Humierres, L. Romagnani, P. Audebert, and J. Fuchs, Phys. Rev. Lett. **107**, 135005 (2011).

18. A. Debayle, J. J. Honrubia, E. d'Humieres, and V. T. Tikhonchuk, Phys. Rev. E **82**, 036405 (2010).

19. D. J. Strozzi, M. Tabak, D. J. Larson, L. Divol, A. J. Kemp, C. Bellei, M. M. Marinak, and M. H. Key, Phys. Plasmas **19**, 072711 (2012).

20. L. Van Woerkom, K. U. Akli, T. Bartal, F. N. Beg, S. Chawla, C. D. Chen, E. Chowdhury, R. R. Freeman, D. Hey, M. H. Key, J. A. King, A. Link, T. Ma, A. J. Mackinnon, A. G. MacPhee, D. Offermann, V. Ovchinnikov, P. K. Patel, D. W. Schumacher, R. B. Stephens, and Y. Y. Tsui, Phys. Plasmas **15**, 056304 (2008).

21. W. B. Pei, Commun. Comput. Phys. **2**, 255 (2007).

22. L. H. Cao, T. Q. Chang, W. B. Pei, Z. J. Liu, M. Li, and C. Y. Zheng, Plasma Sci. Tech. **10**, 18 (2008).

23. L. H. Cao, W. B. Pei, Z. J. Liu, T. Q. Chang, B. Li, and C. Y. Zheng, Plasma Sci. Tech. **8**, 269 (2006).

24. C. T. Zhou, X. T. He, and M. Y. Yu, Appl. Phys. Lett. **92**, 071502 (2008).

25. Z. Y. Mo, A. Q. Zhang, X. L. Cao, Q. K. Liu, X. W. Xu, H. B. An, W. B. Pei, and S. P. Zhu, Front. Comput. Sci. China. **4**, 480 (2010).

26. H. B. Cai, S. P. Zhu, M. Chen, S. Z. Wu, X. T. He, and K. Mima, Phys. Rev. E **83**, 036408 (2011).

27. H. B. Cai, S. P. Zhu, X. T. He, S. Z. Wu, M. Chen, C. T. Zhou, W. Yu, and H. Nagatomo, Phys. Plasmas **18**, 023106 (2011).

28. H. B. Cai, S. P. Zhu, X. T. He, and K. Mima, EPJ Web of Conference **59**, 17017 (2013).

29. S. Z. Wu, C. T. Zhou, and S. P. Zhu, Phys. Plasmas **17**, 063103 (2010).

30. S. Fujioka, Z. Zhang, K. Ishihara, K. Shigemori, Y. Hironaka, T. Johzaki, A. Sunahare, N. Yamamoto, H. Nakashima, T. Watanabe, H. Shiraga, H. Nishimura, and H. Azechi, Sci. Rep. **3**, 1170 (2013); K. Mima Reports in 12th International Workshop on Fast Ignition of Fusion Target Napa Valley, CA, USA 2012.

31. H. B. Cai, S. P. Zhu, and X. T. He, Phys. Plasmas. **20**, 072701 (2013).

32. W. W. Wang, H. B. Cai, Q. Jia, and S. P. Zhu, Phys. Plasmas **20**, 012703 (2013).

33. Q. Jia, H. Cai, W. W. Wang, S. P. Zhu, Z. M. Sheng, and X. T. He, Phys. Plasmas. **20**, 032113 (2013).

34. S. D. Baton, M. Koenig, J. Fuchs, A. Benuzzi-Mounaix, P. Guillou, B. Loupias, T. Vinci, L. Gremillet, C. Rousseaux, M. Drouin, E. Lefebvre, F. Dorchies, C. Fourment, J.J. Santos, D. Batani, A. Morace, R. Redaelli, M. Nakatsutsumi, R. Kodama, A. Nishida, N. Ozaki, T. Norimatsu, Y. Aglitskiy, S. Atzeni, and A. Schiavi, Phys. Plasmas **15**, 042706 (2008).

35. H. B. Cai, K. Mima, A. Sunahara, T. Johzaki, H. Nagatomo, S. P. Zhu, and X. T. He, Phys. Plasmas **17**, 023106 (2010).

36. A. G. MacPhee, L. Divol, A. J. Kemp, K. U. Akli, F. N. Beg, C. D. Chen, H. Chen, D. S. Hey, R. J. Fedosejevs, R. R. Freeman, M. Henesian, M. H. Key, S. Le Pape, A. Link, T. Ma, A. J. Mackinnon, V. M. Ovchinnikov, P. K. Patel, T. W. Phillips, R. B. Stephens, M. Tabak, R. Town, Y. Y. Tsui, L. D. Van Woerkom, M. S. Wei, and S. C. Wilks, Phys. Rev. Lett. **104**, 055002 (2010).

37. K. U. Akli, C. Orban, D. Schumacher, M. Storm, M. Fatenejad, D. Lamb, and R. R. Freeman, Phys. Rev. E **86**, 065402 (2012).

38. H. B. Cai, K. Mima, W. M. Zhou, T. Jozaki, H. Nagatomo, A. Sunahara, and R. J. Mason, Phys. Rev. Lett. **102**, 245001 (2009).

39. D. R. Welch, D. V. Rose, R. E. Clark, T. C. Genoni, and T. P. Hughes, Comput. Phys. Commun. **164**, 183 (2004).

40. J. R. Davies, Phys. Rev. E **65**, 026407 (2002).

41. C. T. Zhou, X. T. He, and L. Y. Chew, Opt. Lett. **36**, 924 (2011).

42. C. T. Zhou and X. T. He, Phys. Plasmas. **15**, 123105 (2008).

43. C. T. Zhou, X. G. Wang, S. C. Ruan, S. Z. Wu, L. Y. Chew, M. Y. Yu, and X. T. He, Phys. Plasmas. **17**, 083103 (2010).

44. C. T. Zhou, X. T. He, and M. Y. Yu, Appl. Phys. Lett. **92**, 151502 (2008).

45. C. T. Zhou, T. X. Cai, W. Y. Zhang, and X. T. He, Laser Part. Beams **30**, 111 (2012).

46. L. S. Brown, D. L. Preston, and R. L. Singleton, Jr. Phys. Rep. **410**, 237 (2005).

47. L. S. Brown and R. L. Singleton, Phys. Rev. E **79**, 066407 (2009).

48. L. H. Cao, M. Chen, X. T. He, W. Yu, and M. Y. Yu, Phys. Plasmas. **19**, 044503 (2012).

49. S. Z. Wu, C. T. Zhou, S. P. Zhu, H. Zhang, and X. T. He, Phys. Plasmas **18**, 022703 (2011).

50. S. Z. Wu, H. Zhang, C. T. Zhou, S. P. Zhu, and X. T. He, EPJ Web of Conferences **59**, 05021 (2013).

51. T. Yokota, Y. Nakao, T. Johzaki, and K. Mima, Phys. Plasmas **13**, 022702 (2006).

52. A. A. Solodov and R. Betti, Phys. Plasmas **15**, 042707 (2008).

18

Irradiation uniformity at the Laser MegaJoule facility in the context of the shock ignition scheme

Mauro Temporal[1], Benoit Canaud[2], Warren J. Garbett[3], Rafael Ramis[4], and Stefan Weber[5]

[1] Centre de Mathématiques et de Leurs Applications, ENS Cachan and CNRS, 61 Av. du President Wilson, Cachan Cedex, France

[2] CEA, DIF, Arpajon Cedex, France

[3] AWE plc, Aldermaston, Reading, Berkshire, United Kingdom

[4] ETSI Aeronáuticos, Universidad Politécnica de Madrid, Madrid, Spain

[5] ELI-Beamlines, Institute of Physics, Academy of Sciences of the Czech Republic, Prague, Czech Republic

Abstract

The use of the Laser MegaJoule facility within the shock ignition scheme has been considered. In the first part of the study, one-dimensional hydrodynamic calculations were performed for an inertial confinement fusion capsule in the context of the shock ignition scheme providing the energy gain and an estimation of the increase of the peak power due to the reduction of the photon penetration expected during the high-intensity spike pulse. In the second part, we considered a Laser MegaJoule configuration consisting of 176 laser beams that have been grouped providing two different irradiation schemes. In this configuration the maximum available energy and power are 1.3 MJ and 440 TW. Optimization of the laser–capsule parameters that minimize the irradiation non-uniformity during the first few ns of the foot pulse has been performed. The calculations take into account the specific elliptical laser intensity profile provided at the Laser MegaJoule and the expected beam uncertainties. A significant improvement of the illumination uniformity provided by the polar direct drive technique has been demonstrated. Three-dimensional hydrodynamic calculations have been performed in order to analyse the magnitude of the azimuthal component of the irradiation that is neglected in two-dimensional hydrodynamic simulations.

Keywords: inertial confinement fusion; shock ignition; laser system

1. Introduction

One of the main goals of inertial confinement fusion (ICF)[1–3] concerns the ignition of the thermonuclear fusion reactions in a mixture of deuterium–tritium (DT) nuclear fuel. After the ignition phase, it is expected that propagation of a thermonuclear burn wave – dominated by the fusion reaction with larger cross section $D + T => \alpha + n + 17.6\,\text{MeV}$ – throughout the compressed fuel should generate a large energy gain $G = E_{\text{fus}}/E_{\text{in}}$ (ratio between thermonuclear fusion and the invested energy). To this aim two schemes have been proposed, namely: direct drive and indirect drive. In both cases a spherical capsule containing the DT nuclear fuel is considered. In the direct drive[4] scheme the spherical capsule is irradiated by a large number of laser beams, whilst in the indirect drive[1] scheme the laser energy is first converted into an x-ray field (confined into a high-Z casing; see *hohlraum*) that irradiates the

capsule. The energy deposited in the external capsule shell provides a series of strong shock waves that induces the capsule implosion. In the classical central ignition scheme[5] the DT fuel is accelerated to high implosion velocities (hundreds of km s^{-1}) before stagnating to produce a high-density (hundreds of g cm^{-3}) shell that confines a fraction of the DT fuel (hundreds of μg). The ignition conditions require that the central mass, called the hot-spot, is heated to high temperature (10 keV) and confined into a volume with an areal density comparable with the α-particle range (≈ 0.3 g cm^{-2}).

A crucial issue concerns the uniformity of the capsule irradiation. A successful capsule implosion requires a very uniform irradiation and capsule target; otherwise, the imploding shell suffers the growth of dangerous hydrodynamic instabilities (Richtmyer–Meshkov[6, 7] and Rayleigh–Taylor (RT)[8, 9]) and shell deformations that could even destroy the hot-spot. A way to reduce the growth of RT instability consists of compressing the capsule fuel at low implosion velocity V. This causes a detriment of the energy gain G which scales[10] as $\varphi/(V^{5/4}I^{1/4})$, where $\varphi = \rho R/(\rho R + 7)$

Correspondence to: Mauro Temporal, Centre de Mathématiques et de Leurs Applications, ENS Cachan and CNRS, 61 Av. du President Wilson, Cachan Cedex, France. Email: mauro.temporal@hotmail.com

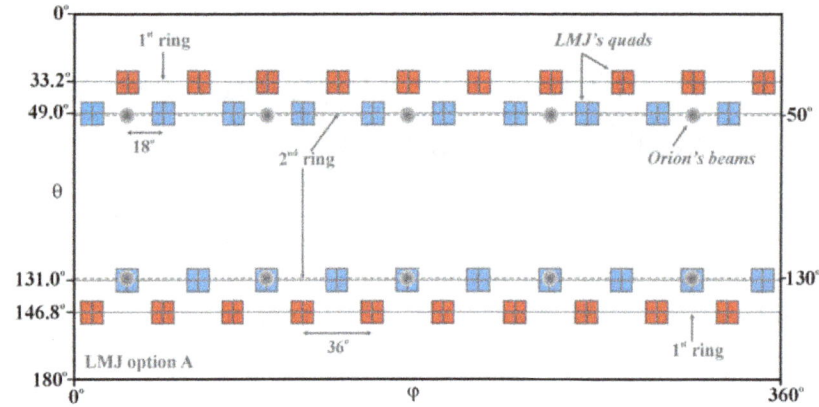

Figure 1. Angular coordinate of the 40 quads (blue and red boxes) distributed to the first and second ring of the LMJ facility. The gray circles represent the polar coordinates of the 10 long-pulse beams of the Orion facility.

is the fractional burn-up[1, 3] with ρR the fuel areal density, and I the incident laser intensity.

Alternative schemes are currently under study, such as fast ignition induced by laser accelerated electrons[11, 12], protons[13–15], or heavier ions[16, 17]. More recently, the shock ignition (SI) scheme[18, 19] has been proposed as an alternative to the classical central ignition in the context of the inertial confinement fusion scenario. In this case, the capsule is directly irradiated by the laser beams providing the compression of the DT fuel. The implosion velocity of the compressed shell is set under the ignition threshold ($V < 2$–3×10^7 cm s^{-1}) and does not allow for the generation of an efficient hot-spot. In the SI scheme, a second high-power (hundreds of TW) laser pulse irradiates the capsule and drives a strong shock wave that reaches the compressed shell providing the fuel ignition. The SI pulse must be carefully tuned in time to synchronize the strong shock wave with the compression shock rebounded from the centre after stagnation. This new scheme promises higher gain[18–21] in comparison to central ignition, and the separation between the compression and the ignition phase allows for less stringent conditions in terms of irradiation uniformity[22, 23]. Moreover, this two-step irradiation would benefit also from the *zooming* technique[24, 25] in order to increase the laser–capsule coupling efficiency. Nevertheless, caution is necessary due to the uncertainties related to laser–plasma instabilities such as stimulated Raman scattering (SRS)[26], stimulated Brillouin scattering (SBS)[27], and the two-plasmon decay (TPD)[28] expected at the high laser intensities $I\lambda^2 > 10^{15}$ W cm^{-2} μm^2[29] provided during the shock ignition pulse. These dangerous instabilities act to reduce the energy deposition efficiency and generate high-energetic ($\approx$10–40 keV) electrons[30–32]. The uncertainties concerning the laser–plasma interaction correlated to the shock ignition scheme have also motivated great interest in experimental activities[33–39]. Moreover, large laser facilities such as the National Ignition Facility (NIF)[40, 41] in the USA and the Laser MegaJoule (LMJ) facility[42, 43] in France as

well as the smaller Orion facility[44] in the UK – all of them devoted to the indirect drive scheme – could be used to test relevant aspects inherent to the shock ignition scheme.

Due to its indirect drive design, the LMJ facility does not provide a favourable laser beam configuration for direct drive irradiation. Nevertheless, this large facility is very attractive for direct drive studies because of its large available energy. In this context, this paper aims to chart a path starting from the current characteristics of the LMJ facility and exploring the potential of the shock ignition scheme. After summarizing the main characteristics of the LMJ facility in Section 2, the paper analyses two different aspects: the requirement in the maximum power on the shock ignition scheme together with possible consequences due by laser–plasma instabilities in Section 3, and the study of the uniformity of the irradiation during the foot pulse of the imprint phase in Section 4, while in Section 5 some three-dimensional (3D) aspects of the hydrodynamics of the implosion are discussed. Then, conclusions are drawn in Section 6.

2. The Laser MegaJoule configuration

The configuration of the Laser MegaJoule facility considered in this paper consists of 176 high-power laser beams. These beams are grouped in 44 identical quads, each one composed by four beams. 40 quads are distributed into the spherical experimental chamber in four axial symmetric rings, and the two hemispheres are rotated by 18°. The two rings closer to the polar axis have an angle of 33.2° and 146.8° with each having 10 quads; another 20 quads are located in the rings at 49° and 131°, as shown in Figure 1. The last four quads, not shown in the figure, will be located in two additional rings at 59.5° and 120.5°. The LMJ architecture is designed to provide a maximum laser energy (power) of 7.5 kJ (2.5 TW) for each beam. Consequently, this corresponds to a total energy (power) of 30 kJ (10 TW) per quad, and each pair of rings will provide a maximum energy (power) of 600 kJ

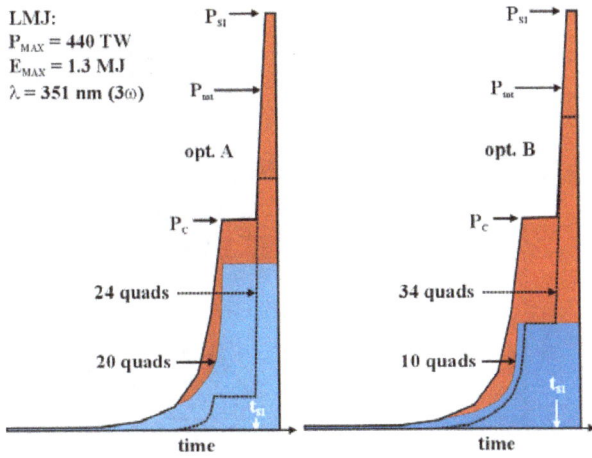

Figure 2. Sketch of the temporal power profile partition for the two LMJ options, A (left) and B (right), in the shock ignition scheme.

(200 TW). This makes the LMJ a large laser facility, able to drive a total energy of about 1.3 MJ with a maximum power of 440 TW delivered by laser beams with a wavelength of $\lambda = 351$ nm (3ω)[45]. The facility design energy is appropriate for indirect drive central ignition, but both direct drive and shock ignition schemes are expected to lower the energy threshold for ignition. Thus the total available energy at the LMJ facility largely exceeds the needs in the direct drive approach.

The polar coordinates of the quads have been optimized for the indirect drive scheme. Nevertheless, this laser beam distribution could be helpful also in the direct drive shock ignition scheme. Indeed, as already mentioned, this scheme involves two laser pulses: one for the capsule compression and a second one for the fuel ignition. Thus, there are several options in order to use the LMJ facility as a direct drive facility in the context of the SI scheme. Hereafter, we will consider two options: option A, where 20 quads of the second ring are devoted to the compression of the capsule and with the other 24 quads driving the high-power shock ignition pulse; and option B, with only 10 quads of the second ring devoted to the foot pulse while together with the remaining 34 quads contributing to the drive and the igniting pulse. The main difference between the two options concerns the role of the different quads in the partition of laser power during the low-power foot pulse, the main drive of the compression phase, and the shock ignition phase.

Of course, the choice of the irradiation configuration also has consequences on the irradiation uniformity. Details of these configurations are given in the temporal power pulse sketched in Figure 2. In option A, the whole 600 kJ (200 TW) of the laser beams of the second ring are available for the compression phase and quads of the first ring are almost entirely associated to the shock ignition pulse. In option B, only 300 kJ (100 TW) – half the laser beams of the second ring – are devoted to the foot pulse and part of the compression phase, while part of the drive and the shock

ignition pulse operate with the 34 quads located in the three rings. In both cases the maximum available power for the shock ignition pulse will be the totality of the 440 TW.

The division of tasks among the different quads also allows implementing in a natural way both the polar direct drive (PDD)[46, 47] and the *zooming* technique. The purpose of the PDD is to adapt a non-optimal configuration of laser beams in order to optimize the uniformity of the capsule irradiation. To this purpose, in the PDD technique the laser beam directions are modified in order to optimize the direct drive capsule irradiation[48–51]. The use – as a direct drive – of the quads in the second rings of the LMJ facility leads to an over-irradiation of the capsule polar regions, whilst the equatorial area is under-irradiated. Thus, applying the PDD by displacing the quad toward the equator improves the capsule illumination uniformity considerably.

It is worth noting that the Orion facility in the UK is composed of 12 beams: two laser beams provide 500 J each at 1ω (1054 nm) in a short pulse of 0.5 ps and the other ten provide a total energy of 5 kJ (3ω, $\lambda = 351$ nm) in 1–5 ns long pulses. The angular positions of these ten beams are indicated by gray circles in Figure 1. These 5 + 5 beams are located in two rings at 50° and 130° with respect to the polar angle. This beam distribution is very similar to the one provided by the second ring of the LMJ facility. This makes the Orion facility the natural choice to test relevant aspects inherent to the LMJ facility such as the laser absorption and the improvement of the irradiation uniformity promised by the PDD technique. Indeed, despite the relatively small dimensions, 5 kJ in 5 ns for 1 TW, this facility is perfectly matched to the requirements of experiments dedicated to the study of the imprint phase, where the first shock wave is driven by a low-power ($\approx$TW) foot pulse.

3. Shock ignition calculations

A relatively large direct drive capsule characterized by an initial aspect ratio $A = 3$ has been considered. This capsule is part of a family of capsules that have been recently studied[52]. This spherical capsule has an external radius of 815 μm and contains a DT fuel mass of 300 μg. The cryogenic nuclear fuel ($\rho_{DT} = 0.25$ g cm^{-3}) is surrounded by a thin (24 μm) shell of plastic ($\rho_{CH} = 1.07$ g cm^{-3}) devoted to the laser energy absorption. Detailed parametric studies have been performed showing that the self-ignition threshold in the implosion velocity is about 3×10^7 cm s^{-1}[53] and that the maximum energy gain is $G = 44$ with an incident energy of about 500 kJ and laser peak power of 230 TW[54]. In the considered LMJ design, the available energy (1.3 MJ) is above that needed by the capsule, which means we retain some energy margin against constraints mainly related to the required irradiation uniformity.

Here, the capsule has been used in the context of the shock ignition scheme, and a series of 1D numerical calculations has been performed with the hydro-radiative MULTI code[55, 56]. In this version, the MULTI code takes into account the tabulated equation of state, heat conduction (Spitzer–Harm flux limited 8%) and a 3D ray-tracing package that manage the laser energy absorption via an inverse bremsstrahlung mechanism. A laser beam characterized by a Gaussian intensity profile with a full-width at half-maximum of 1356 μm – the intensity is reduced to $1/e$ at the initial capsule radius – has been considered. The laser pulse is composed of a low-power foot pulse ($\approx$7 ns at $\approx$TW) followed by the main pulse with a maximum power $P_C = 180$ TW (see the shadow area in Figure 3) which drives the capsule compression. This laser pulse does not provide self-ignition but only serves to compress the DT fuel. Indeed, ignition is achieved by the action of an additional high-power shock ignition spike. In this case, the igniting SI pulse starts at time t_{SI} and the power grows linearly, reaching the maximum power P_{SI} in 100 ps; the maximum power holds for 300 ps and then goes down to zero in another 100 ps.

In Figure 3 are shown details of a numerical calculation performed with a compression pulse which is maintained until time $t_{SI} = 12.6$ ns, where the shock ignition pulse began. The maximum incident laser power during the compression phase is $P_C = 180$ TW, which then grows to $P_{SI} = 350$ TW during the shock ignition pulse. In this case, the calculation provides a fusion energy of $E_{FUS} = 24.3$ MJ while the total incident laser energy is $E_{INC} = 560$ kJ, of which 360 kJ are invested in the compression phase ($t < t_{SI}$) and 200 kJ in the shock ignition pulse ($t > t_{SI}$). The energy gain is $G = E_{FUS}/E_{INC} = 43$. The incident and absorbed laser powers are shown by the two shadowed areas (that correspond to the linear scale) in Figure 3. As can be seen, the absorbed power is almost the half of the incident laser power, and the total energy absorption fraction is $\eta = 59\%$. It is worth noticing that this capsule provides similar performance when is used in the central ignition scheme. Nevertheless, the shock ignition scheme benefits from a larger tolerance with respect to the fuel compression uniformity, which is the drawback at the LMJ facility.

The radial position r_c where the density is equal to the critical value ρ_c [g cm^{-3}] $= 1.865 \times 10^{-3}(A/Z)/\lambda^2$ [μm] has also been calculated, and it is shown by the dashed red curve in Figure 3. The surface associated with the radius r_c has been used to estimate the maximum incident laser intensity $I_{INC} = P_{INC}/(4\pi r_c^2)$, shown by the blue curve. As usual in the shock ignition scheme, a very high laser intensity I_{INC} is needed during the shock ignition pulse, but there is some concern that the laser–plasma interaction at high intensities ($I\lambda^2 > 10^{15}$ W cm^{-2} μm^2) is dominated by laser–plasma instabilities that considerably modify the absorption mechanism[57] and negatively impact coupling of the ignition pulse energy to the capsule. In our case, the maximum intensity is larger than 10^{16} W cm^{-2}, although the true value may not be quite this high, since not all the incident power P_{INC} reaches the critical density surface, also

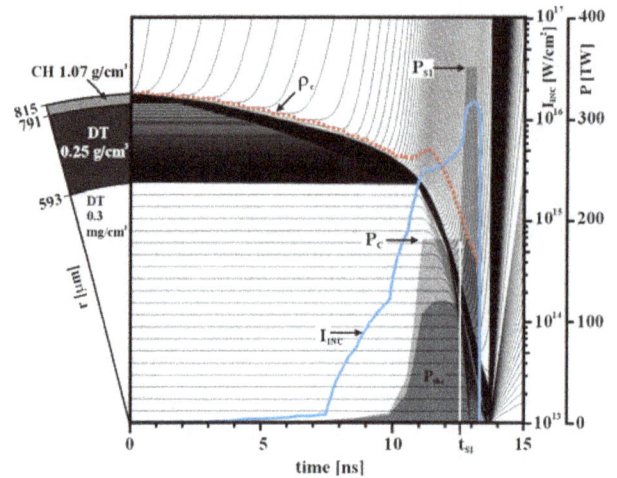

Figure 3. Capsule dimensions and temporal evolution of the Lagrangean radii. The temporal profile of the incident and absorbed power are shown by the two shadowed areas. The position of the critical density (ρ_c) and evolution of the maximum incident laser intensity (I_{INC}) are also shown as a function of time.

due to beam refraction. Nevertheless, these intensities are still in excess of the thresholds for laser plasma instabilities. It is likely that a large part ($\approx$50%) of the photon energy is converted into energetic electrons ($\approx$30 keV) and the laser light does not penetrate until the classical critical density, ρ_c, but reaches only $\rho_c/4$[58–61] where the laser light is absorbed by collective effects. The physics involving these high laser intensities and electronic transport are not included in our hydrodynamic code. Nevertheless, we tried to mimic the reduced critical density assuming a laser wavelength (λ_{SI}) twice the nominal value ($\lambda_{SI} = 2\lambda = 702$ nm) during the shock ignition pulse, i.e., for $t > t_{SI}$. Of course, this is just an attempt to evaluate the effect of the reduction by a factor of four ($\rho_c\alpha\lambda^{-2}$) in the maximum density reached by the photons. Detailed hydrodynamic calculations that also include the high-energetic electron transport will be needed to give a more complete treatment of the problem.

Two parametric studies have been performed, varying the starting time, t_{SI}, of the shock ignition pulse and the maximum incident power, P_{SI}. In a first case, we used the usual laser wavelength $\lambda = 351$ nm during the whole calculations and for each couple of parameters, t_{SI} and P_{SI}, the final energy gain G has been calculated. In a second set of calculations the laser wavelength has been doubled during the shock ignition pulse – i.e., when $t < t_{SI}$ – providing the gain $G^*(t_{SI}, P_{SI})$. The colour maps in Figure 4 shown the gain G and G^* as a function of the two parameters t_{SI} and P_{SI}. In the same figures the white contour curves show the total absorbed energy fraction, η. The case of the gain G – Figure 4(a), evaluated using always the same wavelength λ – shows two regions with high gain. In the first maximum, at smaller parameter t_{SI} and characterized by a lower gain, the high-power laser pulse arrives too early

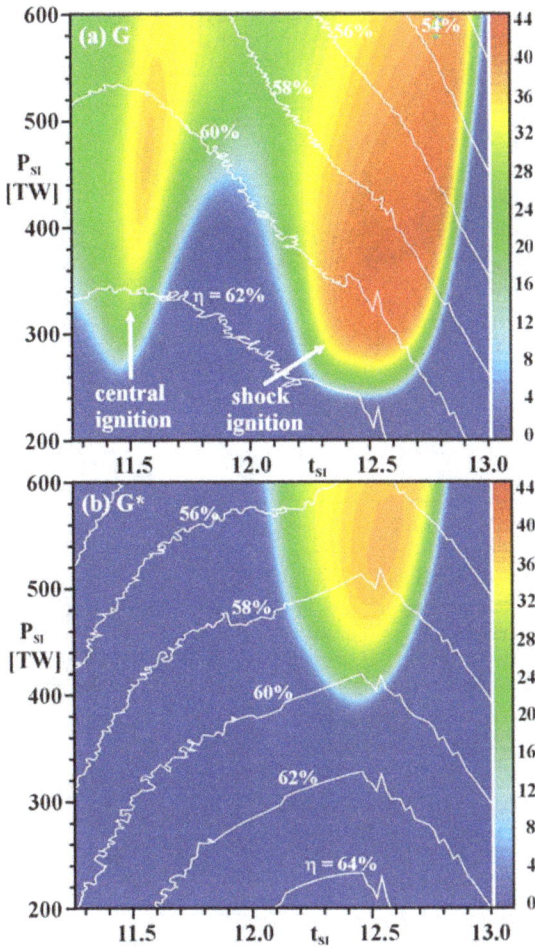

Figure 4. Gain as a function of the starting time t_{SI} and of the maximum power P_{SI} of the shock ignition pulse. (*a*) Gain G, calculated with $\lambda_{SI} = \lambda$; (*b*) Gain G^*, calculated assuming $\lambda_{SI} = 2\lambda$. The white curves represent isovalues of the absorption, η [%].

and generates a Kidder-like exponential laser pulse[62] that induces the classical central ignition. In contrast, the shock ignition mechanism is responsible for the second stronger signal at larger times, t_{SI}. It is found that for this specific laser–capsule configuration the threshold in the power P_{SI} is about 250 TW. The ignition region is reduced in the case of the gain G^* when the laser wavelength has been artificially doubled during the shock ignition pulse. In this case, the threshold moves to higher powers at around 400 TW. In both cases, G and G^*, the energy absorption is around 60%, and the modification in the threshold comes from the fact that the laser energy is deposited far from the compressed fuel. We are aware that our calculations do not deal with the correct laser–plasma interaction mechanisms; thus these results are not conclusive and just indicate a trend. It is also worth noting that the energetic electrons, neglected in our calculations, can transport some energy[63, 64] between the deposition region ($\rho < \rho_C/4$) and the ablation front that should favourably reduce the incident power threshold. In fact the high-energy electrons may not be as detrimental as in

the central ignition scheme, and it is possible they may even contribute positively towards driving the ignition shock[65].

4. Illumination non-uniformity

The shock ignition scheme is less demanding than the central ignition one with respect to the uniformity of the irradiation[66, 67]. However, the spike power needed to ignite the target is sensitive to the uniformity of the fuel assembly[59], and it is necessary to control the irradiation uniformity during the whole duration of the laser pulse. This is in general also difficult, because the plasma corona evolves during this time, and laser parameters optimized at the beginning of the irradiation could be no longer appropriate later during the implosion[68]. Nevertheless, special care must be paid to minimize the initial irradiation non-uniformity that generates the first shock wave of the implosion and dominates the so-called imprint phase.

In this section, we analyse some of the behaviour of the irradiation by using the illumination model[69, 70]. In the model, the capsule is assumed stationary – expansion of the plasma corona is neglected – and is characterized only by the external radius r_0. For a given number of incident laser beams characterized by an arbitrary laser intensity profiles, the model calculates the intensity of the illumination $I(\theta, \phi)$ over the spherical surface. It is thus assumed that laser parameters that optimize the illumination uniformity also minimize the non-uniformity transmitted to the first shock wave[71]. Generally, the quality of the illumination is measured by the root-mean-square deviation, σ_0, associated to the function $I(\theta, \phi)$; this is given by

$$\sigma_0 = \left\{ \frac{1}{4\pi} \int_0^{2\pi} \int_0^{\pi} [I(\theta, \phi) - \langle I \rangle]^2 \sin(\theta) d\theta d\phi \right\}^{1/2} \Big/ \langle I \rangle,$$

where $\langle I \rangle$ is the average intensity over the surface of the spherical target. The intrinsic non-uniformity σ_0 is a characteristic of a given laser–capsule configuration and assumes perfectly ideal laser beams not affected by any imperfections.

In reality, laser beams suffer from unavoidable errors such as beam-to-beam power imbalance σ_{PI}, laser pointing error σ_{PE}, and error in the target positioning σ_{TP}. These errors are statistical quantities that in the case of the LMJ facility are estimated by the standard deviations: $\sigma_{PI} = 10\%$ (beam-to-beam), $\sigma_{PE} = 50$ μm, and $\sigma_{TP} = 20$ μm. In the LMJ facility, the laser beams are grouped in quads; thus the power imbalance benefits from a statistical factor which reduces it to $\sigma_{PI} = 5\%$ (quad-to-quad). The illumination non-uniformity, evaluated taking into account these beam uncertainties, is usually measured as an average value $\langle \sigma \rangle$ estimated over a large number of calculations[72–77]. In these calculations, each of the three parameters (laser power, laser pointing, and target position) varies randomly and follows a Gaussian distribution centred to their nominal values and characterized by the corresponding standard deviation σ_{PI}, σ_{PE}, or σ_{TP}.

As has been already said, the LMJ facility is devoted to the indirect drive scheme. This means that the laser beam directions as well as their intensity profile fit with the *hohlraum* requirements. As a consequence, the geometrical shape of the laser intensity profile is elliptical. Thus, the intensity profile has been parameterized by the super-Gaussian function: $I(x, y) = I_0 \exp -[(x/\Delta_a)^2 + (y/\Delta_b)^2]^{m/2}$, characterized by the parameters Δ_a and Δ_b (half width at $1/e$) and by the exponent m. In our calculations we considered an elliptical laser intensity profile characterized with $\Delta_a = 2\Delta_b$, $\Delta_b = 320 \, \mu$m, and an exponent $m = 4$. Because of specific needs inherent to the indirect drive scheme in the LMJ facility (the same applies also for the Orion facility), the minor axis (Δ_b) of the elliptical intensity profile is located in the meridian defined by the polar and beam axes.

In the first set of calculations we considered the non-uniformity provided by the 20 quads located in the second ring of the LMJ facility (option A in Section 2). The results are shown in the Figure 5 as a function of the target radius r_0. The cloud of dots in the figure represents the results obtained with the elliptical profile for a large number of calculations assuming a random Gaussian distribution for the power imbalance, pointing error, and target positioning. The continuous red curve is the average non-uniformity $\underline{\sigma}$, while the blue curve shows the intrinsic non-uniformity σ_0 evaluated neglecting any beam–capsule uncertainties. As can be seen, for small capsule radii, some configurations with capsule centre, laser powers, and pointing randomly assigned can provide better results with respect to the intrinsic values. For the given elliptical LMJ laser intensity profile, an optimum capsule radius of $r_0 = 320 \, \mu$m is found for which the average non-uniformity assumes the minimum value $\underline{\sigma} = 6.2\%$, while the minimum intrinsic non-uniformity, evaluated neglecting any beam uncertainties, is $\sigma_0 = 4.6\%$.

It has been already shown[78, 79] that, in the case of axis-symmetric beam distributions as in the LMJ or Orion facilities, the elliptical laser intensity profile allows for better non-uniformities with respect to circular shapes. For the sake of comparison we have also shown in Figure 5 the average (red dashed curve) and the intrinsic non-uniformities (blue dashed curve) calculated using a circular intensity profile. In this case it is assumed that $\Delta_a = \Delta_b = 450 \, \mu$m in such a way as to have the same focal spot surface at I_0/e ($450^2 = 320 \times 640$). It is found that for the capsule radius $r_0 = 320 \, \mu$m the non-uniformity provided by the circular profile is almost double that of the elliptical case. These results confirm that the elliptical profile provides better results than the circular one for capsule radius $r_0 < 450 \, \mu$m.

The specific configurations given by the 10 or 20 quads located in the second ring (49°) of the LMJ facility are not

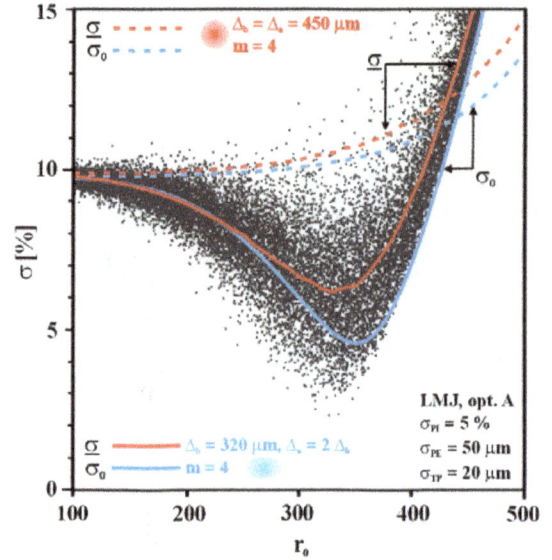

Figure 5. Average illumination non-uniformities $\underline{\sigma}$ (red curves) and intrinsic non-uniformities σ_0 (blue curves) as a function of the capsule radius r_0 evaluated for the LMJ configuration (option A). Continuous and dashed curves refer to the elliptical and circular laser intensity profile, respectively.

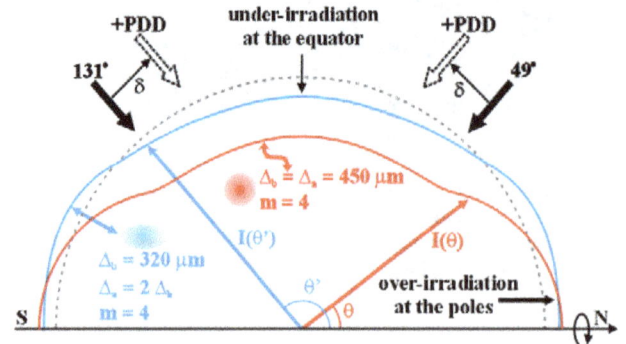

Figure 6. Polar plot of the intensity profile $I(\theta)$ provided by two axis-symmetric laser beams illuminating a capsule of radius $r_0 = 320 \, \mu$m. The laser intensity profiles are elliptical (red) and circular (blue), while the dashed circle is the reference of a perfectly uniform irradiation.

optimized for direct drive irradiation. Nevertheless, it is worth noting that the polar angle of 49° is relatively close to the optimum value, $\theta_S = 54.7°$, as found by Schmitt[70] for optimization of a two-ring configuration assuming that the energy deposition is given by a $\cos^2(\theta)$ distribution. Indeed, the LMJ configuration provides an over-irradiation of the two polar caps in detriment of the under-irradiation of the equatorial band. This is shown in the polar plot of Figure 6, where the radial distance – which has been set proportionally to the intensity $I(\theta)$ – is shown as a function of capsule latitude $\theta \epsilon [0, \pi]$. The calculations have been performed for an axis-symmetric beam distribution for the elliptical and circular laser intensity profiles and a capsule radius $r_0 = 320 \, \mu$m. Both intensity profiles cause an under-irradiation

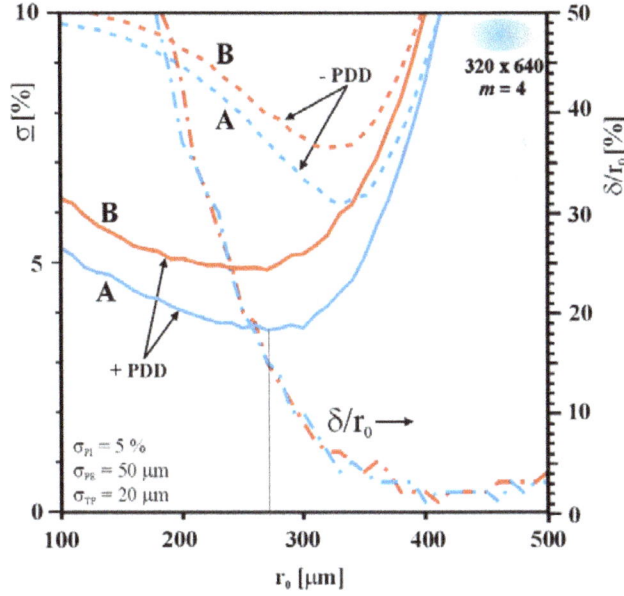

Figure 7. Average irradiation non-uniformity $\bar{\sigma}$ as a function of the capsule radius r_0 for the LMJ options A (blue) and B (red) with (continuous) and without (dashed) applying PDD. In the cases applying PDD, the optimum PDD parameter δ/r_0 is also shown.

Figure 8. Variation of the average non-uniformity with respect to the laser–capsule uncertainties. Continuous (dashed) curves refer to LMJ option A (B).

of the equatorial area but – due to geometrical factors – the elliptical laser spot provides a more uniform radial intensity that better approximates the perfectly spherical symmetry which is represented by the dashed circle.

As previously mentioned, to improve the laser–capsule coupling, the polar direct drive technique has been proposed. In this case, the laser beams are re-directed towards the equator by a quantity δ in order to balance the irradiation between polar and equatorial areas. The displacement δ is also indicated in the sketch of Figure 6. A parametric study varying the PDD parameter between 0 and 100 μm has been performed looking for the optimal PDD parameter δ that minimizes the average illumination non-uniformity, $\underline{\sigma}$. These non-uniformities $\underline{\sigma}$ are shown as a function of the capsule radius r_0 in Figure 7 for the two cases A and B. The dashed curves refer to the calculations without PDD ($\delta = 0$), whilst the continuous curves account for the PDD optimization; in these last cases, the corresponding optimal PDD parameter δ/r_0 is also shown. In both configurations, the PDD technique improves the results and reduces the minimum non-uniformities by about 40%. The minimum illumination non-uniformities $\underline{\sigma}_A = 3.6\%$ and $\underline{\sigma}_B = 4.9\%$ are reached at the capsule radius $r_0 = 270$ μm, for which the associated optimum PDD parameter is $\delta/r_0 = 15\%$.

Another set of calculations has been performed to evaluate the sensitivity of the illumination non-uniformity with respect to a variation of the beam uncertainties σ_{PI}, σ_{PE}, and σ_{TP}. These calculations use the laser intensity profile envisaged for the LMJ facility ($\Delta_a = 640$ μm, $\Delta_b = 320$ μm), a capsule radius $r_0 = 270$ μm, and a PDD

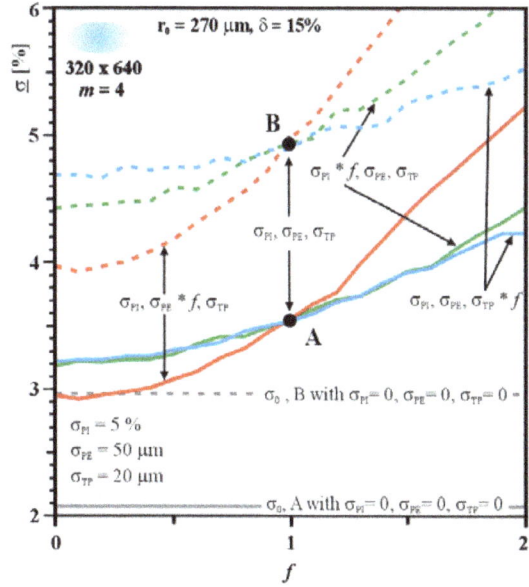

parameter $\delta/r_0 = 15\%$. The average non-uniformities $\underline{\sigma}$ are shown in Figure 8 as a function of an uncertainty scaling parameter f, which varies between 0 and 2. Three series of calculations have been done: (I) keeping constant pointing error ($\sigma_{PE} = 50$ μm) and target positioning ($\sigma_{TP} = 20$ μm) while varying the power imbalance from zero to double the nominal value ($\sigma_{PI} = 5\%$), i.e., considering $f\sigma_{PI}$, σ_{PE}, and σ_{TP}; (II) varying only the pointing error (σ_{PI}, $f\sigma_{PE}$, σ_{TP}); and (III) with only variation of the target positioning (σ_{PI}, σ_{PE}, $f\sigma_{TP}$). The ensemble of the results is shown in Figure 8. As a comparison the two intrinsic values σ_0 (horizontal gray lines) evaluated for the LMJ options A and B have been added. The largest gradient of the average non-uniformity $\underline{\sigma}$ is associated with the variation of the pointing error. This makes these two laser–capsule configurations more sensitive to the pointing error (σ_{PE}) rather than the other two error sources (power imbalance and target positioning).

A final detailed parametric study has been performed to evaluate the sensitivity of the average illumination non-uniformity to a variation of the PDD parameter δ and of the super-Gaussian exponent m of the laser intensity profile. As in the previous case, the capsule radius has been set to $r_0 = 270$ μm and the elliptical intensity profile is characterized by the widths $\Delta_b = 320$ μm and $\Delta_a = 2\Delta_b$. The average non-uniformity, which takes into account the beam uncertainties, is shown as a function of the parameters δ and m in Figure 9 for LMJ option A (top frame) and B (bottom). The shadowed areas indicate the regions where the non-uniformities are within 10% closer to their minimum values $\underline{\sigma}_{min}$ (3.6% for option A and 4.9% for option B). It is thus shown that both systems tolerate a relatively large variation of the super-Gaussian exponent $3 < m < 5$ and a variation of about

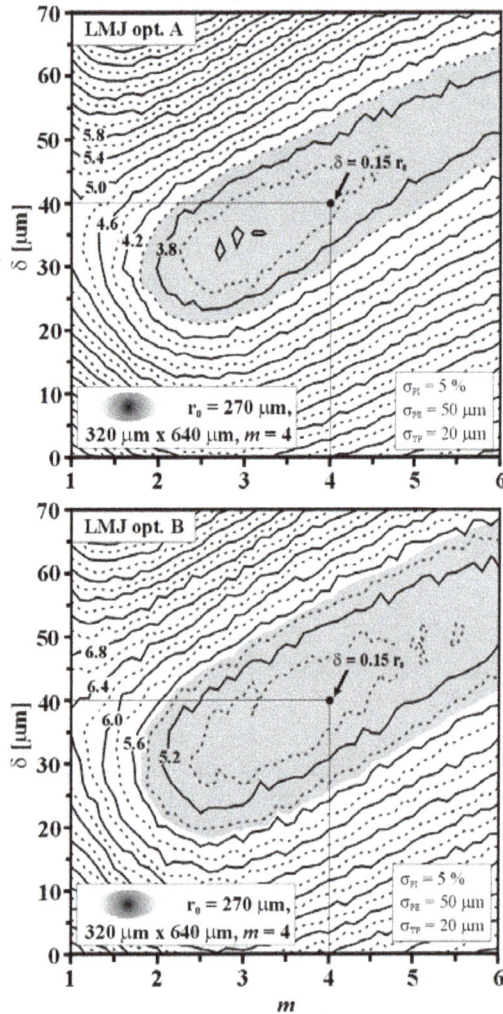

Figure 9. Average non-uniformity as a function of the PDD parameter δ and of the super-Gaussian exponent m of the laser intensity profile.

± 10 μm of the PDD parameter while still providing a non-uniformity smaller than $1.1\underline{\sigma}_{min}$.

5. 3D hydrodynamic simulations

Detailed two-dimensional (2D) hydrodynamic numerical simulations are usually employed to analyse the irradiation, compression, ignition, and thermonuclear burn wave propagation in an ICF capsule. Nevertheless, most actual laser–target configurations are intrinsically three-dimensional (3D) systems, and these have motivated the development of 3D hydrodynamic numerical tools[80–85]. Three-dimensional aspects can play a role also in the present cases considered in this paper that use a limited number of quads. This is especially true in option B, where only five beams are located in each axis-symmetric ring. Recently, using the 3D version of the MULTI[55] code, the uniformity of the irradiation provided by the LMJ facility has been analysed. Here, we

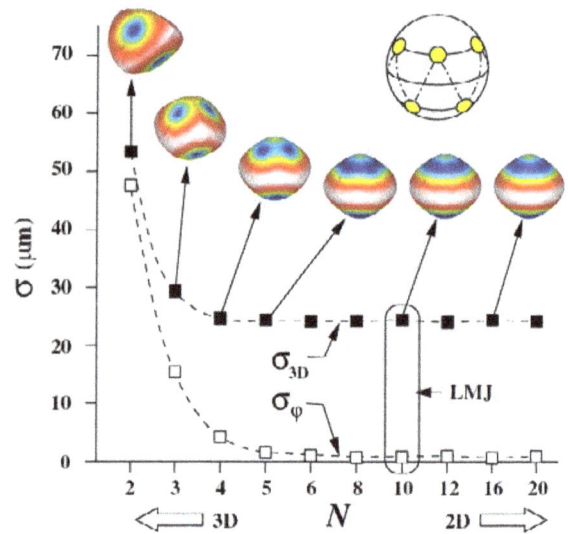

Figure 10. σ_{3D} (black squares, ■) and σ_{φ} (white squares, □) at $t = 12$ ns, as a function of the number of quads, N. Rings of opposite hemispheres are rotated against each other by an angle of $180°/N$.

report only on a few results of a much larger and detailed analysis[86]. The 3D version of the MULTI code assumes a non-structured Lagrangian mesh (tetrahedral elements) and accounts for flux-limited (10%) thermal heat conduction, tabulated equations of state, and a 3D ray-tracing package for the laser energy deposition.

A first issue is when a 3D configuration can be correctly described as a 2D axis-symmetric problem. For this purpose, a configuration with a number N of laser beams in each ring has been considered. In these calculations, the spherical capsule described in Section 3 is illuminated by beams from rings at 49° and 131°, aligned to the target centre ($\delta = 0$), and with a Gaussian radial shape characterized by a full width at half maximum (FWHM) of 1356 μm. The DT shell is followed in time and a mean radial position is defined as

$$R(\theta, \varphi, t) = \int_{DT} \rho(r, \theta, \varphi, t) r^3 dr \Big/ \int_{DT} \rho(r, \theta, \varphi, t) r^2 dr.$$

For a pure axis-symmetric problem, R does not depend on azimuthal angle φ. Here, it is assumed that departure from perfectly sphericity of the surface defined by the radius R is representative of the non-uniformity produced by the laser energy deposition. This surface has been decomposed in spherical harmonics providing the corresponding time-dependent coefficients. Then, these coefficients have been used to measure the azimuthal $\sigma_{\varphi}(t)$ and the polar $\sigma_{\theta}(t)$ components of the total root-mean-square non-uniformity $\sigma_{3D}^2 = \sigma_{\varphi}^2 + \sigma_{\theta}^2$ associated to this interface.

The results of these 3D calculations are summarized in Figures 10 and 11, where we show the values of σ_{3D} and σ_{ϕ} evaluated at $t = 12$ ns, approximately the time when the shell radius reduces to one half of its initial value. Figure 10 corresponds to laser beams arrangements where

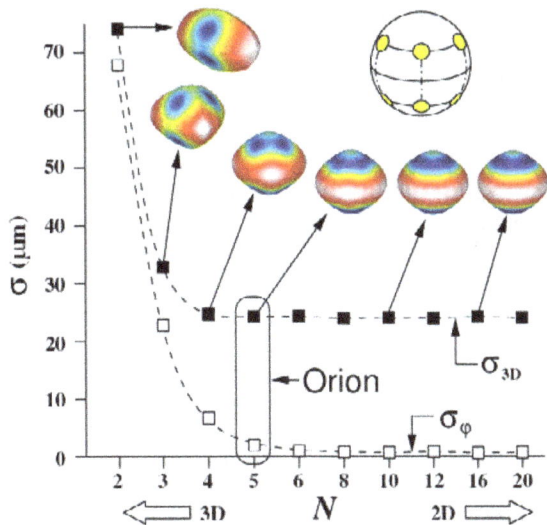

Figure 11. σ_{3D} (■) and σ_φ (□) evaluated at $t = 12$ ns, as a function of the number of quads, N. Rings of opposite hemispheres are symmetric with respect to the equatorial plane.

the rings of opposite hemispheres are rotated each other by an angle $18°/N$ (e.g., the LMJ facility with $N = 10$), while Figure 11 corresponds to symmetric arrangements (e.g., the Orion facility, $N = 5$). The insets in the figures show the shape of the DT–ablator interface evaluated at $t = 12$ ns. The colours indicate the distortion – inversely related to the driver pressure – in terms of radius (white/blue for large/small values). For extremely small number of beams, there is a clear triaxiality. In Figure 10 one can recognize tetrahedral ($N = 2$) and hexahedral ($N = 3$) shapes. In Figure 11, for $N = 2$, the four quads are in the same meridian plane, and the compressed shape is elongated along the perpendicular direction. In all these cases, polar, azimuthal, and total distortions are of the same order. For large values of N, the configuration converges to an axis-symmetric one, the same for both types of laser arrangement. These calculations have been done without PDD correction ($\delta = 0$), and as a consequence a polar overpressure appears. The small residual value $\sigma_\varphi \leqslant 1$ μm, for large N, is due to the spatial discretization and the numerical noise associated to the Monte Carlo nature of the ray-tracing algorithm. It is noteworthy that the transition between three-dimensionality and two-dimensionality occurs at relatively small laser beam numbers, N. For $N = 5$, the values of σ_ϕ are 1.55 μm and 2.0 μm, just above the numerical noise. For $N \geqslant 8$, the results are no longer distinguishable. This fast approach to the 2D axial symmetry justifies the use of 2D codes to treat accurately option A and, in an approximate but reasonable way, option B. These conclusions hold when spots of adjacent beams have enough overlapping, provided that uncertainties in beam power balance and pointing accuracy can be neglected. It must be mentioned that, for configurations where the beam size has been reduced

(FWHM ≈ 1000 μm), 3D effects occur, and azimuthal distortions can becomes significant even for $N \approx 5$[86].

6. Conclusions

The Laser MegaJoule facility has been considered in the context of the shock ignition scheme. Two laser beam configurations have been analysed. A first option (A) uses 20 quads – 80 laser beams (600 kJ, 200 TW) locate at the second ring of the LMJ facility – for the compression of the capsule, making available the remaining 24 quads – 96 laser beams (720 kJ, 240 TW) – for the additional shock ignition pulse. A second option (B) envisages the possibility to use only 10 quads for the compression phase and 34 quads for the compression and SI phases. The total available laser power is 440 TW at 3ω ($\lambda = 351$ nm).

A classical ICF capsule – devoted to the central ignition scheme – has been used in the context of the shock ignition scheme. A set of mono-dimensional numerical simulations has been performed to enlighten some aspect of the shock ignition scheme. For this specific capsule it is found that the threshold power in the shock ignition pulse is about 250 TW. Nevertheless, assuming that all this power is incident to the surface of the critical density provides incident intensity larger than 10^{16} TW cm^{-2}. At these large intensities ($I\lambda^2 > 10^{15}$ W cm^{-2} μm^2) we expect saturation of dangerous laser–plasma instabilities (SRS, SBS, and TPD) that modify the laser energy deposition mechanism. In this new regime, a large fraction of the laser energy is transferred to high-energetic electrons, and the photon penetration depth is limited to a quarter of the critical density ($\rho_c/4$), instead of the classical limit, ρ_c. These physical mechanisms are not included in our numerical tools; however, we performed some calculations to estimate the effect caused by limiting the deposition of the laser energy in the region at lower density ($\rho < \rho_c/4$). To mimic this effect, the light wavelength during the shock ignition pulse has been artificially doubled ($\lambda_{SI} = 2\lambda$); thus, because $\rho_c \alpha \lambda^{-2}$, the critical density becomes a quarter. As expected, this affects negatively the power threshold in the shock ignition pulse that now increases to about 400 TW. This should be considered as a pessimistic estimation. In fact, none of the positive effects associated with the high-energetic electrons are included in our calculations.

The second issue addressed in the paper concerns the irradiation uniformity provided during the first few ns of the foot pulse. First, it has been shown that the elliptical laser intensity profile of the LMJ facility provides better results in comparison to the usually circular profile. The two LMJ options A and B have been considered, taking into account beam uncertainties such as quad-to-quad power imbalance ($\sigma_{PI} = 5\%$), pointing error ($\sigma_{PE} = 50$ μm), and target positioning ($\sigma_{TP} = 20$ μm). Both of these configurations cause an over-irradiation of the capsule polar regions in detriment to the equatorial area. In order to improve these schemes, the polar direct drive technique has been applied to optimize the irradiation uniformity. It has been found that for the elliptical laser intensity profile ($\Delta_b = 320$ μm, $\Delta_a = 2 \Delta_b$, $m = 4$) expected at the LMJ facility the optimal capsule radius is $r_0 = 270$ μm, and this provides an average illumination non-uniformity of $\underline{\sigma} = 3.6\%$ and 4.9% in case A and case B, respectively. These minimum non-uniformities correspond to the use of a PDD parameter $\delta/r_0 = 15\%$. This capsule radius is relatively small in comparison to the available LMJ energy and the requirements for typical ignition capsule designs; however, bigger capsules could be envisaged assuming larger focal spots provided by either defocusing of the laser beams or using an alternative set of phase plates.

A 3D version of the code MULTI has been used to perform a set of preliminary hydrodynamic calculations. The LMJ options A and B have been considered in these calculations, and the laser irradiation uniformity has been split into the azimuthal and polar components by means of decomposition in spherical harmonics. For the analysed laser–capsule configuration – with the laser intensity profile that reduces to $1/e$ at the initial capsule radius – it is found that the azimuthal component is negligible in the case of option A (ten beams per hemisphere). This encouraging result seems indicates that a 2D analysis is appropriate in option A, while in the second case, option B, it may not be. Of course these conclusions depend on the beam and capsule sizes, and further investigations are needed for specific configurations.

Finally, the two LMJ options A and B involve the use of 10 or 20 quads located in the second rings characterized by the polar angles 49° and 131°. These options are in many aspects similar to the configuration already available at the Orion facility, where ten laser beams are located at 50° and 130°. In addition, these ten ns-long laser beams operate at the wavelength $\lambda = 0.351$ μm (3ω) as in the LMJ facility. The similarity between the two installations motivates us to stress the opportunity to perform Orion's experiments addressed to PDD issues of interest also for future direct drive LMJ campaigns. Indeed, although of relatively small energy – 5 kJ in few ns for the ten long-pulse Orion beams – this installation is fully adequate for direct drive experiments that may explore the laser–capsule coupling as well as the uniformity and timing of the first shock wave generated during the low-power ($\approx$TW) ns-long foot pulse needed to control the initial imprint phase of an ICF implosion, thus helping to underwrite modelling of polar direct drive implosions.

Acknowledgements

M. T. and B. C. express their thanks to Daniel Bouche for the support given to this work. R. R. was partially supported by the EURATOM/CIEMAT association in the framework of the 'IFE Keep-in-Touch Activities'. S. W. acknowledges support from the Czech Science Foundation (Project No. CZ.1.07/2.3.00/20.0279) and from ELI (Project No. CZ.1.05/1.1.00/02.0061).

References

1. J. Lindl, Phys. Plasmas **2**, 3933 (1995).
2. J. D. Lindl, P. Amendt, R. L. Berger, S. G. Glendinning, S. H. Glenzer, S. W. Haan, R. L. Kauffman, O. L. Landen, and L. J. Suter, Phys. Plasmas **11**, 339 (2004).
3. S. Atzeni and J. Meyer-ter-Vehn, *The Physics of Inertial Fusion* (Oxford University Press, Oxford, 2004).
4. S. E. Bodner, D. G. Colombant, J. H. Gardner, R. H. Lehmberg, S. P. Obenschain, L. Phillips, A. J. Schmitt, J. D. Sethian, R. L. McCrory, W. Seka, C. P. Verdon, J. P. Knauer, B. B. Afeyan, and H. T. Powell, Phys. Plasmas **5**, 1901 (1998).
5. J. H. Nuckolls, L. Wood, A. Thiessen, and G. B. Zimmermann, Nature **239**, 129 (1972).
6. R. D. Richtmyer, Commun. Pure Appl. Math. **13**, 297 (1960).
7. E. E. Meshkov, Fluid Dyn. **4**, 101 (1969).
8. L. Rayleigh, *Scientific Papers*. Vol. II (Dover, New York, 1965).
9. G. I. Taylor, Proc. R. Soc. London, Ser. A **201**, 192 (1950).
10. R. Betti and C. Zhou, Phys. Plasmas **12**, 110702 (2005).
11. N. G. Basov, S. Yu. Gus'kov, and L. P. Feoktistov, J. Sov. Laser Res. **13**, 396 (1992).
12. M. Tabak, J. Hammer, M. E. Glinsky, W. L. Kruer, S. C. Wilks, J. Woodworth, E. M. Campbell, M. D. Perry, and R. J. Mason, Phys. Plasmas **1**, 1626 (1994).
13. M. Roth, T. E. Cowan, M. H. Key, S. P. Hatchett, C. Brown, W. Fountain, J. Johnson, D. M. Pennington, R. A. Snavely, S. C. Wilks, K. Yasuike, H. Ruhl, F. Pegoraro, S. V. Bulanov, E. M. Campbell, M. D. Perry, and H. Powell, Phys. Rev. Lett. **86**, 436 (2001).
14. S. Yu. Gus'kov, Quantum Electronics **31**, 885 (2001).
15. M. Temporal, J. J. Honrubia, and S. Atzeni, Phys. Plasmas **9**, 3098 (2002).
16. B. M. Hegelich, B. J. Albright, J. Cobble, K. Flippo, S. Letzring, M. Paffett, H. Ruhl, J. Schreiber, R. K. Schulze, and J. C. Fernandez, Nature **439**, 441 (2006).
17. J. C. Fernandez, J. J. Honrubia, B. J. Albright, K. A. Flippo, D. C. Gautier, B. M. Hegelich, M. J. Schmitt, M. Temporal, and L. Yin, Nucl. Fusion **49**, 065004 (2009).
18. V. A. Shcherbakov, Sov. J. Plasma Phys. **9**, 240 (1983).
19. R. Betti, C. D. Zhou, K. S. Anderson, L. J. Perkins, W. Theobald, and A. A. Solodov, Phys. Rev. Lett. **98**, 155001 (2007).
20. A. J. Schmitt, J. W. Bates, S. P. Obenschain, S. T. Zalesak, and D. E. Fyfe, Phys. Plasmas **17**, 042701 (2010).
21. M. Lafon, X. Ribeyre, and G. Schurtz, Phys. Plasmas **17**, 052704 (2010).
22. L. Hallo, M. Olazabal-Loume, X. Ribeyre, V. Drean, G. Schurtz, J. L. Feugeas, J. Breil, Ph. Nicolai, and P. H. Maire, Plasma Phys. Control. Fusion **51**, 014001 (2009).
23. X. Ribeyre, G. Schurtz, M. Lafon, S. Galera, and S. Weber, Plasma Phys. Control. Fusion **51**, 015013 (2009).
24. R. Lehmberg and J. Goldhar, Fusion Tech. **11**, 532 (1987).
25. B. Canaud and F. Garaude, Nucl. Fusion **45**, L43 (2005).
26. H. A. Rose, D. F. DuBois, and B. Bezzerides, Phys. Rev. Lett. **58**, 2547 (1987).
27. K. Estabrook, J. Harte, E. Campbell, F. Ze, D. Phillion, M. Rosen, and J. Larsen, Phys. Rev. Lett. **46**, 724 (1981).
28. R. Yan, A. Maximov, C. Ren, and F. Tsung, Phys. Rev. Lett. **103**, 175002 (2009).
29. O. Klimo, V. T. Tikhonchuk, X. Ribeyre, G. Schurtz, C. Riconda, S. Weber, and J. Limpouch, Phys. Plasmas **18**, 082709 (2011).

30. R. Yan, A. Maximov, and C. Ren, Phys. Plasmas **17**, 052701 (2010).
31. H. Vu, D. DuBois, D. Russell, and J. Myatt, Phys. Plasmas **17**, 072701 (2010).
32. O. Klimo, S. Weber, V. T. Tikhonchuk, and J. Limpouch, Plasma Phys. Control. Fusion **52**, 055013 (2010).
33. W. Theobald, R. Betti, C. Stoeckl, K. S. Anderson, J. A. Delettrez, V. Yu. Glebov, V. N. Goncharov, F. J. Marshall, D. N. Maywar, R. L. McCrory, D. D. Meyerhofer, P. B. Radha, T. C. Sangster, W. Seka, D. Shvarts, V. A. Smalyuk, A. A. Solodov, B. Yaakobi, C. D. Zhou, J. A. Frenje, C. K. Li, F. H. Séguin, R. D. Petrasso, and L. J. Perkins, Phys. Plasmas **15**, 056306 (2008).
34. S. D. Baton, M. Koenig, E. Brambrink, H. P. Schlenvoigt, C. Rousseaux, G. Debras, S. Laffite, P. Loiseau, F. Philippe, X. Ribeyre, and G. Schurtz, Phys. Rev. Lett. **108**, 195002 (2012).
35. W. Theobald, R. Nora, M. Lafon, A. Casner, X. Ribeyre, K. S. Anderson, R. Betti, J. A. Delettrez, J. A. Frenje, V. Yu. Glebov, O. V. Gotchev, M. Hohenberger, S. X. Hu, F. J. Marshall, D. D. Meyerhofer, T. C. Sangster, G. Schurtz, W. Seka, V. A. Smalyuk, C. Stoeck, and B. Yaakobi, Phys. Plasmas **19**, 102706 (2012).
36. D. Batani, L. A. Gizzi, P. Koester, L. Labate, J. Honrubia, L. Antonelli, A. Morace, L. Volpe, J. J. Santos, G. Schurtz, S. Hulin, X. Ribeyre, P. Nicolai, B. Vauzour, F. Dorchies, W. Nazarov, J. Pasley, M. Richetta, K. Lancaster, C. Spindloe, M. Tolley, D. Neely, M. Kozlov'a, J. Nejdl, B. Rus, J. Wolowski, and J. Badziak, Nukleonika **57**, 3 (2012).
37. P. Koester, L. Antonelli, S. Atzeni, J. Badziak, F. Baffigi, D. Batani, C. A. Cecchetti, T. Chodukowski, F. Consoli, G. Cristoforetti, R. De Angelis, G. Folpini, L. A. Gizzi, Z. Kalinowska, E. Krousky, M. Kucharik, L. Labate, T. Levato, R. Liska, G. Malka, Y. Maheut, A. Marocchino, P. Nicolai, T. O'Dell, P. Parys, T. Pisarczyk, P. Raczka, O. Renner, Y. J. Rhee, X. Ribeyre, M. Richetta, M. Rosinski, L. Ryc, J. Skala, A. Schiavi, G. Schurtz, M. Smid, C. Spindloe, J. Ullschmied, J. Wolowski, and A. Zaras, Plasma Phys. Control. Fusion **55**, 124045 (2013).
38. M. Hohenberger, W. Theobald, S. X. Hu, K. S. Anderson, R. Betti, T. R. Boehly, A. Casner, D. E. Fratanduono, M. Lafon, D. D. Meyerhofer, R. Nora, X. Ribeyre, T. C. Sangster, G. Schurtz, W. Seka, C. Stoeckl, and B. Yaakobi, Phys. Plasmas **21**, 022702 (2014).
39. C. Goyon, S. Depierreux, V. Yahia, G. Loisel, C. Baccou, C. Courvoisier, N. G. Borisenko, A. Orekhov, O. Rosmej, and C. Labaune, Phys. Rev. Lett. **111**, 235006 (2013).
40. G. H. Miller, E. I. Moses, and C. R. Wuest, Nucl. Fusion **44**, 228 (2004).
41. E. I. Moses, R. N. Boyd, B. A. Remington, C. J. Keane, and R. Al-Ayat, Phys. Plasmas **16**, 041006 (2009).
42. C. Cavailler, N. Fleurot, T. Lonjaret, and J. M. Di-Nicola, Plasma Phys. Control. Fusion **46**, B135 (2004).
43. C. Lion, Journal of Physics: Conference Series **244**, 012003 (2010).
44. N. Hopps, C. Danson, S. Duffield, D. Egan, S. Elsmere, M. Girling, E. Harvey, D. Hillier, M. Norman, S. Parker, P. Treadwell, D. Winter, and T. Bett, Appl. Opt. **52**, 3597 (2013).
45. S. Jacquemot, F. Amiranoff, S. D. Baton, J. C. Chanteloup, C. Labaune, M. Koenig, D. T. Michel, F. Perez, H. P. Schlenvoigt, B. Canaud, C. Cherfils Clerouin, G. Debras, S. Depierreux, J. Ebrardt, D. Juraszek, S. Lafitte, P. Loiseau, J. L. Miquel, F. Philippe, C. Rousseaux, N. Blanchot, C. B. Edwards, P. Norreys, S. Atzeni, A. Schiavi, J. Breil, J. L. Feugeas, L. Hallo, M. Lafon, X. Ribeyre, J. J. Santos, G. Schurtz, V. Tikhonchuk, A. Debayle, J. J. Honrubia, M. Temporal, D. Batani, J. R. Davies, F. Fiuza, R. A. Fonseca, L. O. Silva, L. A. Gizzi, P. Koester, L. Labate, J. Badziak, and O. Klimo, Nucl. Fusion **51**, 094025 (2011).

46. S. Skupsky, J. A. Marozas, R. S. Craxton, R. Betti, T. J. B. Collins, J. A. Delettrez, V. N. Goncharov, P. W. McKenty, P. B. Radha, J. P. Knauer, F. J. Marshall, D. R. Harding, J. D. Kilkenny, D. D. Meyerhofer, T. C. Sangster, and R. L. McCrory, Plasma Phys. **11**, 2763 (2004).
47. R. S. Craxton, F. J. Marshall, M. J. Bonino, R. Epstein, P. W. McKenty, S. Skupsky, J. A. Delettrez, I. V. Igumenshchev, D. W. Jacobs-Perkins, J. P. Knauer, J. A. Marozas, P. B. Radha, and W. Seka, Phys. Plasmas **12**, 056304 (2005).
48. B. Canaud, X. Fortin, F. Garaude, C. Meyer, and F. Philippe, Laser Part. Beam **22**, 109 (2004).
49. B. Canaud, F. Garaude, P. Ballereau, J. L. Bourgade, C. Clique, D. Dureau, M. Houry, S. Jaouen, H. Jourdren, N. Lecler, L. Masse, A. Masson, R. Quach, R. Piron, D. Riz, J. Van der Vliet, M. Temporal, J. A. Delettrez, and P. W. McKenty, Plasma Phys. Control. Fusion **49**, B601 (2007).
50. P. B. Radha, J. A. Marozas, F. J. Marshall, A. Shvydky, T. J. B. Collins, V. N. Goncharov, R. L. McCrory, P. W. McKenty, D. D. Meyerhofer, T. C. Sangster, and S. Skupsky, Phys. Plasmas **19**, 082704 (2012).
51. P. B. Radha, F. J. Marshall, J. A. Marozas, A. Shvydky, I. Gabalski, T. R. Boehly, T. J. B. Collins, R. S. Craxton, D. H. Edgell, R. Epstein, J. A. Frenje, D. H. Froula, V. N. Goncharov, M. Hohenberger, R. L. McCrory, P. W. McKenty, D. D. Meyerhofer, R. D. Petrasso, T. C. Sangster, and S. Skupsky, Phys. Plasmas **20**, 056306 (2013).
52. V. Brandon, B. Canaud, M. Primout, S. Laffite, and M. Temporal, Laser Part. Beam **31**, 141 (2013).
53. V. Brandon, B. Canaud, M. Temporal, and R. Ramis, Low Initial Aspect-Ratio Direct-Drive target designs for shock- or self-ignition in the context of the laser Megajoule, submitted to Nucl Fusion (2014).
54. M. Temporal, V. Brandon, B. Canaud, J. P. Didelez, R. Fedosejevs, and R. Ramis, Nucl. Fusion **52**, 103011 (2012).
55. R. Ramis, R. Schmaltz, and J. Meyer-ter-Vehn, Comput. Phys. Commun. **49**, 475 (1988).
56. R. Ramis, K. Eidmann, J. Meyer-ter-Vehn, and S. Hüller, Comput. Phys. Commun. **183**, 637 (2012).
57. W. L. Kruer, *The Physics of Laser–Plasma Interactions* (Addison-Wesley, Reading, MA, 1988).
58. C. Riconda, S. Weber, V. T. Tikhonchuk, and A. Heron, Phys. Plasmas **18**, 092701 (2011).
59. S. Weber, C. Riconda, O. Klimo, A. Heron, and V. T. Tikhonchuk, Phys. Rev. E. **85**, 016403 (2012).
60. O. Klimo and V. T. Tikhonchuk, Plasma Phys. Control. Fusion **55**, 095002 (2013).
61. W. Theobald, R. Nora, M. Lafon, A. Casner, X. Ribeyre, K. S. Anderson, R. Betti, J. A. Delettrcz, J. A. Frenje, V. Yu. Glebov, O. V. Gotchev, M. Hohenberger, S. X. Hu, F. J. Marshall, D. D. Meyerhofer, T. C. Sangster, G. Schurtz, W. Seka, V. A. Smalyuk, C. Stoeckl, and B. Yaakobi, Phys. Plasmas **19**, 102706 (2012).
62. R. E. Kidder, Nucl. Fusion **16**, 405 (1976).
63. A. Bell and M. Tzoufras, Plasma Phys. Control. Fusion **53**, 045010 (2011).
64. S. Gus'kov, X. Ribeyre, M. Touati, J. L. Feugeas, Ph. Nicolai, and V. Tikhonchuk, Phys. Rev Lett. **109**, 255004 (2012).
65. X. Ribeyre, S. Gus'kov, J. L. Feugeas, Ph. Nicolai, and V. T. Tikhonchuk, Phys. Plasmas **20**, 062705 (2013).
66. B. Canaud, S. Laffite, V. Brandon, and M. Temporal, Laser Part. Beam **30**, 183 (2012).
67. S. Atzeni, A. Schiavi, and A. Marocchino, Plasma Phys. Control. Fusion **53**, 035010 (2011).
68. M. Temporal and B. Canaud, Eur. Phys. J. D. **55**, 139 (2009).

69. S. Skupsky and K. Lee, J. Appl. Phys. **54**, 3662 (1983).

70. A. J. Schmitt, Appl. Phys. Lett. **44**, 399 (1984).

71. M. Temporal, B. Canaud, W. J. Garbett, and R. Ramis, Comparison between illumination model and hydrodynamic simulation for a Direct Drive laser irradiated capsule (2014), in preparation.

72. M. Murakami, Appl. Phys. Lett. **66**, 1587 (1995).

73. M. Murakami, K. Nishihara, and H. Azechi, J. Appl. Phys. **74**, 802 (1993).

74. B. Canaud, X. Fortin, N. Dague, and J. Bocher, Phys. Plasmas **9**, 4252 (2002).

75. M. Temporal, B. Canaud, and B. J. Le Garrec, Phys. Plasmas **17**, 022701 (2010).

76. M. Temporal, B. Canaud, S. Laffite, B. J. Le Garrec, and M. Murakami, Phys. Plasmas **17**, 064504 (2010).

77. M. Temporal, R. Ramis, B. Canaud, V. Brandon, S. Laffite, and B. J. Le Garrec, Plasma Phys. Control. Fusion **53**, 124008 (2011).

78. M. Temporal, B. Canaud, W. J. Garbett, and R. Ramis, Phys. Plasmas **21**, 012710 (2014).

79. M. Temporal, B. Canaud, W. J. Garbett, F. Philippe, and R. Ramis, Eur. Phys. J. D. **67**, 205 (2013).

80. R. Ramis, Phys. Plasmas **20**, 082705 (2013).

81. M. M. Marinak, G. D. Kerbel, J. M. Koning, M. V. Patel, S. M. Sepke, M. S. McKinley, M. J. O'Brien, R. J. Procassini, and D. Munro, EPJ Web of Conferences **59**, 06001 (2013).

82. M. Gittings, R. Weaver, M. Clover, T. Betlach, N. Byrne, R. Coker, E. Dendy, R. Hueckstaedt, K. New, W. R. Oakes, D. Ranta, and R. Stefan, Comput. Sci. Disc. **1**, 015005 (2008).

83. B. van der Holst, G. Toth, I. V. Sokolov, K. G. Powell, J. P. Holloway, E. S. Myra, Q. Stout, M. L. Adams, J. E. Morel, S. Karni, B. Fryxell, and R. P. Drake, Astrophys. J. Suppl. **194**, 23 (2011).

84. B. Fryxell, K. Olson, P. Ricker, F. X. Timmes, M. Zingale, D. Q. Lamb, P. Macneice, R. Rosner, J. W. Truran, and H. Tufo, Astrophys. J. Suppl. **131**, 273 (2000).

85. Jinghong Lia, Chuanlei Zhai, Shuanggui Li, Xin Li, Wudi Zheng, Heng Yong, Qinghong Zeng, Xudeng Hang, Jin Qi, Rong Yang, Juan Cheng, Peng Song, Peijun Gu, Aiqing Zhang, Hengbin An, Xiaowen Xu, Hong Guo, Xiaolin Cao, Zeyao Mo, Wenbing Pei, Song Jiang, and Shao-ping Zhu, EPJ Web of Conferences **59**, 06002 (2013).

86. R. Ramis, M. Temporal, B. Canaud, V. Brandon, and S. Laffite, Three-dimensional symmetry analysis of a Direct Drive irradiation scheme for the Laser MegaJoule facility (2014), in preparation.

19

Ultrafast ignition with relativistic shock waves induced by high power lasers

Shalom Eliezer[1], Noaz Nissim[2], Shirly Vinikman Pinhasi[2], Erez Raicher[2,3], and José Maria Martinez Val[1]

[1]*Nuclear Fusion Institute, Polytechnic University of Madrid, Madrid, Spain*

[2]*Applied Physics Division, Soreq NRC, Yavne, Israel*

[3]*Hebrew University of Jerusalem, Jerusalem, Israel*

Abstract

In this paper we consider laser intensities greater than 10^{16} W cm^{-2} where the ablation pressure is negligible in comparison with the radiation pressure. The radiation pressure is caused by the ponderomotive force acting mainly on the electrons that are separated from the ions to create a double layer (DL). This DL is accelerated into the target, like a piston that pushes the matter in such a way that a shock wave is created. Here we discuss two novel ideas. Firstly, the transition domain between the relativistic and non-relativistic laser-induced shock waves. Our solution is based on relativistic hydrodynamics also for the above transition domain. The relativistic shock wave parameters, such as compression, pressure, shock wave and particle flow velocities, sound velocity and rarefaction wave velocity in the compressed target, and temperature are calculated. Secondly, we would like to use this transition domain for shock-wave-induced ultrafast ignition of a pre-compressed target. The laser parameters for these purposes are calculated and the main advantages of this scheme are described. If this scheme is successful a new source of energy in large quantities may become feasible.

Keywords: relativistic hydrodynamics; shock waves; laser; plasma; nuclear fusion; fast ignition

1. Introduction

Inertial fusion energy (IFE) is based on high compression[1–3]. The reasoning is that it is energetically cheaper to compress rather than to heat and the nuclear reactions are proportional to the density squared. IFE of deuterium–tritium (DT) requires high compression (>1000) and, in particular, the aneutronic fusion[4–6] of proton–boron 11 needs extremely high compression (>10 000). The high compression is achieved by shock waves and the accumulation of matter during stagnation of the implosion of the target shell.

Shock waves in laser plasma interactions[7] have played an important role in the study of IFE. In 1974 the first direct observation of a laser-driven shock wave was reported[8]. A planar solid hydrogen target was irradiated with a 10 J, 5 ns, Nd laser and a pressure of approximately 2 Mbar was measured. Twenty years after this experiment, the Nova laser from Livermore created a pressure of 750 ± 200 Mbar[9]. This was achieved in a collision between two gold foils,

Correspondence to: Shalom Eliezer, Nuclear Fusion Institute, C. Jose Gutierrez Abascal 2, Madrid 28006, Spain. Email: shalom.eliezer@gmail.com

where the flyer (Au foil) was accelerated by a high intensity X-ray flux created by the laser–plasma interaction.

In order to achieve nuclear fusion ignition, a mega-joule (MJ) laser with a few nanoseconds pulse duration has been constructed in the USA[10]. The central spark ignition of DT is expected in the near future. In this scheme the target and the driver pulse shape are designed in such a way that only a spark at the centre of the compressed fuel is heated and ignited[11, 12]. The rest of the fuel is heated by α particles produced in the DT reactions.

In order to ignite a DT target with significantly less than a few MJ of energy, it was suggested[13, 14] to separate the drivers that compress the target from those that heat the target. This idea is called fast ignition (FI), and triggers not in a central spark, but in a secondary interaction of an igniting driver of a very short duration, such as a multi-Petawatt (PW) laser beam. The PW laser is supposed to form a channel for a few picoseconds in the plasma atmosphere and to ignite a part of the fuel at the stagnation point of the implosion. For this purpose it is estimated that ignition requires of the order of a few tens of kilo-joule of laser energy for a duration of approximately 10 ps with an irradiance of the order of 10^{20} W cm^{-2}. The FI problem is that the laser pulse does not penetrate directly into the compressed target, which has an electron density of the order of 10^{24} cm^{-3}. Therefore many schemes of FI have been suggested: (1) the laser energy is converted into electrons that ignite the target[15]. (2) The laser energy is converted into protons that ignite the target[16]. (3) Since the heating in the previous proposals is not confined and furthermore it is necessary to avoid preheating, a gold cone[17] (Au density/solid DT density $\sim$100) is stuck in the spherical pellet. (4) FI is induced by plasma jets[18] that are induced by the same laser system that compresses the pellet. (5) The FI is achieved by a plasma flow created from a thin exploding pusher foil[19, 20]. (6) Plasma blocks for FI have also been suggested[21, 22]. (7) Murakami et al.[23] revived the old idea of impact fusion with the help of the cone. (8) The use of clusters[24] was also suggested to ignite the compressed pellet. (9) Furthermore, the use of an extra laser-induced shock wave created by the same lasers that compressed the target in order to ignite the target was suggested[25]. (10) Alternatively the FI shock wave from a laser-accelerated impact foil[26, 27] was proposed. The shock wave ignition schemes are actually based on heating by collision of two shock waves using tailored laser pulses that had already been suggested[28] even before the idea of FI was explicitly published[13, 14].

It is well known that the interaction of a high power laser (HPL) with a planar target creates a one dimensional (1D) shock wave[29, 30]. The theoretical basis for laser-induced shock waves analysed and measured experimentally so far is based on plasma ablation. For laser intensities in the range 10^{12} W cm^{-2} $< I_L <$ 10^{16} W cm^{-2} and nanosecond pulse durations a hot plasma is created. This plasma exerts a high pressure on the surrounding material, leading to the formation of an intense shock wave moving into the interior of the target. The momentum of the out-flowing plasma balances the momentum imparted to the compressed medium behind the shock front, similar to a rocket effect.

For $I_L < 10^{16}$ W cm^{-2} the ablation pressure is dominant. For $I_L > 10^{16}$ W cm^{-2} the radiation pressure is the dominant pressure at the solid–vacuum interface and the ablation pressure is negligible. In this last case the ponderomotive force drives the shock wave. For laser irradiances $I_L >$ 10^{21} W cm^{-2} one gets a relativistic laser-induced shock wave[31]. The theoretical foundation of relativistic shock waves is based on relativistic hydrodynamics[32] and is first analysed by Taub[33]. Relativistic shock waves may be of importance in intense stellar explosions or in collisions of extremely high energy nuclear particles. Furthermore, relativistic shock waves may be a new route for FI nuclear fusion.

In Section 2 the formalism of relativistic shock waves is given for further consideration. In Section 3 the laser-induced shock wave equations are explicitly written and solved numerically without approximation for the first time. In a recent publication[31] the solution is given only for very strong relativistic shocks, whereas in this paper the transition

between relativistic and nonrelativistic laser-induced shock waves is obtained. It turns out that this transition domain is important and relevant for the FI scheme as described in Section 4. The paper is concluded with a short summary and discussion.

2. Relativistic shock waves

The relativistic 1D (or non-relativistic[34]) shock wave is described by five variables: the particle density n (or the density $\rho = Mn$, where M is the particle mass), the pressure P, the energy density e, the shock wave velocity u_s, and the particle flow velocity u_p, assuming that we know the initial condition of the target: n_0 (or ρ_0), P_0, e_0, and the particle flow velocity u_0, before the shock arrival. The four equations relating the shock wave variables are the three Hugoniot relations describing the conservation laws of energy, momentum, and particles, and the equation of state[35, 36] connecting the thermodynamic variables of the state under consideration. In order to solve the problem an extra equation is required, which in our case we derive from a laser–plasma interaction model.

The relativistic hydrodynamic starting point is the energy momentum 4-tensor $T_{\mu\nu}$ given by

$$T_{\mu\nu} = (e + P)U_\mu U_\nu + Pg_{\mu\nu}. \tag{1}$$

U_μ ($\mu = 0, 1, 2, 3$) is the dimensionless 4-velocity, where the subscript 0 indicates the time component and subscripts 1, 2, and 3 indicate the space x, y, and z components, respectively, and $g_{\mu\nu}$ is the metric tensor,

$$u_\mu = cU_\mu = (\gamma c, \gamma v_1, \gamma v_2, \gamma v_3),$$

$$g_{\mu\nu} : g_{00} = -1, g_{11} = g_{22} = g_{33} = 1, g_{\mu\nu} = 0 \text{ if } \mu \neq \nu,$$

$$\gamma = \frac{1}{\sqrt{1 - \beta^2}}; \quad \beta = \frac{v}{c}; \quad v = \sqrt{v_1^2 + v_2^2 + v_3^2}, \tag{2}$$

where c is the speed of light in vacuum and v is the 3-dimensional fluid particle velocity. Since our equation (1) is the starting point we write it more explicitly as

$$T_{00} = \gamma^2(e + P) - P$$

$$T_{0i} = T_{i0} = -\gamma^2(e + P)\left(\frac{v_i}{c}\right) \quad \text{for } i = 1, 2, 3 \tag{3}$$

$$T_{ij} = \gamma^2(e + P)\left(\frac{v_i}{c}\right)\left(\frac{v_j}{c}\right) + P\delta_{ij} \quad \text{for } i, j = 1, 2, 3.$$

In our 1D model one has for the velocity vector $\mathbf{v} = (v, 0, 0)$ and the Lorentz transformation is

$$\text{Lorentz transformation} = \begin{pmatrix} \gamma & -\gamma\beta & 0 & 0 \\ -\gamma\beta & \gamma & 0 & 0 \\ 0 & 0 & 1 & 0 \\ 0 & 0 & 0 & 1 \end{pmatrix}. \tag{4}$$

Energy–momentum conservation, particle number conservation, and the equation of state (EOS) are given accordingly (Einstein summation is assumed from 0 to 3 for identical indices)

$$\frac{\partial T^\nu_\mu}{\partial x^\nu} \equiv \partial_\nu T^\nu_\mu = 0 \quad \text{for} \quad \mu = 0, 1, 2, 3$$

$$\frac{\partial (nU^\mu)}{\partial x^\mu} \equiv \partial_\mu (nU^\mu) = 0 \tag{5}$$

$$P = P(e, n).$$

We use equations (1), (3) and (5) for the conservation of energy density flux $c[T_{0x}]_0 = c[T_{0x}]_1$, the conservation of momentum density flux $[T_{xx}]_0 = [T_{xx}]_1$, and the conservation of particle number flux $[nU_x]_0 = [nU_x]_1$ along the shock wave singularity, with the subscripts 0 and 1 respectively denoting the domains before and after shock arrival, to obtain the following equations

$$\gamma_0^2 \beta_0 (e_0 + P_0) = \gamma_1^2 \beta_1 (e_1 + P_1)$$

$$\gamma_0^2 \beta_0^2 (e_0 + P_0) + P_0 = \gamma_1^2 \beta_1^2 (e_1 + P_1) + P_1 \tag{6}$$

$$\gamma_0 \beta_0 n_0 = \gamma_1 \beta_1 n_1,$$

where γ_i and $\beta_i = v_i/c$ for the domains 0 and 1 are defined in equation (2), where v_0 and v_1 are the inflow and outflow onto the shock wave singularity. Figure 1 describes the fluid flow velocities v_0 and v_1 as seen in the shock wave singularity frame of reference S_1 and the shock wave velocity u_{s1} and the particle flow velocities u_{p1} and $u_{p0} = u_0$ as seen in the laboratory frame of reference.

From equations (6) the velocities v_0 and v_1 are obtained

$$\frac{v_0}{c} \equiv \beta_0 = \sqrt{\frac{(P_1 - P_0)(e_1 + P_0)}{(e_1 - e_0)(e_0 + P_1)}}$$

$$\frac{v_1}{c} \equiv \beta_1 = \sqrt{\frac{(P_1 - P_0)(e_0 + P_1)}{(e_1 - e_0)(e_1 + P_0)}}, \tag{7}$$

and the relativistic Hugoniot equation is derived[33],

$$\frac{(e_1 + P_1)^2}{n_1^2} - \frac{(e_0 + P_0)^2}{n_0^2}$$

$$= (P_1 - P_0)\left[\frac{(e_0 + P_0)}{n_0^2} + \frac{(e_1 + P_1)}{n_1^2}\right]. \tag{8}$$

Assuming that in the laboratory the target is initially at rest, $u_0 = 0$, the shock wave velocity u_s and the particle flow velocity u_p in the laboratory frame of reference are related to the flow velocities v_0 and v_1 in the shock wave rest frame of reference by

$$u_s = -v_0,$$

$$u_p = \frac{v_1 - v_0}{1 - \frac{v_0 v_1}{c^2}}. \tag{9}$$

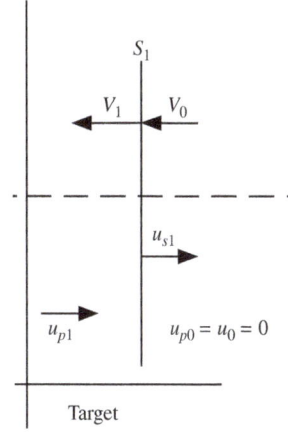

Figure 1. The fluid flow velocities v_0 and v_1 as seen in the shock wave singularity frame of reference S_1 and the shock wave velocity u_{s1} and the particle flow velocities u_{p1} and $u_{p0} = u_0$ as seen in the laboratory frame of reference.

The EOS taken here in order to calculate the shock wave parameters is the ideal gas EOS

$$e = \rho c^2 + \frac{P}{\Gamma - 1}, \tag{10}$$

where Γ is the specific heat ratio and v_0 and v_1 are given in equations (7).

3. Laser-induced shock waves

This paper analyses the shock wave created in a planar target by the ponderomotive force induced by very high laser irradiance. In this domain of laser intensities the force acts on the electrons that are accelerated and the ions that follow accordingly. This model describes our piston model[37, 38] as summarized schematically in Figure 2: Figure 2(a) shows the capacitor model for laser irradiances I_L, where the ponderomotive force dominates the interaction. In Figure 2(b) the system of the negative and positive layers is called a double layer (DL), n_e and n_i are the electron and ion densities respectively, E_x is the electric field, λ_{DL} is the distance between the positive and negative DL charges, and δ is the solid density skin depth of the foil. The DL is geometrically followed by a neutral plasma where the electric field decays within a skin depth and a shock wave is created. The shock wave description in the laboratory frame of reference is given in Figure 2(c). This DL acts as a piston driving a shock wave[39, 40]. This model is supported in the literature by particle in cell (PIC) simulation[39, 41] and independently by hydrodynamic two-fluid simulations[21, 22, 42]. The relativistic shock wave parameters, such as compression, pressure, shock wave and particle flow velocities, and temperature are calculated here for any compression $\kappa = \rho/\rho_0 > 1$ for the first time in the context of relativistic hydrodynamics. In a recent previous paper this was solved only for $\kappa = \rho/\rho_0 > 4$ with $\Gamma = 5/3$.

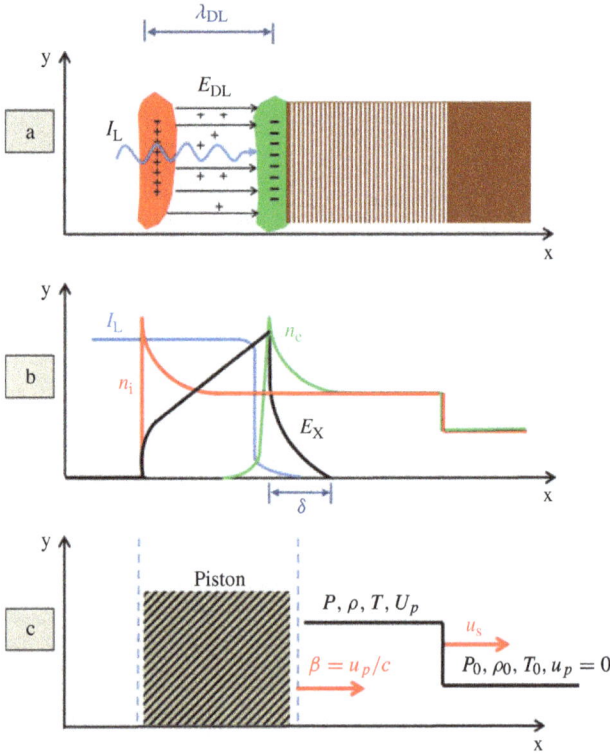

Figure 2. (a) The capacitor model for laser irradiances I_L where the ponderomotive force dominates the interaction. (b) The parameters that define our capacitor model: n_e and n_i are the electron and ion densities accordingly, E_x is the electric field, λ_{DL} is the distance between the positive and negative DL charges. The DL is geometrically followed by a neutral plasma where the electric field decays within a skin depth δ and a shock wave is created. (c) The shock wave description in the piston model.

For $I_L < 10^{16}$ W cm^{-2} the ablation pressure P_a is dominant and scales with the laser irradiance I_L like $P_a \sim I_L^{\alpha}$, where α is of the order of 2/3 in a 1D model[7]. For $I_L > 10^{16}$ W cm^{-2} the radiation pressure is the dominant pressure at the solid–vacuum interface and the ablation pressure is negligible. In this last case the ponderomotive force drives the shock wave. The equations forrelativistic hydrodynamics with the ideal gas EOS in the laboratory frame of reference are

$$\text{(i)} \quad \frac{u_{p1}}{c} = \sqrt{\frac{(P_1 - P_0)(e_1 - e_0)}{(e_0 + P_1)(e_1 + P_0)}}$$

$$\text{(ii)} \quad \frac{u_{s1}}{c} = \sqrt{\frac{(P_1 - P_0)(e_1 + P_0)}{(e_1 - e_0)(e_0 + P_1)}}$$

$$\text{(iii)} \quad \frac{(e_1 + P_1)^2}{\rho_1^2} - \frac{(e_0 + P_0)^2}{\rho_0^2}$$
$$= (P_1 - P_0)\left[\frac{(e_0 + P_0)}{\rho_0^2} + \frac{(e_1 + P_1)}{\rho_1^2}\right] \quad (11)$$

$$\left.\begin{matrix}\text{(iv)}\\ \text{(v)}\end{matrix}\right\} e_j = \rho_j c^2 + \frac{P_j}{\Gamma - 1}; \; j = 0, 1.$$

We have to solve these five equations together with our piston model equation[31, 38].

$$\text{(vi)} \; P_1 = \frac{2 I_L}{c}\left(\frac{1 - \beta}{1 + \beta}\right); \quad \beta \equiv \frac{u_{p1}}{c}. \quad (12)$$

Equations (11) and (12) describe six equations with six unknowns: u_s, u_{p1}, P_1, ρ_1, e_1, and e_0, assuming that we know I_L, ρ_0, P_0, Γ, and $u_o = 0$. We take the ideal gas EOS with $\Gamma = 5/3$. The calculations are conveniently done in the dimensionless units defined by

$$\Pi_L \equiv \frac{I_L}{\rho_0 c^3}; \quad \kappa \equiv \frac{\rho_1}{\rho_0};$$
$$\kappa_0 \equiv \frac{\Gamma + 1}{\Gamma - 1}; \quad \Pi = \frac{P_1}{\rho_0 c^2}; \quad \Pi_0 = \frac{P_0}{\rho_0 c^2}. \quad (13)$$

It is important to emphasize that if we take $P_0 = 0$ then we get only the $\kappa > \kappa_0$ solutions[31], therefore in order to see the behaviour at the transition between the relativistic and nonrelativistic domains one has to take $P_0 \neq 0$. In our numerical estimations we take $P_0 = 1$ bar $= 10^6$ in cgs units. For example, the Hugoniot equation (11)$_{(iii)}$ together with the EOS equations (11)$_{(iv)+(v)}$ yield

$$\frac{P_0}{P_1} = \frac{\Pi_0}{\Pi} = \mathbf{0}$$
$$\Rightarrow \begin{cases} \Pi = -B(\Pi_0 = 0) = \frac{(\Gamma - 1)^2}{\Gamma}\kappa(\kappa - \kappa_0) \\ \kappa \equiv \frac{\rho_1}{\rho_0} \geqslant \kappa_0, \end{cases} \quad (14)$$

$$\frac{P_0}{P_1} = \frac{\Pi_0}{\Pi} \neq \mathbf{0} \Rightarrow \begin{cases} \Pi^2 + B\Pi + C = 0 \\ \kappa \equiv \frac{\rho_1}{\rho_0} \geqslant 1 \end{cases}$$
$$\Pi = \left(\tfrac{1}{2}\right)\left(-B \pm \sqrt{B^2 - 4C}\right) \quad (15)$$
$$B = \frac{(\Gamma - 1)^2}{\Gamma}(\kappa_0 \kappa - \kappa^2) + \Pi_0(\Gamma - 1)(1 - \kappa^2)$$
$$C = \frac{(\Gamma - 1)^2}{\Gamma}(\kappa - \kappa_0 \kappa^2)\,\Pi_0 - \kappa^2 \Pi_0^2.$$

The compression κ as a function of the dimensionless pressure $\Pi = P_1/(\rho_0 c^2)$ is given in Figure 3 for $\kappa_0 = 4(\Gamma = 5/3)$. Although P_0/P_1 is extremely small one cannot neglect it in the very near vicinity of κ_0 and in this domain one has to solve equation (15) numerically. Furthermore, in order to see the transition between the relativistic and nonrelativistic approximations (see appendix A) one has to solve the relativistic equations with equation (15) in order to see transition effects such as the one shown in Figure 3. However for $\kappa > \kappa_0$, for $(\kappa - \kappa_0)/\kappa_0 > 10^{-3}$, the approximation of equation (14) is very good for calculating the shock wave variables as a function of the dimensionless laser irradiance Π_L.

The numerical solutions of equations (11) and (12) are shown in Figures 4 and 5. Figure 4 gives the dimensionless

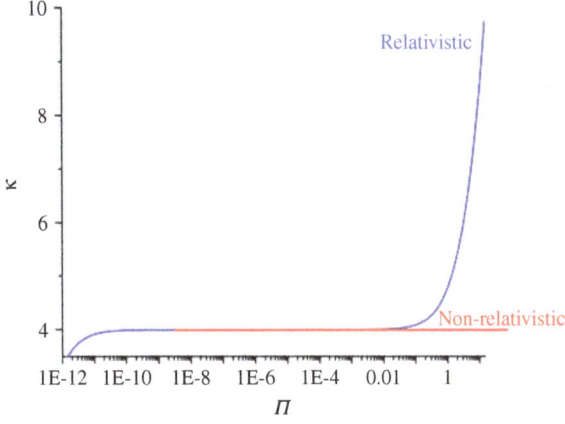

Figure 3. The compression $\kappa = \rho/\rho_0$ as a function of the shock wave dimensionless pressure $\Pi = P/(\rho_0 c^2)$. The numerical values are obtained for $\Gamma = 5/3$.

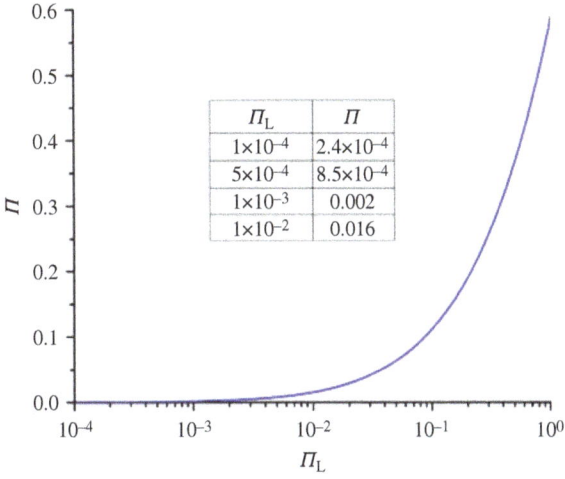

Figure 4. The dimensionless shock wave pressure $\Pi = P/(\rho_0 c^2)$ versus the dimensionless laser irradiance $\Pi_L = I_L/(\rho_0 c^3)$ in the range $10^{-4} < \Pi_L < 1$. For a better understanding of this graph the inserted table shows numerical values in the range $10^{-4} < \Pi_L < 10^{-2}$.

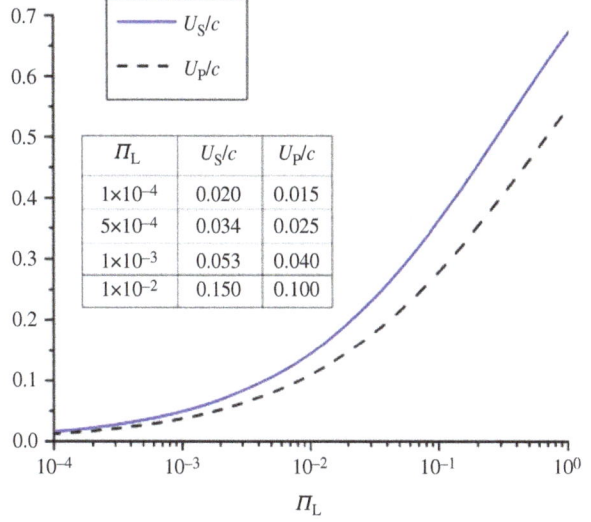

Figure 5. The dimensionless shock wave velocity u_s/c and the particle velocity u_p/c in the laboratory frame of reference versus the dimensionless laser irradiance $\Pi_L = I_L/(\rho_0 c^3)$ in the range $10^{-4} < \Pi_L < 1$. For a better understanding of this graph the inserted tables show numerical values in the range $10^{-4} < \Pi_L < 10^{-2}$.

The relativistic speed of sound c_S for an ideal gas EOS is

$$\frac{c_S}{c} = \sqrt{\left(\frac{\partial P}{\partial e}\right)_S} = \left(\frac{\Gamma P}{e + P}\right)^{1/2} = \left[\frac{\Gamma(\Gamma - 1)\Pi}{\Gamma \Pi + (\Gamma - 1)\kappa}\right]^{1/2}. \tag{16}$$

In the shocked medium the characteristic velocity of a disturbance from the piston to the shock wave front, equal to the rarefaction wave in the shocked medium c_{rw}, is given by

$$c_{rw} = \frac{c_S + u_p}{1 + \left(\frac{c_S u_p}{c^2}\right)}. \tag{17}$$

Figures 6(a) and 6(b) respectively describe the speed of sound in units of the speed of light, c_S/c, and the ratio of shock velocity to the rarefaction velocity, u_s/c_{rw} as a function of the dimensionless laser irradiance $\Pi_L = I_L/(\rho_0 c^3)$ in the range $10^{-4} < \Pi_L < 1$. The inserted tables show numerical values in the range $10^{-4} < \Pi_L < 10^{-2}$.

We now analyse the temperature problem. The partial pressures of an ideal gas that contains electrons and ions with appropriate densities n_e and n_i and temperatures T_e and T_i are P_e and P_i, and can be described by

$$P_e = n_e k_B T_e; \quad P_i = n_i k_B T_i. \tag{18}$$

If the associated photons in this system are in thermal equilibrium then a radiation temperature T_r can be defined, with a radiation pressure P_r given by[35]

shock wave pressure $\Pi = P_1/(\rho_0 c^2)$ versus the dimensionless laser irradiance $\Pi_L = I_L/(\rho_0 c^3)$ in the range $10^{-4} < \Pi_L < 1$. For a better understanding of this graph and for the practical proposal in the next section, the inserted table shows numerical values in the range $10^{-4} < \Pi_L < 10^{-2}$. Figure 5 describes the dimensionless shock wave velocity u_s/c and the particle velocity u_p/c in the laboratory frame of reference versus the dimensionless laser irradiance $\Pi_L = I_L/(\rho_0 c^3)$ in the range $10^{-4} < \Pi_L < 1$ while the inserted table shows numerical values in the range $10^{-4} < \Pi_L < 10^{-2}$. As a numerical example we take a target (liquid DT) with initial density $\rho_0 = 0.2$ g cm^{-3} irradiated by a laser with intensity $I_L = 5 \times 10^{22}$ W cm^{-2}, namely $\Pi_L = 9.26 \times 10^{-2}$. In this case our relativistic equations yield a compression $\kappa = \rho/\rho_0 = 4.09$, a pressure $P = 2 \times 10^{13}$ bar, a shock wave velocity $u_s = 0.35c$ and a particle velocity $u_p = 0.27c$, where c is the speed of light.

$$P_r = (1/3)aT_r^4;$$

$$a = \left(\frac{1}{15}\right)\left(\frac{k_B^4}{h^3c^3}\right) = 7.56 \times 10^{-15} \text{ [erg (cm}^{-3}\text{ K}^{-4})].}$$

(19)

For a plasma in local thermal equilibrium satisfying $T_e = T_i = T_r = T$, where the ions have an ionization Z and an atomic number A, implying a ion mass of Am_p, where m_p is the proton mass, the plasma pressure is given by

$$P = P_i + P_e + P_r = (Z+1)n_i k_B T + (\tfrac{1}{3})aT^4. \quad (20)$$

If the ion density satisfies

$$n_i \text{ [cm}^{-3}] \ll 1.56 \times 10^{27}\left(\frac{k_B T}{m_e c^2}\right), \quad (21)$$

then the radiation pressure is dominant and the temperature is given by

$$T \approx \left(\frac{3P}{a}\right)^{1/4}. \quad (22)$$

It is conceivable to assume that electrons and ions are in thermal equilibrium, i.e., $T_e = T_i$, however the shocked area is not optically thick for the energetic photons. In this case the energetic photons created by bremsstrahlung leave the system, implying $T_r \ll T_e$, or one can have a situation where radiation temperature is not defined at all. Therefore if the photon radiation in equation (19) is negligible then one has

$$k_B T = m_p c^2 \left(\frac{A}{Z+1}\right)\left(\frac{\Pi}{\kappa}\right). \quad (23)$$

Therefore in general we can write that the plasma temperature is constrained to the following range

$$\left(\frac{m_p c^2}{k_B}\right)\left(\frac{A}{Z+1}\right)\left(\frac{\Pi}{\kappa}\right) > T > \left(\frac{3P}{a}\right)^{1/4}. \quad (24)$$

Taking the example given above for liquid DT with $A = 2.5, Z = 1, m_p = 938.3$ MeV/c^2 and initial density $\rho = 0.2$ g cm^{-3} irradiated by a laser with intensity $I_L = 5 \times 10^{22}$ W cm^{-2}, namely $\Pi_L = 9.26 \times 10^{-2}$, we get $\Pi = 0.11, \kappa = 4.09$ and a temperature in the range 26.2 keV < $k_B T$ < 31.6 MeV. However, for $k_B T > 1$ MeV we have electron–positron pair production[43, 44] and new physics is required here for the temperature calculations. It is out of the scope of this paper to analyse this exotic case here.

4. An ultrafast ignition solution to the energy problem

In order to solve the energy problem of future generations scientists have considered using controlled nuclear fusion energy. One of the approaches is the well-known inertial confinement fusion driven by HPLs where the physics is based on compressing and igniting rather than confining the

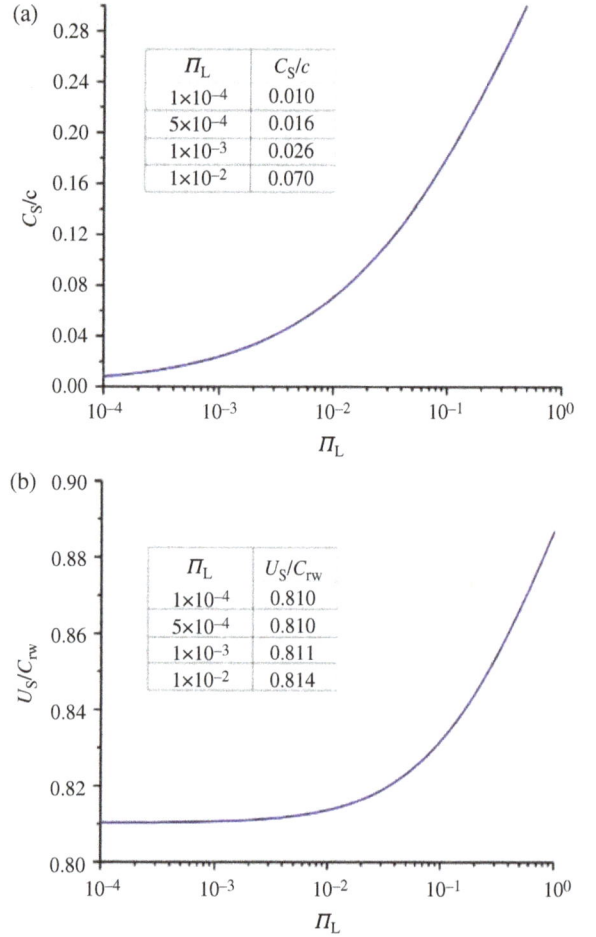

Figure 6. The speed of sound c_S is given in units of the speed of light c in (a) and the ratio of the shock velocity to the rarefaction velocity, u_S/c_{rw} is shown in (b) as function of the dimensionless laser irradiance $\Pi_L = I_L/(\rho_0 c^3)$ in the range $10^{-4} < \Pi_L < 1$. The inserted tables show numerical values in the range $10^{-4} < \Pi_L < 10^{-2}$.

fuel[1, 2]. In order to ignite the fuel with less energy it was suggested to separate the drivers that compress the target from whose that ignite the target[13, 14]. First the fuel is compressed, then a second driver ignites a small part of the fuel while the alpha particles created in the DT interaction heat the rest of the target. This idea is called FI. The problem with FI is that the laser pulse does not penetrate directly into the compressed target; therefore many alternative schemes have been suggested[45].

The laser solution of the energy problem requires very sophisticated high power laser science and engineering (HPLSE). In a recent paper[46] the various HPLSE optimizations and design constraints for a laser fusion power plant are beautifully summarized and analysed. From the many possible proposals to solve the energy problem with HPLs we consider three criteria for choosing the best candidate (present or future): (i) *Understanding the physics*. In HPL–target interactions there are many scientific problems not yet fully understood, such as laser–plasma instabilities,

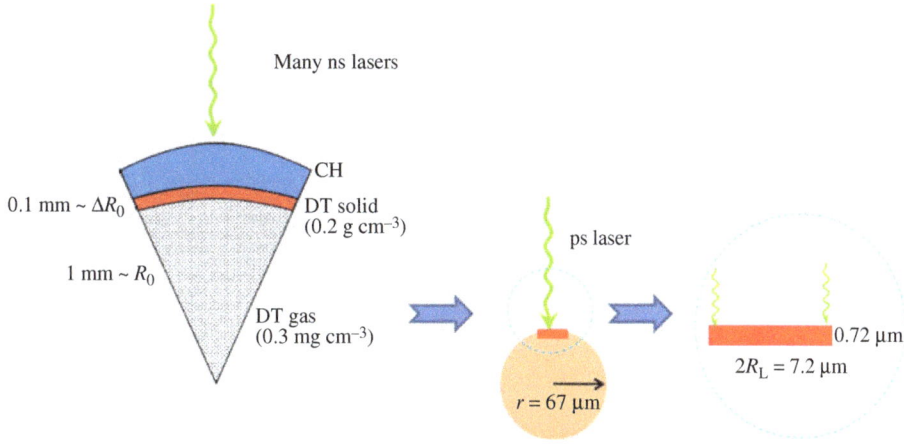

Figure 7. The FI scheme suggested in this paper. As a numerical example an initial pellet with radius $R_0 = 1$ mm and DT fuel of density 0.2 g cm^{-3} with thickness 0.1 mm (i.e., an aspect ratio of 10) is compressed to a density of $\rho_0 = 10^3$ g cm^{-3} by nanosecond lasers with a radius of 67 μm. The picosecond fast igniter laser with a 7.2 μm beam diameter creates a shock wave pulse with a thickness of 0.72 μm and can be considered a 1D shock wave to a reasonable approximation.

hydrodynamic instabilities, equations of state, nonlinear transport issues, non-local thermodynamic equilibrium, even neglecting energy conservation! (ii) *Engineering simplicity.* The IFE project is extremely complicated technologically and therefore a major effort is required in choosing the laser system, target design, etc., from all possible proposals by physicists. Technological simplicity must be seriously taken into account. For example, IFE requires 10^8 or more laser shots per year; therefore complicated target designs (such as inserting a golden cone inside a pellet) are not realistic. (iii) Last, but not least, IFE is supposed to be *economically practical.* This implies the required gain, defined as the nuclear energy output divided by the laser input per shot, be larger than 100 and that the cost of a target should not be more expensive than 0.1 US$.

Taking into account these three criteria it looks as though: (a) direct drive is simpler than indirect drive. (b) FI needs significantly less energy (approximately 0.3 MJ instead of 3 MJ). Therefore direct drive FI has the potential to be the best route to achieve nuclear fusion as an energy source. (c) From all presently known FI schemes the simplest FI seems to be by means of an 'extra shock' wave. We suggest a novel shock wave ignition scheme requiring less energy (in comparison with the present shock wave ignition scheme[25]) and free of laser–plasma instabilities (no more than $I_L \lambda_L^2 = 10^{14}$ (W cm^{-2})μm^2 in the laser compression pulses). In this proposal the ignition shock wave is created by a high irradiance laser and the shock wave is induced by a ponderomotive force in the intermediate domain between the relativistic and non-relativistic hydrodynamics. For this case the relativistic shock wave formalism has to be considered as developed in our previous section. We call our scheme ultrafast, since the laser pulse duration for the ignition process is significantly smaller, by one to two orders of magnitude.

The shock wave ignition criteria for DT nuclear fuel are

(i) $\rho R'' = \kappa \rho_0 (u_s - u_p)\tau_L \geqslant 0.3$ [g cm^{-2}]

(ii) $T \geqslant 10$ keV.
$$(25)$$

For the DT fusion one has $A = 2.5$ and $Z = 1$, therefore equation (23) for 10 keV temperatures and a compression of $\kappa = 4$ implies a minimum dimensionless pressure $\Pi_{min} = 3.4 \times 10^{-5}$. According to our solution the dimensionless laser irradiance satisfies $\Pi_L > \Pi_{L,min} = 1.8 \times 10^{-5}$. Π_{min} and $\kappa = 4$ imply a minimum shock velocity and particle velocity $u_s/c = 0.59 \times 10^{-2}$ and $u_p/c = 0.44 \times 10^{-2}$ respectively. Using these values in equation (25)$_{(i)}$, one gets a laser pulse duration of $\tau_L = 1.6$ ps. Assuming a pre-compression of $\rho_0 = 10^3$ g cm^{-3}, the $\Pi_{L,min} = 1.8 \times 10^{-5}$ requires $I_L = 4.8 \times 10^{22}$ W cm^{-2}. The shock wave thickness turn out to be $l_s = (u_s - u_p)\tau_L = 0.72$ μm. In order to have a 1D shock wave to a reasonable approximation we require a laser focal spot radius $R_L = 5l_s$, implying a laser cross section of $S = \pi R_L^2 = 4.0 \times 10^{-7}$ cm^2. In this case the laser energy W_L and power P_L are 30 kJ and 19 PW respectively. This example was taken to describe our concept in Figure 7. As a numerical example in this figure we take an initial pellet with radius $R_0 = 1$ mm and a DT fuel of density 0.2 g cm^{-3} with thickness 0.1 mm (i.e., an aspect ratio of 10) that is compressed to a density of $\rho_0 = 10^3$ g cm^{-3} (with a radius of 67 μm) by the nanosecond lasers. The picosecond fast igniter laser with a 7.2 μm in diameter creates a shock wave pulse with a thickness of 0.72 μm, which can be consider a 1D shock wave to a reasonably good approximation. The compressed pellet has a radius much larger than $\sqrt{S} \gg l_s$ in order to have a 1D shock wave. In Table 1 we show how larger values of Π_L change the laser and shock wave parameters.

The compression of a typical pellet as discussed in the literature[12, 47] requires between 100 and 300 kJ of energy,

Table 1. The laser is defined by its irradiance I_L, pulse duration τ_L, energy W_L and power P_L. This laser creates a shock wave with a compression κ in a pre-compressed target with an initial density ρ_0. The shock wave thickness ($= (u_s - u_p)\tau_L$, where u_s and u_p are the shock wave velocity and the particle velocity respectively) and its cross section are l_s and S, respectively, satisfying $\sqrt{S} \gg l_s$ in order to have a 1D shock wave.

Π_L	ρ_0 (g cm^{-3})	I_L (W cm^{-2})	κ	$(u_s - u_p)/c$	τ_L (ps)	l_s (μm)	S (cm^2)	W_L (kJ)	P_L (PW)
1.8×10^{-5}	10^3	4.8×10^{22}	4	0.15×10^{-2}	1.6	0.72	4.0×10^{-7}	30	19
1×10^{-4}	10^3	2.7×10^{23}	4	0.5×10^{-2}	0.5	0.75	4.4×10^{-7}	60	120
1×10^{-3}	10^3	2.7×10^{24}	4	1.3×10^{-2}	0.2	0.78	4.8×10^{-7}	260	1300

depending on the EOS, target design and the final required density. The FI in our case needs approximately 30 kJ of energy. Such a laser is under development and may be available in the near future.

5. Summary and discussion

Recently[31] it was suggested that relativistic shock waves with a shock wave velocity of more than 50% light speed can be created in the laboratory with HPLs which are recently under development. In this paper we discuss two novel ideas. The first is the transition domain between the relativistic and non-relativistic laser-induced shock waves. The second is the use of this transition domain for shock-wave-induced ultrafast ignition of a pre-compressed target. The laser parameters for these purposes are calculated and the main advantages of this scheme are described. The many laser beams with few nanosecond pulses that compresses the target do not require $I_L\lambda_L^2 = 10^{15}$ W cm^{-2} μm^2 as in the previously proposed shock wave ignition scheme[25], thus disturbing laser plasma instabilities do not occur. Furthermore, in the present scheme less energy is required in the main laser pulses where a picosecond laser with very high power ($\sim$30 PW) is required for the ultrafast ignition with the shock wave in the intermediate domain between the relativistic and non-relativistic hydrodynamics.

Presently existing petawatt lasers (see appendix B) might be used to start relativistic experimental research in the laboratory. Recent and future developments of HPLs in the multi-petawatt domain could be important for relativistic shock waves in the laboratory with pressures of 10^{15} atmospheres or energy densities of the order of 10^{14} J cm^{-3}. Such pressures or energy densities have been suggested so far only in astrophysical objects.

The ultrafast ignition scheme suggested in this paper appears advantageous in comparison with the many FI proposals, as given in our introduction section. It is based on the following merit criteria: (i) *Understanding the physics*, (ii) *Engineering simplicity*, and (iii) *Economically practical*. We think that shock wave FI is the best choice and the model suggested here between the relativistic and non-relativistic domain has significant advantages and should be taken seriously into account.

Finally we must mention a very recent FI proposal[48] using a laser system similar to our laser parameters estimated in Section 4. In particular this scheme requires a temporally tailored pulse with an energy of 65 kJ of duration 1.48 ps with a maximum intensity of 4×10^{22} W cm^{-2}. This model is based on the use of a hole–boring[39, 49, 50] phenomenon that enables the HPL beam to penetrate beyond the critical density. As early as 1971 it was analytically calculated[49] that the condition for a laser to propagate in non-uniform plasma with an electron density n_e and density gradient scale length L_n beyond the electron critical density n_c is given by

$$\frac{1}{4\pi}\left(\frac{L_n}{\lambda_L}\right)\left(\frac{n_c}{n_e}\right)^2 a^2 > 1$$

$$a \equiv \frac{eE_L\lambda_L}{2\pi m_e c^2} \approx 840\left[I_L\lambda_L^2/10^{24}\left(\frac{W \cdot \mu m^2}{cm^2}\right)\right]^{1/2}.$$

(26)

PIC simulations[39, 48, 50] derived $L_n \sim 20\lambda_L$, implying a laser penetration up to approximately 10^{24} cm^{-3} for a one micron wavelength laser with an intensity $I_L = 5 \times 10^{22}$ W cm^{-2}. The quasi mono-energetic ions[48] generated by the tailored laser pulses penetrate beyond this density to ignite the pre-compressed pellet. The *in situ* accelerated ions are the driver of FI.

In our model the laser-induced relativistic shock wave induces the ignition. The preliminary compressed fusion target for ICF is usually spherically symmetric and the density increases very rapidly towards the core of the target when our shock wave model is applied as described in Figure 7. As long as the thickness of the shock wave l_s is much smaller than the density gradient L_n, i.e., $L_n \gg l_s$, we can assume that the target is uniform in the shock wave domain. Looking at our Table 1 one gets $l_s \sim 1$ μm, which is much smaller than the density gradient $L_n \sim 20\lambda_L$ derived in PIC simulations[48, 50].

The solution suggested in this paper, like all other solutions to the energy problem, is extremely difficult scientifically, and a lot of money and enormous optimism is required for a positive solution. HPLSE is complex, complicated but possible. If civilization is to survive we need large new sources of energy. To quote Mark Twain (1835–1910): 'And what is a man without energy? Nothing – nothing at all'.

Appendix A

For convenience we write the nonrelativistic Hugoniot equations and the ideal gas EOS:

$$\text{(i) } u_{p1} = [P_1 - P_0]^{1/2} \left(\frac{1}{\rho_0} - \frac{1}{\rho_1} \right)^{1/2}$$

$$\text{(ii) } u_s = \left(\frac{1}{\rho_0} \right) \frac{[P_1 - P_0]^{1/2}}{\left(\frac{1}{\rho_0} - \frac{1}{\rho_1} \right)^{1/2}}$$

(A 1)

$$\text{(iii) } E_1 - E_0 = \left(\frac{1}{2} \right) [P_1 + P_0] \left(\frac{1}{\rho_0} - \frac{1}{\rho_1} \right)$$

$$\left. \begin{array}{c} \text{(iv)} \\ \text{(v)} \end{array} \right\} E_j = \left(\frac{1}{\Gamma - 1} \right) \left(\frac{P_j}{\rho_j} \right) \quad \text{for } j = 0, 1.$$

The equations are obtained from the relativistic equations (11) by using $e = \rho c^2 + \rho E$, P and ρE are much smaller than ρc^2 and $v/c \ll 1$.

Appendix B

In this appendix we give a list of petawatt lasers that are in use in different laboratories at the end of the year 2013. The following data are not officially confirmed – however, this was used in the literature and conferences according to our knowledge.

USA
Michigan University, Ann Arbor: 10J/30fs
Texas University, Austin: 186J/167fs
Berkeley National Laboratory: 40J/40fs
Rochester University, Rochester: 1 kJ/1ps
LLNL, Livermore: 600J/500fs
CHINA
Beijing National Laboratory: 32J/28fs
Shanghai Institute of Optics and Fine Mechanics: 35J/27fs
EUROPE
Central Laser Facility, UK: 500J/500fs & 15J/30fs
Jena, Germany: 120J/120fs
GSI Darmstadt, Germany: 500J/500fs
JAPAN
Osaka University: 500J/500fs
KOREA
Gwangju University: 34J/30fs

References

1. J. H. Nuckolls, L. Wood, A. Thiessen, and G. B. Zimmermann, Nature **239**, 139 (1972).
2. G. Velarde and N. Carpintero-Santamaria, *Inertial Confinement Nuclear Fusion: A Historical Approach by its Pioneers* (Foxwell and Davies Publ., 2007).
3. K. Mima, M. Murakami, S. Nakai, and S. Eliezer, *Applications of Laser–Plasma Interactions* S. Eliezer and K. Mima, eds. (CRC Press, Boca Raton, 2009).
4. J. M. Martinez Val, S. Eliezer, M. Piera, and G. Velarde, Phys. Lett. A **216**, 142 (1996).
5. S. Eliezer and J. M. Martinez Val, Laser Part. Beams **16**, 581 (1998).
6. S. Son and N. J. Fish, Phys. Lett. A **329**, 76 (2004).
7. S. Eliezer, *The Interaction of High Power Lasers with Plasmas* (CRC Press, Boca Raton, 2002).
8. C. G. M. Van Kessel and R. Sigel, Phys. Rev. Lett. **33**, 1020 (1974).
9. R. Cauble, D. W. Phillion, T. J. Hoover, N. C. Holmes, J. D. Kilkenny, and R. W. Lee, Phys. Rev. Lett. **70**, 2102 (1993).
10. E. I. Moses, Nucl. Fusion **49**, 104022 (2009).
11. J. D. Lindl, *Inertial Confinement Fusion: The Quest for Ignition and High Gain Using Indirect Drive* (Springer, New York, 1997).
12. S. Atzeni and J. Meyer-ter-Vehn, *The Physics of Inertial Fusion* (Clarendon Press, Oxford, 2004).
13. N. G. Basov, S. Y. Guskov, and L. P. Feoktistov, J. Sov. Laser Res. **13**, 396 (1992).
14. M. Tabak, J. Hammer, M. E. Glinsky, W. L. Kruer, S. C. Wilks, J. Woodworth, E. M. Campbell, M. D. Perry, and R. J. Mason, Phys. Plasmas **1**, 1626 (1994).
15. P. A. Norreys, R. Allot, R. J. Clarke, J. Colliers, D. Neely, S. J. Rose, M. Zepf, M. Santala, A. R. Bell, K. Krushelnick, A. E. Dangor, N. C. Woolsey, R. G. Evans, H. Habara, T. Norimatsu, and R. Kodama, Phys. Plasmas **7**, 3721 (2000).
16. M. Roth, E. T. Cowan, M. H. Key, S. P. Hatchett, C. Brown, W. Fountain, J. Johnson, D. M. Pennington, R. A. Snavely, S. C. Wilks, K. Yasuike, H. Ruhl, F. Pegoraro, S. V. Bulanov, E. M. Campbell, M. D. Perry, and H. Powell, Phys. Rev. Lett. **86**, 436 (2001).
17. R. Kodama, P. A. Norreys, K. Mima, A. E. Dangor, R. G. Evans, H. Fujita, Y. Kitagawa, K. Krushelnick, T. Miyakoshi, N. Miyanaga, T. Norimatsu, S. J. Rose, T. Shozaki, K. Shigemori, A. Sunahara, M. Tampo, K. A. Tanaka, Y. Toyama, and M. Zepf, Nature **412**, 798 (2001).
18. J. M. Martinez Val and M. Piera, Fusion Tech. **32**, 131 (1997).
19. A. Caruso and C. Strangio, Laser Part. Beams **19**, 295 (2001).
20. S. Y. Guskov, Quantum Electron. **31**, 885 (2001).
21. P. Lalousis, I. B. Foldes, and H. Hora, Laser Part. Beams **30**, 233 (2012).
22. P. Lalousis, H. Hora, S. Eliezer, J. M. Martinez Val, S. Moustaizis, G. H. Miley, and G. Mourou, Phys. Lett. A **377**, 885 (2013).
23. M. Murakami, H. Nagatomo, H. Azechi, F. Ogando, M. Perlado, and S. Eliezer, Nucl. Fusion **46**, 9 (2006).
24. S. Eliezer, J. M. Martinez Val, and C. Deutsch, Laser Part. Beams **13**, 43 (1995).
25. R. Betti, C. D. Zhou, K. S. Anderson, L. J. Perkins, W. Theobald, and A. A. Sokolov, Phys. Rev. Lett. **98**, 155001 (2007).
26. S. Eliezer and J. M. Martinez Val, Laser Part. Beams **29**, 175 (2011).
27. S. Eliezer and S. V. Pinhasi, High Power Laser Sci. Eng. **1**, 44 (2013).
28. S. Jackel, D. Saltzmann, A. Krumbein, and S. Eliezer, Phys. Plasmas **26**, 3138 (1983).
29. V. E. Fortov and I. V. Lomonosov, Shock Waves **20**, 53 (2010).
30. S. Eliezer, *Laser–Plasma Interactions and Applications*, P. D. McKenna, P. D. Neely, R. Bingham and D. A. Jaroszynski, eds. 68th Scottish Universities Summer School in Physics, p. 49 (Springer Publication, Heidelberg, 2013).

31. S. Eliezer, N. Nissim, E. Raicher, and J. M. Martinez Val, Laser Part. Beams **32**, 243 (2014).

32. L. D. Landau and E. M. Lifshitz, *Fluid Mechanics* 2nd edition (Pergamon Press, Oxford, 1987).

33. A. H. Taub, Phys. Rev. **74**, 3 (1948).

34. Y. B. Zeldovich and Y. P. Raizer, *Physics of Shock Waves and High Temperature Hydrodynamic Phenomena* (Academic Press Publications, New York, 1966).

35. S. Eliezer, A. Ghatak, and H. Hora, *Fundamental of Equation of State* (World Scientific, Singapore, 2002).

36. S. Eliezer and R. A. Ricci, *High Pressure Equation of State: Theory and Application*, Enrico Fermi School CXIII 1989, (North Holland Publication, Amsterdam, 1991).

37. S. Eliezer and H. Hora, Phys. Rep. **172**, 339 (1989).

38. S. Eliezer, N. Nissim, J. M. Martinez Val, K. Mima, and H. Hora, Laser Part. Beams **32**, 211 (2014).

39. N. Naumova, T. Schlegel, V. T. Tikhonchuk, C. Labaune, I. V. Sokolov, and G. Mourou, Phys. Rev. Lett. **102**, 025002 (2009).

40. S. Eliezer, J. M. Martinez Val, and S. V. Pinhasi, Laser Part. Beams **31**, 113 (2013).

41. T. Esirkepov, M. Borghesi, S. V. Bulanov, G. Mourou, and T. Tajima, Phys. Rev. Lett. **92**, 175003 (2004).

42. H. Hora, P. Lalousis, and S. Eliezer, Phys. Rev. Lett. **53**, 1650 (1984).

43. A. Di Piazza, C. Muller, K. Z. Hatsagortsyan, and C. H. Keitel, Rev. Modern Phys. **84**, 1177 (2012).

44. R. Ruffini, G. Vereshchagin, and S. Xue, Phys. Rep. **487**, 1 (2010).

45. S. Y. Guskov, Plasma Phys. Rep. **39**, 1 (2013).

46. S. E. Bodner, A. J. Schmitt, and J. D. Sethian, High Power Laser Sci. Eng. **1**, 2 (2013).

47. S. Eliezer, M. Murakami, and J. M. Martinez Val, Laser Part. Beams **25**, 585 (2007).

48. S. M. Weng, M. Murakami, H. Azechi, J. W. Wang, N. Tasoko, M. Chen, Z. M. Sheng, P. Mulser, W. Yu, and B. F. Shen, Phys. Plasmas **21**, 012705 (2014).

49. C. Max and F. Perkins, Phys. Rev. Lett. **20**, 1342 (1971).

50. A. Pukhov and J. Meyer-ter-Vehn, Phys. Rev. Lett. **79**, 2686 (1997).

20

Ion motion effects on the generation of short-cycle relativistic laser pulses during radiation pressure acceleration

W. P. Wang, X. M. Zhang, X. F. Wang, X. Y. Zhao, J. C. Xu, Y. H. Yu, L. Q. Yi, Y. Shi, L. G. Zhang, T. J. Xu, C. Liu, Z. K. Pei, and B. F. Shen

State Key Laboratory of High Field Laser Physics, Shanghai Institute of Optics and Fine Mechanics, Chinese Academy of Sciences, P. O. Box 800-211, Shanghai 201800, China

Abstract

The effects of ion motion on the generation of short-cycle relativistic laser pulses during radiation pressure acceleration are investigated by analytical modeling and particle-in-cell simulations. Studies show that the rear part of the transmitted pulse modulated by ion motion is sharper compared with the case of the electron shutter only. In this study, the ions further modulate the short-cycle pulses transmitted. A 3.9 fs laser pulse with an intensity of 1.33×10^{21} W cm^{-2} is generated by properly controlling the motions of the electron and ion in the simulations. The short-cycle laser pulse source proposed can be applied in the generation of single attosecond pulses and electron acceleration in a small bubble regime.

Keywords: radiation pressure acceleration; short-cycle pulses; particle-in-cell simulations.

With the rapid development of laser technology[1–3], the use of ultra-intense laser irradiation on ultra-thin foils has been studied for various fields ranging from fast ignition for inertial confinement fusion[4, 5], medical therapy[6, 7], and laboratory astrophysics[8], to the generation of high-energy particle sources[1]. However, prepulses or laser pulses with slowly increasing fronts may lead to premature ionization of the target, and significant expansion of the plasma sheet occurs before the amplitude peak of the pulse arrives[9, 10]. These events jeopardize the relativistic interaction of the ultra-thin target. Thus, high-contrast[9–20] and short-duration laser pulses[21, 22] are needed.

Plasma mirrors may be a feasible method by which to solve these problems. Using a double plasma mirror[23], the laser contrast may be improved to $\sim 10^{-12}$ at the picosecond time scale[12]. A 30 nm thick carbon foil irradiated by such a high-contrast laser at an intensity of $\sim 7 \times 10^{19}$ W cm^{-2} can produce a 185 MeV carbon ion beam. A few-cycle laser pulse with an intensity up to 3×10^{20} W cm^{-2} may also be generated when a laser irradiates an ultra-thin foil[22]. In this method, the pulse duration and intensity mainly depend on the laser profile and foil conditions. Generally, the intensity of few-cycle laser pulses is limited in conventional optical methods[21] because of the relatively low damage threshold of the optical components and other problems. However, no laser intensity due to material damage would be limited by this method, because this process only involves laser–plasma interactions, like the plasma grating[24].

Ref. [22] discusses the generation of a short-cycle laser pulse. It is also necessary to discuss an explicit explanation of the difference and the progress. Ref. [22] considers in detail the transmission of the incident laser pulse in constant conditions. There the electron layer is assumed to be at rest, and the ion motion effects are not considered. In fact the ion motion can significantly modulate the transmission of the laser. For example, the Doppler effect on the mirror is enhanced because the velocities of the ions initially in the middle of the foil are higher than that of the compressed electron layer (CEL) during the hole-boring stage. The rear part of the incident laser pulse can be reflected in this case.

In this letter, the effects of ion motion on the generation of short-cycle relativistic laser pulses are investigated by analytical modeling and particle-in-cell (PIC) simulations. The generation of a near single-cycle laser pulse has been obtained in the simulations, and the corresponding theoretical analysis has been discussed[22]. However, all the solutions are based on static conditions, where the motions of the CEL and ions are not included. In fact the ion motion can

Correspondence to: Wenpeng Wang, State Key Laboratory of High Field Laser Physics, Shanghai Institute of Optics and Fine Mechanics, Chinese Academy of Sciences, Shanghai 201800, China. Email: wwpvin@hotmail.com

significantly modulate the transmission of the laser. For example, the Doppler effect on the electron layer mirror is enhanced because the velocities of the ions initially in the middle of the foil are higher than that of the CEL during the hole-boring stage. At the end, the incident laser pulse can be further reflected. One-dimensional (1D) PIC simulations show that the ion motion can further modulate the transmitted short-cycle pulse compared with the case of the electron shutter only. No transmitted pulse is generated when the ions are accelerated together with the CEL at the end of the hole-boring stage. The dynamics of the electrons and ions during laser–plasma interaction are investigated using a simple model. A 3.9 fs laser pulse with an intensity of 1.33×10^{21} W cm^{-2} is generated by properly controlling the electrons and ions in two-dimensional (2D) PIC simulations. Such a short-cycle high-intensity laser pulse has important applications in single attosecond pulse generation[25] and electron acceleration in a small bubble regime[26, 27].

Figure 1 shows the interaction model. As an intense circularly polarized (CP) laser pulse irradiates a thin foil, the laser pressure[28–36] quickly sweeps all electrons forward in a compressed layer. Initially, the protons are left behind because their mass ($m_i = 1836m_e$) is much larger than the electron mass m_e. As the CEL is further pushed forward, the maximum charge separation field at the surface of the compressed layer increases with the depletion distance d. When the CEL reaches the back side of the target, as shown in Figure 1, the charge separation field, $E_0 = 4\pi e n_0 d$, becomes larger. At this stage, the front portion of the pulse with the smaller intensity is reflected by the foil, while the portion with the larger intensity begins to propagate through the foil[37]. As the ions catch up with the electron layer, the transmission is reduced and the rear portion of the pulse is reflected by the foil. Thus, only the part of the laser pulse with the highest intensity is transmitted with a much shorter duration than the incident pulse.

1D PIC simulations are used to study the effects of ion motion on the generation of a short-cycle relativistic laser pulse. A CP laser pulse with wavelength $\lambda = 1$ μm is incident on the target from the left boundary. The laser front arrives at the front surface of the target at $t = 20T_0$, where $T_0 = \lambda/c$ is the laser cycle and c is the speed of light. The laser pulse has a trapezoidal shape profile (linear growth–plateau–linear decrease) with a duration of $9.66T_0$ ($4.08T_0 - 1.5T_0 - 4.08T_0$). Here, the short width of the flat top ($1.5T_0$) is used to reduce the content of high frequencies. The frequency of the trapezoidal laser pulse used in this case is mainly at the base frequency c/λ. The laser amplitude gradient is $a_0/t_{up} = 7.35$ ($a_0 = 30$ and $t_{up} = 4.08T_0$). Here, $a_0 = eE_L/m_e\omega_L c$ is the normalized amplitude, where m_e and e are the electron mass and charge, respectively, E_L is the laser electric field, ω_L is the laser frequency, and t_{up} is the rising time of the laser pulse. The front surface of the target is located at $x = 20\lambda$. The foil density is $n_0 = 8n_c$ and the foil thickness is $l_0 = 1.03\lambda$. Here, $n_c = \omega_L^2 m_e/4\pi e^2$ is

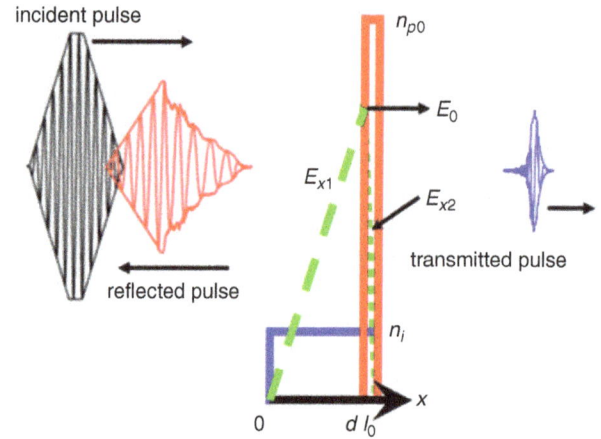

Figure 1. Scheme for generating nearly single-cycle laser pulses. The incident pulse irradiates a thin foil, producing an ultra-short transmitted pulse and a reflected pulse. Electrostatic fields E_{x1} (green dashed line) and E_{x2} (green dotted line) are produced at both sides of the surface (at $x = d$) of the CEL (red solid line) at the initial stage of the interaction. Ions (blue solid line) remain at rest. The distribution of the electrons corresponds to the case where the CEL just reaches the back side of the target. The CEL then oscillates and disperses, as shown in Figures 2c and 2d.

the critical density. Low-density plasma with a step density profile is used to simplify the model and reduce the simulation time. The longitudinal length of the 1D simulation box is $x = 60\lambda$. The mesh size is $\lambda/200$. Each cell contains 100 numerical macro particles in the plasma region.

Figure 2a shows the evolution of electrons and ions. The laser front arrives at the front surface of the foil at $t = 20T_0$, and the electrons are pushed forward in a thin compressed layer. The peak density of the layer increases with time. Both the CEL and the amplitude peak of the incident laser pulse arrive at the back side of the foil ($x = 21.03\lambda$) at $t \sim 25.3T_0$. The peak density of the CEL is $n = 192.5n_c$ at this time. Then, the electron layer begins to disperse in the vacuum, and the density decreases sharply to $n = 143.9n_c$ [Figure 2a]. The incident pulse begins to transmit. However, the electron density increases again up to $n = 406 \cdot 4n_c$ at $t \sim 26T_0$. The ions catch up with the CEL, forming a double layer with the electrons. The remainder part of the incident pulse is then reflected and the transmitted field drops sharply from its peak value. Figure 2b shows that a short-cycle transmitted laser pulse with a duration of $\sim 0.9T_0$ and amplitude peak $a_t = 16$ is produced. Here, we consider the cases that the shortening of the incident laser pulse does not seriously depend on the harmonic generation. The central frequency is barely changed, meaning that there is almost no frequency shift compared with the incident laser pulse. The number of lightwave cycles is indeed reduced by the pulse shortening in our case. The frequency broadening should be taken into account when the duration of the transmitted pulse is much shorter than the laser period, which is not considered in this paper. Figures 2c and 2d show the case of ions at rest. The duration of the transmitted laser pulse increases to $2.8T_0$

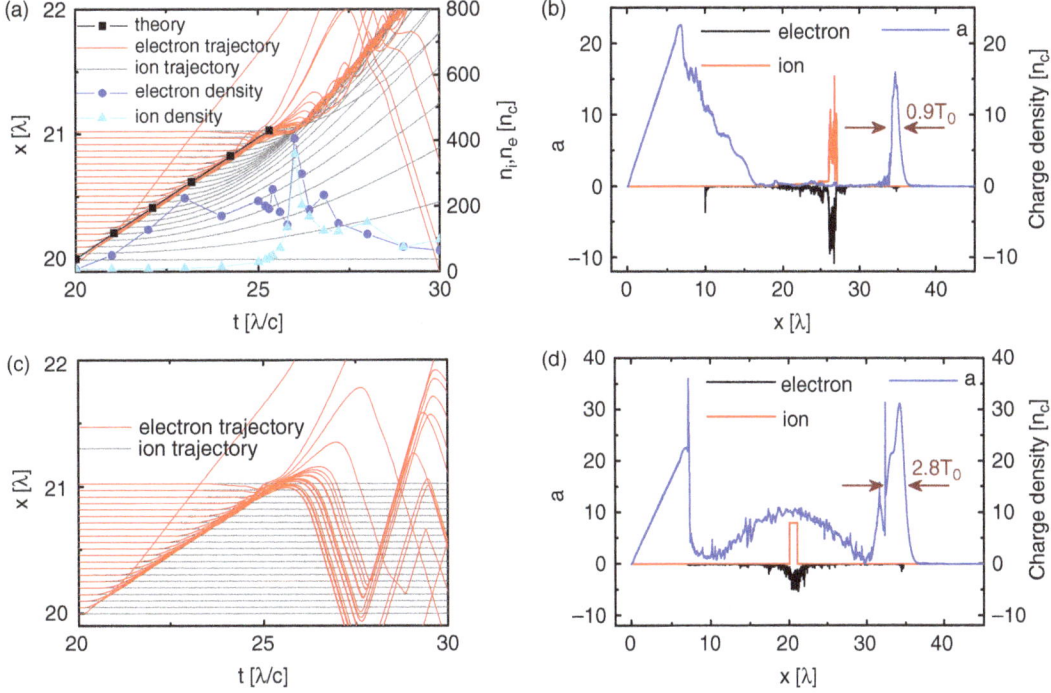

Figure 2. 1D PIC simulation results for $a_0/t_{up} = 7.35$ ($a_0 = 30$ and $t_{up} = 4.08T_0$), $n_0 = 8n_c$, and $l_0 = 1.03\lambda$. (a) Electron and proton trajectories and their density peaks versus time. (b) Laser profile (blue solid line) and charge density of electrons (black solid line) and ions (red solid line) at $t = 40T_0$ for the case of moving ions. (c) Electron and proton trajectories and (d) laser profile (blue solid line) and charge density of electrons (black solid line) and ions (red solid line) at $t = 40T_0$ for the case of ions at rest.

at this point [Figure 2d]. Compared with Figures 2b and 2d, the ion motion has an important role in modulating the rear part of the incident laser pulse.

The simulations above show that the ion motion is important for the generation of short-cycle laser pulses. The main reason is that the re-entering of the ions into the CEL at the back side of the foil can further reflect the rear part of the laser pulse. In the case of Figure 2a, the laser pulse begins to transmit the foil as the CEL slightly disperses at the back side of the foil. Then the ions re-enter into the CEL, and a double-layer reflecting mirror is produced. In this stage, the Doppler effect on the mirror is enhanced because the velocities of the ions initially in the middle of the foil are higher than that of the CEL during the hole-boring stage. At the end, the rear part of the incident laser pulse is reflected.

The dynamics of the electrons and ions are investigated to obtain insights into the generation of short-cycle lasers. The action of the electric field in the CEL E_{x2} (Figure 1) is initially neglected for ion acceleration because the velocity of the CEL is large[35]. Moreover, the ions do not catch the CEL when this layer arrives at the back side of the target[35]. Here, the unrelativistic interaction of the laser pulse with the linearly increasing front in the hole-boring stage is considered to simplify our study. A time-dependent theory model can be used to deal with the relativistic interaction of the laser with any pulse profile in Ref. [38].

For electrons, a uniform velocity of the CEL can be obtained for a laser with a linearly increasing front in

simulations[35]. The electrons are first accelerated by the ponderomotive force of the CP laser pulse within a very short time, and then a balance between the laser pressure and the electrostatic forces appears. Then the swept electrons pile up to form a dense skin-depth layer by the laser pressure and electrostatic forces. We define this skin-depth layer as the CEL. The equation of motion of a perfectly reflecting mirror, $dp/dt = Pd\sigma$, can be used to describe the motion of the CEL, where p is the momentum of the mirror element, $d\sigma$ is the element of area of the mirror and normal to the mirror surface, and P is the radiation pressure[39]. In this paper, we focus on the case that the ions move after the electrons during the hole-boring stage because the laser-pulse front is very sharp. The laser energy is largely contained in the charge separation field during this stage. The velocities of the electron layer and ions are not relativistic. Simulations have shown that the surface of the CEL moves roughly at a constant velocity; if the laser-pulse front is linearly increasing and short enough, there must be a balance between the electrostatic force and the laser pressure force[35],

$$2(\pi n_0 v_{CEL} t)^2 = 2a^2 \frac{1 - v_{CEL}}{1 + v_{CEL}}, \qquad (1)$$

where the normalized amplitude of the laser on the moving CEL surface $a = (a_0/t_{up})(t - v_{CEL}t)$, and $l = v_{CEL}t$. Here, v_{CEL}, t_{up}, and n_0 are normalized by c, λ/c, and $n_c = \omega_L^2 m_e/4\pi e^2$, respectively. Although the linear amplitude increases, a part of the electrons is moving while another part is stationary; most electrons move around a averaged velocity during the hole-boring stage. Such averaged velocity of the skin-depth reflecting mirror has been verified in the simulations[38]. Equation (1) describes the velocity of the CEL for the case that the laser is totally reflected by the plasma. The transmission of the laser pulse is not considered during the hole-boring stage in our case. Based on Equation (1), a uniform velocity of the CEL ($v_{CEL} \approx 0.194c$) is obtained for $a_0 = 30$, $t_{up} = 4.08T$, $n_0 = 8n_c$, and $l_0 = 1.03\lambda$, which is consistent with the simulation result in Figure 2a.

For ions, the velocity for the ion initially at rest at x_{10} ($0 < x_{10} < d$) can be approximated by[35]

$$v_{i1} = \frac{2\pi q_i E_0}{m_i} \frac{x_{10}}{d} t_1, \qquad (2)$$

where x_{10}, x_{i1}, and d are normalized by λ, and the ion mass m_i and charge q_i are normalized by m_e and e, respectively. In addition, the action time t_1 is normalized by λ/c. $E_0 = 2\pi n_0 d$ is the maximum electrostatic field of charge depletion, which is normalized by $e/m_e\omega_L c$. $E_0 = 2\pi n_0 l_0$ is obtained when the CEL is assumed to be superthin ($d \approx l_0$)[35]. For the case shown in Figure 2a, $E_0 \approx 52$ for $n_0 = 8n_c$, and $l_0 = 1.03\lambda$.

The velocity of the ions initially at rest in the middle of the foil ($x_{10} = d/2$) $v_{i1} \sim 0.24c$ is obtained using Equation (2). Here, $t_1 \sim 0.5d/v_{CEL}$ is used. Ions initially resting in the middle of the foil run fast during the sharp-front laser interaction at the end of the hole-boring stage[35]. The CEL quickly catches up to these ions after the hole-boring stage because the electron mass is much smaller than the ion mass. Therefore, we can assume that $v_{CEL} \sim 0.24c$ at this point, which is larger than the velocity of the CEL ($\sim 0.194c$) during the hole-boring stage. The Doppler effect parameter $(c - v_{CEL})/(c + v_{CEL})$ is thus reduced by the ion motion. Moreover, the reflection of the laser pulse is enhanced. The rear part of the transmitted pulse can be further modulated by the ion motion compared with the case of the electron shutter only, which is verified by the simulations in Figures 2a–2d.

From the discussion above, two conditions are required for the generation of an intense short-cycle transmitted pulse. First, the peak of the incident pulse must arrive at the back surface when the CEL disperses, thereby generating a short-duration transmitted pulse with higher amplitude. This condition is simply an approximation because the exact amplitude of the transmitted pulse is not considered. Second, the ions must not catch up to the CEL during the hole-boring stage. Otherwise all of the laser pulses may be reflected.

For the first condition, the laser and foil parameters are as follows:

$$a_{open} > \frac{a_0}{t_{up}} \left(\frac{l_0}{v_{CEL}} - l_0 \right), \qquad (3)$$

where a_{open} is the peak amplitude of the incident laser pulse that may open the electron shutter. For a certain laser amplitude $a_{open} = a_0$ and gradient a_0/t_{up}, the incident laser can transmit if $l_0 < t_{up} \cdot v_{CEL}/(1 - v_{CEL})$ according to Equation (3). Here, v_{CEL} is calculated using Equation (1). The laser peak cannot reach the back side of the target when $l_0 \gg t_{up} \cdot v_{CEL}/(1 - v_{CEL})$. In this case, the incident laser is completely reflected, and no transmitted laser pulse can be generated. For the case shown in Figure 2, $l_0 < 0.98\lambda$ is obtained for $a_0/t_{up} = 7.35$ ($a_0 = 30$ and $t_{up} = 4.08T_0$), and $n_0 = 8n_c$, which is consistent with the simulation results (Figure 3). Figure 3 shows that the transmitted laser pulse with $a_0 > 30$ is generated for $l_0 < 0.98\lambda$. A nearly single-cycle laser pulse is generated for $l_0 = 1.03\lambda$, as shown in Figure 3.

For the second condition, the laser and foil parameters are[35]

$$l_0 < \frac{1}{\pi} \sqrt{\frac{2m_i}{q_i n_0}} v_{CEL}, \qquad (4)$$

where $l_0 = \sqrt{2m_i/q_i n_0} v_{CEL}/\pi$ corresponds to the case where ions are initially in the middle of the foil and the CEL reaches the back surface of the foil at the same time. The incident laser is completely reflected if $l_0 \geqslant \sqrt{2m_i/q_i n_0} v_{CEL}/\pi$[35], which is verified by the simulation results in Figure 3. $l_0 \geqslant 1.32\lambda$ is obtained for $a_0/t_{up} = 7.35$ ($a_0 = 40.4$ and $t_{up} = 5.5T_0$), and $n_0 = 8n_c$. Figure 3b shows that no transmitted laser pulse is generated for $l_0 = 1.32\lambda$. A thinner foil ($l_0 = 1.2\lambda$) is simulated for the same laser and foil parameters to verify the condition of Equation (4). Figure 3c shows that a quasi-single-cycle transmitted pulse is generated, where the ions does not catch up the CEL as the CEL leaves the back surface of the foil. This indicates that ion motion indeed affects the transmitting of the laser pulse. Here, a higher density ($n_0 = 200n_c$) of the foil is also theoretically estimated. The velocity of the CEL $v_{CEL} \sim 0.07c$ is obtained for an ultra-intense laser pulse ($a_0/t_{up} \sim 54$) based on Equation (1). Then $l_0 < 0.1\lambda$ can be obtained according to Equation (4). And $a_{open} > 57.4$ is obtained for $l_0 = 0.08\lambda$ according to Equation (3). In Figure 3c, the duration of the transmitted pulse is much shorter than the laser period when the foil thickness further increases ($l_0 > 1\lambda$). The frequency broadening should be taken into account when the duration of the transmitted pulse is much shorter than the laser period, which is not considered in this paper.

Different simulations of carbon ions are also performed to verify the theory. The velocity of the CEL $v_{CEL} \sim 0.2c$ is obtained for $a_0/t_{up} \sim 15$ and $n_0 = 15n_c$ according to Equation (1). $l_0 < 1.4\lambda$ is calculated based on Equation (4) for C^{6+} ions, where $m_i = 12$ and $q_i = 6$. Here, the foil thickness $l_0 \sim 1.25\lambda$ is used in the simulation. And $a_0 = 80$ is chosen to satisfy the condition $a_{open} > 75$ according to Equation (3). Figure 4 shows that a near single-cycle laser pulse is generated, indicating that the conditions [Equations (3) and (4)] can be used to deal with different ion masses. However, a smaller value of m_i/q_i is suggested for the generation of a short-cycle transmitted laser pulse. The

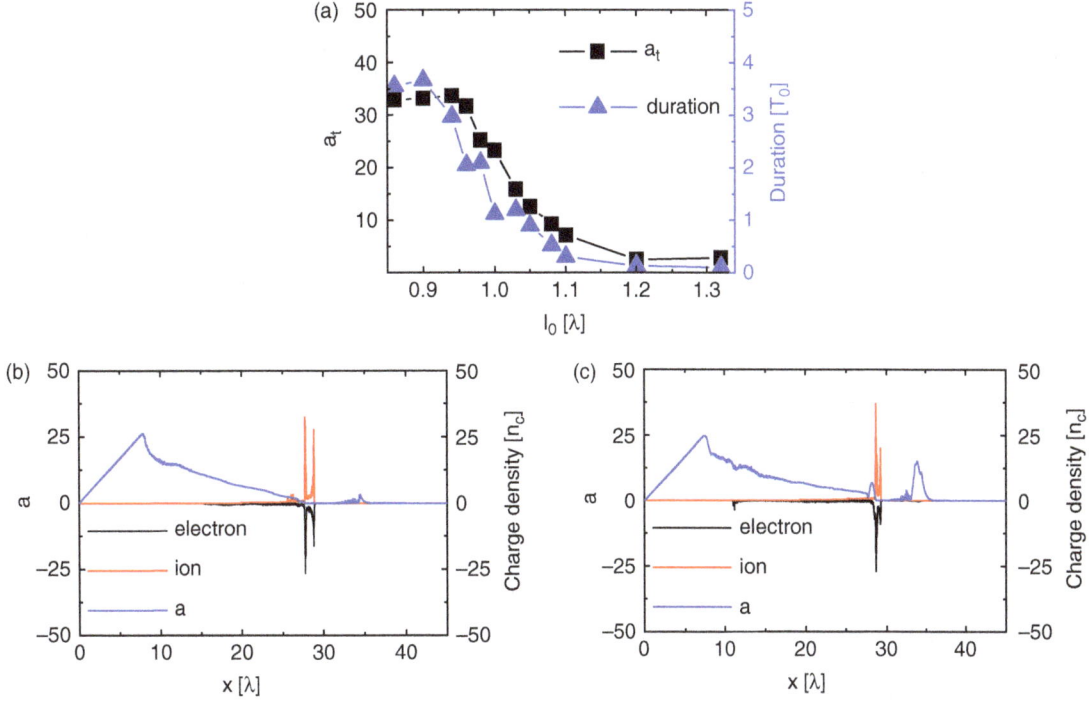

Figure 3. (a) Amplitude peak a_t (black square) and duration (blue triangle) of transmitted pulses versus foil thickness l_0. For the incident laser, $a_0/t_{up} = 7.35$ ($a_0 = 30$ and $t_{up} = 4.08T_0$). The foil density is $n_0 = 8n_c$. Laser profile (blue solid line) and charge density of electrons (black solid line) and ions (red solid line) for $a_0/t_{up} = 7.35$ ($a_0 = 40.4$ and $t_{up} = 5.5T_0$) and (b) $l_0 = 1.32\lambda$, (c) 1.2λ.

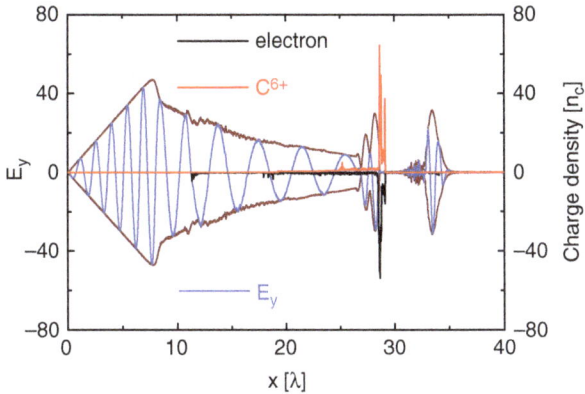

Figure 4. Laser profile E_y (blue solid line) and charge density of electrons (black solid line) and C^{6+} ions (red solid line) for $a_0/t_{up} = 15$ ($a_0 = 80$ and $t_{up} = 6T_0$). The foil density is $n_0 = 15n_c$, and the foil thickness is $l_0 = 1.25\lambda$.

main reason is that a thinner foil is obtained according to Equation (4), and some multidimensional effects and instabilities can be reduced to a certain extent for a thinner foil.

A clean laser pulse is used to simplify the model in this work. In fact, the effect of a laser prepulse is always critical for a thin foil and it may alter the conditions [see Equations (3) and (4)]. A density gradient at the front or back side of the foil produced by the prepulse must be considered if the laser contrast is not ultra-high[9, 10]; such a condition is not considered in this work.

To verify our theoretical model, we also carried out 2D PIC simulations. The same parameters as in the 1D PIC simulations [see Figures 2a and 2b], which include $a_0 = 30$, $t_{up} = 4.08T_0$, $n_0 = 8n_c$, and $l_0 = 1.03\lambda$, are used in the 2D case. The laser pulse has a trapezoidal shape profile ($4.08T_0 - 1.5T_0 - 4.08T_0$) in time and is a transverse four-order super Gaussian, $I \sim \exp[-(r/r_0)^4]$, where $r_0 = 10\lambda$. The simulation region is a $50\lambda \times 50\lambda$ box (3000 cells along the laser axis x, and 3000 cells transversely along axis y). The plasma foil occupies the region from $x = 20\lambda$ to 21.03λ and from $x = -22.5\lambda$ to 22.5λ. Ten macroparticles are available in each cell. Absorbing boundaries are used for both electromagnetic waves and macroparticles.

Figure 5 shows the electron and ion distributions and the laser profile in the (x, y) plane at $t = 40T_0$. From the laser axial profiles at $y = 0\lambda$, a nearly single-cycle transmitted pulse is produced. The laser duration is about 3.9 fs (FWHM), and the amplitude peak is $a_t \sim 22$ ($I_0 = 2a_t^2 \times 1.37 \times 10^{18}$ W cm^{-2} $= 1.33 \times 10^{21}$ W cm^{-2}). Compared with the results from 1D simulations in Figure 2b, the duration and amplitude peak are larger in the 2D case, which is mainly attributed to foil deformation and self-focusing of the laser in the 'hole-boring' stage. Figure 5 shows that the transmitted laser amplitude decreases sharply from its peak because ions catch the slightly dispersed electron layer at the end of the hole-boring stage. Here, the ion motion further modulates the rear part of the incident laser pulse. In addition, the transverse width is also reduced when the

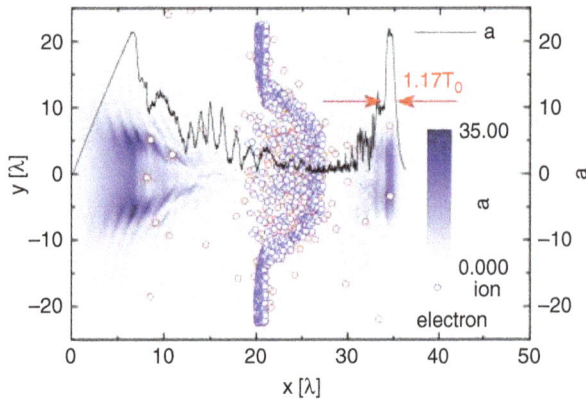

Figure 5. Electron (red circle) and ion (blue circle) distribution and the laser profiles in the (x, y) plane for foils with $n_0 = 8n_c$ and $l_0 = 1.03$ irradiated by CP laser pulses with $a_0/t_{up} = 7.35$ ($a_0 = 30$ and $t_{up} = 4.08T_0$) at $t = 40T_0$. The axial laser profiles at $y = 0$ are denoted by the black solid lines.

laser mainly passes through the foil. Figure 4 further shows that the transverse width is reduced from 10λ (FWHM) to about 6λ (FWHM). The front and back portions of the pulse are reflected after laser–foil interaction. About 92% of the energy of the incident laser pulse is reflected, as shown in Figure 5.

In conclusion, the effects of ion motion on the generation of a short-cycle relativistic laser pulse are investigated by analytical modeling and PIC simulations. 1D PIC simulations show that the ion shutter can further modulate the transmitted short-cycle pulse compared with the case of the electron shutter only. Two conditions are theoretically proposed to generate short-cycle transmitted laser pulses, which are proven by the simulations. A near single-cycle (3.9 fs) laser pulse with an intensity of 1.33×10^{21} W cm^{-2} is generated by properly controlling the electron and ion shutters in 2D PIC simulations.

Acknowledgements

This work was supported by the 973 Program (No. 2011CB808104), the National Natural Science Foundation of China (Nos. 11335013, 10834008, 11125526, 60921004, and 11305236), the International S&T Cooperation Program of China (No. 2011DFA11300), and Shanghai Natural Science Foundation (No. 13ZR1463300).

References

1. G. A. Mourou, T. Tajima, and S. V. Bulanov, Rev. Modern Phys. **78**, 309 (2006).
2. V. Chvykov, P. Rousseau, S. Reed, G. Kalinchenko, and V. Yanovsky, Op. Lett. **31**, 1456 (2006).
3. V. Yanovsky, V. Chvykov, G. Kalinchenko, P. Rousseau, T. Planchon, T. Matsuoka, A. Maksimchuk, J. Nees, G. Cheriaux, G. Mourou, and K. Krushelnick, Opt. Express **16**, 2109 (2008).
4. M. Tabak, J. Hammer, M. E. Glinsky, W. L. Kruer, S. C. Wilks, J. Woodworth, E. M. Campbell, M. D. Perry, and R. J. Mason, Phys. Plasmas **1**, 1626 (1994).
5. N. Naumova, T. Schlegel, V. T. Tikhonchuk, C. Labaune, I. V. Sokolov, and G. Mourou, Phys. Rev. Lett. **102**, 025002 (2009).
6. S. V. Bulanov, T. Z. Esirkepov, V. S. Khoroshkov, A. V. Kuznetsov, and F. Pegoraro, Phys. Lett. A **299**, 240 (2002).
7. S. V. Bulanov and V. S. Khoroshkov, Plasma Phys. Rep. **28**, 453 (2002).
8. B. A. Remington, Science **284**, 1488 (1999).
9. W. P. Wang, H. Zhang, B. Wu, C. Y. Jiao, Y. C. Wu, B. Zhu, K. G. Dong, W. Hong, Y. Q. Gu, B. F. Shen, Y. Xu, Y. X. Leng, R. X. Li, and Z. Z. Xu, Appl. Phys. Lett. **101**, 214103 (2012).
10. W. P. Wang, B. F. Shen, H. Zhang, Y. Xu, Y. Y. Li, X. M. Lu, C. Wang, Y. Q. Liu, J. X. Lu, Y. Shi, Y. X. Leng, X. Y. Liang, R. X. Li, N. Y. Wang, and Z. Z. Xu, Appl. Phys. Lett. **102**, 224101 (2013).
11. A. Mackinnon, Y. Sentoku, P. Patel, D. Price, S. Hatchett, M. Key, C. Andersen, R. Snavely, and R. Freeman, Phys. Rev. Lett. **88**, 215006 (2002).
12. A. Henig, D. Kiefer, K. Markey, D. C. Gautier, K. A. Flippo, S. Letzring, R. P. Johnson, T. Shimada, L. Yin, B. J. Albright, K. J. Bowers, J. C. Fernández, S. G. Rykovanov, H. C. Wu, M. Zepf, D. Jung, V. K. Liechtenstein, J. Schreiber, D. Habs, and B. M. Hegelich, Phys. Rev. Lett. **103**, 045002 (2009).
13. A. Henig, S. Steinke, M. Schnürer, T. Sokollik, R. Hörlein, D. Kiefer, D. Jung, J. Schreiber, B. M. Hegelich, X. Q. Yan, J. Meyer-ter-Vehn, T. Tajima, P. V. Nickles, W. Sandner, and D. Habs, Phys. Rev. Lett. **103**, 245003 (2009).
14. D. Neely, P. Foster, A. Robinson, F. Lindau, O. Lundh, A. Persson, C. G. Wahlstrom, and P. McKenna, Appl. Phys. Lett. **89**, 021502 (2006).
15. T. Ceccotti, A. Lévy, H. Popescu, F. Réau, P. D'Oliveira, P. Monot, J. Geindre, E. Lefebvre, and P. Martin, Phys. Rev. Lett. **99**, 185002 (2007).
16. S. A. Gaillard, T. Kluge, K. A. Flippo, M. Bussmann, B. Gall, T. Lockard, M. Geissel, D. T. Offermann, M. Schollmeier, Y. Sentoku, and T. E. Cowan, Phys. Plasmas **18**, 056710 (2011).
17. A. Zigler, T. Palchan, N. Bruner, E. Schleifer, S. Eisenmann, M. Botton, Z. Henis, S. A. Pikuz, A. Y. Faenov, Jr, D. Gordon, and P. Sprangle, Phys. Rev. Lett. **106**, 134801 (2011).
18. K. Ogura, M. Nishiuchi, A. S. Pirozhkov, T. Tanimoto, A. Sagisaka, T. Z. Esirkepov, M. Kando, T. Shizuma, T. Hayakawa, H. Kiriyama, T. Shimomura, S. Kondo, S. Kanazawa, Y. Nakai, H. Sasao, F. Sasao, Y. Fukuda, H. Sakaki, M. Kanasaki, A. Yogo, S. V. Bulanov, P. R. Bolton, and K. Kondo, Optics Lett. **37**, (2012).
19. M. Kaluza, J. Schreiber, M. I. K. Santala, G. D. Tsakiris, K. Eidmann, J. Meyer-ter-Vehn, and K. J. Witte, Phys. Rev. Lett. **93**, 045003 (2004).
20. Y. Xu, J. Wang, Y. Huang, Y. Li, X. Lu, and Y. Leng, High Power Laser Sci. Eng. **1**, 98 (2013).
21. F. Tavella, Y. Nomura, L. Veisz, V. Pervak, A. Marcinkevicius, and F. Krausz, Optics Lett. **32**, 2227 (2007).
22. L. L. Ji, B. F. Shen, X. M. Zhang, F. C. Wang, Z. Y. Jin, C. Q. Xia, M. Wen, W. P. Wang, J. C. Xu, and M. Y. Yu, Phys. Rev. Lett. **103**, 215005 (2009).
23. T. Wittmann, J. P. Geindre, P. Audebert, R. S. Marjoribanks, J. P. Rousseau, F. Burgy, D. Douillet, T. Lefrou, K. Ta Phuok, and J. P. Chambaret, Rev. Sci. Instrum. **77**, 083109 (2006).

24. H.-C. Wu, Z.-M. Sheng, and J. Zhang, Appl. Phys. Lett. **87**, 201502 (2005).

25. T. Brabec and F. Krausz, Rev. Modern Phys. **72**, 545 (2000).

26. A. Pukhov and J. Meyer-ter-Vehn, Appl. Phys. B **74**, 355 (2002).

27. K. Schmid, L. Veisz, F. Tavella, S. Benavides, R. Tautz, D. Herrmann, A. Buck, B. Hidding, A. Marcinkevicius, U. Schramm, M. Geissler, J. Meyer-ter-Vehn, D. Habs, and F. Krausz, Phys. Rev. Lett. **102**, 124801 (2009).

28. B. F. Shen and Z. Z. Xu, Phys. Rev. E **64**, 056406 (2001).

29. T. Esirkepov, M. Borghesi, S. V. Bulanov, G. Mourou, and T. Tajima, Phys. Rev. Lett. **92**, 175003 (2004).

30. A. Macchi, F. Cattani, T. V. Liseykina, and F. Cornolti, Phys. Rev. Lett. **94**, 165003 (2005).

31. X. Q. Yan, C. Lin, Z. M. Sheng, Z. Y. Guo, B. C. Liu, Y. R. Lu, J. X. Fang, and J. E. Chen, Phys. Rev. Lett. **100**, 135003 (2008).

32. X. Zhang, B. Shen, X. Li, Z. Jin, F. Wang, and M. Wen, Phys. Plasmas **14**, 123108 (2007).

33. A. P. L. Robinson, M. Zepf, S. Kar, R. G. Evans, and C. Bellei, New J. Phys. **10**, 013021 (2008).

34. B. Qiao, M. Zepf, M. Borghesi, and M. Geissler, Phys. Rev. Lett. **102**, 145002 (2009).

35. W. P. Wang, B. F. Shen, X. M. Zhang, L. L. Ji, M. Wen, J. C. Xu, Y. H. Yu, Y. L. Li, and Z. Z. Xu, Phys. Plasmas **18**, 013103 (2011).

36. T.-P. Yu, A. Pukhov, G. Shvets, and M. Chen, Phys. Rev. Lett. **105**, 065002 (2010).

37. V. A. Vshivkov, N. M. Naumova, F. Pegoraro, and S. V. Bulanov, Phys. Plasmas **5**, 2727 (1998).

38. W. P. Wang, B. F. Shen, X. M. Zhang, L. L. Ji, Y. H. Yu, L. Q. Yi, X. F. Wang, and Z. Z. Xu, Phys. Rev. Special Topics - Accelerators and Beams **15**, 081302 (2012).

39. F. Pegoraro and S. V. Bulanov, Phys. Rev. Lett. **99**, 065002 (2007).

21

Conceptual designs of a laser plasma accelerator-based EUV-FEL and an all-optical Gamma-beam source

Kazuhisa Nakajima

Center for Relativistic Laser Science, Institute for Basic Science (IBS), Gwangju 500-712, Republic of Korea

Abstract

Recently, intense research into laser plasma accelerators has achieved great progress in the production of high-energy, high-quality electron beams with GeV-level energies in a cm-scale plasma. These electron beams open the door for broad applications in fundamental, medical, and industrial sciences. Here we present conceptual designs of an extreme ultraviolet radiation source for next-generation lithography and a laser Compton Gamma-beam source for nuclear physics research on a table-top scale.

Keywords: high peak high average power lasers; laser wakefield accelerators

1. Introduction

To date, intense research has been carried out on laser plasma acceleration concepts[1] to achieve high-energy, high-quality electron beams with GeV energies in a cm-scale plasma[2–6], a 1%-level energy spread[7], a 1 mm mrad level transverse emittance[8], and a 1 fs level bunch duration[9], ensuring that the stability of reproduction is as high as that of present high-power ultra-short-pulse lasers[10]. Recently, staged laser plasma acceleration[11–13] has been successfully demonstrated in conjunction with ionization-induced injection[14–16] and phase-locking acceleration[17]. Relativistic electron beams from ultraintense laser plasma interactions can be conceived to be compact particle accelerators, inspiring a wide range of applications of unique particle beam and radiation sources, such as THz[18, 19] and X-ray/Gamma-ray radiation[20–25].

Here we present an extreme ultraviolet (EUV) radiation source for next-generation lithography and a laser Compton Gamma-beam source for nuclear physics research. EUV lithography with wavelengths below 13.5 nm is capable of providing resolution below 30 nm in semiconductor manufacturing. We propose a self-amplified spontaneous emission (SASE) free electron laser (FEL) driven by relativistic electron beams from laser plasma accelerators. For example, this FEL system, capable of generating an average EUV power of 1 kW at 13.5 nm, comprises a fiber-based chirped pulse amplification (CPA) laser delivering a 1 MW average laser power, a 5 cm gas cell-type plasma accelerator producing a 660 MeV electron beam with a 1.6% relative energy spread and a 0.5 nC charge, and a 1 m long undulator with a 15 mm period and a 1.4 T peak magnetic field.

High-quality Gamma beams generated from inverse Compton scattering off relativistic electron beams interacting with an intense laser pulse have aroused interest in photonuclear physics and nuclear astrophysics research, the characterization of nuclear materials or radioactive waste and so on. We present a table-top all-optical laser plasma accelerator-based Gamma-beam source comprising a high-power laser system with synchronous dual outputs, a laser plasma accelerator producing 300–900 MeV electron beams, and scatter optics whereby the laser pulse is focused onto the electron beam to generate a Gamma beam via Compton scattering with photon energies of 2–20 MeV.

2. Design of laser plasma accelerators for driving electron beams

2.1. Accelerator stage

Most of the laser plasma acceleration experiments that have successfully demonstrated the production of quasi-monoenergetic electron beams with a narrow energy spread have been elucidated in terms of self-injection and acceleration mechanisms in the bubble regime[26, 27], where a drive laser pulse with wavelength λ_L, peak power P_L, intensity I_L, and focused spot radius r_L is characterized by a normalized vector potential $a_0 \gg 1$ with respect to the electron rest energy mc^2, given for the linear polarization as

Correspondence to: K. Nakajima, Center for Relativistic Laser Science, Institute for Basic Science (IBS), Gwangju 500-712, Republic of Korea. Email: naka115@dia-net.ne.jp

$$a_0 = \left(\frac{2e^2\lambda_L^2 I_L}{\pi m^2 c^5} \right)^{1/2}$$

$$\cong 8.55 \times 10^{-10} \sqrt{I_L(\text{W cm}^{-2})} \lambda_L \, (\mu\text{m})$$

$$\approx 6.82 \sqrt{P_L(\text{TW})} \frac{\lambda_L}{r_L}. \tag{1}$$

In these experiments, electrons are self-injected into a non-linear wake, often referred to as a bubble, i.e., a cavity void of plasma electrons consisting of a spherical ion column surrounded by a narrow electron sheath, formed behind the laser pulse instead of a periodic plasma wave in the linear regime. The phenomenological theory of nonlinear wakefields in the bubble (blowout) regime[26] describes the accelerating wakefield $E_z(\xi)/E_0 \approx (1/2)k_p\xi$ in the bubble frame moving in a plasma with velocity v_B, i.e., $\xi = z - v_B t$, where $k_p = \omega_p/c = (4\pi r_e n_e)^{1/2}$ is the plasma wavenumber evaluated with a plasma frequency ω_p, an unperturbed on-axis electron density n_e and the classical electron radius $r_e = e^2/mc^2$, and $E_0 = mc\omega_p/e$ is the non-relativistic wave-breaking field, approximately given by $E_0 \approx 96 \, (\text{GV m}^{-1})(n_e/10^{18} \, (\text{cm}^{-3}))^{1/2}$. In the bubble regime for $a_0 \geqslant 2$, since an electron-evacuated cavity shape is determined by balancing the Lorentz force of the ion sphere exerted on the electron sheath with the ponderomotive force of the laser pulse, the bubble radius R_B is approximately given as $k_p R_B \approx 2\sqrt{a_0}$[27]. Thus, the maximum accelerating field is given by $E_{z0}/E_0 = (1/2)\alpha k_p R_B$, where α represents a factor taking into account the difference between the theoretical estimation and the accelerating field reduction due to the beam loading effects.

Here we consider the self-guided case, where a drive laser pulse propagates in a homogeneous density plasma. The equations of longitudinal motion of an electron with normalized energy $\gamma = E_b/mc^2$ and longitudinal velocity $\beta_z = v_z/c$ are written approximately as[28]

$$\frac{d\gamma}{dz} = \frac{1}{2}\alpha k_p^2 R_B \left(1 - \frac{\xi}{R_B} \right),$$

$$\frac{d\xi}{dz} = 1 - \frac{\beta_B}{\beta_z} \approx 1 - \beta_B \approx \frac{3}{2\gamma_g^2}, \tag{2}$$

where $\xi = z - v_B t$ $(0 \leqslant \xi \leqslant R_B)$ is the longitudinal coordinate of the bubble frame moving at a velocity of $v_B = c\beta_B \approx v_g - v_{\text{etch}}$, taking into account diffraction at the laser pulse front that etches back at a velocity $v_{\text{etch}} \sim ck_p^2/k^2$[27] with laser wavenumber k, and $\gamma_g = (1-\beta_g^2)^{-1/2} \approx k/k_p \gg 1$ is assumed. Integrating Equations (2), the energy and phase of the electron can be calculated as[28]

$$\gamma(z) = \gamma_0 + \frac{1}{3}\alpha\gamma_g^2 k_p^2 R_B \xi(z) \left(1 - \frac{1}{2}\frac{\xi(z)}{R_B} \right),$$

$$\xi(z) = \frac{3}{2}\frac{z}{\gamma_g^2}, \tag{3}$$

where $\gamma_0 = \gamma(0)$ is the injection energy. Hence, the maximum energy gain is obtained at $\xi = R_B$ as

$$\Delta\gamma_{\max} = \gamma_{\max} - \gamma_0 \approx \frac{1}{6}\alpha\gamma_g^2 k_p^2 R_B^2 \approx \frac{2}{3}\alpha a_0 \gamma_g^2$$

$$= \frac{2}{3}\alpha\kappa_{\text{self}} a_0 \frac{n_c}{n_e}, \tag{4}$$

where κ_{self} is the correction factor of the relativistic factor for the group velocity in a uniform plasma for a self-guided pulse, i.e., $\gamma_g^2 = (1-\beta_g^2)^{-1} \approx \kappa_{\text{self}} k^2/k_p^2 = \kappa_{\text{self}} n_c/n_e$, obtained from[28]

$$\kappa_{\text{self}} = \frac{a_0^2}{8} \left(\sqrt{1 + a_0^2/2} - 1 - \ln\frac{\sqrt{1 + a_0^2/2} + 1}{2} \right)^{-1}, \tag{5}$$

and $n_c = m\omega_L^2/4\pi e^2 = \pi/(r_e\lambda_L^2) \approx 1.115 \times 10^{21} \, (\text{cm}^{-3})$ $(\lambda_L/1 \, \mu\text{m})^{-2}$ is the critical plasma density. The dephasing length L_{dp} for the self-guided bubble regime is given by

$$k_p L_{\text{dp}} \approx \frac{2}{3}k_p R_B \gamma_g^2 = \frac{4}{3}\sqrt{a_0}\kappa_{\text{self}}\frac{n_c}{n_e}. \tag{6}$$

For a given energy gain E_b, the operating plasma density is determined from Equation (4) as

$$n_e = \frac{2}{3}\alpha\kappa_{\text{self}} a_0 \frac{n_c}{\Delta\gamma_{\max}}$$

$$\approx 1.9 \times 10^{18} \, (\text{cm}^{-3})\kappa_{\text{self}} a_0$$

$$\times \left(\frac{1 \, \mu\text{m}}{\lambda_L} \right)^2 \left(\frac{200 \, \text{MeV}}{E_b/\alpha} \right). \tag{7}$$

The accelerator length equal to the dephasing length becomes

$$L_{\text{acc}} = L_{\text{dp}} \approx \sqrt{\frac{3}{2}} \frac{(\Delta\gamma_{\max}/\alpha)^{3/2}}{\pi\kappa_{\text{self}}^{1/2} a_0} \lambda_L$$

$$\approx \frac{3.1 \, (\text{mm})}{\kappa_{\text{self}}^{1/2} a_0} \left(\frac{\lambda_L}{1 \, \mu\text{m}} \right) \left(\frac{E_b/\alpha}{200 \, \text{MeV}} \right)^{3/2}, \tag{8}$$

while the pump depletion length due to pulse-front erosion is given by

$$L_{\text{pd}} \approx c\tau_L \frac{n_c}{n_e} = \frac{3}{2}\frac{c\tau_L \Delta\gamma_{\max}/\alpha}{\kappa_{\text{self}} a_0}$$

$$\approx \frac{5 \, (\text{mm})}{\kappa_{\text{self}} a_0} \left(\frac{\tau_L}{30 \, \text{fs}} \right) \left(\frac{E_b/\alpha}{200 \, \text{MeV}} \right). \tag{9}$$

The dephasing length should be less than the pump depletion length, i.e., $L_{\text{pd}} \geqslant L_{\text{dp}}$. Thus, the required pulse duration for self-guiding of the drive laser pulse is given by

$$\tau_L \geqslant 18 \text{ (fs)} \kappa_{\text{self}}^{1/2} \left(\frac{\lambda_L}{1\,\mu\text{m}} \right) \left(\frac{E_b/\alpha}{200\,\text{MeV}} \right)^{1/2}. \qquad (10)$$

The matched spot radius becomes

$$r_m \approx 3.9 \text{ (}\mu\text{m)} \frac{R_m}{\sqrt{\kappa_{\text{self}}} a_0} \left(\frac{\lambda_L}{1\,\mu\text{m}} \right) \left(\frac{E_b/\alpha}{200\,\text{MeV}} \right)^{1/2}, \quad (11)$$

where $R_m \equiv k_p r_L$ is the dimensionless matched spot radius given by[28]

$$R_m = \left\{ \frac{\ln(1 + a_0^2/2)}{\sqrt{1 + a_0^2/2} - 1 - 2\ln[(\sqrt{1 + a_0^2/2} + 1)/2]} \right\}^{1/2}. \qquad (12)$$

The corresponding matched power is calculated as

$$P_L = \frac{k_p^2 r_L^2 a_0^2}{32} P_c \approx 0.312 \text{ (TW)} \frac{a_0 R_m^2}{\kappa_{\text{self}}} \left(\frac{E_b/\alpha}{200\,\text{MeV}} \right). \qquad (13)$$

The required pulse energy becomes

$$U_L = P_L \tau_L \geqslant 5.62 \text{ (mJ)} \frac{a_0 R_m^2}{\kappa_{\text{self}}^{1/2}} \left(\frac{\lambda_L}{1\,\mu\text{m}} \right) \left(\frac{E_b/\alpha}{200\,\text{MeV}} \right)^{3/2}. \qquad (14)$$

2.2. Beam loading effects

In laser wakefield acceleration, an accelerated electron beam induces its own wakefield and cancels the laser-driven wakefield. Assuming the beam loading efficiency $\eta_b \equiv 1 - E_z^2/E_M^2$ defined by the fraction of plasma wave energy absorbed by particles of the bunch with a root mean square (r.m.s.) radius σ_b, the beam-loaded field is given by $E_z = \sqrt{1 - \eta_b} E_M = \alpha E_M$, where E_M is the accelerating field without beam loading, given by $E_M \approx a_0^{1/2} E_0$ for the bubble regime $a_0 \geqslant 2$. Thus, a loaded charge is calculated as[29]

$$Q_b \simeq \frac{e}{4 k_L r_e} \frac{\eta_b k_p^2 \sigma_b^2}{1 - \eta_b} \frac{E_z}{E_0} \left(\frac{n_c}{n_e} \right)^{1/2}$$

$$\approx 76 \text{ (pC)} \frac{\eta_b a_0^{1/2} k_p^2 \sigma_b^2}{\sqrt{1 - \eta_b}} \left(\frac{n_e}{10^{18}\,\text{cm}^{-3}} \right)^{-1/2}. \qquad (15)$$

Using the plasma density Equation (7), the loaded charge is given by

$$Q_b \approx 55 \text{ (pC)} \frac{1 - \alpha^2}{\alpha^{3/2}} \frac{k_p^2 \sigma_b^2}{\kappa_{\text{self}}^{1/2}} \left(\frac{\lambda_L}{1\,\mu\text{m}} \right) \left(\frac{E_b}{200\,\text{MeV}} \right)^{1/2}. \qquad (16)$$

Therefore, the field reduction factor α for accelerating charge Q_b up to energy E_b is obtained by solving the equation

$$\alpha^2 + C\alpha^{3/2} - 1 = 0, \qquad (17)$$

where the coefficient C is defined as

$$C \equiv \frac{Q_b}{55\text{ (pC)}} \frac{\kappa_{\text{self}}^{1/2}}{k_p^2 \sigma_b^2} \left(\frac{1\,\mu\text{m}}{\lambda_L} \right) \left(\frac{200\,\text{MeV}}{E_b} \right)^{1/2}. \qquad (18)$$

2.3. Injector stage

Electron beams can be produced and accelerated in the injector stage driven by the same laser pulse as that in the accelerator stage, relying on a self-injection mechanism such as the expanding bubble self-injection mechanism[30] or an ionization-induced injection scheme with a short mixed gas cell[14–16, 31], where tunnel ionization leads to electron trapping near the centre of the laser wakefield. Here we consider the ionization-induced injection scheme. According to theoretical considerations in ionization-induced injection[31], for trapping electrons ionized at the peak of the laser electric field, the minimum laser intensity is given by $1 - \gamma_g^{-1} \leqslant 0.64 a_{\min}^2$. At a plasma density $n_{\text{inj}} = 10^{18}$ cm^{-3} in the injector, the required minimum laser field is $a_{\min} \geqslant 1.23$. The maximum number of trapped electrons saturates at approximately $N_{e\,\max} \sim 5 \times 10^6$ μm^{-2} at a gas length $L_{\text{inj}} \approx 1000\lambda_L$ for a plasma density $n_{\text{inj}} = 0.001 n_c$ with a nitrogen concentration of $\alpha_N = 1\%$ and laser parameters of $a_0 = 2$ and $c\tau_L \approx 15\lambda_L$ due to the beam loading effects and initially trapped particle loss from the separatrix in the phase space. From the particle-in-cell (PIC)-simulation results[31], the trapped electron density scales as

$$N_e \text{ (}\mu\text{m}^{-2}) \sim 8 \times 10^7 \alpha_N k_p L_{\text{inj}} \left(\frac{n_{\text{inj}}}{n_c} \right)^{1/2}$$

$$\approx 5 \times 10^8 \alpha_N \left(\frac{L_{\text{inj}}}{\lambda_L} \right) \left(\frac{n_{\text{inj}}}{n_c} \right). \qquad (19)$$

The energy spread is also proportional to both the mixed gas length and the nitrogen concentration. In a injector with gas length L_{inj}, the electron charge Q_b trapped inside a bunch with radius $r_b = 1/k_p \approx 5.3$ (μm) at $n_{\text{inj}} = 10^{18}$ cm^{-3} is estimated as

$$Q_b \sim \frac{k_p^2 r_b^2}{4 r_e n_{\text{inj}}} e N_e \approx 6.4 \text{ (pC)} \alpha_N k_p^2 r_b^2 \left(\frac{\lambda_L}{1\,\mu\text{m}} \right) \left(\frac{L_{\text{inj}}}{1\,\mu\text{m}} \right). \qquad (20)$$

An electron charge of 500 pC will be trapped via the ionization-induced injection mechanism in an injector gas cell with a 2 mm length and a nitrogen concentration of $\alpha_N = 4\%$.

2.4. Design of a SASE FEL

In the SASE FEL process, coupling the electron bunch with a co-propagating undulator radiation field induces an energy modulation of electrons that yields current modulation of the bunch due to the dispersion of the undulator dipole fields, known as microbunching. It means that the electrons are grouped into small bunches separated by a fixed distance that resonantly coincides with the wavelength of the radiation field. Consequently, the radiation field can be amplified coherently. In the absence of an initial resonant radiation field, a seed may build up from spontaneous incoherent emission in the SASE process.

The design of the FEL-based EUV light source is carried out using one-dimensional FEL theory as follows[32]. The FEL amplification takes place in an undulator with undulator period λ_u and peak magnetic field B_u at a resonant wavelength λ_X given by

$$\lambda_X = \frac{\lambda_u}{2\gamma^2}\left(1 + \frac{K^2}{2}\right), \qquad (21)$$

where $\gamma = E_b/m_e c^2$ is the relativistic factor of the electron beam energy E_b, and $K_u = 0.934 B_u$ (T)λ_u (cm) $= \gamma\theta_e$ is the undulator parameter, which is related to the maximum electron deflection angle θ_e. In the high-gain regime required for the operation of a SASE FEL, an important parameter is the Pierce parameter ρ_{FEL}, given by

$$\rho_{FEL} = \frac{1}{2\gamma}\left[\frac{I_b}{I_A}\left(\frac{\lambda_u K_u A_u}{2\pi\sigma_b}\right)^2\right]^{1/3}, \qquad (22)$$

where I_b is the beam current, $I_A = 17$ kA is the Alfven current, σ_b is the r.m.s transverse size of the electron bunch, and the coupling factor is $A_u = 1$ for a helical undulator and $A_u = J_0(\Xi) - J_1(\Xi)$ for a planar undulator, where $\Xi = K_u^2/[4(1 + K_u^2/2)]$ and J_0 and J_1 are Bessel functions of the first kind. Another important dimensionless parameter is the longitudinal velocity spread Λ of the beam normalized by the Pierce parameter:

$$\Lambda^2 = \frac{1}{\rho_{FEL}^2}\left[\left(\frac{\sigma_\gamma}{\gamma}\right)^2 + \left(\frac{\varepsilon\lambda_u}{4\lambda_X\beta}\right)^2\right]$$
$$= \frac{1}{\rho_{FEL}^2}\left[\left(\frac{\sigma_\gamma}{\gamma}\right)^2 + \left(\frac{\varepsilon_n^2}{2\sigma_b^2(1 + K_u^2/2)}\right)^2\right], \qquad (23)$$

where σ_γ/γ is the relativistic r.m.s. energy spread, ε is the r.m.s. transverse emittance, $\beta = \sigma_b^2/\varepsilon$ is the beta function provided by the guiding field (undulator plus external focusing) and ε_n is the normalized emittance, defined as $\varepsilon_n \equiv \gamma\varepsilon$, assuming that the beta function is constant along the length of the undulator. The e-folding gain length L_{gain} over which

the power grows exponentially according to $\exp(2s/L_{gain})$ is given by

$$L_{gain} = \frac{\lambda_u}{4\pi\sqrt{3}\rho_{FEL}}(1 + \Lambda^2). \qquad (24)$$

In order to minimize the gain length, one needs a large Pierce parameter ρ_{FEL} and a normalized longitudinal velocity spread Λ sufficiently low compared to unity, which means a sufficiently small energy spread σ_γ/γ and ε. This expression applies to a moderately small beam size σ_b such that the diffraction parameter $B \gg 1$, where B is defined as

$$B = \frac{16\pi^2 A_u \sigma_b^2}{\lambda_X\lambda_u}\left[\frac{K_u^2/2}{\gamma(1 + K_u^2/2)}\frac{I_b}{I_A}\right]^{1/2}. \qquad (25)$$

The saturation length L_{sat} required to saturate the amplification can be expressed as

$$L_{sat} = L_{gain}\ln\left[\left(\frac{\Lambda^2 + 3/2}{\Lambda^2 + 1/6}\right)\frac{P_{sat}}{P_{in}}\right], \qquad (26)$$

where P_{in} and P_{sat} are the input power and the saturated power, which are related to the electron beam power P_b according to

$$P_b = \gamma I_b m_e c^2 = I_b E_b,$$
$$P_{sat} \cong 1.37 \rho_{FEL} P_b \exp(-0.82\Lambda^2),$$
$$P_{in} \cong 3\sqrt{4\pi}\rho_{FEL}^2 P_b[N_{\lambda_X}\ln(N_{\lambda_X}/\rho_{FEL})]^{-1/2}, \qquad (27)$$

where N_{λ_X} is the number of electrons per wavelength, given by $N_{\lambda_X} = I_b\lambda_X/(ec)$.

For an EUV light source based on a FEL, a planar undulator comprising alternating dipole magnets is used, e.g., a pure permanent magnet (PPM) undulator with $Nd_2Fe_{14}B$ (Nd–Fe–B) blocks or a hybrid undulator comprising PPMs and ferromagnetic poles, e.g., a high saturation cobalt steel such as Vanadium Permendur or a simple iron. For a hybrid undulator, the thickness of the pole and magnet is optimized in order to maximize the peak field. The peak field B_u of the gap is estimated in terms of the gap g and period λ_u according to $B_u = a$ (T) $\exp[b(g/\lambda_u) + c(g/\lambda_u)^2]$ for a gap range $0.1 < g/\lambda_u < 1$, where $a = 3.694$ T, $b = -5.068$ and $c = 1.520$ for the hybrid undulator with Vanadium Permendur. Table 1 summarizes design examples for a fiber laser-driven laser plasma accelerator-based FEL-produced EUV radiation source at 13.5 nm wavelength using undulators with periods 5 mm (Case A), 10 mm (Case B), 15 mm (Case C), 20 mm (Case D), and 25 mm (Case E), all cases of which have the same gap:period ratio 0.2, e.g., $g = 1$ mm (Case A), 2 mm (Case B), 3 mm (Case C), 4 mm (Case D), and 5 mm (Case E), respectively. The bunch duration of the electron beam in the injector stage at a plasma density of $n_e \approx 10^{18}$ cm^{-3} is assumed to be ~10 fs full-width at half-maximum (FWHM), based on a

Table 1. Parameters for laser plasma accelerator-based EUV FEL light sources.

Case	A	B	C	D	E
Laser					
Laser wavelength (μm)	1	1	1	1	1
Average laser power (MW)	1.63	1.24	1.19	1.22	1.27
Repetition rate (MHz)	1.22	0.515	0.315	0.223	0.168
Laser energy per pulse (J)	1.34	2.40	3.79	5.52	7.57
Peak power (TW)	29	43	59	75	93
Pulse duration (fs)	46	56	65	73	82
Matched spot radius (μm)	19	23	27	30	34
Laser plasma accelerator					
Electron beam energy (MeV)	243	427	659	937	1257
Plasma density (10^{17} cm^{-3})	8.3	5.6	4.2	3.2	2.6
Accelerator length (mm)	18	32	51	74	102
Charge per bunch (nC)	0.5	0.5	0.5	0.5	0.5
Field reduction factor α	0.223	0.267	0.302	0.325	0.364
Bunch duration (fs)	10	10	10	10	10
Energy spread (%)	~1.1	~1.5	~1.6	~1.6	~1.6
Transverse beam size (μm)	25	25	25	25	25
Peak current (kA)	50	50	50	50	50
Average beam power (kW)	148	110	104	104	105
Efficiency of laser to beam (%)	9.1	8.9	8.7	8.5	8.3
Free electron laser					
Undulator period (mm)	5	10	15	20	25
Radiation wavelength (nm)	13.5	13.5	13.5	13.5	13.5
Gap (mm)	1	2	3	4	5
Peak magnetic field (T)	1.425	1.425	1.425	1.425	1.425
Undulator parameter K_u	0.666	1.33	2.00	2.66	3.33
Pierce parameter (%)	1.12	1.51	1.60	1.60	1.57
Gain length (mm)	41	61	86	115	146
Saturation length (mm)	499	721	1016	1355	1723
Number of periods	100	72	68	68	69
Spectral bandwidth (%)	1.0	1.4	1.5	1.5	1.5
R.m.s. radiation cone angle (μrad)	116	97	82	71	63
Input power (MW)	0.94	3.03	5.26	7.48	9.72
Saturated power (GW)	82	194	317	451	596
Duration of EUV pulse (fs)	10	10	10	10	10
Average EUV power (kW)	1	1	1	1	1
Efficiency of EUV generation (%)	0.061	0.081	0.084	0.082	0.079

measurement of the electron bunch duration in a recent laser wakefield acceleration experiment[33]. The relative energy spread of the accelerated electron beam with an injection energy of $0.1E_b$, where E_b is the final beam energy in the accelerator stage, is assumed to be of the order of 10% in the injector stage. After acceleration up to 10 times higher energy in the accelerator stage, the relative energy spread at the final beam energy is reduced to $\Delta E/E_b \sim 1\%$ due to adiabatic damping in the longitudinal beam dynamics. The transverse beam size is tuned by employing a beam focusing system. Figure 1 shows a schematic of the EUV light source based on a compact FEL driven by a fiber laser-based plasma accelerator.

2.5. Design of all-optical Gamma-beam source

The design of a Gamma-beam source based on inverse Compton scattering is carried out by using a result of quantum electrodynamics on photon–electron interactions, namely, the Klein–Nishina formula, which gives the differential cross section of photons scattered from a single electron in the lowest order of quantum electrodynamics. In Compton scattering of a laser photon with energy $\hbar\omega_L$ ($\hbar\omega_L$ (eV) $= 1.240/\lambda_L$ (μm) for laser wavelength λ_L (μm)) off a beam electron, the maximum energy of the scattered photon is given by $E_{\gamma\,\max} = 4\gamma_e^2 a\hbar\omega_L$, where $\gamma_e = E_b/m_ec^2$ is the relativistic factor for an electron beam energy E_b with electron rest mass $m_ec^2 \simeq 0.511$ MeV and the factor $a = [1 + 4\gamma_e(\hbar\omega_L/m_ec^2)]^{-1}$. In the laboratory frame, the differential cross section of Compton scattering[34] is given by

$$\frac{d\sigma}{d\kappa} = 2\pi a r_e^2 \left\{ 1 + \frac{\kappa^2(1-a)^2}{1-\kappa(1-a)} + \left[\frac{1-\kappa(1+a)}{1-\kappa(1-a)} \right]^2 \right\}, \tag{28}$$

where $\kappa = E_\gamma/E_{\gamma\,\max}$ is the energy of a scattered photon normalized by the maximum photon energy and $r_e^2 \simeq 79.4$ mb (1 barn $= 10^{-24}$ cm^2) with the classical electron radius r_e. In the laboratory frame, the scattering angle θ of the photon is given by $\tan\theta = \gamma_e^{-1}\sqrt{(1-\kappa)/a\kappa}$. Integrating the differential cross section over $0 \leqslant \kappa \leqslant 1$, the total cross section of Compton scattering becomes

$$\sigma_{\text{total}} = \pi r_e^2 a \left[\frac{2a^2 + 12a + 2}{(1-a)^2} + a - 1 \right. $$
$$\left. + \frac{6a^2 + 12a - 2}{(1-a)^3} \ln a \right]. \tag{29}$$

This total cross section leads to a cross section of Thomson scattering $\sigma_{\text{Thomson}} = 8\pi r_e^2/3 = 665$ mb for an electron beam energy $E_b \to 0$. The fractional cross section for the photon energy range $E_{\gamma\,\max} - \Delta E_\gamma \leqslant E_\gamma \leqslant E_{\gamma\,\max}$ is given by

$$\Delta\sigma = 2\pi a r_e^2 \Delta\kappa \left[\left(\frac{1+a}{1-a} \right)^2 + \frac{4}{(1-a)^2} \right. $$
$$\times \left(1 + \frac{1-a}{a}\Delta\kappa \right)^{-1} + (a-1)\left(1 + \frac{\Delta\kappa}{2} \right) $$
$$\left. + \frac{1-6a-3a^2}{(1-a)^3\Delta\kappa} \ln\left(1 + \frac{1-a}{a}\Delta\kappa \right) \right], \tag{30}$$

with $\Delta\kappa = \Delta E_\gamma/E_{\gamma\,\max} \ll 1$. All photons in this energy range are scattered in the forward direction within a half-cone angle $\theta \sim \gamma_e^{-1}\sqrt{\Delta\kappa/a}$. For an electron beam interacting with a laser pulse at an angle of α_{int} in the horizontal

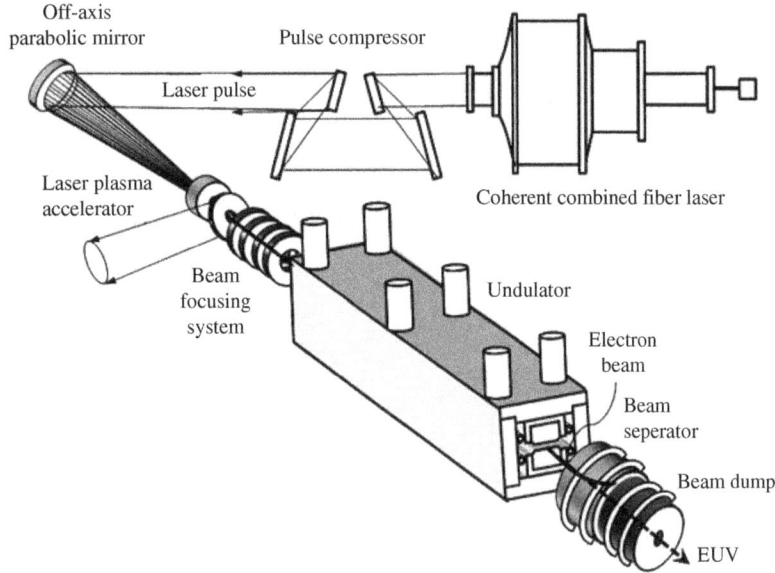

Figure 1. Schematic of the EUV light source based on a compact FEL driven by a fiber laser-based plasma accelerator.

plane (x-plane), the luminosity representing the probability of collisions between electron and laser beams per unit cross section per unit time is obtained by L (mb^{-1}s^{-1}) = $N_e N_L f_L / 2\pi \Sigma$, where N_e is the number of electrons contained in the electron bunch, N_L is the number of photons per laser pulse, f_L is the repetition rate of laser pulses, and Σ is the area where the two beams overlap, given by

$$\Sigma = (\sigma_{ey}^2 + \sigma_{Ly}^2)^{1/2}[\cos^2(\alpha_{\text{int}}/2)(\sigma_{ex}^2 + \sigma_{Lx}^2) + \sin^2(\alpha_{\text{int}}/2)(\sigma_{ez}^2 + \sigma_{Lz}^2)]^{1/2}, \quad (31)$$

where σ_{ex} and σ_{ey} are the r.m.s. horizontal and vertical sizes of the electron beam, σ_{ez} is the r.m.s. bunch length of the electron beam, σ_{Lx} and σ_{Ly} are the r.m.s. horizontal and vertical spot sizes of the laser beam, and σ_{Lz} is the r.m.s. pulse length of the laser beam. For a head-on collision providing efficient Gamma-beam production, the crossing angle between the electron and laser beams is chosen to be $\alpha_{\text{int}} = 0$. Tuning the beam focusing system and the interaction optics so as to give $\sigma_{ex} \approx \sigma_{ey} \approx \sigma_{Lx} \approx \sigma_{Ly}$, the luminosity turns out to be $L = N_e N_L f_L / (4\pi r_{\text{int}}^2)$, where r_{int} is the laser spot radius at the interaction point. Using $N_e = 1.6022 \times 10^{10}(Q_e/1 \text{ nC})$ and $N_L = U_{LS}/\hbar\omega_L = 5.0334 \times 10^{18} U_{LS}$ (J)λ_L (μm), where Q_e is the charge of the electron bunch and $U_{LS} = P_{LS}\tau_{LS}$ is the energy of a scatter pulse with peak power P_{LS} and duration τ_{LS}, the luminosity is calculated as

$$L \text{ (mb}^{-1}\text{s}^{-1}) = Q_e I_{\text{int}} f_L \tau_{LS}/(8e\hbar\omega_L)$$
$$\approx 1.0 \times 10^{-14} f_L \text{ (s}^{-1})Q_e \text{ (nC)}$$
$$\times I_{\text{int}}(\text{W cm}^{-2})\tau_{LS}(\text{fs})\lambda_L \text{ (μm)}, \quad (32)$$

where I_{int} is the focused intensity of the scatter pulse at the interaction point. Thus the Gamma-beam flux is given by

$$N_\gamma \text{ (s}^{-1}) = L\sigma_{\text{tot}} \approx 1 \times 10^{-14}\sigma_{\text{tot}} \text{ (mb)}f_L \text{ (s}^{-1})Q_e \text{ (nC)}$$
$$\times I_{\text{int}} \text{ (W cm}^{-2})\tau \text{ (fs)}\lambda_L(\text{μm}). \quad (33)$$

The fractional Gamma-beam flux with photon energy spread $\Delta\kappa = \Delta E_\gamma / E_{\gamma \max}$ is estimated as

$$\Delta N_\gamma \text{ (s}^{-1}) = L\Delta\sigma \approx 1 \times 10^{-14}\Delta\sigma \text{ (mb)}f_L \text{ (s}^{-1})Q_e \text{ (nC)}$$
$$\times I_L(\text{W cm}^{-2})\tau(\text{fs})\lambda_L \text{ (μm)}. \quad (34)$$

Table 2 summarizes design examples for an all-optical laser plasma accelerator-based Gamma-beam source at photon energies 2.5 MeV (Case A), 5 MeV (Case B), 10 MeV (Case C), 15 MeV (Case D), and 20 MeV (Case E), respectively. Figure 2 is a schematic illustration of the Gamma-beam source based on inverse Compton scattering off relativistic electron beams driven by a laser plasma accelerator.

3. Conclusion

We present methods for producing EUV light at a wavelength of 13.5 nm from a SASE FEL generated by electron beams from a laser plasma accelerator driven by a fiber-based CPA laser and also for producing a Gamma beam with photon energies of 1–20 MeV via inverse Compton scattering off relativistic electron beams from a laser plasma accelerator. For these practical applications of laser plasma accelerators, it is essential to employ high average power,

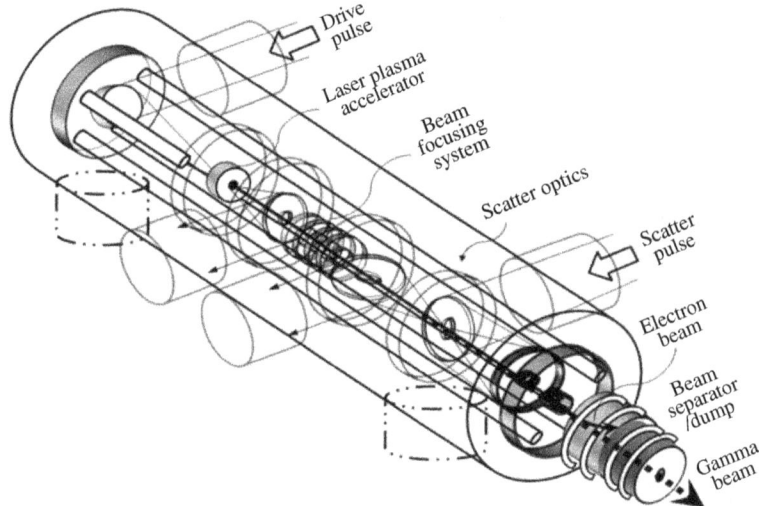

Figure 2. Schematic illustration of the Gamma-beam source based on inverse Compton scattering off relativistic electron beams driven by a laser plasma accelerator.

Table 2. Parameters for all-optical laser plasma accelerator-based Gamma-beam sources.

Case	A	B	C	D	E
Laser plasma accelerator					
Laser wavelength (μm)	0.8	0.8	0.8	0.8	0.8
Repetition rate (Hz)	10	10	10	10	10
Laser energy per pulse (J)	1.78	2.56	3.68	4.55	5.31
Peak power (TW)	41	52	66	77	85
Pulse duration (fs)	43	49	55	59	62
Matched spot radius (μm)	18	20	23	24	26
Electron beam energy (MeV)	326	461	654	802	928
Plasma density (10^{17} cm^{-3})	9.2	7.3	5.7	4.9	4.5
Accelerator length (mm)	24	34	50	61	72
Charge per bunch (nC)	0.5	0.5	0.5	0.5	0.5
Bunch duration (fs)	~10	~10	~10	~10	~10
Transverse beam size (μm)	25	25	25	25	25
Compton scatter					
Photon energy (MeV)	2.5	5	10	15	20
Laser peak power (TW)	10	10	10	10	10
Pulse duration (fs)	250	250	250	250	250
Pulse energy (J)	2.5	2.5	2.5	2.5	2.5
Laser spot radius (μm)	25	25	25	25	25
Focused intensity (10^{18} W cm^{-2})	1	1	1	1	1
Repetition rate (Hz)	10	10	10	10	10
Luminosity (10^6 mb^{-1} s^{-1})	10	10	10	10	10
Total cross section (mb)	660	658	655	653	651
Total photon flux (10^9 s^{-1})	6.60	6.58	6.55	6.53	6.51
Spectral bandwidth (%)	1.0	1.0	1.0	1.0	1.0
Scattering angle within 1% BW (μrad)	313	222	157	128	111
Cross section within 1% BW (mb)	9.80	9.77	9.73	9.69	9.66
Photon flux within 1% BW (10^8 s^{-1})	0.980	0.977	0.973	0.969	0.966

source can produce a high-quality photon flux of 3×10^{12} s^{-1} at 10 MeV energy within a 1% bandwidth. One such high average power laser is a coherent combining fiber laser system[35], comprising a plurality of amplifying fibers wherein an initial laser pulse is distributed and amplified to a 1 mJ level, intended for grouping together the elementary pulses amplified in the fiber in order to form a single amplified global laser pulse with a 1 J level energy.

In both radiation sources, beam transport and imaging from the laser plasma accelerator to the undulator or a focal point of the scatter laser pulse is provided by a beam focusing system that comprises Halbach-type permanent quadrupole magnets made of NdFeB-type rare-earth magnets with a high remanent field[36, 37]. According to simulation results on ionization-induced injection at a plasma density $n_e \approx 10^{18}$ cm^{-3}[31], the normalized emittance is assumed to be $\varepsilon_n \approx 1$ μm inside the wakefield. The transverse beam size in the beam transport optics is given by $\sigma_b = \sqrt{\beta \varepsilon_n / \gamma}$, where β is the beta function of the beam optics at the undulator or the scattering point. For Case C in Table 1, the beta function should be set to $\beta = \gamma \sigma_b^2 / \varepsilon_n \approx 80$ cm inside the undulator. The electron beam, after passing through the undulator or being scattered by the scatter laser pulse, is bent by the dipole field of a permanent magnet (a beam separator) made of NdFeB material and dumped to a copper beam dump with a water cooling element, while the EUV radiation or the Gamma beam is extracted from a beam separator and directed to an EUV lithography scanner or a photon beam irradiation system.

high efficiency drive lasers operating at high repetition pulse rates (of the order of 300 kHz); the corresponding average power of 1 MW means that the EUV FEL is capable of producing an average radiation power of 1 kW at a wavelength of 13.5 nm and the all-optical Gamma beam

Acknowledgements

The work was supported by the National Natural Science Foundation of China (Project No. 51175324). The author was supported by IZEST, Ecole Polytechnique, France,

Shanghai Jiao Tong University, Institute of Physics, CAS, China, and the Center for Relativistic Laser Science, Institute for Basic Science (IBS), Korea.

References

1. T. Tajima and J. M. Dawson, Phys. Rev. Lett. **43**, 267 (1979).
2. W. P. Leemans, B. Nagler, A. J. Gonsalves, C. Toth, K. Nakamura, C. G. R. Geddes, E. Esarey, C. B. Schroeder, and S. M. Hooker, Nat. Phys. **2**, 696 (2006).
3. C. E. Clayton, J. E. Ralph, F. Albert, R. A. Fonseca, S. H. Glenzer, C. Joshi, W. Lu, K. A. Marsh, S. F. Martins, W. B. Mori, A. Pak, F. S. Tsung, B. B. Pollock, J. S. Ross, L. O. Silva, and D. H. Froula, Phys. Rev. Lett. **105**, 105003 (2010).
4. H. Lu, M. Liu, W. Wang, C. Wang, J. Liu, A. Deng, J. Xu, C. Xia, W. Li, H. Zhang, X. Lu, C. Wang, J. Wang, X. Liang, Y. Leng, B. Shen, K. Nakajima, R. Li, and Z. Xu, Appl. Phys. Lett. **99**, 091502 (2011).
5. X. Wang, R. Zgadzaj, N. Fazel, Z. Li, S. A. Yi, X. Zhang, W. Henderson, Y. Y. Chang, R. Korzekwa, H. E. Tsai, C. H. Pai, H. Quevedo, G. Dyer, E. Gaul, M. Martinez, A. C. Bernstein, T. Borger, M. Spinks, M. Donovan, V. Khudik, G. Shvets, T. Ditmire, and M. C. Downer, Nat. Commun. **4**, 1988 (2013).
6. H. T. Kim, K. H. Pae, H. J. Cha, I. J. Kim, T. J. Yu, J. H. Sung, S. K. Lee, T. M. Jeong, and J. Lee, Phys. Rev. Lett. **111**, 165002 (2013).
7. T. Kameshima, W. Hong, K. Sugiyama, X. Wen, Y. Wu, C. Tang, Q. Zhu, Y. Gu, B. Zhang, H. Peng, S.-I. Kurokawa, L. Chen, T. Tajima, T. Kumita, and K. Nakajima, Appl. Phys. Exp. **1**, 066001 (2008).
8. S. Karsch, J. Osterhoff, A. Popp, T. P. Rowlands-Rees, Z. Major, M. Fuchs, B. Marx, R. Hörlein, K. Schmid, L. Veisz, S. Becker, U. Schramm, B. Hidding, G. Pretzler, D. Habs, F. Grüner, F. Krausz, and S. M. Hooker, New J. Phys. **9**, 415 (2007).
9. O. Lundh, J. Lim, C. Rechatin, L. Ammoura, A. Ben-Ismail, X. Davoine, G. Gallot, J. P. Goddet, E. Lefebvre, V. Malka, and J. Faure, Nat. Phys. **7**, 219 (2011).
10. N. A. M. Hafz, T. M. Jeong, I. W. Choi, S. K. Lee, K. H. Pae, V. V. Kulagin, J. H. Sung, T. J. Yu, K.-H. Hong, T. Hosokai, J. R. Cary, D.-K. Ko, and J. Lee, Nat. Photon. **2**, 571 (2008).
11. J. S. Liu, C. Q. Xia, W. T. Wang, H. Y. Lu, C. Wang, A. H. Deng, W. T. Li, H. Zhang, X. Y. Liang, Y. X. Leng, X. M. Lu, C. Wang, J. Z. Wang, K. Nakajima, R. X. Li, and Z. Z. Xu, Phys. Rev. Lett. **107**, 035001 (2011).
12. B. B. Pollock, C. E. Clayton, J. E. Ralph, F. Albert, A. Davidson, L. Divol, C. Filip, S. H. Glenzer, K. Herpoldt, W. Lu, K. A. Marsh, J. Meinecke, W. B. Mori, A. Pak, T. C. Rensink, J. S. Ross, J. Shaw, G. R. Tynan, C. Joshi, and D. H. Froula, Phys. Rev. Lett. **107**, 045001 (2011).
13. W. T. Wang, W. T. Li, J. S. Liu, C. Wang, Q. Chen, Z. J. Zhang, R. Qi, Y. X. Leng, X. Y. Liang, Y. Q. Liu, X. M. Lu, C. Wang, R. X. Li, and Z. Z. Xu, Appl. Phys. Lett. **103**, 243501 (2013).
14. A. Pak, K. A. Marsh, S. F. Martins, W. Lu, W. B. Mori, and C. Joshi, Phys. Rev. Lett. **104**, 025003 (2010).
15. C. McGuffey, A. G. R. Thomas, W. Schumaker, T. Matsuoka, V. Chvykov, F. J. Dollar, G. Kalintchenko, V. Yanovsky, A. Maksimchuk, K. Krushelnick, V. Y. Bychenkov, I. V. Glazyrin, and A. V. Karpeev, Phys. Rev. Lett. **104**, 025004 (2010).
16. C. Xia, J. Liu, W. Wang, H. Lu, W. Cheng, A. Deng, W. Li, H. Zhang, X. Liang, Y. Leng, X. Lu, C. Wang, J. Wang, K. Nakajima, R. Li, and Z. Xu, Phys. Plasmas **18**, 113101 (2011).
17. W. T. Li, J. S. Liu, W. T. Wang, Z. J. Zhang, Q. Chen, Y. Tian, R. Qi, C. H. Yu, C. Wang, T. Tajima, R. X. Li, and Z. Z. Xu, Appl. Phys. Lett. **104**, 093510 (2014).
18. W. Leemans, C. G. R. Geddes, J. Faure, Cs. Tóth, J. van Tilborg, C. B. Schroeder, E. Esarey, and G. Fubiani, Phys. Rev. Lett. **91**, 074802 (2003).
19. Z. D. Hu, Z. M. Sheng, W. M. Wang, L. M. Chen, Y. T. Li, and J. Zhang, Phys. Plasmas **20**, 080702 (2013).
20. M. Fuchs, R. Weingartner, A. Popp, Zs. Major, S. Becker, J. Osterhoff, I. Cortrie, B. Benno Zeitler, R. Rainer Hörlein, G. D. Tsakiris, U. Schramm, T. P. Rowlands-Rees, S. M. Hooker, D. Habs, F. Krausz, S. Karsch, and F. Grüner, Nat. Phys. **5**, 826 (2009).
21. S. Kneip, C. McGuffey, J. L. Martins, S. F. Martins, C. Bellei, V. Chvykov, F. Dollar, R. Fonseca, C. Huntington, G. Kalintchenko, A. Maksimchuk, S. P. D. Mangles, T. Matsuoka, S. R. Nagel, C. A. J. Palmer, J. Schreiber, K. Ta Phuoc, A. G. R. Thomas, V. Yanovsky, L. O. Silva, K. Krushelnick, and Z. Najmudin, Nat. Phys. **6**, 980 (2010).
22. W. T. Yan, L. M. Chen, D. Z. Li, L. Zhang, N. A. M. Hafz, J. Dunn, Y. Ma, K. Hung, L. Su, M. Chen, Z. M. Sheng, and J. Zhang, Proc. Natl. Acad. Sci. **111**, 5825 (2014).
23. S. Cipiccia, M. R. Islam, B. Ersfeld, R. P. Shanks, E. Brunetti, G. Vieux, X. Yang, R. C. Issac, S. M. Wiggins, G. H. Welsh, M.-P. Anania, D. Maneuski, R. Montgomery, G. Smith, M. Hoek, D. J. Hamilton, N. R. C. Lemos, D. Symes, P. P. Rajeev, V. O. Shea, J. M. Dias, and D. A. Jaroszynski, Nat. Phys. **7**, 867 (2011).
24. K. Ta Phuoc, S. Corde, C. Thaury, V. Malka, A. Tafzi, J. P. Goddet, R. C. Shah, S. Sebban, and A. Rousse, Nat. Photon. **6**, 308 (2012).
25. N. D. Powers, I. Ghebregziabher, G. Golovin, C. Liu, S. Chen, S. Banerjee, J. Zhang, and D. P. Umstadter, Nat. Photon. **8**, 28 (2014).
26. I. Kostyukov, A. Pukhov, and S. Kiselev, Phys. Plasmas **11**, 5256 (2004).
27. W. Lu, M. Tzoufras, C. Joshi, F. S. Tsung, W. B. Mori, J. Vieira, R. A. Fonseca, and L. O. Silva, Phys. Rev. ST Accel. Beams **10**, 061301 (2007).
28. K. Nakajima, H. Lu, X. Zhao, B. Shen, R. Li, and Z. Xu, Chin. Opt. Lett. **11**, 013501 (2013).
29. K. Nakajima, A. Deng, X. Zhang, B. Shen, J. Liu, R. Li, Z. Xu, T. Ostermayr, S. Petrovics, C. Klier, K. Iqbal, H. Ruhl, and T. Tajima, Phys. Rev. ST Accel. Beams **14**, 091301 (2011).
30. S. Kalmykov, A. Yi, V. Khudik, and G. Shvets, Phys. Rev. Lett. **103**, 135004 (2009).
31. M. Chen, E. Esarey, C. B. Schroeder, C. G. R. Geddes, and W. P. Leemans, Phys. Plasmas **19**, 033101 (2012).
32. P. Elleaume, J. Chavanne, and B. Faatz, Nucl. Instrum. Methods Phys. Res. A **455**, 503 (2000).
33. A. Buck, M. Nicolai, K. Schmid, C. M. S. Sears, A. Sävert, J. M. Mikhailova, F. Krausz, M. C. Kaluza, and L. Veisz, Nat. Phys. **7**, 543 (2011).
34. H. A. Tolhokk, Rev. Mod. Phys. **28**, 277 (1956).
35. G. Mourou, B. Brocklesby, T. Tajima, and J. Limpert, Nat. Photon. **7**, 258 (2013).
36. J. K. Lim, P. Frigola, G. Travish, J. B. Rosenzweig, S. G. Anderson, W. J. Brown, J. S. Jacob, C. L. Robbins, and A. M. Tremaine, Phys. Rev. ST Accel. Beams **8**, 0072401 (2005).
37. K. Nakajima, A. H. Deng, H. Yoshitama, N. A. M. Hafz, H. Y. Lu, B. F. Shen, J. S. Liu, R. X. Li, and Z. Z. Xu, *Free Electron Laser* Chap. 5, S. Varró, ed. p. 119 (InTech, Rijeka, Croatia, 2012).

22

Nonlinear wake amplification by an active medium in a cylindrical waveguide using a modulated trigger bunch

Zeev Toroker, Miron Voin, and Levi Schächter

Department of Electrical Engineering, Technion, Israel Institute of Technology, Haifa 32000, Israel

Abstract

Cerenkov wake amplification can be used as an accelerating scheme, in which a trigger bunch of electrons propagating inside a cylindrical waveguide filled with an active medium generates an initial wake field. Due to the multiple reflections inside the waveguide, the wake may be amplified significantly more strongly than when propagating in a boundless medium. Sufficiently far away from the trigger bunch the wake, which travels with the same phase velocity as the bunch, reaches saturation and it can accelerate a second bunch of electrons trailing behind.

For a CO_2 gas mixture our numerical and analytical calculations indicate that a short saturation length and a high gradient can be achieved with a large waveguide radius filled with a high density of excited atoms and a trigger bunch that travels at a velocity slightly above the Cerenkov velocity. To obtain a stable level of saturated wake that will be suitable for particle acceleration, it is crucial to satisfy the single-mode resonance condition, which requires high accuracy in the waveguide radius and the ratio between the electron phase velocity and the Cerenkov velocity. For single-mode propagation our model indicates that it is feasible to obtain gradients as high as $GV\,m^{-1}$ in a waveguide length of cm.

Keywords: laser–plasma interaction; novel optical material and device

1. Introduction

Currently, high electron energies of tens of GeV are achieved with radio-frequency (RF) linear accelerators that operate in the GHz frequency range with a typical length of a few kilometers[1]. In idealized conditions, breakdown[2] limits the accelerating electric field to the order of a few hundreds of MV m^{-1}. In practice, gradients reach values of 25 MV m^{-1} when operating at room temperature[1] and 35 MV m^{-1} in their superconductive counterpart[3].

In the past two decades, with the immense progress in laser technology, laser plasma accelerators have become able generate hundreds of GV m^{-1}[4–6]. In this scheme, high intensity focused laser pulses with lengths of the order of the plasma wavelength generate an intense wake. The plasma wake, which trails behind the laser pulse with the same group velocity, can accelerate electrons from the plasma itself. Work is in progress to accelerate electrons that do not originate in the plasma.

Based on the chirped pulse amplification (CPA) technique[7], pulsed laser technology facilitates focus on plasma target pulses with intensities as high as 10^{18} W cm^{-2}

Correspondence to: Zeev Toroker, Department of Electrical Engineering, Technion, Israel Institute of Technology, Haifa 32000, Israel. Email: ztoroker@tx.technion.ac.il

and duration of the order of femtoseconds (10^{-15} s) for a laser wavelength of 1 μm that is optimized to a plasma density of 10^{18} cm^{-3}. In a series of experiments reported in 2004[8–10], quasi-monoenergetic e-beams with energies of the order of 100 MeV have been demonstrated. More recently, several groups have demonstrated quasi-monoenergetic e-beams with energies of up to 2 GeV[11].

An intense wake may also develop by replacing the laser pulse with an energetic e-beam in plasma, and it is shown experimentally[12] that an initial bunch of 40 GeV can generate an intense wake of the order of 50 GV m^{-1}, which results in acceleration of a trailing bunch to an energy of about 80 GeV with about 16% energy spread. The total number of injected electrons in the bunch is 10^{10} and the spot size is 10 μm, whereas the number of electrons accelerated to 80 GeV is about 240×10^6. For comparison, in a dielectric loaded waveguide, a 60 MeV bunch of electrons can generate Cerenkov gradients of 250 MV m^{-1}[13]. The total number of injected electrons in the bunch is 3×10^6 and the spot size is 10 μm, whereas the number of accelerated electrons is about 70×10^3.

In the paradigm analyzed here, the gradients are more modest (order of 1 GV m^{-1}) and it is conceptually closer to a conventional two-beam acceleration scheme. It relies on transferring energy stored from the active medium to a train of electron bunches – see Refs. [14, 15]. In contrast to previously mentioned schemes, where the energy for the acceleration comes from either energetic laser pulses or an energetic electron bunch, in this scheme the energy comes from excited atoms. In the first approach of this scheme, an electron bunch injected into an active medium generates a wake comprised of a broadband spectrum of evanescent waves. Since the active medium is a resonant medium, only the fraction of the wake spectrum that is close to the resonance frequency of the medium will be amplified. Thus, it is proposed to inject a spatially modulated bunch with periodicity equal to the resonance wavelength so that a large portion of the wake spectrum lies in the vicinity of the resonance frequency of the medium. As a result, the amplified wake accelerates directly the injected bunch. This approach has been demonstrated in an experiment performed at Brookhaven National Laboratory Accelerator Test Facility (BNL-ATF)[16], in which a density modulated bunch with an energy of 45 MeV gained energy of 200 keV from an active CO_2 gas mixture.

For a description of the paradigm, we start with a general description of the system followed by a simplified model used to investigate the essential phenomena involved. A trigger bunch propagates in a vacuum channel surrounded by a low loss dielectric layer which is thick enough to sustain the 8–10 atm pressure of the CO_2 mixture consisting of the active medium. The latter in turn is confined by a Bragg waveguide which facilitates excitation of the active medium on the one hand and allows full confinement of a resonant Cherenkov wake. One of the eigenmodes is amplified by the active medium and many wavelengths behind the trigger bunch the former accelerates a trailing bunch. For the sake of simplicity, the analysis that follows relies on a metallic waveguide that contains the gaseous active medium.

Initially, a triggering bunch of electrons generates a Cerenkov wake with electric field in the longitudinal direction, which in turn is amplified by the active medium - stimulated emission process. As the wake field is amplified, the population inversion in the active medium is reduced and, as a result, the spatial gain is also reduced. When the spatial gain is zero, the wake reaches saturation and it can accelerate a second bunch of electrons trailing behind the triggering bunch.

Previous study of this scheme used a linear model[15, 17–19] and a simplified nonlinear model[20]. In the linear model, where a constant population inversion density (PID) is assumed, the wake is exponentially amplified. Since the propagating wake in the cylindrical waveguide propagates in an oblique angle, it reflects multiple times from the boundaries. As a result, the effective propagation path is longer than if the wake were propagating in a boundless medium. For a structure of a given length, the effective gain of the wake is enhanced. The *simplified* nonlinear model assumes the propagation of one electromagnetic (EM) mode[20]; it does not account for the multiple reflections

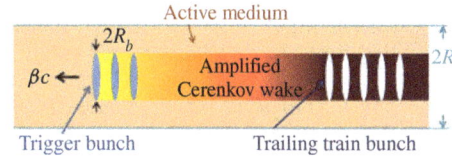

Figure 1. Schematic description of the accelerating structure. A trigger bunch propagates in a cylindrical metallic waveguide of radius R filled with an active medium. This bunch is injected into the structure with velocity βc larger than the Cerenkov velocity $c/\sqrt{\varepsilon_r}$ and generates an entire manifold of TM modes which propagate behind. One of the eigenmodes is amplified by the active medium and many wavelengths behind the trigger bunch the former accelerates a trailing train bunch.

of the wake from the waveguide boundary and only the dynamics of the polarization field is considered.

In this paper we extend the previous linear model[18, 19] and include the nonlinear dynamics of the active medium. The extended model can describe the wake saturation level and the interval of time in which the wake reaches saturation. While the previously mentioned approaches[18–20] predict the saturation process qualitatively, the present approach is far more quantitative.

This paper is organized as follows. Section 2 presents the dynamics equations that describe the Cerenkov wake amplification by the active medium. It also describes the dynamics of the polarization and the population inversion density. In Section 3 the single-mode resonance condition, the value of the saturated wake and the saturation length are calculated analytically. In addition, we calculate for a single mode the width or the spot size of the longitudinal wake. In Section 4 we show numerically for a modulated trigger bunch the dynamics of the wake and the population inversion density for a CO_2 gas mixture. We conclude (Section 5) with discussion and conclusions.

2. Formulation of the problem

The structure of interest consists of a cylindrical metallic waveguide of radius R filled with an active medium (see Figure 1). Far from the resonance of the active medium it is assumed that the dielectric coefficient of the medium is ε_r and it is frequency independent. A bunch of electrons, the trigger bunch, is injected into the structure with velocity βc larger than the Cerenkov velocity $c/\sqrt{\varepsilon_r}$, where c is the speed of light in vacuum. Assuming azimuthal symmetry, the bunch excites an entire manifold of transverse magnetic (TM) modes which propagate behind[21]. Each mode consists of longitudinal and radial electric field components (E_z, E_r) as well as an azimuthal magnetic field (H_ϕ).

This superposition of TM modes, also known as the wake field, travels at the same speed as the trigger bunch. Among this infinite manifold of TM modes only those with frequency equal to or close enough to the resonance frequency of the active medium, ω_0, will be amplified through the stimulated emission process. Thus, the EM field has the following generic form

$$\psi(T, r) = \frac{1}{2} \sum_{s} J_{\nu}(k_s r) \psi_s(T) e^{i\omega_0 T} + \text{c.c.}, \qquad (2.1)$$

where each field ψ ($\psi \in \{E_z, E_r, H_\phi\}$) depends on the radial coordinate, r, and a variable that follows the e-beam, $T \equiv t - z/\beta c$. Specifically, for field $\psi = E_z$ the index of the Bessel function of the first kind is $\nu = 0$ and for $\psi = \{E_r, H_\phi\}$ the index is $\nu = 1$. In addition, the radial wavenumber is $k_s \equiv p_s/R$, where p_s is defined through $J_0(k_s R) = 0$ ($s = 1, 2, 3, \ldots$) since $E_z(T, r = R) = 0$. Finally, the dependence of each mode on T is given by $\psi_s(T)$.

Similarly to the EM fields, the current density of the trigger bunch is assumed to be of the form

$$J_z(T, r) = \frac{1}{2} I(T) \frac{\theta(R_b - r)}{\pi R_b^2} e^{i\omega_0 T} + \text{c.c.}, \qquad (2.2)$$

where $\theta(x)$ is the Heaviside step function and $I(T)$ is the e-beam longitudinal current envelope. Assuming that the spatially modulated bunch profile is given by $f(T)$ then the current envelope is $I(T) = I_0 f(T)$, where the modulated current is $I_0 = Q_b \beta c / L_b$. Here, the total charge is $Q_b = -eN_b$, where $-e$ is the electron charge and N_b is the number of macro-particles that comprise the bunch; L_b is the bunch length.

Having in mind that the growth rate is much smaller than the resonance frequency, the dynamics of the fields can be separated into two major time scales: the 'fast' and the 'slow' time scales. The 'fast' time scale is the medium resonance period of time $1/\omega_0$ whereas the 'slow' time scale associated with the growth rate of the field envelopes, $\psi_s(T)$, is $1/\omega_p$, where $\omega_p = \omega_0 \sqrt{\frac{2N_0 \mu^2}{\varepsilon_0 \hbar \omega_0}} \ll \omega_0$ is the 'plasma frequency' of the active medium. Here, N_0 is the initial PID, $\mu = \mu_{12}/\sqrt{3}$ is the average dipole moment and μ_{12} is the dipole moment. The parameter $\hbar$ is the reduced Planck constant and ε_0 is the vacuum permittivity. Thus, the dynamics of the wake and the active medium can be described by the slowly varying envelope approximation. Consequently, the dynamics of the normalized EM fields derived from the Ampere and Faraday laws read

$$\frac{\partial \bar{E}_{z,s}}{\partial \tau} = -i\bar{\omega}_0 \bar{E}_{z,s} + \frac{\bar{k}_s}{2\sqrt{\varepsilon_r}} \bar{E}_{+,s} + \frac{\bar{k}_s}{2\sqrt{\varepsilon_r}} \bar{E}_{-,s}$$
$$+ \frac{2}{\varepsilon_r} \bar{P}_{z,s} - \frac{2}{\varepsilon_r} \bar{J}_s f, \qquad (2.3)$$

$$\frac{\partial \bar{E}_{+,s}}{\partial \tau} = -i\bar{\omega}_0 \bar{E}_{+,s} - \frac{\bar{k}_s}{\sqrt{\varepsilon_r} \Delta \varepsilon_-} \bar{E}_{z,s} + \frac{2}{\varepsilon_r \Delta \varepsilon_-} \bar{P}_{r,s}, \qquad (2.4)$$

$$\frac{\partial \bar{E}_{-,s}}{\partial \tau} = -i\bar{\omega}_0 \bar{E}_{-,s} - \frac{\bar{k}_s}{\sqrt{\varepsilon_r} \Delta \varepsilon_+} \bar{E}_{z,s} - \frac{2}{\varepsilon_r \Delta \varepsilon_+} \bar{P}_{r,s}. \qquad (2.5)$$

In our model the time T is normalized by $1/\omega_A$ such that $\tau = T\omega_A$, where $\omega_A = \omega_p/(2\sqrt{\varepsilon_r})$ and, as already indicated, ε_r is the dielectric constant of the medium excluding the population inversion dynamics. Also, the normalized electric field envelopes, $\bar{E}_{z,s}$ and $\bar{E}_{r,s}$, are normalized with $E_0 = \frac{1}{J_1(k_s R)} \sqrt{\frac{\hbar \omega_0 N_0}{2\varepsilon_0}}$ and the magnetic field envelope, $\bar{H}_s$, is normalized with $E_0/(\mu_0 c)$. In addition, $\bar{E}_{\pm,s} = \frac{\bar{H}_s}{\sqrt{\varepsilon_r}} \pm \bar{E}_{r,s}$, $\Delta \varepsilon_\pm = 1 \pm \frac{1}{\beta \sqrt{\varepsilon_r}}$, $\bar{\omega}_0 = \omega_0/\omega_A$ and $\bar{k}_s = k_s c/\omega_A$.

The expression $\frac{2}{\varepsilon_r} \bar{J}_s f$ is the normalized bunch current, where $\bar{J}_s = \frac{I_0}{\pi R^2} \frac{J_c(k_s R_b)}{J_1(k_s R)} \frac{\sqrt{\varepsilon_r}}{\omega_0 \mu N_0}$, $J_c(x) \equiv 2J_1(x)/x$ and $f = f(\tau)$ describes the electron bunch profile in the longitudinal direction. In this study, the bunch injected at $\tau = \tau_0$ with a length of $\bar{L}_b = L_b \omega_A / \beta c$ has a profile of $f(\tau_0 < \tau < \tau_1) = 1$, $f(\tau = \tau_0) = 1/2$, $f(\tau = \tau_1) = 1/2$ and zero otherwise, where $\tau_1 = \tau_0 + \bar{L}_b$.

The active medium is modeled semi-classically as a two-level system within the framework of the dipole approximation[22, 23]. In addition, it is assumed that only stimulated emission can reduce the population inversion density and collisions of the second kind are neglected here. The response of the active medium to the wake is through the normalized polarization fields $\bar{P}_z$ and $\bar{P}_r$,

$$\frac{\partial \bar{P}_{z,s}}{\partial \tau} + \frac{\Delta \bar{\omega}}{2} \bar{P}_{z,s} = \varepsilon_r \bar{N} \bar{E}_{z,s}, \qquad (2.6)$$

$$\frac{\partial \bar{P}_{r,s}}{\partial \tau} + \frac{\Delta \bar{\omega}}{2} \bar{P}_{r,s} = \frac{1}{2} \varepsilon_r \bar{N} \bar{E}_{+,s} - \frac{1}{2} \varepsilon_r \bar{N} \bar{E}_{-,s}, \quad (2.7)$$

where the polarization envelopes are normalized with $i\frac{\mu N_0}{\sqrt{\varepsilon_r} J_1(k_s R)}$ and $\bar{N} = \bar{N}(\tau)$ is the normalized population inversion density measured in units of N_0. Also, it is assumed for simplicity that $\bar{N}$ is radially independent. Radial variations will be considered elsewhere.

The dynamics of the PID, $\bar{N}$, reads

$$\frac{\partial \bar{N}}{\partial \tau} + \bar{A}_{21}(\bar{N} - \bar{N}^e) = -\frac{1}{4} \sum_{s} [2\bar{E}_{z,s}^* \bar{P}_{z,s} + 2\bar{E}_{z,s} \bar{P}_{z,s}^*$$
$$+ (\bar{E}_{+,s} - \bar{E}_{-,s}) \bar{P}_{r,s}^* + (\bar{E}_{+,s}^* - \bar{E}_{-,s}^*) \bar{P}_{r,s}], \quad (2.8)$$

where $\bar{A}_{21} = A_{21}/\omega_A$ is the normalized Einstein coefficient associated with the spontaneous emission time $\tau_{spon} = 1/\bar{A}_{21}$ and $\bar{N}^e$ is the PID in thermal equilibrium. In this study, we consider an active medium with a long spontaneous emission time compared with the order of the amplification time of $1/\omega_p$, which results in neglecting the second term on the left-hand side of Equation (2.8) associated with the spontaneous emission effect.

Finally, the set of equations introduced in Equations (2.3)–(2.8) conserves energy,

$$\frac{\partial}{\partial \tau} \bar{W}_{tot} = 0, \qquad (2.9)$$

where the total energy is

$$\bar{W}_{tot} = \bar{W}_N + \sum_{s=1}^{\infty}[\bar{W}_s^{(EM,lo)} + \bar{W}_s^{(EM,tr)} + \bar{W}_s^{(B)}], \quad (2.10)$$

$\bar{W}_s^{(EM,lo)} = \frac{\varepsilon_r}{4}|\bar{E}_{z,s}|^2$ is the energy density associated with the longitudinal electric field and $\bar{W}_s^{(EM,tr)} = \frac{\varepsilon_r}{8}(\Delta\varepsilon_-|\bar{E}_{+,s}|^2 + \Delta\varepsilon_+|\bar{E}_{-,s}|^2)$ is the transverse component counterpart. Also, the energy density of the active medium is $\bar{W}_N = \bar{N}$, and the energy of the bunch is denoted by $\bar{W}_s^{(B)} = \bar{J}_s \int_0^{\tau} f(\tau')\frac{1}{2}[\bar{E}_{z,s}(\tau') + \bar{E}_{z,s}^*(\tau')]d\tau'$.

3. Analytical assessments

In this section we determine *analytically* the single-mode resonance condition, the value of the saturated wake and the saturation length. In addition, we calculate for a single mode the spot size of the longitudinal wake.

In the considered structure only the modes with frequencies adjacent to the resonance frequency will be amplified. In our model it is possible to find the modes that will be amplified from the dynamics of $\bar{E}_{z,s}$ as follows. Substitution of $\bar{E}_{+,s}$ and $\bar{E}_{-,s}$ from Equations (2.4) and (2.5), respectively, into Equation (2.3) results in

$$\frac{\partial \bar{E}_{z,s}}{\partial \tau} = -i\Delta\bar{\omega}_s \bar{E}_{z,s} - i\frac{2\bar{k}_s}{\beta\varepsilon_r\sqrt{\varepsilon_r\varepsilon_c}\bar{\omega}_0}\bar{P}_{r,s} + \frac{2}{\varepsilon_r}\bar{P}_{z,s}$$
$$+ i\frac{\bar{k}_s}{2\bar{\omega}_0\sqrt{\varepsilon_r}}\frac{\partial}{\partial\tau}(\bar{E}_{+,s} + \bar{E}_{-,s}) - \frac{2}{\varepsilon_r}\bar{J}_s f, \quad (3.1)$$

where $\Delta\bar{\omega}_s = \bar{\omega}_0(1 - \frac{\bar{k}_s^2}{\varepsilon_r\varepsilon_c\bar{\omega}_0^2})$ is the frequency detuning of mode s. The parameter $\varepsilon_c = \Delta\varepsilon_+\Delta\varepsilon_- = 1 - \frac{1}{\beta^2\varepsilon_r}$ will be referred to as the Cerenkov coefficient. Thus, the modes with frequency detuning smaller than the active medium band will be amplified or $|\Delta\bar{\omega}_s| < \Delta\bar{\omega}$. Therefore, the condition for single-mode propagation at the resonance frequency is $1 - \frac{\bar{k}_s^2}{\varepsilon_r\varepsilon_c\bar{\omega}_0^2} = 0$. In physical units this reads

$$\left(\frac{\omega_0\sqrt{\varepsilon_r}}{c}\right)^2 - \left(\frac{p_s}{R}\right)^2 = \left(\frac{\omega_0}{\beta c}\right)^2. \quad (3.2)$$

Thus, the single resonance occurs when the dispersion relation of the waveguide (left-hand side of Equation (3.2)) coincides with that of the electron (right-hand side of Equation (3.2)) at the resonant frequency (ω_0) of the medium.

The dynamics of the wake can be divided into three parts. In the first part of the amplification, where the PID is weakly depleted i.e., $\bar{N} \simeq 1$, known also as the *linear regime*, the solution of Equations (2.3)–(2.7) assuming a single-mode

longitudinal wake is

$$\bar{E}_{z,s0} = i\frac{\bar{J}_s}{\varepsilon_r}\left[A_1 \int_0^{\tau} e^{i\Omega_1(\tau-\tau')}f(\tau')d\tau' + A_2 \int_0^{\tau} e^{i\Omega_2(\tau-\tau')}f(\tau')d\tau'\right], \quad (3.3)$$

where $A_1 = \frac{i\Omega_1+0.5\Delta\bar{\omega}}{\Omega_1-\Omega_2}$ and $A_2 = i - A_1$. Here, the normalized linear growth and decay rates of the wake are

$$\Omega_{1,2} = i\frac{\Delta\bar{\omega}}{4}\left(1 \pm \sqrt{1 + \frac{16\bar{N}}{\varepsilon_c\Delta\bar{\omega}^2}}\right), \quad (3.4)$$

where it is assumed that $|\Omega_{1,2}| \ll \bar{\omega}_0$ and $\bar{N}$ is constant. In the limit of $\frac{1}{\sqrt{\varepsilon_c}} \gg \Delta\bar{\omega}$ and $\bar{N} = 1$ the growth rate is

$$\Omega_1 \approx -i\left(\frac{1}{\sqrt{\varepsilon_c}} - \frac{\Delta\bar{\omega}}{4}\right), \quad (3.5)$$

and the decay rate is

$$\Omega_2 \approx i\left(\frac{1}{\sqrt{\varepsilon_c}} + \frac{\Delta\bar{\omega}}{4}\right). \quad (3.6)$$

At the limit of $\tau \gg \tau_0$ and $\beta \to 1$ the longitudinal wake is

$$\bar{E}_{z,s0} \simeq -\frac{\bar{J}_s}{2\varepsilon_r}\frac{1 + \frac{\Delta\bar{\omega}}{4}\sqrt{\varepsilon_c}}{\frac{1}{\sqrt{\varepsilon_c}} - \frac{\Delta\bar{\omega}}{4}}e^{\left(\frac{1}{\sqrt{\varepsilon_c}} - \frac{\Delta\bar{\omega}}{4}\right)(\tau-\tau_0)}. \quad (3.7)$$

Clearly, in the linear regime the wake is growing exponentially. Moreover, for a trigger bunch satisfying the Cerenkov condition or $\varepsilon_c \to 0$ the growth rate can be significantly larger than the medium bandwidth $\Delta\bar{\omega}$. This is in contrast to the growth rate of source-free EM pulse propagation in an active medium which is limited by the medium's bandwidth $\Delta\bar{\omega}$.

In the second part of the amplification the PID is significantly depleted ($|\bar{N}| \ll 1$) as a result of the stimulated emission (right-hand side of Equation (2.8)), thus reducing the effective gain of the medium. In this *nonlinear regime* of amplification, where the PID is getting depleted and the wake is intense, both the amplified wake and the PID can experience Rabi oscillations[24] before reaching the third regime of *deep saturation*. In the latter case the PID is completely depleted ($\bar{N} \simeq 0$) and as a result the effective gain is zero. Hence, the medium is transparent to the propagating wake and the interaction reaches full saturation.

The value of the saturated wake can be found from the energy conservation (Equation (2.9)). Assuming that $\bar{N}(\tau = 0) = 1$ and at $\tau = 0$ most of the energy is stored in the active medium rather than the trigger bunch ($\bar{W}_{b,s}(\tau) \ll 1$) then

Equation (2.9) becomes

$$\bar{N}(\tau) + \sum_s \frac{\varepsilon_r}{4}|\bar{E}_{z,s}(\tau)|^2 + \frac{\varepsilon_r}{8}\Delta\varepsilon_-|\bar{E}_{+,s}(\tau)|^2$$
$$+ \frac{\varepsilon_r}{8}\Delta\varepsilon_+|\bar{E}_{-,s}(\tau)|^2 = 1. \qquad (3.8)$$

Now, from Equations (2.4) and (2.5) we have $\bar{E}_{\pm} \approx \frac{ik_s}{\bar{\omega}_0\sqrt{\varepsilon_r\Delta\varepsilon_\mp}}\bar{E}_{z,s}$. In addition, we assume a single propagating mode ($k_{s0}^2 = \varepsilon_r\varepsilon_c\bar{\omega}_0^2$) and a relativistic bunch ($\beta \to 1$). Hence, the energy conservation reads

$$\frac{1}{2}\varepsilon_r|\bar{E}_{z,s0}(\tau)|^2 + \bar{N}(\tau) = 1. \qquad (3.9)$$

In the saturation regime where $\bar{N}(\tau = \tau_{sat}) = 0$ one can obtain from Equation (3.9) that the value of the saturated wake is

$$|\bar{E}_{sat}| = |\bar{E}_{z,s0}(\tau = \tau_{sat})| = \sqrt{\frac{2}{\varepsilon_r}}, \qquad (3.10)$$

and in real units it reads

$$|E_{sat}| = \frac{1}{|J_1(p_{s0})|}\sqrt{\frac{\hbar\omega_0 N_0}{\varepsilon_0\varepsilon_r}}. \qquad (3.11)$$

Clearly, the value of the saturated wake is proportional to the square root of the initial PID, N_0. Moreover, since for a large mode number ($s \gg 1$) $|J_1(p_s \simeq s\pi)| \sim \sqrt{2/\pi^2 s}$, the saturation value is proportional to the square root of the mode number or (from Equation (3.2)) the waveguide radius.

To determine the saturation time we first evaluate the time interval of the quasi-linear regime and then the nonlinear regime. We define the time interval of the quasi-linear regime to be from the point where the trigger bunch ends ($\tau = \tau_1$) until the point where the PID reaches its first time zero, τ_d: $\bar{N}(\tau = \tau_1+\tau_d) = 0$. Motivated by the linear regime (see Equations (3.4) and (3.7)), where the amplitude during $\tau_1 < \tau < \tau_1 + \tau_d$ satisfies

$$\frac{\partial\bar{E}_{z,s0}}{\partial\tau} = \bar{\Omega}\bar{E}_{z,s0}, \qquad (3.12)$$

wherein $\bar{\Omega}$ is virtually constant, in the nonlinear regime this term is assumed to vary according to the PID, namely, $\bar{\Omega} = i\Omega_1\sqrt{\bar{N}} \simeq \frac{\sqrt{\bar{N}}}{\sqrt{\varepsilon_c - \frac{\Delta\bar{\omega}}{4}}}$. Using Equation (3.9) the nonlinear growth rate is $\bar{\Omega}(\tau) = \bar{\Omega}_0\sqrt{1 - \frac{1}{2}\varepsilon_r|\bar{E}_{z,s0}(\tau)|^2}$, where $\bar{\Omega}_0 = \frac{1}{\sqrt{\varepsilon_c}} - \frac{\Delta\bar{\omega}}{4}$. The solution of Equation (3.12) at the first depletion point in real units is

$$\tau_d = \frac{1}{\Omega_0}\ln\left[\frac{2\sqrt{2}}{\sqrt{\varepsilon_r}|\bar{E}_{z,s0}(\tau_1)|}\right], \qquad (3.13)$$

where the wake just behind the trigger bunch (calculated from Equation (3.3)) is

$$\bar{E}_{z,s0}(\tau = \tau_1) = i\frac{\bar{J}_{s0}}{\varepsilon_r}\left\{-\frac{A_1}{i\Omega_1}[1 - e^{i\Omega_1(\tau_1-\tau_0)}]\right.$$
$$\left.-\frac{A_2}{i\Omega_2}[1 - e^{i\Omega_2(\tau_1-\tau_0)}]\right\}. \qquad (3.14)$$

Now, the time interval of the nonlinear regime is defined from the first zero of the PID ($\tau = \tau_1 + \tau_d$) until the excited Rabi oscillation decays ($\tau = \tau_1 + \tau_d + \tau_r$). A simple way to evaluate the time interval of this nonlinear regime is to assume that the strongly amplified wake is virtually constant. By differentiating equation (2.8) with τ and substituting Equations (2.6) and (2.7) into it one obtains

$$\frac{\partial^2\bar{N}}{\partial\tau^2} - \frac{\Delta\bar{\omega}}{2}\frac{\partial\bar{N}}{\partial\tau} + \bar{\omega}_R^2\bar{N} = 0, \qquad (3.15)$$

where $\bar{\omega}_R = \sqrt{\varepsilon_r(|\bar{E}_{z,s0}|^2 + |\frac{\bar{E}_{+,s0}-\bar{E}_{-,s0}}{2}|^2)}$ is the normalized Rabi frequency.

Since $\bar{E}_\pm \approx \frac{ik_{s0}}{\bar{\omega}_0\sqrt{\varepsilon_r\Delta\varepsilon_\mp}}\bar{E}_{z,s0}$, the Rabi frequency in physical units is $\omega_R^2 = \omega_{R,0}^2\frac{J_1^2(p_{s0})}{\varepsilon_c}$, where $\omega_{R,0} = \frac{\mu|E_{z,s0}|}{\sqrt{\varepsilon_r}\hbar}$ is in the same form as the Rabi frequency in a homogeneous medium.

For a relativistic bunch that travels close to the Cerenkov velocity ($\beta \to 1$ and $\varepsilon_c \ll 1$) the Rabi frequency is $\bar{\omega}_R \approx |\bar{E}_{z,s0}|/\sqrt{\varepsilon_c}$. Hence, in the nonlinear regime where $|\bar{E}_{z,s0}| \approx |\bar{E}_{sat}| = \sqrt{2/\varepsilon_r}$ the wake oscillates at

$$\bar{\omega}_R \approx \sqrt{\frac{2}{\varepsilon_r\varepsilon_c}}, \qquad (3.16)$$

and the solution of Equation (3.15) is

$$\bar{N}(\tau \geqslant \tau') = \frac{\bar{N}_0'}{\bar{\omega}_R}e^{-\frac{\Delta\bar{\omega}}{4}(\tau-\tau')}\sin[\bar{\omega}_R(\tau - \tau')], \qquad (3.17)$$

where $\tau' = \tau_1 + \tau_d$ and $\bar{N}_0' = \frac{\partial\bar{N}}{\partial\tau}|_{\tau=\tau'}$. Thus, we define according to Equation (3.17) the time in which the oscillations of the PID relax to be

$$\tau_r = 5\left(\frac{4}{\Delta\bar{\omega}}\right) = \frac{20}{\Delta\bar{\omega}}. \qquad (3.18)$$

Therefore, from Equations (3.13) and (3.18) the saturation time is given by

$$\tau_{sat} = \tau_d + \tau_r = \frac{1}{\Omega_0}\ln\left[\frac{2\sqrt{2}}{\sqrt{\varepsilon_r}|\bar{E}_{z,s0}(\tau_1)|}\right] + \frac{20}{\Delta\bar{\omega}}, \qquad (3.19)$$

and in real units it is

Table 1. Structure parameters of our studied example. Note that the set of parameters used here is the same as in Ref. [22].

Parameter	Symbol	Value
Active medium resonance wavelength	$\lambda_0 = 2\pi c/\omega_0$	10.6 μm
Active medium resonance bandwidth	$\Delta f = \Delta\omega/2\pi$	37 GHz
Active medium plasma frequency	ω_p	1.18×10^{10} rad s^{-1}
Electrical dipole moment	μ_{12}	0.0275 Debye
Initial PID	N_0	1.3×10^{23} m^{-3}
Einstein's coefficient	A_{21}	0.2 s^{-1}
Relative permittivity	ε_r	1.0014
Waveguide radius	R	5.065 cm
E-beam Lorentz factor	γ	600
E-beam total charge	Q_b	$-e10^9$
E-beam length	L_{tr}	$150\lambda_0 = 1.6$ mm
E-beam modulation	M_{tr}	20%
E-beam radius	R_b	4 mm

$$t_{sat} = \frac{1}{\Omega}\ln\left[\frac{2\sqrt{2}}{\sqrt{\varepsilon_r}|\bar{E}_{z,s0}(\tau_1)|}\right] + \frac{20}{\Delta\omega}, \qquad (3.20)$$

where $\Omega = \frac{\omega_p}{2\sqrt{\varepsilon_r\varepsilon_c}} - \frac{\Delta\omega}{4}$ is the linear growth rate for $\beta \to 1$ and $\varepsilon_c \ll 1$. In addition, $\bar{E}_{z,s0}(\tau_1)$ is given in Equation (3.14).

Finally, the theoretical spot size, R_{sp}, of the longitudinal wake can be calculated through $J_0(p_{s0}R_{sp}/R) = 0$. Since the first zero of $J_0(x)$ is ~2.4 and from the single-resonance condition $p_{s0} = R\frac{\omega_0}{c}\sqrt{\varepsilon_r\varepsilon_c}$, we obtain that the spot size of the longitudinal wake is

$$R_{sp} \simeq \frac{2.4}{\frac{\omega_0}{c}\sqrt{\varepsilon_r\varepsilon_c}}. \qquad (3.21)$$

4. Simulations

In this section we show the wake dynamics for an active CO_2 gas mixture with the set of parameters that is given in Table 1.

Since the bunch profile is continuously changing except for a finite number of points where it can have first-order discontinuity, the initial conditions of all the envelopes are set to zero but the initial PID is $\bar{N}(\tau = 0) = 1$.

Figure 2 shows the wake and the medium dynamics on the waveguide axis ($r = 0$). In this example the trigger bunch (Figures 2(a) and (b) dashed curves) which appears at $\tau = \tau_0 = 0.05$ and ends at $\tau = \tau_1 = 0.0813$ (corresponding to a bunch length of $L_b = 300$ μm) generates the initial wake. As seen in Figure 2(a), the wake (solid curve) amplification begins in the linear regime, where the PID (dashed–dotted curve) is weakly depleted ($\bar{N}(\tau) \approx \bar{N}(0) = 1$). In this regime the growth rate of the medium is constant, which results in exponential amplification of the wake (Figure 2(b)).

As the wake is amplified the PID is depleted. In this nonlinear regime of amplification, where the PID is significantly reduced and the wake is intense, Figure 2(a) shows that the

amplified wake and the PID experience Rabi oscillations. Finally, when the PID relaxes to zero, the effective gain of the medium is zero and the wake reaches deep saturation.

Figure 2(a) shows that the normalized wake reaches a saturation value similar to the theoretical calculation (Equation (3.10) with $\varepsilon_r \simeq 1$) of $\bar{E}_{sat} = \sqrt{2}$. Also, the normalized saturation time in this example is $\tau_{sat}^{sim} = \tau_d^{sim} + \tau_r^{sim} = 1.32$, where the depletion time is $\tau_d^{sim} = 0.8448$ and the relaxation time is $\tau_r^{sim} = 0.512$. These time parameters are in good agreement with the theoretical formulas (Equations (3.13) and (3.18)) of $\tau_d^{th} = 0.8448$, $\tau_r^{th} = 0.508$ and $\tau_{sat}^{th} = 1.353$.

Figure 2(b) shows in a logarithmic scale the nonlinear wake amplification in physical units of V m^{-1}. Clearly, in the linear regime the wake (solid curve) grows exponentially and it saturates about 6 cm behind the trigger bunch to a value of 0.7 GV m^{-1}. In addition, Figure 2(b) shows in the linear regime good agreement between the nonlinear wake dynamics (solid curve) and the linear wake dynamics (dots) which is calculated for $\bar{N}(\tau) = 1$.

Similarly to the linear regime approach[18, 19], the waveguide radius is chosen such that only a single TM mode will be in resonance with the active medium. Indeed, Figure 3 shows at $\tau = 1.67$ that among the first 500 modes of $\bar{E}_{z,s}$ that are calculated in the simulation only $s0 = 360$ is the dominant mode.

Figure 4 shows the two-dimensional plot of the longitudinal wake. At $\tau = 0.95$, corresponding to $z - \beta ct = 50$ mm, Figures 4(b) and (c) show that the radius of the spot is about 100 μm, which is smaller than the waveguide radius of $R = 50.6475$ mm. This spot size radius is in good agreement with our theoretical expression (Equation (3.21)) of 107.6 μm. This means that such a wake can accelerate a second bunch of electrons with a beam radius of less than a hundred microns.

The energy balance of the structure shown in Figure 5 reveals that initially most of the energy comes from the active medium (Figure 5(c)) and only a small fraction of it comes from the trigger bunch (Figure 5(b)). In the steady state of the amplification all the stored energy from the active medium is transferred to EM energy (Figure 5(a)) such that the energy of the longitudinal wake, $\bar{W}^{(EM,lo)} = \sum_s \bar{W}_s^{(EM,lo)}$, is the same as that of the transverse wake, $\bar{W}^{(EM,tr)} = \sum_s \bar{W}_s^{(EM,tr)}$. In addition, the stored energy of level 1 in the active medium, $\bar{W}_{N1}(\tau) = \frac{1}{2}[\bar{W}_{tot}(\tau = 0) - \bar{W}_N(\tau)]$, is equal to that of level 2, $\bar{W}_{N2}(\tau) = \frac{1}{2}[\bar{W}_{tot}(\tau = 0) + \bar{W}_N(\tau)]$ (Figure 5(c)), which means that in the steady state the total stored energy in the active medium is zero, $\bar{W}_N = \bar{W}_{N2} - \bar{W}_{N1} = 0$. The deviation from energy conservation $\eta(\%) = 100 \times |\bar{W}_{tot}(\tau) - \bar{W}_{tot}(\tau = 0)|/\bar{W}_{tot}(\tau = 0)$ is shown in Figure 5(d) and can be used for the numerical simulation error which in our example is about $6 \times 10^{-5}\%$.

As mentioned previously, the shape of the single-mode wake is sensitive to the active medium bandwidth, $\Delta\bar{\omega}$, and the Cerenkov parameter, ε_c. These two parameters affect

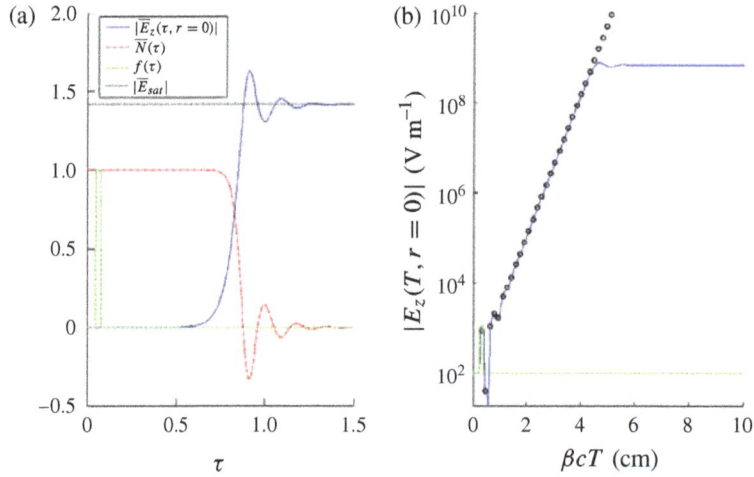

Figure 2. (a) The dynamics of the wake $\bar{E}_z(\tau, r = 0)$ on the axis (solid curve), the PID, $\bar{N}$ (dashed–dotted curve), and the trigger bunch profile, f (dashed curve). The value of the saturated wake $|\bar{E}_{sat}| = \sqrt{2}$ is shown by the dotted curve. (b) A comparison of the nonlinear wake dynamics in real units (solid curve) with the linear wake dynamics (dots). In addition, the profile of the bunch is drawn as a reference (dashed curve).

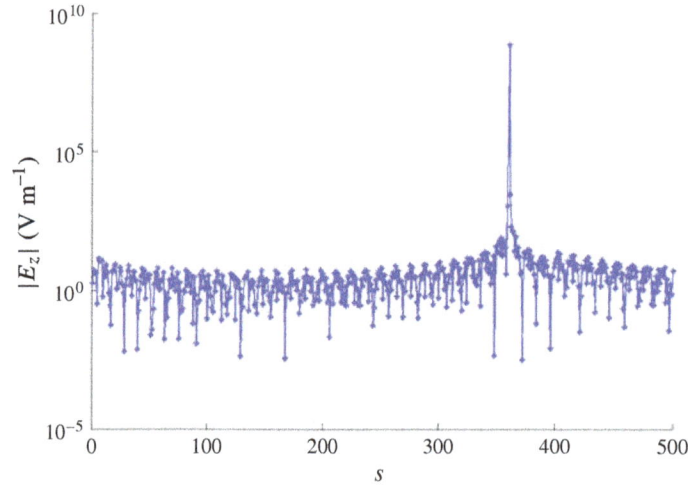

Figure 3. The mode spectrum of the wake $|E_{z,s}(\beta cT = 10\ \mathrm{cm}, r = 0)|$. Here, the single-resonance mode is $s0 = 360$.

the Rabi frequency $\bar{\omega}_R$ and the relaxation time τ_r. Figure 6 shows the dynamics of the wake for different Cerenkov and bandwidth parameters. The solid curve corresponds to the same wake dynamics as in Figure 6(a) with $\bar{\omega}_R = 37.6$ and $\Delta\bar{\omega} = 39.4$. As seen for a Cerenkov parameter five times larger than our former example (dashed curve) the number of Rabi oscillations is significantly lowered. However, the saturation length, τ_{sat}, is greatly increased as a result of increased τ_d—see Equation (3.13). It may be possible to suppress the Rabi oscillation when the medium bandwidth is enlarged. However, increasing the bandwidth can result in multi-mode interaction if $|\Delta\bar{\omega}_s| < \Delta\bar{\omega}$. Figure 6(a) shows (dotted curve) that for a bandwidth that is five times larger the shape of the wake in the steady-state regime has a steady oscillation which results from multi-mode coupling. In this example, $|\Delta\bar{\omega}_{s_0\pm1}| = 75 < 88.06 = \Delta\bar{\omega}$, whereas for the first single-mode propagation we have $|\Delta\bar{\omega}_{s_0\pm1}| =$

$167 > 39.4 = \Delta\bar{\omega}$, where $s_0 = 360$. Figure 6(b) shows the mode spectrum of the wake. As seen for large bandwidth (dotted curve), the adjacent modes ($s = 359$ and 361) participate in the wake dynamics shown in Figure 6(a) (dotted curve).

The two major parameters that characterize our scheme are the value of the wake at the saturation and the saturation length. The sensitivities of these two quantities to the other parameters of the scheme are revealed in Figure 7. Specifically, Figure 7 shows the dependence of the value of the saturated wake (Figure 7(a)) and the saturation time (Figure 7(b)) on the waveguide radius, the dielectric constant, the active medium bandwidth, the bunch energy and the bunch radius. Clearly, the saturation value of the wake is significantly dependent on the waveguide radius, R, the medium bandwidth, $\Delta\omega$, and the parameter $\varepsilon_r - 1$, which is proportional to the Cerenkov parameter ε_c.

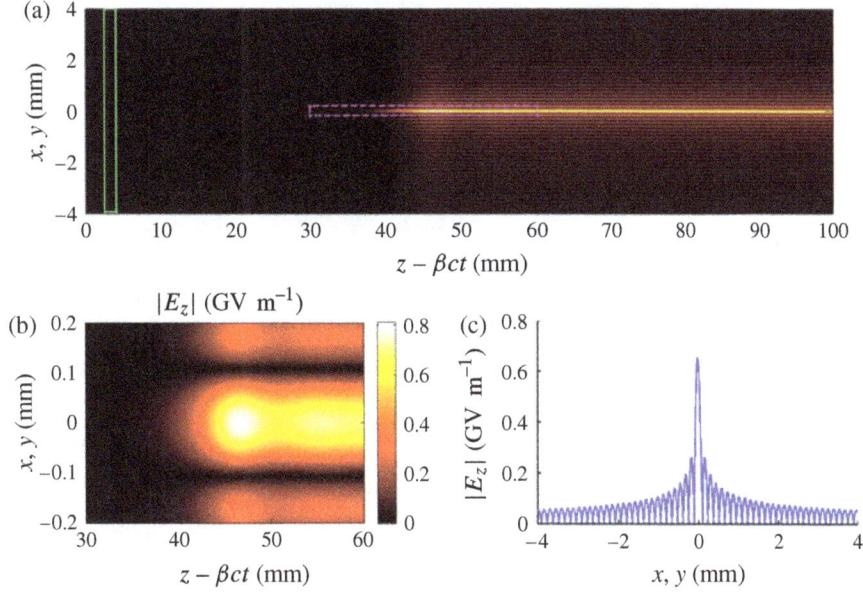

Figure 4. (a) A two-dimensional plot of the longitudinal wake $|E_z|$. The green rectangle is the location of the trigger bunch. (b) The same as (a) but in the region that is marked in magenta in (a). (c) The radial dependence of the wake at $z - \beta ct = 50$ mm.

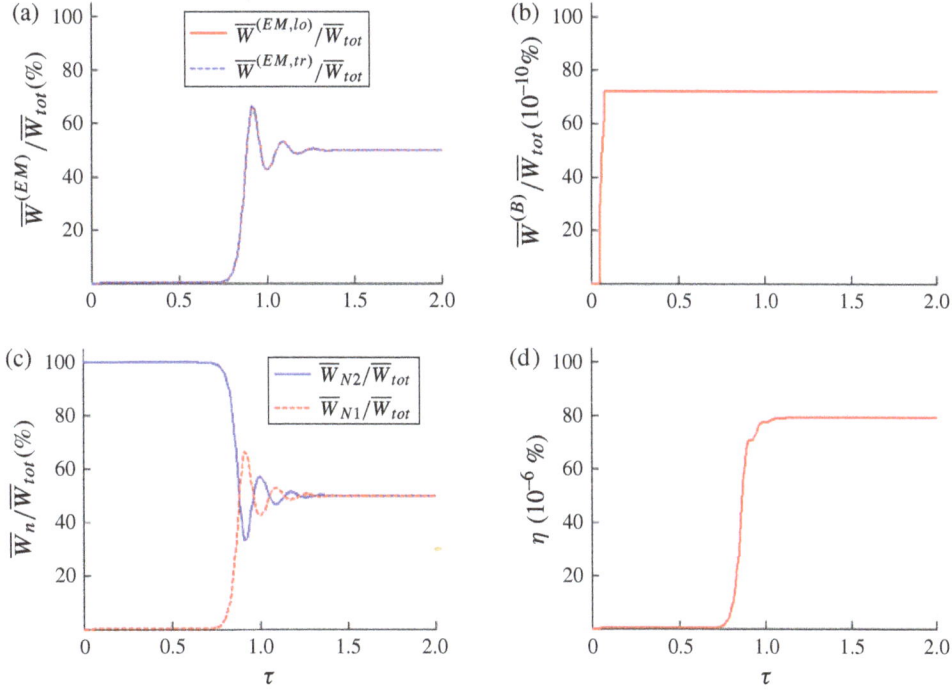

Figure 5. The energy conservation. Here, $\bar{W}_{N1}$ is the energy of the ground state and $\bar{W}_{N2}$ is the energy of the excited state; (d) shows the deviation from energy conservation.

For a relativistic beam the scaling law of the saturated wake on the structure parameters can be explained by our analytical calculations (Equation (3.11)) and the single-resonance condition of $p_s = R(\omega_0/c)\sqrt{\varepsilon_r \varepsilon_c}$ (see Equation (3.2)),

$$|E_{sat}| \propto (\Delta\omega R)^{1/2}(\varepsilon_r - 1)^{1/4}, \qquad (4.1)$$

where it is assumed that $\bar{N} \propto \Delta\bar{\omega}$ (because of $\omega_p = \omega_0 \sqrt{\frac{2N_0\mu^2}{\varepsilon_0 \hbar\omega_0}}$ and $\omega_p = \sqrt{2c\Delta\omega\alpha}$, where α is the small signal gain). This means that the waveguide radius and bandwidth strongly affect the saturation – compared with the Cerenkov parameter. In addition, for a relativistic beam the saturation value does not depend on the beam radius and energy.

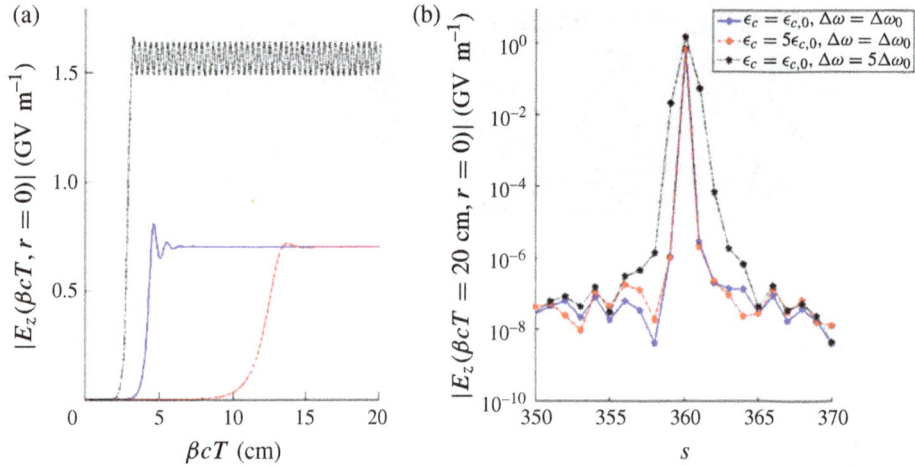

Figure 6. The wake dynamics (a) and the wake spectrum (b) for various Cerenkov parameters and bandwidths. The solid curve corresponds to $\varepsilon_c = \varepsilon_{c,0}$ and $\Delta\omega = \Delta\omega_0$, where $\varepsilon_{c,0}$ and $\Delta\omega_0$ are the Cerenkov and bandwidth parameters as in Figure 2. The dotted curve corresponds to $\varepsilon_c = \varepsilon_{c,0}$ and $\Delta\omega = 5\Delta\omega_0$. The dashed curve corresponds to $\varepsilon_c = 5\varepsilon_{c,0}$ and $\Delta\omega = \Delta\omega_0$.

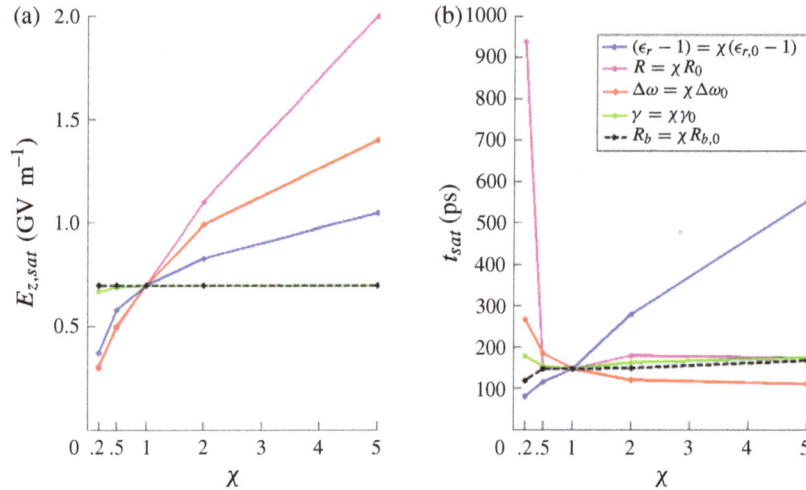

Figure 7. The dependence of the saturation value (a) and saturation length (b) on the waveguide and trigger bunch parameters. Here, the index 0 represents the parameter value as in Figure 2 or Ref. [18].

Similarly, the scaling law of the saturation time on the structure parameters can be explained by our analytical calculations of the saturation time (Equations (3.14) and (3.20)) and the single resonance condition of $p_s = R(\omega_0/c)\sqrt{\varepsilon_r\varepsilon_c}$ (see Equation (3.2)),

$$t_{sat} \propto (\ldots)\sqrt{\varepsilon_c}\ln[\ldots(\varepsilon_r\varepsilon_c)^{1/2}(RR_b)^{3/2}] + \frac{1}{\Delta\omega}. \quad (4.2)$$

For a single resonant mode, the saturation time is strongly dependent on the Cerenkov parameter and the medium bandwidth but weakly dependent on the waveguide radius and beam spot size. Indeed, Figure 7(b) confirms the scaling law in Equation (4.2), but not for the waveguide radius at $\chi = 0.2$.

At this point the saturation time is significantly large because the deviation from satisfying the resonance condition

is larger than for a larger waveguide radius ($\chi > 0.2$). Specifically, for $R = 0.2R_0$ the deviation from the nearest resonance mode is $\Delta\bar{\omega}_s(s = 72) = 167$, whereas for $R = 0.5R_0$ the deviation is $\Delta\bar{\omega}_s(s = 180) = 42$. When the waveguide radius is close to satisfying a single resonance, then there is a small dependence on the saturation time, as confirmed by the logarithmic dependence in Equation (4.2). Otherwise the wake growth rate is significantly reduced.

5. Discussion and conclusions

In this paper we present the nonlinear aspects of a new scheme of electron acceleration by an active medium. In contrast to plasma-based accelerators, where the energy for generating the intense wake originates in the intense laser pulse, in our scheme the energy source is stored in excited atoms/molecules. Most importantly, we have shown that a centimeter-sized cylindrical waveguide filled with a CO_2 gas mixture can generate gradients of the order of GV m^{-1} traveling at the trigger bunch velocity.

We analyzed the dynamics of the wake and the active medium both analytically and numerically. Our numerical simulations indicate that the dynamics of the wake can be divided into three regimes. In the first regime, known as the *linear regime*, the PID is nearly constant ($\bar{N} \simeq 1$). Hence, the gain of the medium is constant which results in exponential growth of the wake. The second regime starts when the amplified wake reaches *high intensity*, and the PID is significantly reduced ($\bar{N} \ll 1$) by the stimulated emission effect. In this regime both the PID and the wake can experience Rabi oscillations. The Rabi frequency in our confined structure can be significantly larger than in a boundless medium. More specifically, this frequency is inversely proportional to the square root of both the Cerenkov parameter and the mode number. Finally, in the third regime, when the Rabi oscillations are relaxed, the PID reaches complete depletion ($\bar{N} \simeq 0$) and the wake reaches *deep saturation*.

Energy conservation proves that most of the initial energy originates from the active medium, and in the deep saturation regime half of the initial stored energy is transferred to the longitudinal EM field component and the other half is transferred to the transverse EM field components.

To obtain maximum performance from our studied structure we should fulfill the following constraints. First, it is desired to design the structure for single-resonance-mode operation to obtain a constant value of the wake in the saturation regime. In order to avoid multi-mode propagation of adjacent frequencies, the active medium bandwidth should be narrow enough. Second, the spot size of the wake should be large enough in order to accelerate most of the trailing train bunch. We found that for single-mode propagation a large spot size is achieved for a small Cerenkov parameter. The last requirement is obviously the generation of a high gradient in a short saturation length.

On one hand, we have found that for a relativistic trigger bunch the *saturation value* is strongly dependent on the waveguide radius and the medium bandwidth but weakly dependent on the Cerenkov parameter. On the other hand, we have found that for single-mode propagation and a relativistic trigger bunch the *saturation length* is strongly dependent on the Cerenkov parameter and the medium bandwidth but weakly dependent on the waveguide and electron beam radius. Consequently, optimal performance may be achieved with a large waveguide radius filled with a high density of excited atoms and a trigger bunch that travels at a velocity slightly above the Cerenkov velocity. Interestingly, the short saturation length of the wake may be used as a seed pulse for backward Raman amplification in plasma[25].

For a proper perspective, it is important to comment that the high value of the gradient is limited by self-focusing and ionization effects[23], which are not analyzed here, but which may be overcome by using a vacuum channel, as previously mentioned in Section 1.

In spite of the relatively modest gradient, compared with laser plasma accelerators, our paradigm may benefit from a few aspects. First, our scheme can support staging[26, 27] in a fairly natural way. Second, our setup, in principle, does not seem to suffer from instabilities that may evolve in plasma at a high repetition rate. A typical repetition rate for laser plasma accelerators is less than 1 Hz for 100 J laser pulses with fs duration in a laser plasma accelerator[6]. The various laser wake acceleration schemes in plasma take advantage of the fact that on the fs time scale all ion instabilities are far below threshold; at high repetition rate these instabilities may develop[28]. Their suppression may require a dramatic reduction of the laser power.

Regarding our paradigm as a new source of coherent radiation excited by a trigger bunch, it possesses important features that distinguish it from a conventional laser driven by spontaneous emission. The most remarkable ones are the excitation of a specific mode number and wake propagation at the same phase and group velocity as the trigger bunch.

Acknowledgement

This work was supported by the Bi-National Science Foundation (BSF).

References

1. R. B. Neal, *The Stanford Two-Mile Accelerator* (Benjamin, New York, 1968).
2. T. P. Wangler, *RF Linear Accelerators* 2nd edition (Wiley–VCH, 2008).
3. T. Behnke, J. E. Brau, B. Foster, J. Fuster, M. Harrison, J. McEwan Paterson, M. Peskin, M. Stanitzki, N. Walker, and H. Yamamoto, 'The International Linear Collider', *Technical Design Report*, volume 1: Executive summary, (2013).
4. E. Esarey, C. B. Schroeder, and W. P. Leemans, Rev. Mod. Phys. **81**, 1229 (2009).
5. V. Malka, Phys. Plasmas **19**, 055501 (2012).
6. S. M. Hooker, Nature Photon. **7**, 775 (2013).
7. D. Strickland and G. Mourou, Opt. Commun. **56**, 219 (1985).
8. S. P. D. Mangles, C. D. Murphy, Z. Najmudin, A. G. R. Thomas, J. L. Collier, A. E. Dangor, E. J. Divall, P. S. Foster, J. G. Gallacher, C. J. Hooker, D. A. Jaroszynski, A. J. Langley, W. B. Mori, P. A. Norreys, F. S. Tsung, R. Viskup, B. R. Walton, and K. Krushelnick, Nature **535**, 535 (2004).
9. C. G. R. Geddes, Cs. Toth, J. van Tilborg, E. Esarey, C. B. Schroeder, D. Bruhwiler, C. Nieter, J. Cary, and W. P. Leemans, Nature **431**, 538 (2004).
10. J. Faure, Y. Glinec, A. Pukhov, S. Kiselev, S. Gordienko, E. Lefebvre, J.-P. Rousseau, F. Burgy, and V. Malka, Nature **431**, 541 (2004).

11. X. Wang, R. Zgadzaj, N. Fazel, Z. Li, S. A. Yi, X. Zhang, W. Henderson, Y.-Y. Chang, R. Korzekwa, H.-E. Tsai, C.-H. Pai, H. Quevedo, G. Dyer, E. Gaul, M. Martinez, A. C. Bernstein, T. Borger, M. Spinks, M. Donovan, V. Khudik, G. Shvets, T. Ditmire, and M. C. Downer, Nature Commun. **4**, 1988 (2013).

12. I. Blumenfeld, C. E. Clayton, F.-J. Decker, M. J. Hogan, C. Huang, R. Ischebeck, R. Iverson, C. Joshi, T. Katsouleas, N. Kirby, W. Lu, K. A. Marsh, W. B. Mori, P. Muggli, E. Oz, R. H. Siemann, D. Walz, and M. Zhou, Nature **445**, 741 (2007).

13. E. A. Peralta, K. Soong, R. J. England, E. R. Colby, Z. Wu, B. Montazeri, C. McGuinness, J. McNeur, K. J. Leedle, D. Walz, E. B. Sozer, B. Cowan, B. Schwartz, G. Travish, and R. L. Byer, Nature **503**, 91 (2013).

14. L. Schächter, Phys. Lett. A **205**, 65 (2000).

15. L. Schächter, Phys. Rev. Lett. **83**, 92 (1999).

16. S. Banna, V. Berezovsky, and L. Schächter, Phys. Rev. Lett. **13**, 134801 (2006).

17. L. Schächter, Phys. Lett. A **277**, 65 (2000).

18. M. Voin, W. D. Kimura, and L. Schächter, Nucl. Instrum. Methods Phys. Res. A **740**, 117 (2013).

19. M. Voin and L. Schächter, Phys. Rev. Lett. **112**, 05480 (2014).

20. L. Schächter, E. Colby, and R. H. Siemann, Phys. Rev. Lett. **87**, 134802 (2001).

21. L. Schächter, Beam–Wave Interaction in Periodic and Quasi-Periodic Structures (Springer, New York, 2011).

22. R. H. Pantell and H. E Puthoff, Fundamentals of Quantum Electronics (Wiley, New York, 1969).

23. R. W. Boyd, Nonlinear Optics (Academic Press, San Diego, CA, 2008).

24. A. E. Siegman, Lasers (University Science Books, Mill Valley, CA, 1986).

25. V. M. Malkin, G. Shvets, and N. J. Fisch, Phys. Rev. Lett. **82**, 4448 (1999).

26. W. D. Kimura, A. van Steenbergen, M. Babzien, I. Ben-Zvi, L. P. Campbell, D. B. Cline, C. E. Dilley, J. C. Gallardo, S. C. Gottschalk, P. He, K. P. Kusche, Y. Liu, R. H. Pantell, I. V. Pogorelsky, D. C. Quimby, J. Skaritka, L. C. Steinhauer, and V. Yakimenko, Phys. Rev. Lett. **86**, 4041 (2001).

27. M. Dunning, E. Hemsing, C. Hast, T. O. Raubenheimer, S. Weathersby, and D. Xiang, Phys. Rev. Lett. **110**, 244801 (2013).

28. W. L. Kruer, The Physics of Laser Plasma Interactions (Addison-Wesley, Reading, MA, 1988).

23

Optimal laser intensity profiles for a uniform target illumination in direct-drive inertial confinement fusion

Mauro Temporal[1], Benoit Canaud[2], Warren J. Garbett[3], and Rafael Ramis[4]

[1] *Centre de Mathématiques et de Leurs Applications, ENS Cachan and CNRS, 61 Av. du President Wilson, F-94235 Cachan Cedex, France*

[2] *CEA, DIF, F-91297, Arpajon Cedex, France*

[3] *AWE plc, Aldermaston, Reading, Berkshire RG7 4PR, United Kingdom*

[4] *ETSI Aeronáuticos, Universidad Politécnica de Madrid, 28040 Madrid, Spain*

Abstract

A numerical method providing the optimal laser intensity profiles for a direct-drive inertial confinement fusion scheme has been developed. The method provides an alternative approach to phase-space optimization studies, which can prove computationally expensive. The method applies to a generic irradiation configuration characterized by an arbitrary number N_B of laser beams provided that they irradiate the whole target surface, and thus goes beyond previous analyses limited to symmetric configurations. The calculated laser intensity profiles optimize the illumination of a spherical target. This paper focuses on description of the method, which uses two steps: first, the target irradiation is calculated for initial trial laser intensities, and then in a second step the optimal laser intensities are obtained by correcting the trial intensities using the calculated illumination. A limited number of example applications to direct drive on the Laser MegaJoule (LMJ) are described.

Keywords: direct drive; inertial confinement fusion; laser system

1. Introduction

In the direct-drive (DD) inertial confinement fusion (ICF)[1, 2] context a spherical capsule containing the deuterium–tritium (DT) nuclear fuel is irradiated by laser beams. The final goal is to generate energy gain via a nuclear fusion reaction: $D + T \rightarrow \alpha + n + 17.6$ MeV. The external shell of the capsule absorbs a fraction of the incoming laser energy producing a plasma; the plasma temperature ($\approx$keV) increase provides the outward expansion of the low-density corona and launches a series of inward shock waves. These shock waves compress the DT payload in a high-density shell that implodes and reaches stagnation. In the classical central ignition scheme, the high-density shell confines a small amount ($\approx$10 µg) of DT fuel – called a hot-spot – which is heated to high temperature ($\approx$10 keV) and compressed to areal densities comparable with the α-particle range ($\int \rho \, dr \approx 0.3$ g cm^{-2}), thus providing the ignition of the thermonuclear fusion reactions. Recently, the new shock ignition (SI) scheme[3] has been proposed. In the SI scheme the fuel is first compressed by the usual DD technique, then a high-power laser pulse ($\approx$hundreds of TW) is used to launch a strong shock wave which provides the fuel ignition.

In all ICF schemes the capsule irradiation must be very uniform in order to inhibit growth of dangerous hydrodynamic instabilities that can prevent a successful fuel compression. In the promising SI scheme the requirements in terms of irradiation uniformity are less stringent in comparison with the classical central ignition scheme. Nevertheless, the irradiation uniformity represents one of the major constraints in ICF and a great deal of effort has been dedicated to its optimization.

In this paper we propose a numerical method to calculate the optimal laser intensity profiles of a generic number N_B of laser beams irradiating a spherical target. Analytical optimization of the laser intensity profiles has been already performed for configurations based on the geometry of the Platonic solids[4–7]. These analyses always provide axially symmetric laser intensity profiles where all the N_B laser intensities are equal. These solutions can be applied to laser configurations such as Gekko XII[8] ($N_B = 12$) or Omega[9] (60 beams) but are not suitable for laser configurations like the National Ignition Facility (NIF)[10], the Laser

Correspondence to: M. Temporal, Centre de Mathematiques et de Leurs Applications, ENS Cachan and CNRS, 61 Av. du President Wilson, Cachan Cedex, France. Email: mauro.temporal@hotmail.com

MegaJoule (LMJ)[11] or the smaller Orion[12] facility where the locations of the beams are optimized for the indirect-drive[13] ICF scheme. Indeed, in these latter cases the optimal laser intensity profiles must be adapted to the laser configuration. As a consequence, the laser intensity profiles are not necessarily equal to each other or axially symmetric.

2. Numerical method to optimize the intensity profiles

The proposed numerical method allows us to find the laser intensity profiles that optimize the illumination uniformity for a given laser configuration. These calculations are performed within an illumination model in which laser refraction is neglected, photons propagate linearly and the results only apply to the low-power foot-pulse that characterizes the first few ns of an ICF irradiation, the so-called imprint phase. Thus, the solution guarantees the uniformity of the first shock wavefront[14].

In the past, optimizing methods have usually been based on analytical or numerical parametric studies looking for the laser parameters that minimize the illumination nonuniformity. In contrast, in the present case a sort of predictor–corrector method is used: in a first step, trial laser intensity profiles are used to evaluate the – imperfect – irradiation of the spherical target; in a second step, the laser intensities are recalculated using the results of the first step in order to provide perfect illumination uniformity.

The model problem is characterized by a spherical target of radius r_0 irradiated by N_B laser beams. The target centre is located at the origin of a Cartesian coordinate system $O(x, y, z)$ and the laser beam directions are characterized by the unitary vector r_n defined by their polar angles θ_n and φ_n (see the details of the geometry in Figure 1). Each given elementary surface element of the target, $ds = r_0^2 \, d\Omega$, is associated with a vector direction $r(\theta, \varphi)$ of co-latitude $\theta \in [0 - \pi]$ and longitude $\varphi \in [0 - 2\pi]$. The laser intensity profile $g_n(x', y')$ of the N_B beams is defined in the planes orthogonal to the beam directions r_n. In these planes we define a secondary Cartesian coordinate system $O'(x'_n, y'_n)$ where the orthogonal axes are given by the two versors: $x'_n = (r_n \wedge z)/|r_n \wedge z|$ and $y'_n = (x'_n \wedge r_n)/|x'_n \wedge r_n|$. In this way the y'_n-axis is located in the meridian plane containing the nth laser beam axis (r_n), while the x'-axis is orthogonal to both y'_n and r_n. Therefore, there is a one-to-one correspondence between a position $r_0 r(\theta, \varphi)$ over the target surface and the corresponding coordinate $x'_n = r_0 r \cdot x'_n$ and $y'_n = r_0 r \cdot y'_n$, on the focal plane of the nth laser beam.

The elementary surface ds, located at the polar coordinates (θ, φ), is irradiated by a given number $(\leqslant N_B)$ of laser beams and receives a total laser intensity $I(\theta, \varphi)$. This intensity is given by the contributions of all the incoming intensities associated with the laser profiles $g_n(x', y')$ multiplied by the scalar product $(r \cdot r_n)$ to account for projection of the surface area. Thus, for the N_B laser beams the irradiation of the

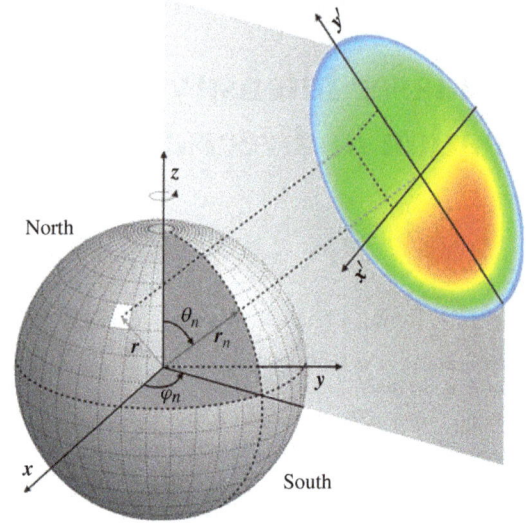

Figure 1. Spherical target and main coordinate system [O]; vector direction r of a generic surface element and versor of the nth laser beam, r_n; coordinate system [x', y'] for the nth laser intensity profile.

spherical target surface is given by

$$I(\theta, \varphi) = \sum_n g_n(x', y')(r \cdot r_n)^\beta, \qquad (1)$$

where the scalar product $(r \cdot r_n)$ is set to zero when it assumes a negative value. The exponent β must be larger than or equal to one and account for the specific assumption on the laser–capsule coupling, e.g., setting $\beta = 2$ recovers the hypothesis of the laser absorption assumed by Schmitt[6]. Hereafter we use the standard illumination model for which $\beta = 1$. For a given laser intensity profile g_n, direct application of the illumination model provides the laser intensity $I(\theta, \varphi)$ used to calculate the root-mean-square (r.m.s.) deviation σ, which is assumed as a measure of the target irradiation nonuniformity:

$$\sigma = \left\{ \frac{1}{4\pi} \int_0^{2\pi} \int_0^\pi [I(\theta, \varphi) - I_a]^2 \sin(\theta) d\theta d\varphi \right\}^{1/2} \Bigg/ I_a, \tag{2}$$

where I_a is the average intensity calculated over the whole sphere.

Of course, the condition to obtain $\sigma = 0$ is to generate a perfectly uniform irradiation over the whole target surface, i.e., to realize $I(\theta, \varphi) = I_0$, where I_0 is the desired intensity over the target surface. This could be done by a simple re-normalization of the laser intensity profiles, by

$$g'_n(x', y') = g_n(x', y')[I_0/I(\theta, \varphi)]. \tag{3}$$

Now, substituting the old intensity profiles g_n with the new ones g'_n in Equation (1) will provide $I(\theta, \varphi) = I_0$ and therefore $\sigma = 0$. A necessary condition to realize the uniform constant intensity illumination – $I(\theta, \varphi) = I_0$ – over the whole target surface is that each elementary surface of the target must be irradiated by at least one laser beam. Indeed, if some part of the target is not irradiated at all, e.g., for $N_B = 1$, the method fails. Thus, the optimization is obtained in two steps: first, by means of a set of trial laser intensity profiles g_n, the target irradiation $I(\theta, \varphi)$ is calculated, and then these trial intensities are corrected – by using Equation (3) – to provide the optimized profiles g'_n. In these calculations the target surface is subdivided into 180×360 elementary elements and thus the method provides a discrete number of coordinates (x', y') where the laser intensity profiles g'_n are defined. These data are used to estimate the intensity profiles, g'_n, on the focal plane where the spatial resolution has been set to $dx' = dy' = r_0/200$, which provides a total of 400×400 values for the laser intensity profiles. With these values the irradiation nonuniformities σ for the cases calculated in this paper are kept below $\sigma = 10^{-4}$. It is worth noticing that an increase of the resolution on the intensity profile or a reduction of the number of elementary surfaces on the target surface increases the precision of the calculation, providing a smaller σ.

The optimal laser intensity profiles, solution of the coupled Equations (3) and (1), depend on the choice of trial intensities g_n; therefore, the set of solutions is not unique. The trial intensity must be well posed in order to generate reasonably final optimized profiles. Hereafter, we use the trial intensity given by the scalar product: $g_n(x', y') = |\mathbf{r} \cdot \mathbf{r}_n| = [r_0^2 - x'^2 - y'^2]^{1/2}/r_0$; a higher intensity is assigned to the beam centre and it vanishes at the target border; the aim of this choice is to look for a solution g'_n that maximizes the laser–capsule coupling. It is worth noticing that although the trial beams are axially symmetric, the final solution will not be symmetric, as we will see later. Moreover, in these calculations we used the same trial function – the scalar product – for all the g_n; nevertheless, the method applies equally well even if the trial functions are different for each laser beam.

3. Profiles for a two-ring 2D irradiation configuration

As a first example we considered a two-dimensional (2D) axially symmetric laser configuration where the beams can be approximated by two annular rings at the co-latitudes θ_1 and $\theta_2 = \pi - \theta_1$; this is achieved by imposing that $I(\theta, \varphi) \equiv I(\theta) = [\int I(\theta, \varphi)d\phi]/(2\pi)$. The two optimal intensity profiles $g'_1(\theta_1)$ and $g'_2(\theta_2)$ have been calculated for different values of θ_1. In this perfectly symmetric case, the two solutions are equal and are just rotated by 180°, $g'_1(x', y') = g_2(x', -y')$. The laser intensity profiles g'_1 normalized to one and corresponding to the annular ring of the north hemisphere are shown in Figure 2 for different polar angles, θ_1.

In these frames the grey curves show the projection of the equator in the focal plane, while the full grey dots localize the projection of the north pole. In these images

the laser intensity has been normalized to 1 and the colour scale ranges from 0 to 1. For small polar angles, e.g., $\theta_1 = 10°$, the laser beams are closer to the z-axis, thus the surfaces of the polar areas are highly irradiated with a nearly orthogonal angle of incidence; on the contrary, for larger angles, it is the equatorial belt that will be over-irradiated in comparison with the polar regions. To compensate for this unbalanced irradiation the optimal laser intensity profile provides different intensities in correspondence to the equatorial and polar target areas. Specifically, at smaller angles (see, e.g., $\theta_1 = 10°$) a maximum intensity is directed towards the equator, while at larger angles (e.g., $\theta_1 = 80°$) the laser intensity is higher in proximity to the polar areas. It is worth noticing that the laser intensity profile becomes circular (axially symmetric) when θ_1 equals the Schmitt angle $\theta_1 = \theta_S = 54.7°$ (Ref. [6]). This is not surprising; indeed, as shown by Schmitt, the angle θ_S is the best co-latitude if the axisymmetric beam intensity profile is given by $I_0 \cos(\gamma)$ with $\beta = 1$.

4. Optimal profile for some LMJ configurations

An LMJ configuration consisting of 40 quads has been considered in this paper. The quad of the LMJ is composed of a bundle of four laser beams and provides a maximum laser energy (power) of 30 kJ (10 TW) at 3ω ($\lambda = 351$ nm). The polar coordinates of the 40 quads are shown in Figure 3. Here, we assume that each quad can be characterized by a single beam with a given laser intensity profile. Four configurations have been considered: (A) a total of four quads (two in the second ring and two in the third ring), labelled A in Figure 3; (B) five quads in the second ring and five in the third ring (ten quads, labelled B); (C) a total of eight quads, two quads in each of the four rings (red quads for the north hemisphere); (D) five quads in each ring (blue quads for the south hemisphere) for a total of 20 quads.

The optimization method has been applied to the four laser configurations A–D. As above, all the N_B trial intensities are given by the scalar product: $g_n(x', y') = (\mathbf{r} \cdot \mathbf{r}_n)$. The optimal intensity profiles g'_n provided for these configurations are shown in Figure 4. These profiles have been normalized to one by dividing the intensities by their maximum value: $\text{Max}[g']_A = 1.52 I_0$; $\text{Max}[g']_B = 0.62 I_0$; $\text{Max}[g']_C = 0.92 I_0$; $\text{Max}[g']_D = 0.38 I_0$. These images correspond to the beams of the north hemisphere, while those of the south hemisphere are obtained by a rotation of 180°. In these irradiation configurations the optimal laser intensity profiles provide an r.m.s. irradiation nonuniformity σ lower than 10^{-4}.

The configuration A has only two beams per hemisphere, and their optimized intensity profile shows three zones at higher intensity: one located below the equator and the other two closer to the pole. The intensity profile for the ten beams of configuration B is shown in Figure 4 (bottom left image). In this case the higher intensity is situated between

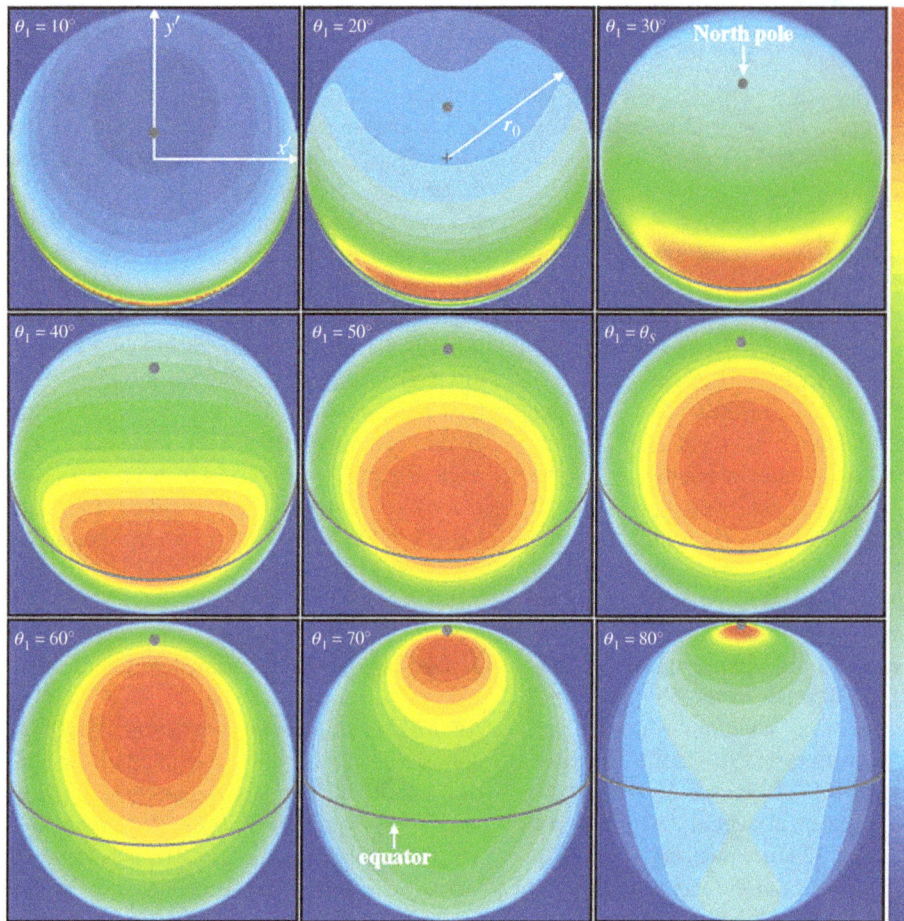

Figure 2. Optimal laser intensity profiles $g'_1(x', y')$ (north hemisphere) for an axially symmetric beam configuration. The intensity profiles have been normalized to one ($g'_1/\text{Max}[g'_1]$) and the scale colour ranges from 0 to 1. Full dots correspond to the north pole and the grey curve is the equator projection on the focal planes.

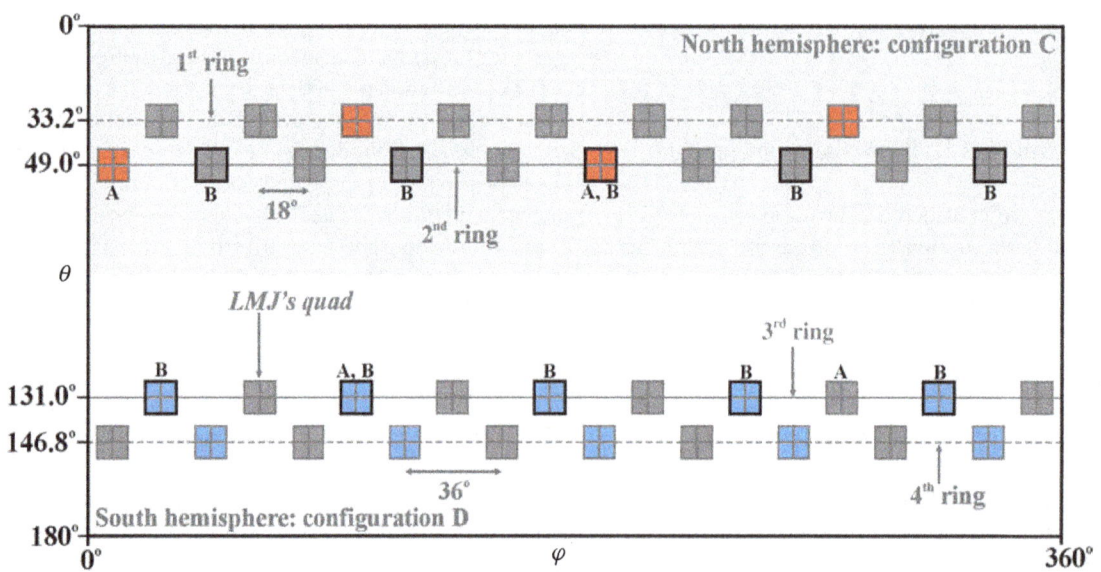

Figure 3. Polar coordinates of 40 quads of the LMJ facility. Quads for the configurations A and B; red quads of the north hemisphere (C) and blue quads of the south hemisphere (D).

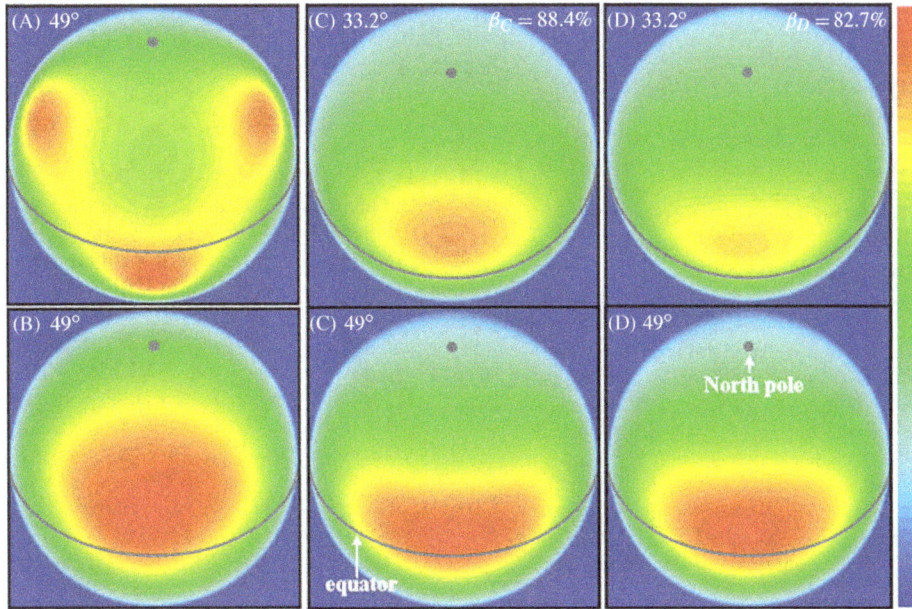

Figure 4. Optimal laser intensity profiles g'_n normalized to one (north hemisphere) for the LMJ configurations A–D. The power imbalance is given by the parameter β and the laser intensity scale colour varies linearly from 0 to 1.

the beam's centre and the equator position and the shape of the laser intensity is not symmetric due to the rotation of 18° between the hemispheres. In the configurations C and D the numerical method provides two intensity profiles: one for the beams of the first ring and a second for those of the second ring. In these two cases the calculations provide similar shapes. Moreover, the numerical results naturally introduce a power imbalance ratio between the maximum beam intensities: $\beta_C = I_{\mathrm{Max}}(49°)/I_{\mathrm{Max}}(33.2°) = 88.4\%$ and $\beta_D = 82.7$. It is worth noticing that these optimum intensities are located off-centre. This could be regarded as the application of the polar direct-drive[15] (PDD) technique where a centred laser intensity profile moves towards the equator to compensate the over-irradiation of the polar zones[16]. In addition, it has been shown[17, 18] that for the configurations C and D elliptical intensity profiles provide a more uniform irradiation than circular ones. The current calculations confirm this trend and show that the optimal intensity profiles are closer to an elliptical shape – characteristic of an indirect-drive installation – rather than circular profiles.

The NIF configuration has been also analysed, providing four optimal laser intensity profiles. In this case the 48 quads of the NIF facility are located at four rings in each hemisphere: four quads at $\theta_1 = 23.5°$, four at $\theta_2 = 30.0°$, eight at $\theta_3 = 44.5°$ and eight at $\theta_4 = 50.0°$. The method of optimization, initialized with the trial intensity $g_n(x', y') = (r \cdot r_n)$, produces intensity profiles similar to those found for configuration D. These calculations also supply the optimal power imbalances $\beta_1 = 69.3\%$, $\beta_2 = 77.2\%$, $\beta_3 = 94\%$, while the maximum power ($\beta_4 = 1$) is assigned to the laser beams located at the larger angle $\theta_4 = 50.0°$.

5. Conclusions

In conclusion, we developed a general method to calculate the optimal laser intensity profiles that optimize the illumination nonuniformity of a spherical target. The method can be used for any DD laser configuration accounting for a general number N_B of laser beams, provided that the beams irradiate the whole target surface. In some sense this is a kind of predictor–corrector method that consists of two steps: firstly, initialized by a set of N_B trial laser intensity profiles, the imperfect surface irradiation is calculated; then, the beam profiles are recalculated in order to correct the previously estimated nonuniform illumination.

A set of four laser configurations based on the LMJ facility has been considered. In these cases, the optimal intensity profiles have been individuated using axially symmetric trial profiles. The resulting optimal intensity profiles are not axially symmetric and their shapes look like to those envisaged by the PDD technique; in addition, these calculations also predict the optimal beam-to-beam power imbalance. These results assume perfect beam-to-beam power imbalance, neglecting laser pointing errors and target positioning uncertainties; deviation from these idealized assumptions would damage the uniformity of the target illumination.

Acknowledgements

M.T. and B.C. express their thanks to Daniel Bouche for the support given to this work. R.R. was partially supported by the EURATOM/CIEMAT association in the framework of the 'IFE Keep-in-Touch Activities'.

References

1. J. Lindl, Phys. Plasmas **2**, 3933 (1995).
2. S. Atzeni and J. Meyer-ter-Vehn, *The Physics of Inertial Fusion* (Oxford University Press, Oxford, 2004).
3. R. Betti, C. D. Zhou, K. S. Anderson, L. J. Perkins, W. Theobald, and A. A. Solodov, Phys. Rev. Lett. **98**, 155001 (2007).
4. M. Murakami, Appl. Phys. Lett. **66**, 1587 (1995).
5. S. Skupsky and K. Lee, J. Appl. Phys. **54**, 3662 (1983).
6. A. J. Schmitt, Appl. Phys. Lett. **44**, 399 (1984).
7. M. Murakami, K. Nishihara, and H. Azechi, J. Appl. Phys. **74**, 802 (1993).
8. H. Azechi, T. Jitsuno, T. Kanabe, M. Katayama, K. Mima, N. Miyanaga, M. Nakai, S. Nakai, H. Nakaishi, M. Nakatsuka, A. Nishiguchi, P. A. Norrays, Y. Setsuhara, M. Takagi, M. Yamanaka, and C. Yamanaka, Laser Particle Beams **9**, 193 (1991).
9. T. R. Boehly, D. L. Brown, R. S. Craxton, R. L. Keck, J. P. Knauer, J. H. Kelly, T. J. Kessler, S. A. Kumpan, S. J. Loucks, S. A. Letzring, F. J. Marshall, R. L. McCrory, S. F. B. Morse, W. Seka, J. M. Soures, and C. P. Verdon, Opt. Commun. **133**, 495 (1997).
10. E. I. Moses, R. N. Boyd, B. A. Remington, C. J. Keane, and R. Al-Ayat, Phys. Plasmas **16**, 041006 (2009).
11. J. Ebrardt and J. M. Chaput, J. Phys.: Conf. Ser. **244**, 032017 (2010).
12. N. Hopps, C. Danson, S. Duffield, D. Egan, S. Elsmere, M. Girling, E. Harvey, D. Hillier, M. Norman, S. Parker, P. Treadwell, D. Winter, and T. Bett, Appl. Opt. **52**, 3597 (2013).
13. J. D. Lindl, P. Amendt, R. L. Berger, S. G. Glendinning, S. H. Glenzer, S. W. Haan, R. L. Kauffman, O. L. Landen, and L. J. Suter, Phys. Plasmas **11**, 339 (2004).
14. M. Temporal, B. Canaud, W. J. Garbett, and R. Ramis, Laser and Particle Beams (in press) (2014) available on CJO2014.
15. S. Skupsky, J. A. Marozas, R. S. Craxton, R. Betti, T. J. B. Collins, J. A. Delettrez, V. N. Goncharov, P. W. McKenty, P. B. Radha, J. P. Knauer, F. J. Marshall, D. R. Harding, J. D. Kilkenny, D. D. Meyerhofer, T. C. Sangster, and R. L. McCrory, Plasma Phys. **11**, 2763 (2004).
16. M. Temporal, B. Canaud, W. J. Garbett, and R. Ramis, High Power Laser Sci. Eng. **2**, e8 (2014).
17. M. Temporal, B. Canaud, W. J. Garbett, F. Philippe, and R. Ramis, Eur. Phys. J. D **67**, 205 (2013).
18. M. Temporal, B. Canaud, W. J. Garbett, and R. Ramis, Phys. Plasmas **21**, 012710 (2014).

24

Generation of high-energy neutrons with the 300-ps-laser system PALS

J. Krása[1], D. Klír[2], A. Velyhan[1], E. Krousky[1], M. Pfeifer[1], K. Řezáč[2], J. Cikhardt[2], K. Turek[4], J. Ullschmied[3], and K. Jungwirth[1]

[1]*Institute of Physics, AS CR, 182 21 Prague 8, Czech Republic*

[2]*Czech Technical University in Prague, FEE, 166 27 Prague, Czech Republic*

[3]*Institute of Plasma Physics, AS CR, 182 00 Prague 8, Czech Republic*

[4]*Nuclear Physics Institute, AS CR, 180 00 Prague 8, Czech Republic*

Abstract

The laser system PALS, as a driver of a broad-beam ion source, delivered deuterons which generated neutrons with energies higher than 14 MeV through the ^{7}Li(d, n)^{8}Be reaction. Deuterons with sub-MeV energy were accelerated from the front surface of a massive CD_2 target in the backward direction with respect to the laser beam vector. Simultaneously, neutrons were emitted from the primary CD_2 target and a secondary LiF catcher. The total maximum measured neutron yield from ^{2}D(d, n)^{3}He, ^{7}Li(d, n)^{8}Be, ^{12}C(d, n)^{13}N reactions was $\sim 3.5(\pm 0.5) \times 10^8$ neutrons/shot.

Keywords: beam–target fusion; deuterons; laser ion sources; lithium; neutrons

1. Introduction

Recent rapid development of laser plasma accelerators has made it possible to accelerate protons and deuterons to high energies of approximately 70 and 170 MeV, respectively[1, 2]. These beams hitting a secondary target can create high-energy neutrons through, for example, D(d, n)^{3}He, ^{7}Li(p, n)^{7}Be, ^{7}Li(d, xn)^{8}Be, ^{9}Be(p, n)$^5_9$B, and ^{9}Be(d, n)$^5_{10}$B nuclear reactions[2–15]. Fast neutrons with energies in excess of 10 MeV resulting from the ^{7}Li(d, xn)^{8}Be reaction ($Q = 15.03$ MeV) have been reported by several authors[6–11]. Acceleration of deuterons is mostly reported in experiments using lasers with intensities of 10^{19} W cm^{-2}. The deuterons are accelerated in focal spots on thin-film targets through either the target-normal sheath acceleration (TNSA) mechanism or the newly recognized break-out afterburner (BOA) mechanism[2]. There is another laser ignition of fusion (LIF) scheme for applications, based on the combination of ultra-high laser nonlinear force driven plasma blocks and the relativistic acceleration of ion blocks, which has shown how 70 MeV D$^+$ and T$^+$ ions can be produced using of ps-laser pulses[16]. The laser-driven bright sources of neutrons can be used in fusion material research[2, 11, 17] and proton beams in medical disciplines as hadron therapy for the treatment of cancer[18].

In contrast to ultra-short high-intensity lasers which allow the generation of beams of protons and deuterons possessing kinetic energies $\gg 1$ MeV, sub-nanosecond lasers of the kJ-class capable of delivering a moderate intensity onto a target make it possible to accelerate ions up to MeV energies per nucleon[19–22]. The clear-cut evidence that the fastest protons accelerated by the laser system PALS (1.315 μm, 300 ps, 3×10^{16} W cm^{-2}) have energies up to ~ 4 MeV[20, 21] creates a way to accelerate a high number of deuterons from the front side of a target and exploit them in the production of high-energy (~ 15 MeV) neutrons through the ^{7}Li(d, xn) nuclear reaction even if the mean kinetic energy of the bunch of deuterons is < 1 MeV. In addition, under these conditions up to 2×10^8 neutrons per laser shot have be generated with the laser system PALS through the D(d, n)^{3}He reaction, which is scalable with energy of other laser systems[22]. This scheme is applicable to newly developed high-energy-class cryogenically cooled Yb^{3+}:YAG multi-slab laser systems, allowing the production of a plasma with intensity $I\lambda^2 > 1 \times 10^{16}$ μm^2 cm^{-2}[23].

A common phenomenon observed in experiments is the acceleration of protons coming from a contaminant layer on the irradiated surface of the target. The protons accelerated by the laser system PALS have a broad energy spectrum around a mean value of ~ 2.5 MeV. Although the total cross section for neutron production through the ^{7}Li(p, n) reaction is only about three times lower than that for the

Correspondence to: Email: krasa@fzu.cz

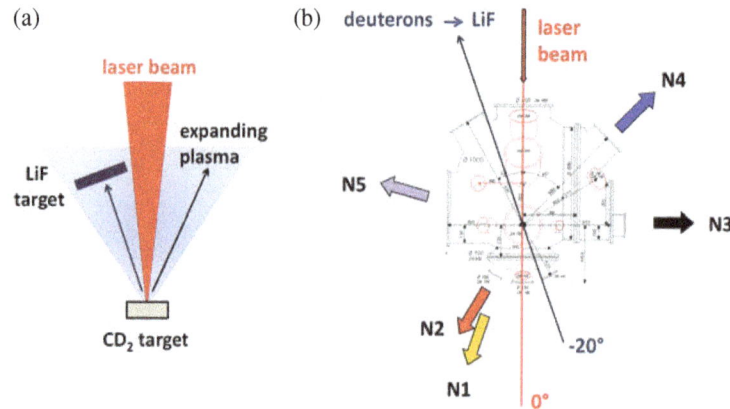

Figure 1. (a) Diagram of the dual target configuration. (b) Configuration of scintillation detectors N1 to N5 around the target chamber.

^{7}Li(d, n) reaction[6], the neutron yield from the ^{7}Li(p, n) reaction should be insignificant due to the very low content of protons in the produced plasma. Moreover, the peculiarity of the ^{7}Li(p, n) reaction that for proton kinetic energies just above the production threshold the emitted neutrons are kinematically focused into a cone in the direction of the proton beam makes it difficult to distinguish them among other neutrons because of the very large divergence angle of the proton beam.

In the experiment reported, a single planar deuterated polyethylene target was irradiated with the PALS laser, resulting in the production of neutrons through the D(d, n)^{3}He, ^{12}C(d, n)^{13}N nuclear reactions, as well as in acceleration of deuterons that were capable of initiating the ^{7}Li(d, n)^{8}Be reaction in a secondary LiF catcher target.

2. Experimental arrangement

A deuterated polyethylene target of 0.2 mm in thickness was exposed to a laser intensity of aapproximately 3×10^{16} W cm^{-2} with the fundamental wavelength of 1.315 μm delivered by the laser system PALS. The laser beam struck the CD$_2$ target parallel to the target surface normal. The accelerated deuterons impacted a 1-mm-thick natural LiF slab with a surface area of 17 cm^2, which was positioned 10 cm off the primary (CD$_2$)$_n$ target, as Figure 1 shows. The characteristics of the ions were measured using ion collectors (ICs) and a Thomson parabola spectrometer (TPS) positioned in the far expansion zone, i.e. outside the recombination zone. The emission of neutrons was observed by means of a calibrated temperature-compensated bubble dosimeters (BD-PND) with a sensitivity of ~4 b μSv^{-1} (Bubble Technology Industries, Chalk River, Ontario, Canada KOJ 1J0) and five calibrated neutron time-of-flight (N-TOF) scintillation detectors composed of a fast plastic scintillator of BC408 type and of a photomultiplier tube[24], which were positioned at distances of 1–3 m in various directions, as sketched in Figure 1. The BD-PND detectors were

positioned inside and outside the target chamber to observe anisotropy in the neutron emission. To reduce the scintillator response to the gamma rays, the scintillation detectors were mounted inside a protective housing composed of 10-cm-thick interlocking lead bricks. The scintillation detectors were operated in a current mode.

3. Results and discussion

When a CD$_2$ thick foil target is irradiated with the PALS laser, neutrons are generated via D(d, n)^{3}He and ^{12}C(d, n)^{13}N nuclear reactions[22]. The neutron yield per energy reaches a value up to ~3.5×10^5 neutrons/J for an average laser energy of 550 J. This value equals the highest efficiency values observed at laser–solid interactions driven by high laser intensities ranging from 1×10^{18} to 5×10^{19} W cm^{-2}. Besides the neutrons, the laser-produced CD$_2$ plasma emits fast ions having energies of MeV, as Figure 2(a) shows. The time of flight value of 70 ns at a distance of 1.5 m corresponding to the peak induced by a group of fast protons indicates an energy of 2.4 MeV, while the maximum energy reaches a value of approximately 4 MeV. These protons are followed by deuterons, which impact the LiF catcher target traversing the distance between both the targets. It causes a delay in the emission of the neutrons from the D–Li reaction. The time needed to travel that distance can be determined from the TOF spectrum of ions obtained by transforming the IC signal to the TOF spectrum of the ion charge density:

$$\rho(L, t) = q n_{\text{IC}}(L, t) \propto j_{\text{IC}}(L, t)/v, \qquad (1)$$

where q is the ion charge, $n_{\text{IC}}(L, t)$ is the ion density, $j_{\text{IC}}(L, t)$ is the ion current density observed at a distance L from the primary target and v is the ion velocity. We note that the ion current is commonly a sum of partial currents j_i of all the ionized species $j_{\text{IC}}(L, t) = \Sigma j_i(L, t)$, including deuterons and carbon ions which are fully ionized as well

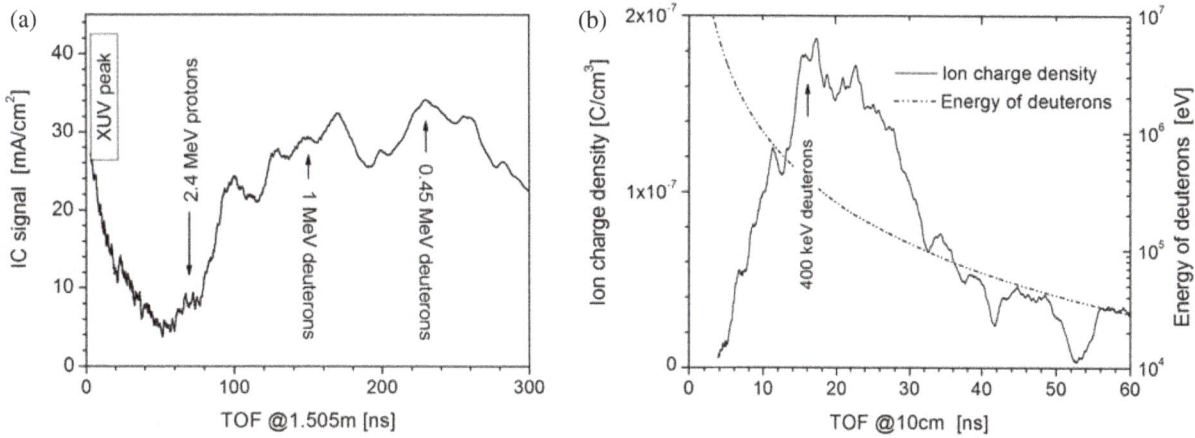

(a)

(b)

Figure 2. (a) Typical time-resolved ion current density observed using an IC positioned at a distance of 1.5 m from a massive CD_2 target in a backward direction at $30°$ with respect to the laser vector. The first peak was induced by XUV radiation. The peak at 70 ns was induced by 2.4 MeV protons. (b) Charge density of ions impacting on a secondary target at a distance of 10 cm for the CD_2 primary target, which was derived from the IC signal using the relationship (2). The dashed line shows the energy of deuterons. The laser irradiance on target was $\sim 3 \times 10^{16}$ W cm^{-2}.

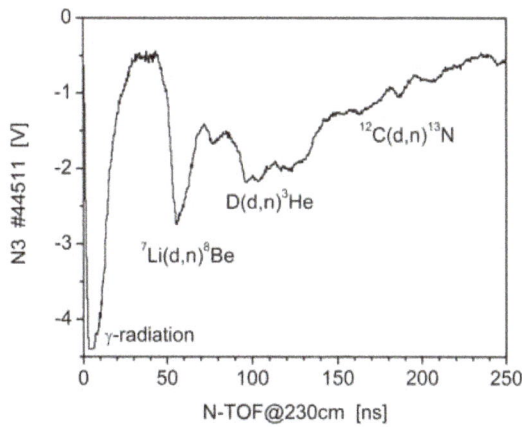

Figure 3. Scintillation detector signal induced by emission of γ radiation and neutrons produced via ^{7}Li(d, n)^{8}Be, D(d, n)^{3}He, and ^{12}C(d, n)^{13}N nuclear reactions. The emission was observed in the radial direction N3 (see Figure 1) at a distance of 230 cm from the target (shot #44511).

as partially recombined. One of the partial currents is that of the protons, which originate from impurities on the front target surface. The partial currents can be revealed using a deconvolution of the TOF spectrum, as has been shown for a polyethylene plasma[22, 25].

Using a similarity relationship for ρ detected by identical detectors placed at two different distances L_1 and L_2 in the same direction[26]

$$\rho(L_1, t_1)L_1^3 = \rho(L_2, t_2)L_2^3, \qquad (2)$$

where $L_1/t_1 = L_2/t_2$, we can determine a time-resolved ion charge density $\rho(L_2, t_2)$ at a chosen distance L_2 when $\rho(L_1, t_1)$ is observed at a distance L_1. The validity of the relationship (2) is limited for distances beyond some critical distance from the target, L_{cr}, where the three-body recombination becomes insignificant due to the plasma rarefaction,

the expanding ions have frozen charge states and their total charge Q_0 can be regarded as constant[27]. The value of L_{cr} depends on the laser irradiance on the target and the target material; in this experiment $L_{cr} < 50$ cm. It is evident that the transformation of $\rho(L, t)$ to a short distance $L_s < L_{cr}$ gives underestimated values of the ion charge density due to the three-body recombination in this area. Although the ICs can hardly be used at a short distance from the target due to harmful effects caused by both the electrical breakdown of IC initiated by the collected current with a density of ~ 100 A cm^{-2} and the strong electromagnetic pulse (EMP) interference[28], the use of a passive detector, e.g., solid state track detector stacks, may partially solve the problem. Nevertheless, the obtained time-resolved density of charges $\rho(L, t)$ impinging on a catcher target provides an acceptable approximation of the flight times of deuterons to the LiF target, as Figure 2(b) shows. Two dominating peaks corresponding to the deuteron energy of ~ 400 and ~ 200 keV are characteristic for the time-resolved charge density of ions.

A typical signal of the scintillation detector that is induced by γ radiation and neutrons coming from nuclear reactions ^{7}Li(d, n)^{8}Be, D(d, n)^{3}He, ^{12}C(d, n)^{13}N is shown in Figure 3.

In contrast to the zero time coordinate for neutrons from the D(d, n)^{3}He and ^{12}C(d, n)^{13}N reactions, which is related to the TOF of gamma radiation, the determination of the start time of neutrons produced via the ^{7}Li(d, n)^{8}Be reaction is encumbered by uncertainty caused by the broadband TOF spectrum of deuterons impinging on the LiF catcher target. A way to minimize this uncertainty consists of calculations of flight times TOF$_D$ of deuterons to the LiF catcher target and flight times N-TOF$_{D-Li}$ of D–Li neutrons to the scintillation detectors. The value of N-TOF$_{D-Li}$ can be determined using formulae describing the kinematics of neutron production in binary collisions between a projectile and an atom in

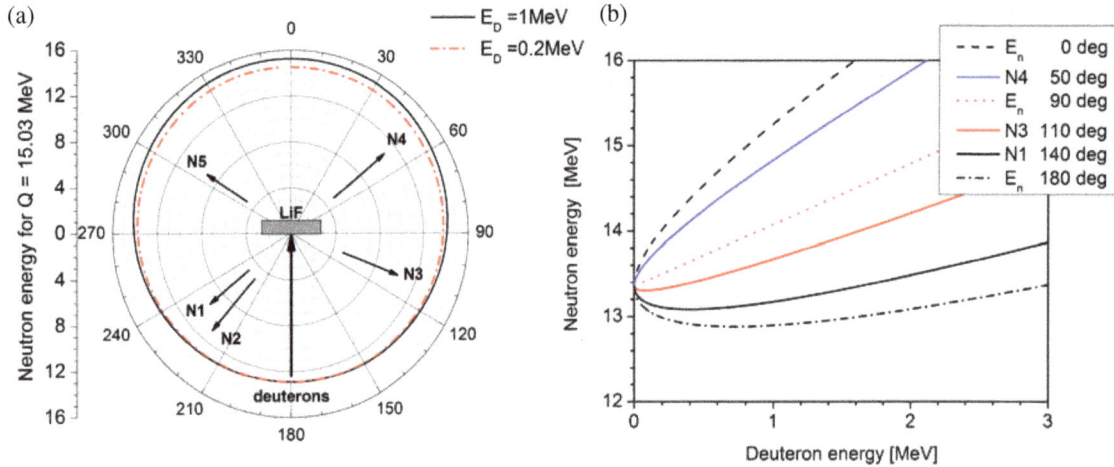

Figure 4. (a) Geometry of a nuclear reaction in the laboratory frame in which an incident deuteron with energy E_D impinges on a Li atom of the stationary LiF target and angular distribution of the neutron energy calculated for a value $Q = 15.03$ MeV of the ^{7}Li(d, n)^{8}Be reaction. (b) Deuteron energy dependence of the energy of d–Li neutrons detected in chosen directions as calculated using formula (1) in [10] describing the kinematics of neutron production in binary collisions between a projectile and an atom in a stationary target.

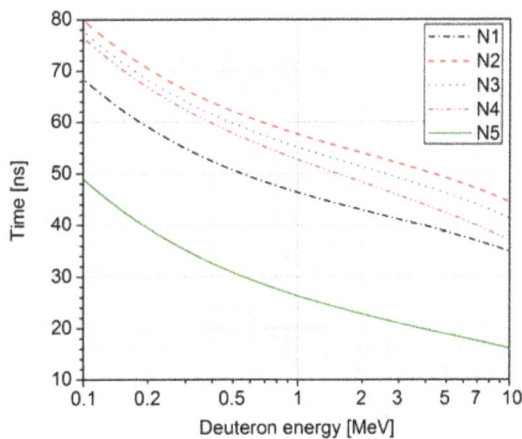

Figure 5. Deuteron energy dependence of the neutron arrival time at detectors N1–N5 (see Figure 4) related to the laser–target interaction. The distances from the catcher LiF target to the scintillation detectors were $L_{N1} = 1.81$ m, $L_{N2-4} = 2.28$–2.41 m, and $L_{N5} = 0.85$ m.

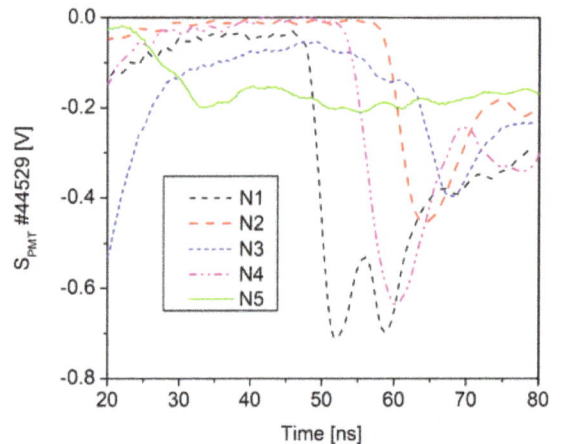

Figure 6. Scintillation detector signals observed in directions N1 to N5 ranging from 50° to 110° with respect to the mean direction of deuterons impinging on the LiF target and at distances from 0.8 to 2.4 m (see Figures 1 and 5).

a stationary target, such as those derived and discussed in[10, 29]. The dependence of neutron energy on the energy of deuterons and on the direction of observation is shown in Figure 4.

Figure 5 shows the arrival time of neutrons, t_N, related to the time of the beam–target interaction, which is the sum of the time of flight TOF$_D$ of deuterons to the catcher LiF target and the time of flight N-TOF$_{D-Li}$ of ^{7}Li(d, n)^{8}Be neutrons from the LiF target to the scintillation detectors:

$$t_N = \text{TOF}_D + \text{N-TOF}_{D-Li}. \tag{3}$$

Figure 6 shows the scintillation detector signals observed at five different distances and directions N1 to N5 with respect to the mean direction of deuterons impinging on the

LiF target (see Figures 1 and 4). Using the dependence of the arrival times of neutrons on the energy of deuterons, E_D, we can interpret individual maxima or partial peaks in the scintillation detector signals. The maxima in the signals shown in figure 6 may be sorted into two groups with respect to the energy of deuterons: $E_{D-1} \cong 400$ keV (N1 – 52 ns, N2 – 64 ns, N4 – 60 ns, N5 – 33 ns) and $E_{D-2} \cong 200$ keV (N1 – 59 ns, N2 – 64 ns, N3 – 68 ns, N5 – 41 ns). The beginnings of all the signals N1 – N5 correspond to a deuteron energy of $\sim$1 MeV. The 400 keV deuterons induce a peak in the time-resolved ion charge density at $\sim$16 ns for a distance of 10 cm, and the second group of deuterons with $E_{D-2} \sim 200$ keV create a peak at $\sim$23 ns, which is the second highest one. The occurrence of protons and deuterons with energy >1.9 and 2.4 MeV, respectively, was confirmed by the tracks observed on a CR-39 track detector protected with an Al foil of 40 μm in thickness. The number of these ions was estimated to be $\sim 10^{12}$ per shot. The contribution of MeV deuterons to the neutron generation was insignificant.

Sorting of the peaks in the scintillation detector signals indicates that the repetitive bursts observed in the emission of fast ions in sub-nanosecond laser–solid experiments[22] may give rise to bursts in emission of fast neutrons generated via the beam–target ^{7}Li(d, n)^{8}Be reaction. Clear evidence is given by the observed well-separated maxima, for example at 52 and 59 ns in the N1 signal, and also at 60 and 78 ns in the N4 signal, that correlate well with the peaks of 200 and 400 keV deuterons appearing in the time dependence of the ion charge density (see Figure 2(b)).

The energy of the observed neutrons depends not only on the kinetic energy of fusing deuterons, but also on the direction of observation. Most of the neutrons emitted in the forward direction with respect to the direction of impinging deuterons have energy >14 MeV. Only a minority of them can reach ~ 16 MeV energy. The maximum neutron yield from both the primary and secondary reactions was 3.5×10^{8} neutrons per shot. The partial contributions of both the D(d, n)^{3}He and ^{7}Li(d, n)^{8}Be reactions can be estimated from the scintillation detector signal, by taking into account the dependence of the scintillation intensity on the energy deposited by neutrons. If we consider the solid angle of the expanding plasma plume containing the fast deuterons[25] reduced by the solid angle of the laser beam, because the deuterons are emitted from the front target surface, the possible maximum yield could be as high as $Y_{\text{D–Li,max}} \sim 2 \times 10^{8}$ neutrons sr^{-1}. This value is more or less comparable to the value of $Y_{\text{D–Li}} = 8 \times 10^{8}$ neutrons sr^{-1} of the ^{7}Li(d, xn) nuclear reaction driven by an intensity of 2×10^{19} W cm^{-2}[7]. In that work the deuterons were produced using the Titan laser (energy 360 J and pulse length 9 ps) and accelerated from the rear surface of the foil through the TNSA mechanism. It follows from this comparison that different power-law relations govern both the experiments. Moreover, the similarity parameter $I\lambda^2$ does not also fully scale the generation of DD neutrons driven by high-power lasers with femtosecond to nanosecond pulse durations[22]. Thus, not only the laser intensity, but also the nonlinear phenomena occurring affect the laser energy absorption by the generated plasma and, thus, the ion acceleration.

Both the analysis of the N-TOF spectra and time-resolved ion charge density of the produced ions have shown that deuterons induced the ^{7}Li(d, n) nuclear reaction while possessing kinetic energy $E_{\text{d}} < 1$ MeV. Under these conditions the neutron source becomes isotropic[6]. A number of neutron detectors used contemporaneously at various directions and distances allow an evaluation of the anisotropy in the neutron emission. Nevertheless, our previous measurement of fusion neutron spectra outside the plasma focus device PF-1000 demonstrated the significant influence of the plasma vessel on the primary spectrum of generated neutrons[30]. Considering also shot-to-shot fluctuations in the ion emission[31], more complex laboratory equipment is needed to measure anisotropy in neutron emission through the beam–target reaction.

4. Conclusions

The deuterons generated on the front side of a thick CD$_2$ foil exposed to an intensity of $\sim 3 \times 10^{16}$ W cm^{-2} using the PALS laser system were exploited as drivers of the ^{7}Li(d, n)^{8}Be nuclear fusion reaction. The deuterons accelerated in the backward direction impinged on a LiF secondary target with an energy of 200–400 keV and produced D–Li neutrons with energies >14 MeV. Since only a small fraction of fast deuterons were hitting the LiF catcher target used, the maximum yield of D–D and D–Li neutrons was $\sim 3.5 \times 10^{8}$ neutrons/shot, which gives the normalized yield of $\sim 5.8 \times 10^{5}$ neutrons/J. This value should still increase when increasing the area of the LiF target catcher.

Acknowledgements

The authors gratefully acknowledge the support of the staff of the PALS laser facility without whose assistance this work would not have been possible. The research leading to these results has received funding from the Czech Science Foundation (Grant No. P205/12/0454), the Czech Republic's Ministry of Education, Youth and Sports (Project No. LM2010014), LASERLAB-EUROPE (grant agreement no 284464, EC's Seventh Framework Programme) and the European Social Fund and state budget of the Czech Republic (Project No. CZ.1.07/2.3.00/20.0279).

References

1. S. A. Gaillard, T. Kluge, K. A. Flippo, M. Bussmann, B. Gall, T. Lockard, M. Geissel, D. T. Offermann, M. Schollmeier, Y. Sentoku, and T. E. Cowan, Phys. Plasmas **18**, 056710 (2011).
2. M. Roth, D. Jung, K. Falk, N. Guler, O. Deppert, M. Devlin, A. Favalli, J. Fernandez, D. Gautier, M. Geissel, R. Haight, C. E. Hamilton, B. M. Hegelich, R. P. Johnson, F. Merrill, G. Schaumann, K. Schoenberg, M. Schollmeier, T. Shimada, T. Taddeucci, J. L. Tybo, F. Wagner, S. A. Wender, C. H. Wilde, and G. A. Wurden, Phys. Rev. Lett. **110**, 044802 (2013).
3. K. L. Lancaster, S. Karsch, H. Habara, F. N. Beg, E. L. Clark, R. Freeman, M. H. Key, J. A. King, R. Kodama, K. Krushelnick, K. W. D. Ledingham, P. McKenna, C. D. Murphy, P. A. Norreys, R. Stephens, C. Stöeckl, Y. Toyama, M. S. Wei, and M. Zepf, Phys. Plasmas **11**, 3404 (2004).
4. J. M. Yang, P. McKenna, K. W. D. Ledingham, T. McCanny, L. Robson, S. Shimizu, R. P. Singhal, M. S. Wei, K. Krushelnick, R. J. Clarke, D. Neely, and P. A. Norreys, J. Appl. Phys. **96**, 6912 (2004).

5. T. Žagar, J. Galy, J. Magill, and M. Kelltt, New J. Phys. **7**, 253 (2005).

6. J. Davis, G. M. Petrov, Tz. Petrova, L. Willingale, A. Maksimchuk, and K. Krushelnick, Control. Fusion **52**, 045015 (2010).

7. D. P. Higginson, J. M. McNaney, D. C. Swift, G. M. Petrov, J. Davis, J. A. Frenje, L. C. Jarrott, R. Kodama, K. L. Lancaster, A. J. Mackinnon, H. Nakamura, P. K. Patel, G. Tynan, and F. N. Beg, Phys. Plasmas **18**, 100703 (2011).

8. G. M. Petrov, D. P. Higginson, J. Davis, Tz. B. Petrova, J. M. McNaney, C. McGuffey, B. Qiao, and F. N. Beg, Phys. Plasmas **19**, 093106 (2012).

9. C. Zulick, F. Dollar, V. Chvykov, J. Davis, G. Kalinchenko, A. Maksimchuk, G. M. Petrov, A. Raymond, A. G. R. Thomas, L. Willingale, V. Yanovsky, and K. Krushelnick, Appl. Phys. Lett. **102**, 124101 (2013).

10. G. M. Petrov, D. P. Higginson, J. Davis, Tz. B. Petrova, C. McGuffey, B. Qiao, and F. N. Beg, Plasma Phys. Control. Fusion **55**, 105009 (2013).

11. J. Davis and G. M. Petrov, Phys. Plasmas **18**, 073109 (2011).

12. D. Jung, K. Falk, N. Guler, O. Deppert, M. Devlin, A. Favalli, J. C. Fernandez, D. C. Gautier, M. Geissel, R. Haight, C. E. Hamilton, B. M. Hegelich, R. P. Johnson, F. Merrill, G. Schaumann, K. Schoenberg, M. Schollmeier, T. Shimada, T. Taddeucci, J. L. Tybo, S. A. Wender, C. H. Wilde, G. A. Wurden, and M. Roth, Phys. Plasmas **20**, 056706 (2013).

13. M. Storm, S. Jiang, D. Wertepny, C. Orban, J. Morrison, C. Willis, E. McCary, P. Belancourt, J. Snyder, E. Chowdhury, W. Bang, E. Gaul, G. Dyer, T. Ditmire, R. R. Freeman, and K. Akli, Phys. Plasmas **20**, 053106 (2013).

14. S. Karsch, S. Düsterer, H. Schwoerer, F. Ewald, D. Habs, M. Hegelich, G. Pretzler, A. Pukhov, K. Witte, and R. Sauerbrey, Phys. Rev. Lett. **91**, 015001 (2013).

15. L. Willingale, G. M. Petrov, A. Maksimchuk, J. Davis, R. R. Freeman, A. S. Joglekar, T. Matsuoka, C. D. Murphy, V. M. Ovchinnikov, A. G. R. Thomas, L. Van Woerkom, and K. Krushelnick, Phys. Plasmas **18**, 083106 (2011).

16. S. Moustaizis, P. Lalousis, and H. Hora, Proc. SPIE **8780**, 878029 (2013).

17. L. J. Perkins, B. G. Logan, M. D. Rosen, M. D. Perry, T. Diaz de la Rubia, N. M. Ghoniem, T. Ditmire, P. T. Springer, and S. C. Wilks, Nucl. Fusion **40**, 1 (2000).

18. R. Banati, H. Hora, P. Lalousis, and S. Moustaizis, J. Intense Pulsed Lasers Appl. Adv. Phys. **4**, 11 (2014).

19. L. Láska, K. Jungwirth, J. Krása, E. Krouský, M. Pfeifer, K. Rohlena, J. Ullschmied, J. Badziak, P. Parys, J. WoLowski, S. Gammino, L. Torrisi, and F. P. Boody, Laser Part. Beams **24**, 175 (2006).

20. D. Margarone, J. Krása, L. Giuffrida, A. Picciotto, L. Torrisi, T. Nowak, P. Musumeci, A. Velyhan, J. Prokůpek, L. Láska, T. Mocek, J. Ullschmied, and B. Rus, J. Appl. Phys. **109**, 103302 (2011).

21. A. Picciotto, D. Margarone, P. Bellutti, S. Colpo, L. Torrisi, J. Krása, A. Velyhan, and J. Ullschmied, Appl. Phys. Express **4**, 126401 (2011).

22. J. Krása, D. Klír, A. Velyhan, D. Margarone, E. Krouský, K. Jungwirth, J. Skála, M. Pfeifer, J. Kravárik, P. Kubeš, K. Řezáč, and J. Ullschmied, Laser Part. Beams **31**, 395 (2013).

23. M. Divoky, P. Sikocinski, J. Pilar, A. Lucianetti, M. Sawicka, O. Slezak, and T. Mocek, Opt. Eng. **52**, 064201 (2013).

24. D. Klir, J. Kravarik, P. Kubes, K. Rezac, E. Litseva, K. Tomaszewski, L. Karpinski, M. Paduch, and M. Scholz, Rev. Sci. Instrum. **82**, 033505 (2011).

25. J. Krása, A. Velyhan, K. Jungwirth, E. Krouský, L. Láska, K. Rohlena, M. Pfeifer, and J. Ullschmied, Laser Part. Beams. **27**, 171 (2009).

26. J. Krása, P. Parys, L. Velardi, A. Velyhan, L. Ryć, D. Delle Side, and V. Nassisi, Laser Part. Beams. **32**, 15 (2014).

27. A. Lorusso, J. Krása, K. Rohlena, V. Nassisi, F. Belloni, and D. Doria, Appl. Phys. Lett. **86**, 081501 (2005).

28. M. De Marco, M. Pfeifer, E. Krousky, J. Krasa, J. Cikhardt, D. Klir, and V. Nassisi, J. Phys. Conf. Series **508**, 012007 (2014).

29. N. Izumi, Y. Sentoku, H. Habara, K. Takahashi, F. Ohtani, T. Sonomoto, R. Kodama, T. Norimatsu, H. Fujita, Y. Kitagawa, K. Mima, K. A. Tanaka, and T. Yamanaka, Phys. Rev. E **65**, 036413 (2002).

30. M. Králík, J. Krása, A. Velyhan, M. Scholz, I. M. Ivanova-Stanik, B. Bienkowska, R. Miklaszewski, H. Schmidt, K. Řezáč, D. Klír, J. Kravárik, and P. Kubeš, Rev. Sci. Instrum. **81**, 113503 (2010).

31. J. Krása, A. Velyhan, D. Margarone, E. Krouský, L. Láska, K. Jungwirth, K. Rohlena, J. Ullschmied, P. Parys, L. Ryć, and J. Wołowski, Rev. Sci. Instrum. **83**, 02B302 (2012).

25

Pulse fidelity in ultra-high-power (petawatt class) laser systems

Colin Danson[1,2], David Neely[3], and David Hillier[2]

[1] Centre for Inertial Fusion Studies (CIFS), Imperial College London, UK

[2] AWE plc, Aldermaston, UK

[3] Central Laser Facility, STFC Rutherford Appleton Laboratory, UK

Abstract

There are several petawatt-scale laser facilities around the world and the fidelity of the pulses to target is critical in achieving the highest focused intensities and the highest possible contrast. The United Kingdom has three such laser facilities which are currently open for access to the academic community: Orion at AWE, Aldermaston and Vulcan & Astra-Gemini at the Central Laser Facility (CLF), STFC (Science and Technology Facilities Council) Rutherford Appleton Laboratory (RAL). These facilities represent the two main classes of petawatt facilities: the mixed OPCPA/Nd:glass high-energy systems of Orion and Vulcan and the ultra-short-pulse Ti:Sapphire system of Astra-Gemini. Many of the techniques used to enhance and control the pulse generation and delivery to target have been pioneered on these facilities. In this paper, we present the system designs which make this possible and discuss the contrast enhancement schemes that have been implemented.

Keywords: petawatt laser; contrast; wavefront correction; plasma mirror

1. Systems description

The facilities highlighted within this paper are operated at AWE, Aldermaston, UK and the Central Laser Facility (CLF), STFC (Science and Technology Facilities Council) Rutherford Appleton Laboratory (RAL), UK and have the following system parameters:

- Orion is the latest facility to be built in the UK and became operational in April 2013[1]. It is a Nd:glass laser system which combines 10 long-pulse beamlines (500 J, 1 ns @ 351 nm) with two synchronized infrared petawatt beams (500 J in 500 fs). Up to 15% of the Orion beam time is available to the UK academic community.

- HELEN is the Orion predecessor and was operational at AWE for nearly 30 years[2]. It had two 527 nm, 500 J long-pulse (nanosecond) beamlines capability with a synchronized 100 TW short-pulse beam. It closed down in April 2009 to allow resources to be switched to the Orion facility during its commissioning.

- Vulcan is a high-power Nd:glass academic user facility[3] which has been operational for over 30 years. It enables a broad range of experiments through a flexible geometry[4, 5]. It has two target areas: one with 6×300 J (1053 nm @1 ns) long pulses combined with two synchronized short-pulse beams and a separate target area with high-energy petawatt capability (500 J in 500 fs) synchronized with a single long-pulse beamline.

- Astra-Gemini is a Ti:Sapphire laser system[6] operating at 800 nm pumped by green pulsed lasers in multi-stage amplifiers. It is operated as an academic user facility. It has two ultra-high-power beamlines delivering 15 J in 30 fs pulses @800 nm, generating focused intensities $>10^{21}$ W cm^{-2} to target.

2. Petawatt generation delivering focused intensities of 10^{21} W cm^{-2}

Short-pulse capabilities on all these systems are based on the technique of chirped pulse amplification (CPA)[7], where an ultra-short pulse is stretched, amplified and then recompressed to overcome nonlinear propagation issues primarily

Correspondence to: Colin Danson, AWE, Aldermaston, Reading, RG7 4PR, UK. Email: c.danson@imperial.ac.uk

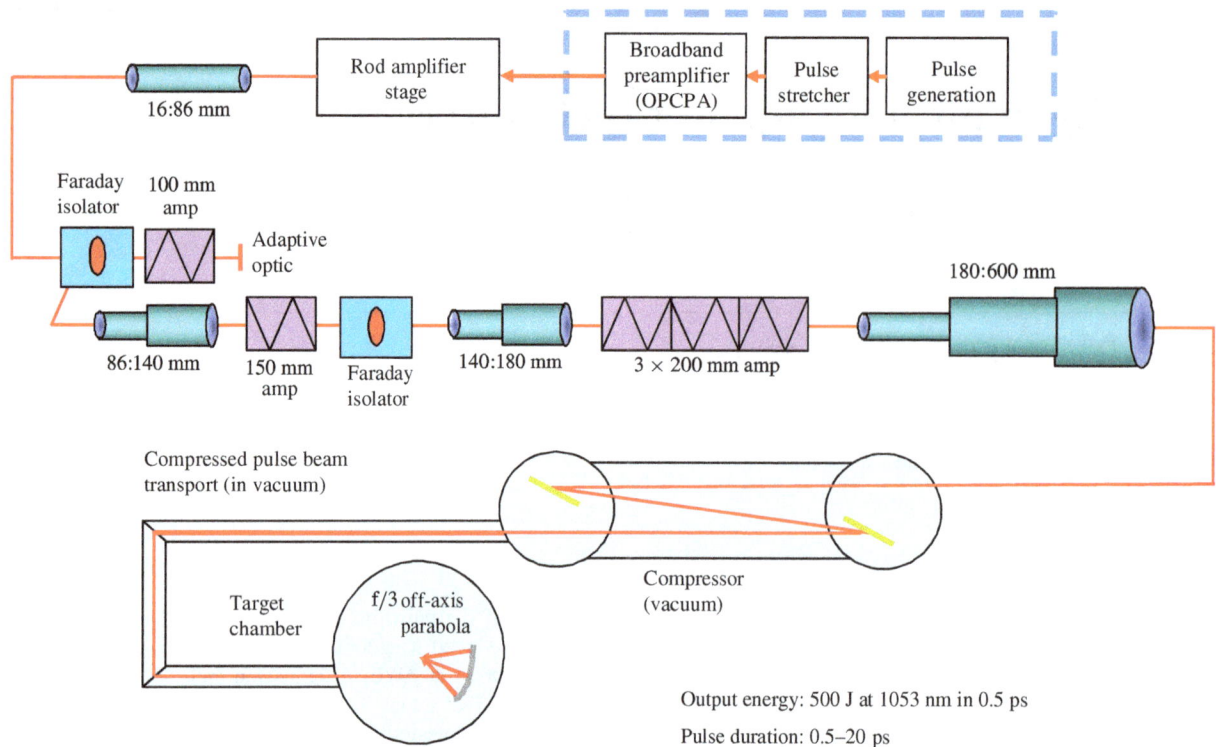

Figure 1. Layout of the Orion/Vulcan petawatt laser systems.

in the glass amplifiers. The first kilojoule class system to be configured to deliver a petawatt was the Nova Facility at LLNL[8]. All of the basic building blocks used on later systems were deployed on Nova, including: broadbandwidth pulse generation; optical pulse stretching; pulse amplification; deformable mirrors; pulse compression and reflective focusing.

2.1. Orion & Vulcan ultra-high-power delivery

The Orion and Vulcan short-pulse petawatt beamlines are based on the master oscillator–power amplifier (MOPA) architecture (see Figure 1). A pulse generated in a commercial Ti:Sapphire oscillator is initially stretched in a double-pass Offner stretcher. Its first stages of amplification are in a broadband three-stage optical parametric chirped pulse amplifier (OPCPA)[9, 10]. Further amplification in mixed glass (phosphate/silicate) takes the energy to the joule level prior to the final amplification in three stages of disc amplifiers (100 mm {double-passed}, 150 and 200 mm {×3} aperture). An adaptive optic mirror, a critical component in generating the ultra-high intensities to target, is positioned as the reflecting mirror between the double passes of the 100 mm disc amplifier. Feedback is provided from a Hartmann sensor positioned in the output diagnostic package.

Following the disc amplifier stages, a beam at the 700 J level is expanded to 600 mm in diameter and compressed in a single-pass geometry using a pair of gold-coated diffraction

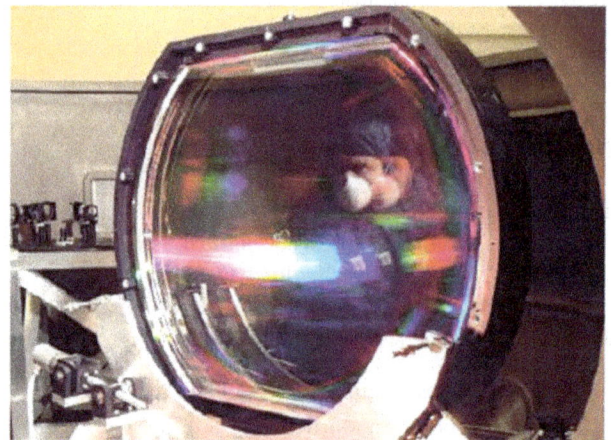

Figure 2. One of the 940 mm aperture compressor gratings installed on the Vulcan Petawatt beamline.

gratings (with 1480 lines mm^{-1}) separated by 13 m and housed in a large vacuum chamber. The resulting 500 J, 0.5 ps beams are then propagated to target in vacuum and focused using an F/3, 1.8 m focal length off-axis parabola (OAP).

There is now vast experience and understanding of the generation of petawatt pulses to target, as detailed in [1, 3, 11]. It can be summarized as:

- Deliver enough energy (>500 J) limited by grating damage (see Figure 2 for the large, 940 mm aperture installed on Vulcan Petawatt)

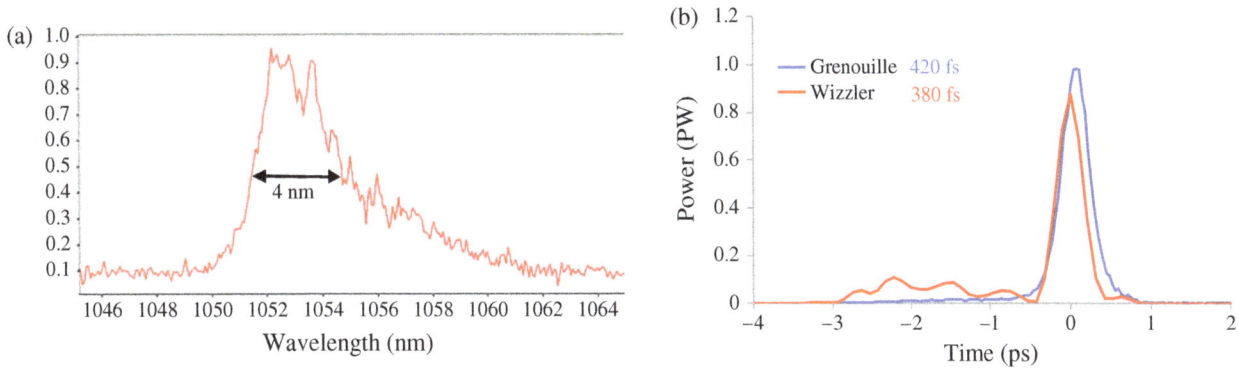

Figure 3. (a) The Orion petawatt output bandwidth. (b) The Orion output temporal profile demonstrating <500 fs pulsewidth.

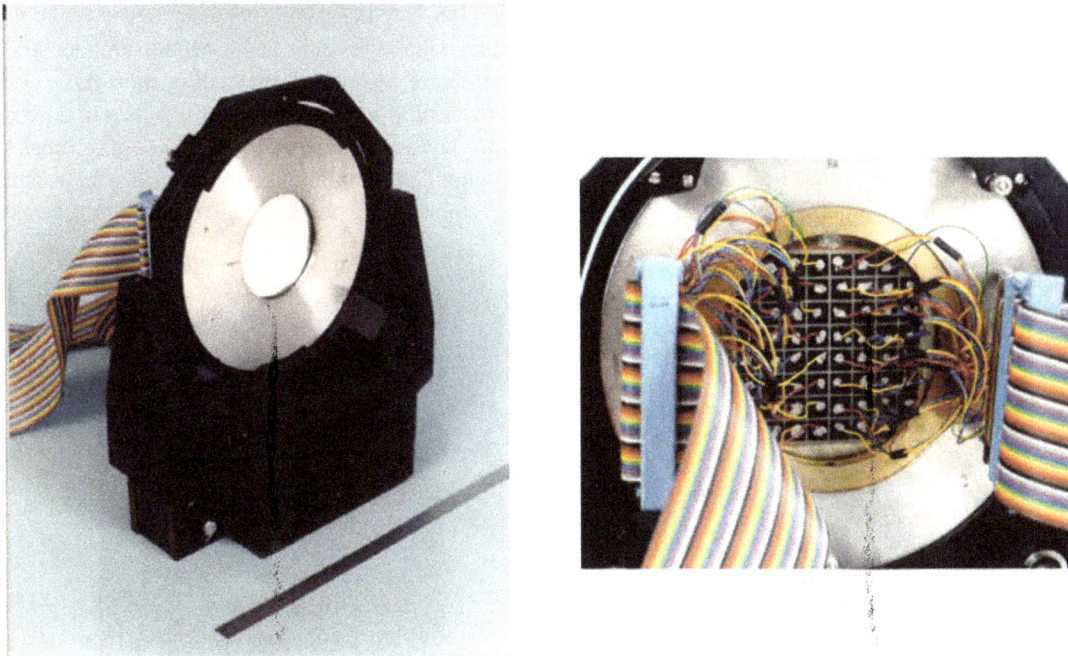

Figure 4. Prototype adaptive mirror as deployed on Vulcan Petawatt.

- Generate sufficient bandwidth (>4 nm; see Figure 3(a))

- Optimize the stretcher/compressor for short-pulse delivery (<500 fs see Figure 3(b)).

Having generated the petawatt beam, the use of adaptive optics holds the key to achieving the spot size and pulse duration required for irradiances on target $>10^{20}$ W cm^{-2}. In CLF tests, an adaptive optics system achieved a cycle rate of ~8 Hz with correction to 0.45 λ. At the time that Vulcan Petawatt was being developed there were no commercial solutions to large aperture wavefront correction; therefore, an in-house prototype '8 × 8 bimorph' mirror was developed at RAL (see Figure 4).

The final output beam to target was found to have residual astigmatism which could be compensated for by optimizing the focal spot using the offset and tilt of the OAP. The spots were first optimized using the Vulcan narrowband CW alignment beam and then optimized at the joule level using rod amplifier shots. Focal spot diameters of ~7 μm were generated to target (see Figure 5).

On full-energy shots, the focal spot on target was monitored using X-ray pinhole camera imaging, resulting in focal spot sizes of ~11 μm FWHM (the X-ray image is likely to be larger than the optical image due to plasma expansion), giving focused intensities to target on high-energy shots of ~10^{21} W cm^{-2}.

On Orion, a commercial wavefront corrector (Imagine Optics/CILAS) could be deployed. With a nominally flat mirror in the deformable mirror location an aberration of 2.5 λ peak-to-valley (mostly astigmatism) could be observed. With the 63-element monomorph deformable mirror in place this is

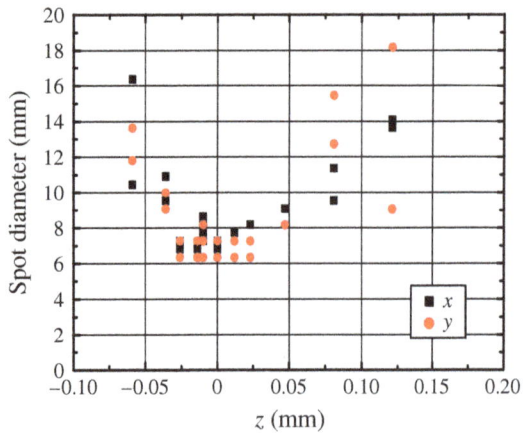

Figure 5. Vulcan Petawatt beam focal spot scan to target using joule-level pulses.

corrected to <0.5 λ peak-to-valley. As with Vulcan, the optimization to target is done using the final focusing OAP. The optical images, taken in the output beam diagnostics prior to the compressors and shown in Figures 6(a) and 6(b), show the quantitative improvement in focusability in moving from the use of the flat mirror to the deformable mirror.

To see the image on higher-energy shots to target, a combination of X-ray pinhole imaging and optical imaging of the plasma self-emission was used. Figure 6(c) shows the time-integrated X-ray emission from the focal spot, recorded by a pinhole camera on an image plate. The deconvolved spot size is 16 μm FWHM with a limiting resolution of 15 μm. The streaked optical transition radiation (OTR) emission gives a central image size of ∼8 μm FWHM (shown in Figure 6(d)) on a 100 TW shot, which gives an irradiance on target of 2×10^{20} W cm^{-2}.

2.2. Astra-Gemini ultra-high-power delivery

The Astra-Gemini system has a MOPA architecture. The seed pulses are generated from an ultra-short-pulse oscillator that provides low-energy, high-quality pulses of around 12 fs duration. These are stretched to ∼7 ps in a glass block before being amplified to millijoule energies in a kHz repetition rate preamplifier. Individual pulses are then selected by a

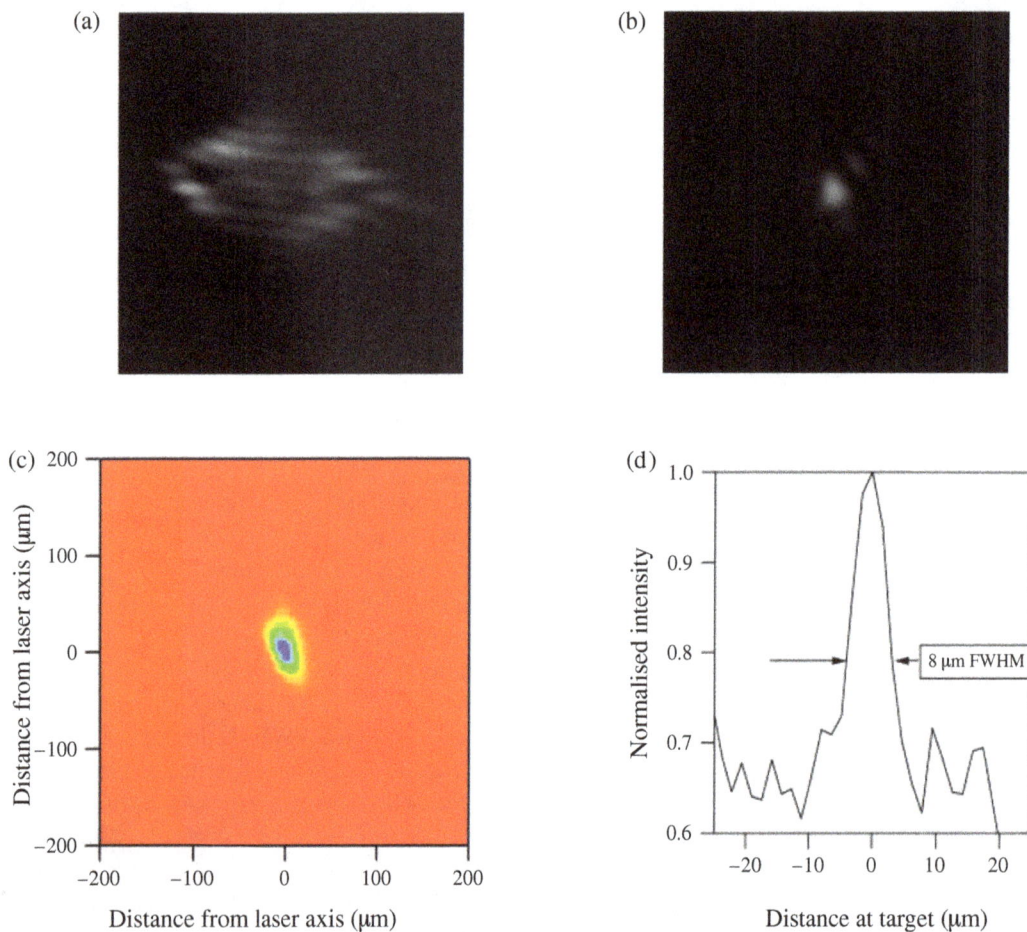

Figure 6. (a) Orion uncorrected petawatt output focal profile. (b) Orion corrected petawatt output focal profile. (c) Soft x-ray emission from an Orion target shot demonstrating an ∼15 μm X-ray spot size. (d) Optical self-emission from an Orion target shot indicating a focal spot profile of ∼8 μm.

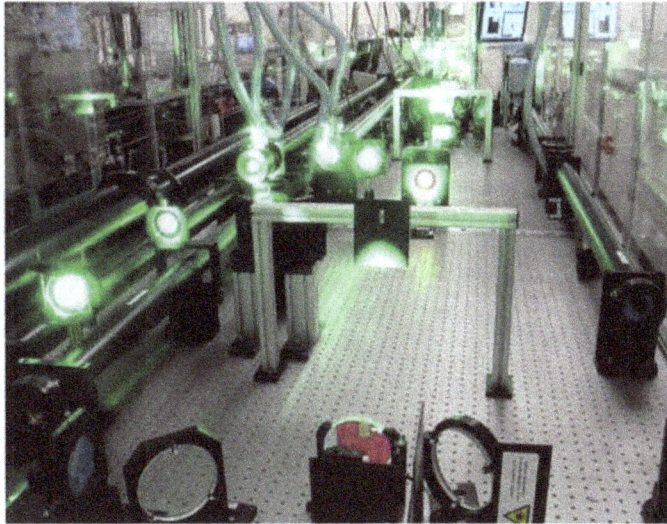

Figure 7. Astra-Gemini laser system.

fast Pockels cell at a repetition rate of 10 Hz. A much greater stretch is then applied, using a standard grating-based pulse stretcher. The pulses are stretched to ~1 ns and then amplified in a series of Ti:Sapphire crystals that are pumped by pulses of green light from another laser. The infrared beam is sent through each crystal several times to increase efficiency. As the energy in the beam increases, both the beam size and the crystals are made larger in successive amplifiers to keep the intensity of the light below the level where damage will occur.

The Astra-Gemini output amplifiers (see Figure 7) each have a Ti:Sapphire crystal 90 mm in diameter and 25 mm thick. The design of the amplifiers calls for a small-signal gain of around 4.2 per pass, to achieve the design output energy of 25 J. Modelling of the performance showed that this output could be achieved with a total of around 60 J of pump energy in a 50 mm diameter beam, while keeping the energy density on the crystal at a safe level. These final amplifiers have a shot rate of one shot every 20 s.

Laser output energies of 25 J are delivered with around 60 J of pump energy, in good agreement with the modelling. At this energy, assuming a transmission efficiency of 60% in the compressor (which would be lower than average) the energy reaching the target area is 15 J.

The main optics of the compressor are the two gratings and a plane mirror. The smaller grating is 320 mm by 205 mm, and the larger is 265 mm by 420 mm. Both are holographically generated with 1480 grooves per millimetre, and coated with gold. The plane mirror is coated with protected silver for increased reflectivity and damage resistance. The output bandwidth of Astra-Gemini is ~35 nm, which overfills the height of the second grating after dispersion by the first grating. The compressor is therefore operated in double pass, so the spatial dispersion of the beam in the first pass is reversed in the second.

The compressed pulse is steered to the final mirror, which sends the beam downwards into a vacuum pipe connected to the interaction chamber. Each beam will deliver 15 J to target in a pulse of 30 fs (i.e., a peak power of 0.5 PW). In the Gemini target chamber, shown in Figure 8, adaptive optics allows the wavefront to be controlled before the beams are focused with parabolic mirrors; with a focal spot as small as two microns, intensities up to 10^{22} W cm^{-2} can be achieved on target.

3. 2ω operation

The first experiments to demonstrate the feasibility of contrast enhancement using second harmonic generation (SHG) were conducted at small aperture and low energy (~1 J) at the Centre for Ultrafast Optical Science, University of Michigan by Gerard Mourou's group[12, 13]. The Chien research describes SHG experiments using 500 fs laser pulses at the joule level, demonstrating efficiencies of 70%–80% at intensities of 400 GW cm^{-2} from a Type I KDP crystal. The Queneuille work uses SHG on a terawatt laser system to demonstrate an intensity contrast of 10^9, making possible, for the first time, the study of laser–matter interactions in the ultra-high-intensity, high-contrast regime.

First large aperture (100 mm) trials were carried out on the Vulcan facility[14, 15]. These experiments used a picosecond pulse comparing the use of 2 mm thick and 4 mm thick Type I KDP crystals. Type I doubling is ideal as it produces the 1ω and 2ω outputs orthogonally polarized such that they can be separated. The experiments showed (see Figure 9) that a 4 mm thick crystal gives optimum conversion efficiency at an irradiance of 150 GW cm^{-2} and the optimum conversion efficiency for a 2 mm thick crystal is at an intensity greater than 250 GW cm^{-2} with efficiencies of ~60%. These experiments also demonstrated that at higher energies the

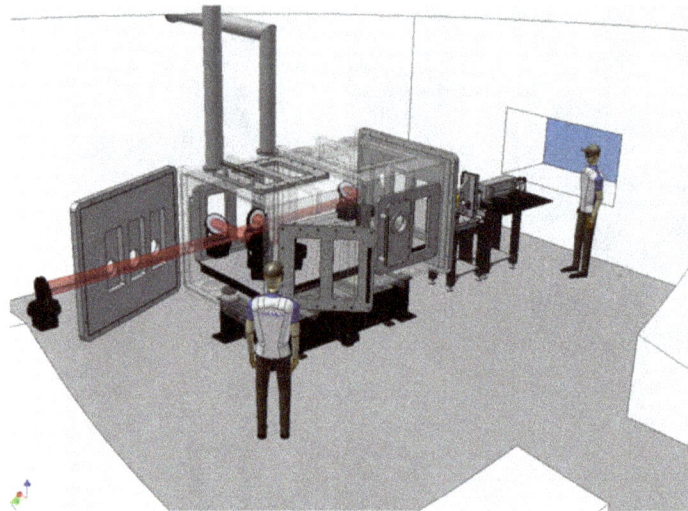

Figure 8. Astra-Gemini target chamber.

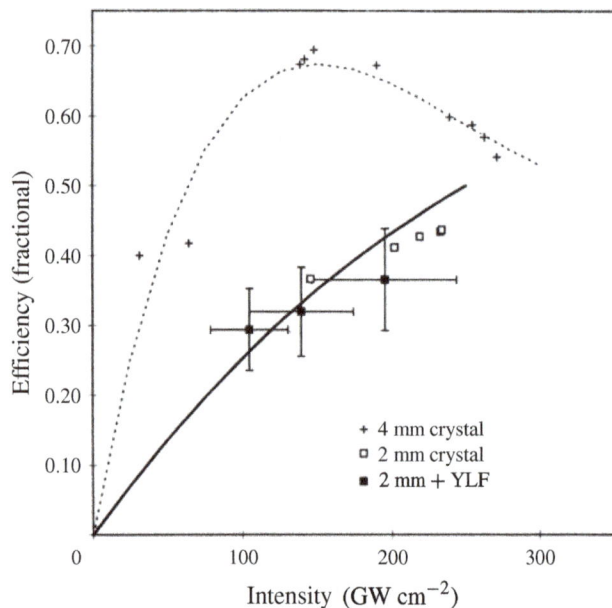

Figure 9. Comparison of the frequency conversion efficiencies of a 2 and a 4 mm Type I KDP frequency-doubling crystal.

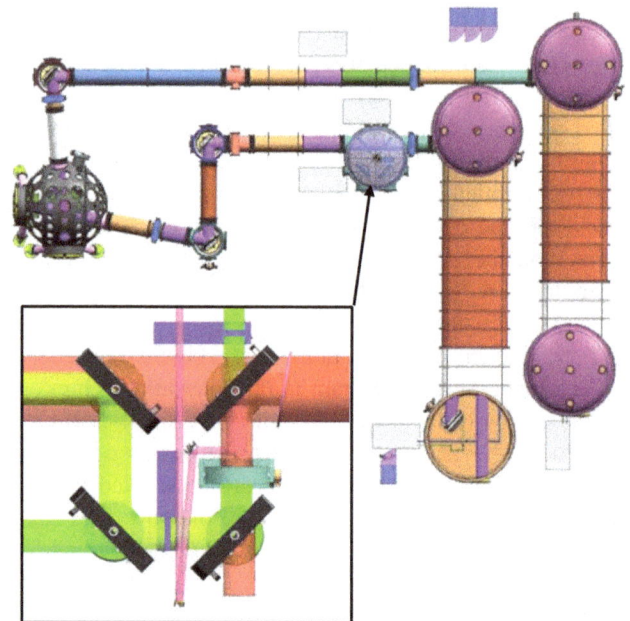

Figure 10. Schematic of the second harmonic option on Orion.

focal spot would break up, drastically reducing the delivered focused intensity.

The frequency-doubling experiments at large aperture on Vulcan showed that a 2 mm Type I KDP crystal was the optimal choice. It was anticipated, and subsequently demonstrated, that 10^{19} W cm^{-2} could be delivered to target in a frequency-doubled beam, with a much enhanced contrast ratio, opening up new plasma physics experiments.

The first frequency-doubling contrast enhancement experiments to be carried out at AWE were conducted on the HELEN facility in 2006 using 200 mm diameter, 2 mm thick Type I KDP. Although no quantitative contrast measurements were made, experimental data from buried-layer experiments suggests a greatly reduced pre-pulse[16]. Conversion efficiencies of approximately 70% were achieved for intensities of over 100 GW cm^{-2} at pulse durations of 0.5 and 2 ps, with delivered target intensities of $\sim 10^{19}$ W cm^{-2}.

Following the early HELEN experiments, Orion was designed from the outset to include the option of frequency doubling on one of the short-pulse petawatt beamlines. The maximum aperture crystal that could be obtained commercially was 300 mm with 3 mm thick Type I KDP. As the output of the Orion petawatt beamline is ~ 600 mm this meant that the beam diameter had to be reduced by a factor of two. A schematic of the second harmonic option as installed on Orion is shown in Figure 10.

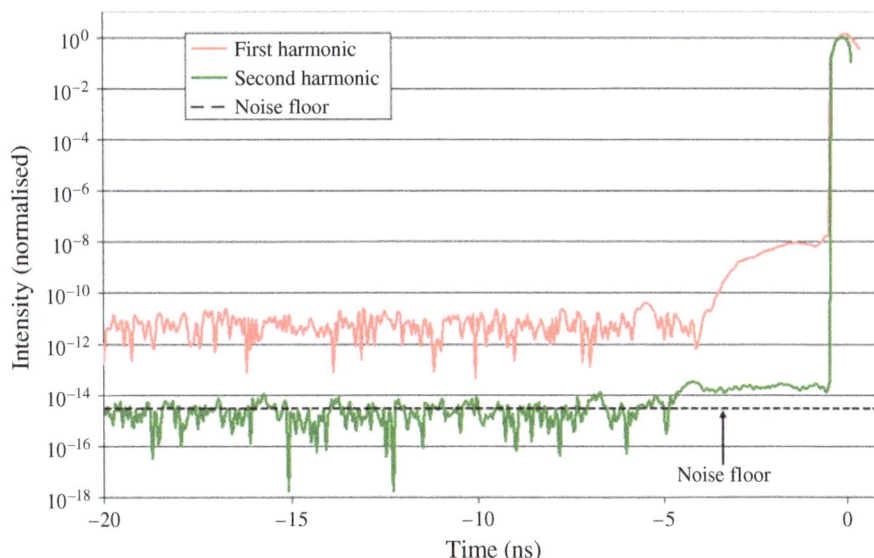

Figure 11. Contrast measurements on Orion of the fundamental and second harmonic.

Figure 10 shows the two Orion short-pulse compressor chambers on the left-hand side. Both of these are 20 m in length with a large aperture grating housed in each of the end chambers. The second harmonic conversion option is housed in one of these beamlines. The detail within this chamber is shown in the expanded square image. The beam is apertured down as it enters the 2ω chamber on the right-hand side. A mirror then directs it in to the 320 mm aperture crystal. There follow three 2ω reflective mirrors and a parabola which reject the residual 1053 nm light and propagate only the 527 nm, frequency-doubled light to target. This option delivers >100 J at the second harmonic (200 TW) with a 70% conversion efficiency. This can be focused down to a spot delivering $\sim 10^{20}$ W cm^{-2}[17].

The temporal contrast after compression was measured in the first harmonic from a leak through from the first mirror after the compressor. In the second harmonic it is measured from a leak through the second mirror after the frequency-doubling crystal. By concatenating photodiode (Newport 818-21A) traces from multiple shots with different attenuation levels, the temporal contrast over a large dynamic range can be measured, as shown in Figure 11. A pedestal shoulder is observed on the infrared pulse at $\sim 10^{-8}$, about 3 ns before the main pulse, which is due to parametric fluorescence in the OPCPA. When the frequency-doubling option is used, a large improvement is observed, generating a contrast to target of $\sim 10^{-14}$.

In both measurements, biased silicon photodiodes were used. To achieve sufficient dynamic range the photodiodes required extensive baffling, spectral filtering, and shielding to prevent any stray reflections or scattered light within the diagnostics enclosure swamping the signal. These measurements were all performed with no protection in front of the

photodiodes (water cells, etc.), which shortened their life and that of some of the optics in the diagnostics line significantly.

4. High-contrast front-end

The concept of using an OPCPA system as a seed for the front-end of a high-power Nd:glass laser system was first proposed in Ref. [10]. This allowed an ultra-short pulse to be amplified in a broad-bandwidth preamplifier before injection in the larger aperture Nd:Glass chain. The first OPCPA front-end system became operational on the Vulcan facility in 1998[18, 19]. In subsequent years, many facilities have implemented these front-end systems [20–22]. The OPCPA technique is an excellent broad-bandwidth amplifier (from nanojoule levels to tens of millijoules), producing fluorescence which generates a nanosecond pre-pulse shoulder on the pulse which gets amplified through to target. This fluorescence produces a contrast limit of $\sim 10^{-8}$ (this contrast level is still far better than that generated when using a traditional Nd:glass regenerative amplifier, where levels of $\sim 10^{-3}$ have been measured[23], or when using multiple Nd:glass phosphate amplifiers, where levels of $\sim 10^{-6}$ have been measured[24]). This fluorescence is present only when the nonlinear crystal is pumped; therefore, by going to picosecond pumping it is possible to enhance the contrast.

Itantini et al., at the University of Michigan first introduced the concept of injecting an ultra-clean picosecond pulse to achieve good amplified contrast[25] where an ultra-short-pulse preamplifier was used for the first three orders of magnitude gain in a Ti:Sapphire system. Other contrast enhancement techniques have been used based on double optical CPA[26], where the pulse is stretched, amplified and

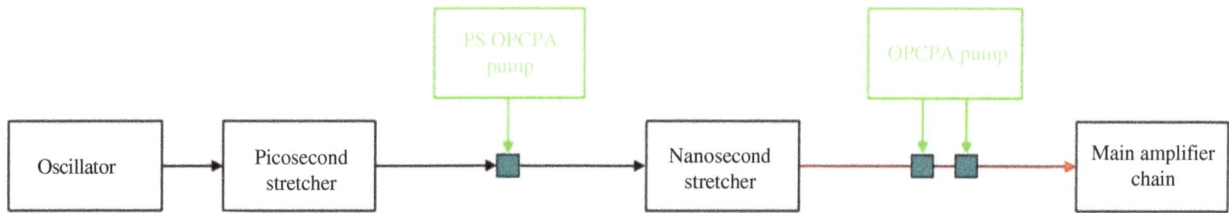

Figure 12. Scheme to introduce an additional picosecond stretcher.

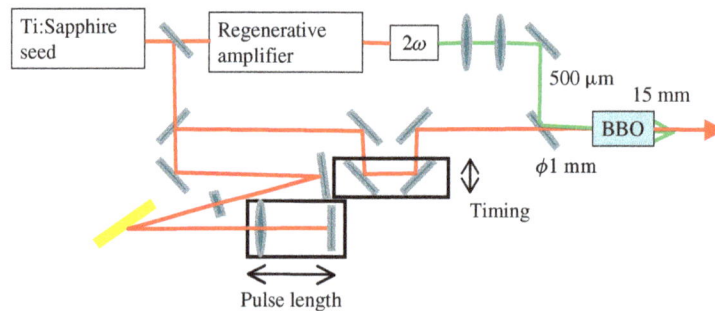

Figure 13. Schematic of the picosecond stretcher used on Vulcan for contrast enhancement.

compressed, then passed through a nonlinear filter before being stretched again and passed to the main beamlines. A range of nonlinear filters are used, depending on the pulse duration, energy and wavelength of the system, these include hollow waveguides, birefringence, cross polarization and low-gain OPA [27–30]. These nonlinear techniques suffer from a degree of complexity in their set-up and operation as they are very sensitive to pulse-to-pulse instabilities. They can also have very low throughput efficiencies, <10%.

An alternative approach first developed for Omega EP[31] was used on the Vulcan facility[32], where a single stage of amplification was introduced with the pulse a few picoseconds before the main stretcher, a schematic of which is shown in Figure 12. This technique reduces the amount of nanosecond gain by using a picosecond OPCPA stage.

The initial seed pulses of 200 fs duration are generated from a commercial Ti:Sapphire oscillator. These pulses are then stretched to 3 ps and amplified in a single-stage OPCPA to take the pulses from 10 pJ to 70 µJ. A schematic of the picosecond stretch and amplification scheme is shown in Figure 13.

In this scheme, a common seed is used for both the signal and pump pulses, thus ensuring they are optically synchronized. The pump seed pulse is fed into the regenerative amplifier and then frequency-doubled to generate pulses of 500 µJ to pump the BBO crystal. The regenerative amplifier uses Nd:YLF as the gain medium, producing gain narrowing as the pulse is amplified, increasing the pulse duration to ∼10 ps. A stretcher in the signal beam enables pulse length matching between the two pulses and a dog-leg timing adjuster ensures synchronization.

Following picosecond amplification, the pulse is stretched to the nanosecond regime before being amplified in two further OPCPA stages and then injected into the Nd:Glass amplifiers. This increased seed energy has led to an improvement of the nanosecond ASE contrast intensity to 10^{10} of the main pulse, without degrading the output performance of the Vulcan Petawatt system.

Following the successful demonstration of the high-contrast front-end system on Vulcan, a very similar configuration was installed on the Orion facility. This also demonstrated a nanosecond ASE contrast improvement by ∼2 orders of magnitude. Figure 14 shows the initial pulse (in red), with a 3 ns shoulder, and the pulse following the deployment of the high-contrast front-end (in blue).

5. Optics fidelity

Typically, in Ti:Sapphire-based petawatt class laser systems there is a 'coherent' contrast feature that frequently appears as an exponentially-rising pedestal within a few tens of picoseconds of the compressed pulse. Contrast measurements[33] were made with several optical configurations of the pulse stretcher, in which different components were eliminated or replaced, thus allowing us to distinguish the contribution to the coherent pedestal from different optics. These results show that scatter from mirrors and dispersion by a prism with polished surfaces did not make a measurable contribution to the coherent pedestal, whereas scatter from the diffraction gratings in the pulse stretcher gave its main contribution.

New higher-quality gratings were installed in the Astra-Gemini pulse stretcher. The new gratings are etched directly

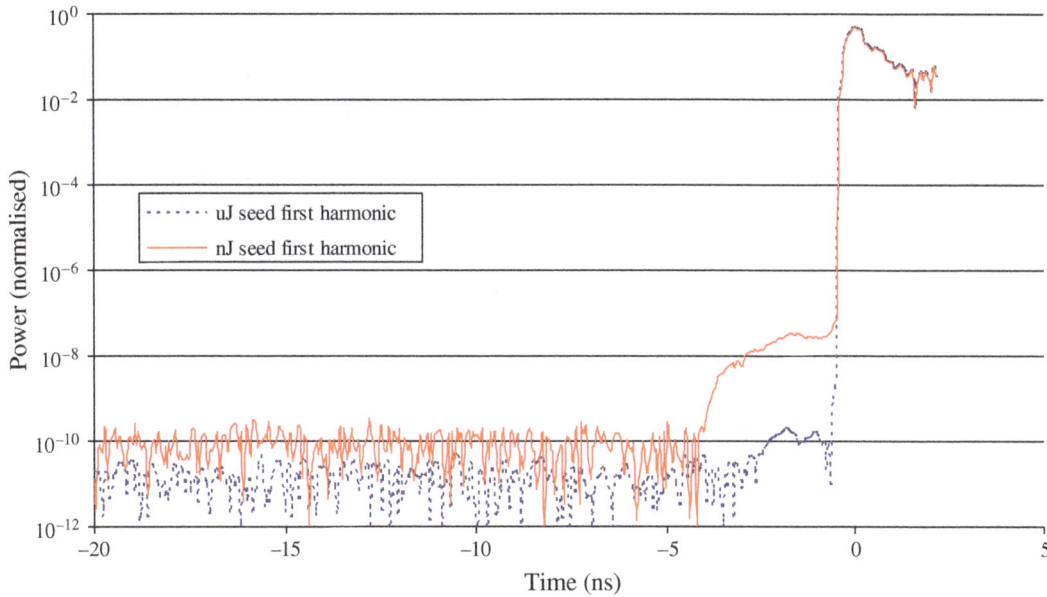

Figure 14. Measurement of contrast improvement on the Orion facility from introducing the high-contrast front-end system.

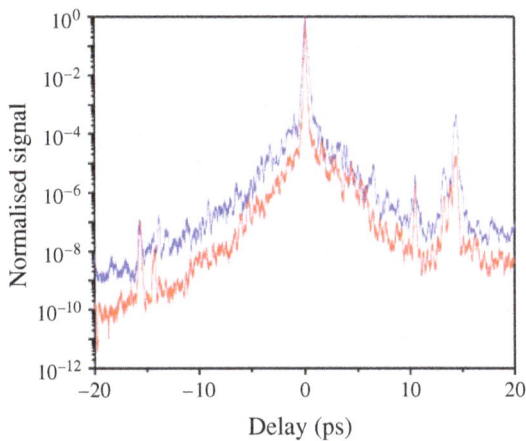

Figure 15. Contrast measurements on the Astra-Gemini system with the original gratings in blue and the replacement gratings in red.

into the fused silica substrates, rather than being formed in a photoresist layer. The contrast was measured both before and after the stretcher gratings were replaced and the results shown in Figure 15. The red line shows the initial contrast measurement and the blue line shows the contrast following grating replacement. This demonstrates an order-of-magnitude reduction in the intensity of the coherent pedestal, giving a significant improvement in the overall contrast of the compressed pulse from the laser.

The etching process may reduce the roughness of the grating surface, giving lower scatter and leading to the observed improvement in contrast. In the ten or more years since the original gratings were made, the technology of grating production has advanced in response to the requirements for high-energy CPA lasers such that gratings made today undoubtedly have better quality.

6. Plasma mirror operation

The use of plasma mirrors close to the target interaction is a technique that can be used to suppress unwanted pre-pulses[34–36] and has been demonstrated to deliver a contrast enhancement of over two orders of magnitude using a single mirror[35, 37]. The operation of a plasma mirror is shown in the visualization in Figure 16. At a certain intensity a laser pulse will form a plasma on the surface of a substrate. Any pre-pulse up to the point of plasma formation will be transmitted through the substrate. At times soon after the formation of the plasma there is no plasma expansion and the rest of the pulse will be reflected with relatively high efficiencies, thereby enhancing the pulse contrast.

In an experiment at the CLF[38] the specular reflectivity of plasma mirrors formed by sub-picosecond pulses from a Ti:Sapphire laser was measured for different angles of incidence and for two different pulse lengths as a function of the laser intensity. Laser pulses with energies up to 250 mJ and pulse durations of 90 and 500 fs were focused onto a flat fused silica substrate. A focusing optic with a large F-number (F/17) and focal length of 1016 mm was used to avoid averaging over a large range of different angles of incidence. The contrast of the pulse was measured at a level of $<10^{-8} \times E_{\mathrm{main}}$. The specular reflectivity was measured for intensities on target between 10^{12} and 4×10^{17} W cm^{-2}.

The target was moved after each shot. The intensity on target was varied either by moving the target out of focus (z-scan) or by decreasing the laser energy with a constant focal spot size. A typical result from the paper, as shown in Figure 17, demonstrates that reflectivities of over 80% can be achieved at low angles (6° in this case) and at intensities on the substrate of 10^{15}–10^{16} W cm^{-2}. For intensities above

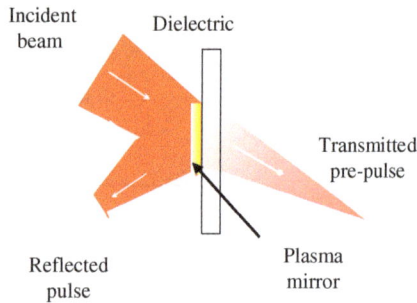

Figure 16. Cartoon of plasma mirror operation.

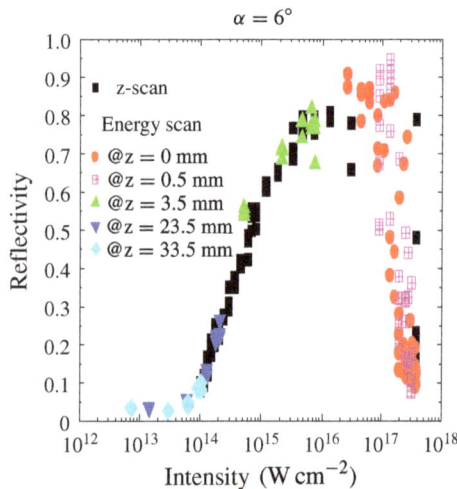

Figure 17. Typical data from the Ziener plasma mirror experiments.

mid 10^{16} W cm^{-2} the specular reflectivity decreases again, to values of only a few percent. The drop-off in specular reflectivity has been attributed to an increase in non-specular reflectivity (scatter, absorption or diffuse reflectivity at the plasma surface).

The relative merits of using 's' or 'p' polarization have been discussed and verified experimentally in many individual publications. As we often use plasma mirrors at large angles, it is observed that good reflectivities can be obtained when using the 's' polarization but efficiencies drop off when using the 'p' polarization, due to higher absorption of the laser light. A more complete characterization of plasma mirrors, both experimentally and theoretically, can be found in the paper by researchers at the CEA Saclay[37].

During the past few years, the routine use of plasma mirrors on Astra-Gemini has moved from being a research project to a properly engineered capability. A system has been installed in the Astra-Gemini target chamber consisting of a 10×2 cm plasma mirror capable of delivering 100 shots driven by a motorized stage before needing to be changed, as shown in Figure 18.

The system is a folded geometry for compactness and consists of two parabolic mirrors which allow an enhanced contrast collimated beam to be delivered to the final focusing

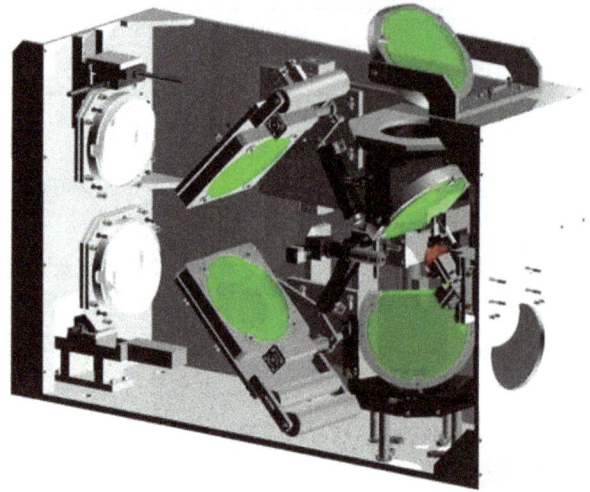

Figure 18. Double plasma mirror system on Astra-Gemini.

optic within the target chamber. A pair of by-pass optics can be inserted into the system for alignment and experimental set-up.

The contrast improvement from employing the double plasma mirror could not be absolutely measured owing to the use of an autocorrelator which had a dynamic range limit of $\sim 10^9$. The real contrast is believed to be 10^{-12} at 1 ns, giving a pre-pulse free relativistic interaction to within 2 ps of the peak of the main interaction pulse.

7. Discussion of contrast enhancement

The techniques discussed can be used together to enhance the contrast further. For instance, the 2ω contrast enhancement has been used on Orion with the high-contrast front-end. A contrast ratio of 10^{-17}–10^{-19} is anticipated, but at the time of writing this had not been measured. The plasma mirror technique has also been used on Vulcan in tandem with its high-contrast front-end operation.

In the early 1990s, as the first CPA systems were coming on line, contrast was not the highest priority and 10^{-6} was seen as the state of the art. When this was first aired, a useful visualization was that this ratio was the same as the ratio of the size of a grain of sand to the height of the Eiffel tower. Having recently measured contrast ratios of 10^{-14} it has been pointed out that this is now the same ratio as the same grain of sand but now to the distance to the Sun (with 10^{-19} being the ratio of a grain of sand to the distance to Alpha Centuari!!!).

8. Conclusion

In the twenty years or so of development of ultra-high-power laser facilities to the petawatt level there have been dramatic improvements in the pulse fidelity delivered to target, critical in achieving the highest focused intensities and the highest possible contrast. This paper has summarized how these developments have been implemented on the petawatt class laser facilities based in the UK which are currently open for access to the academic community: Orion at AWE, Aldermaston and Vulcan & Astra-Gemini at the CLF, STFC RAL.

Acknowledgements

Thanks to the Orion, Astra-Gemini and Vulcan teams who provided input for this paper and in particular to Steve Hawkes, Chris Hooker, Nick Hopps, Ian Musgrave, Rajeev Pattathil and Trevor Winstone for assistance in putting it all together.

References

1. N. Hopps, C. Danson, S. Duffield, D. Egan, S. Elsmere, M. Girling, E. Harvey, D. Hillier, M. Norman, S. Parker, P. Treadwell, D. Winter, and T. Bett, Appl. Opt. **52**, 3597 (2013).
2. M. J. Norman, J. E. Andrew, T. H. Bett, R. K. Clifford, J. E. England, N. W. Hopps, K. W. Parker, K. Porter, and M. Stevenson, Appl. Opt. **41**, 3497 (2002).
3. C. N. Danson, P. A. Brummitt, R. J. Clarke, J. L. Collier, B. Fell, A. J. Frackiewicz, S. Hancock, S. Hawkes, C. Hernandez-Gomez, P. Holligan, M. H. R. Hutchinson, A. Kidd, W. J. Lester, I. O. Musgrave, D. Neely, D. R. Neville, P. A. Norreys, D. A. Pepler, C. J. Reason, W. Shaikh, T. B. Winstone, R. W. W. Wyatt, and B. E. Wyborn, IAEA J. Nucl. Fusion **44**, S239 (2004).
4. C. Hernandez-Gomez, P. A. Brummitt, D. J. Canny, R. J. Clarke, J. Collier, C. N. Danson, A. M. Dunne, B. Fell, A. J. Frackiewicz, S. Hancock, S. Hawkes, R. Heathcote, P. Holligan, M. H. R. Hutchinson, A. Kidd, W. J. Lester, I. O. Musgrave, D. Neely, D. R. Neville, P. A. Norreys, D. A. Pepler, C. J. Reason, W. Shaikh, T. B. Winstone, and B. E. Wyborn, J. Phys. IV France **133**, 555 (2006).
5. I. Musgrave, A. Boyle, D. Carroll, R. Clarke, R. Heathcote, M. Galimberti, J. Green, D. Neely, M. Notley, B. Parry, W. Shaikh, T. Winstone, D. Pepler, A. Kidd, C. Hernandez-Gomez, and J. Collier, Proc. SPIE **8780**, 878003 (2013).
6. C. J. Hooker, S. Blake, O. Chekhlov, R. J. Clarke, J. L. Collier, E. J. Divall, K. Ertel, P. S. Foster, S. J. Hawkes, P. Holligan, B. Landowski, B. J. Lester, D. Neely, B. Parry, R. Pattathil, M. Streeter, and B. E. Wyborn, in *Proceedings of CLEO* JThB2 (2008).
7. D. Strickland and G. Mourou, Opt. Commun. **56**, 219 (1985).
8. M. D. Perry, D. Pennington, B. C. Stuart, G. Tietbohl, J. A. Britten, C. Brown, S. Herman, B. Golick, M. Kartz, J. Miller, H. T. Powell, M. Vergino, and V. Yanovsky, Opt. Lett. **24**, 160 (1999).
9. A. Dubietis, G. Jonusauskas, and A. Piskarskas, Opt. Commun. **88**, 437 (1992).
10. I. N. Ross, P. Matousek, M. Towrie, A. J. Langley, and J. L. Collier, Opt. Commun. **144**, 125 (1997).
11. C. N. Danson, P. A. Brummitt, R. J. Clarke, J. L. Collier, B. Fell, A. J. Frackiewicz, S. Hawkes, C. Hernandez-Gomez, P. Holigan, M. H. R. Hutchinson, A. Kidd, W. J. Lester, I. O. Musgrave, D. Neely, D. R. Neville, P. A. Norreys, D. A. Pepler, C. J. Reason, W. Shaikh, T. B. Winstone, R. W. W. Wyatt, and B. E. Wyborn, Laser Part. Beams **23**, 87 (2005).
12. C. Y. Chien, R. S. Craxton, G. Korn, J. S. Coe, J. Squier, and G. Mourou, Opt. Lett. **20**, 353 (1995).
13. J. Queneuille, F. Druon, A. Maksimchuk, G. Chériaux, G. Mourou, and K. Nemoto, Opt. Lett. **25**, 508 (2000).
14. D. Neely, C. N. Danson, R. Allott, F. Amiranoff, J. L. Collier, A. E. Dangor, C. B. Edwards, P. Flintoff, P. Hatton, M. Harman, M. H. R. Hutchinson, Z. Najmudin, D. A. Pepler, I. N. Ross, M. Salvati, and T. Winstone, Laser Part. Beams **17**, 281 (1999).
15. D. Neely, R. M. Allott, R. L. Clarke, J. L. Collier, C. N. Danson, C. B. Edwards, C. Hernandez-Gomez, M. H. R. Hutchinson, M. Notley, D. A. Pepler, M. Randerson, I. N. Ross, J. Springall, M. Stubbs, T. Winstone, and A. E. Dangor, Laser Part. Beams **18**, 405 (2000).
16. C. R. D. Brown, D. J. Hoarty, S. F. James, D. Swatton, S. J. Hughes, J. W. Morton, T. M. Guymer, M. P. Hill, D. A. Chapman, J. E. Andrew, A. J. Comley, R. Shepherd, J. Dunn, H. Chen, M. Schneider, G. Brown, P. Beiersdorfer, and J. Emig, Phys. Rev. Lett. **106**, 185003 (2011).
17. D. Hillier, C. Danson, S. Duffield, D. Egan, S. Elsmere, M. Girling, E. Harvey, N. Hopps, M. Norman, S. Parker, P. Treadwell, D. Winter, and T. Bett, Appl. Opt. **52**, 4258 (2013).
18. J. Collier, C. Hernandez-Gomez, I. N. Ross, P. Matousek, C. N. Danson, and J. Walczak, Appl. Opt. **38**, 7486 (1999).
19. I. N. Ross, J. L. Collier, P. I. Matousek, C. N. Danson, D. Neely, R. M. Allott, D. A. Pepler, C. Hernandez-Gomez, and K. Osvay, Appl. Opt. **39**, 2422 (2000).
20. V. Bagnoud, I. A. Begishev, M. J. Guardalben, J. Puth, and J. D. Zuegel, Opt. Lett. **30**, 1843 (2005).
21. H. Kiriyama, M. Mori, Y. Nakai, T. Shimomura, M. Tanoue, A. Akutsu, S. Kondo, S. Kanazawa, H. Okada, T. Motomura, H. Daido, T. Kimura, and T. Tajima, Opt. Lett. **33**, 645 (2008).
22. J. Schwarz, P. Rambo, M. Geissel, M. Kimmel, E. Brambrink, B. Atherton, and J. Glassman, Opt. Commun. **281**, 4984 (2008).
23. Y. Kitagawa, H. Fujita, R. Kodama, H. Yoshida, S. Matsuo, T. Jitsuno, T. Kawasaki, H. Kitamura, T. Kanabe, S. Sakabe, K. Shigemori, N. Miyanaga, and Y. Izawa, IEEE J. Quantum Electron. **40**, 281 (2004).
24. C. N. Danson, L. J. Barzanti, Z. Chang, A. E. Damerell, C. B. Edwards, S. Hancock, M. H. R. Hutchinson, M. H. Key, S. Luan, R. R. Mahadeo, L. P. Mercer, P. Norreys, D. A. Pepler, D. A. Rodkiss, I. N. Ross, M. A. Smith, R. A. Smith, P. Taday, W. T. Toner, K. W. M. Wigmore, T. B. Winstone, R. W. W. Wyatt, and F. Zhou, Opt. Commun. **103**, 392 (1993).
25. J. Itatani, J. Faure, M. Nantel, G. Mourou, and S. Watanabe, Opt. Commun. **148**, 70 (1998).
26. M. P. Kalashnikov, E. Risse, H. Schönnagel, and W. Sandner, Opt. Lett. **30**, 923 (2005).
27. D. Homoelle, A. L. Gaeta, V. Yanovsky, and G. Mourou, Opt. Lett. **27**, 1646 (2002).
28. A. Jullien, F. Augé-Rochereau, G. Chériaux, J. Chambaret, P. d'Oliveira, T. Auguste, and F. Falcoz, Opt. Lett. **29**, 2184 (2004).
29. A. Jullien, O. Albert, F. Burgy, G. Hamoniaux, J. Rousseau, J. Chambaret, F. Augé-Rochereau, G. Chériaux, J. Etchepare, N. Minkovski, and S. M. Saltiel, Opt. Lett. **30**, 920 (2005).
30. R. C. Shah, R. P. Johnson, T. Shimada, K. A. Flippo, J. C. Fernandez, and B. M. Hegelich, Opt. Lett. **34**, 2273 (2009).
31. C. Dorrer, I. A. Begishev, A. V. Okishev, and J. D. Zuegel, Opt. Lett. **32**, 2143 (2007).

32. I. Musgrave, W. Shaikh, M. Galimberti, A. Boyle, C. Hernandez-Gomez, K. Lancaster, and R. Heathcote, Appl. Opt. **49**, 6558 (2010).

33. C. Hooker, Y. Tang, O. Chekhlov, J. Collier, E. Divall, K. Ertel, S. Hawkes, B. Parry, and P. P. Rajeev, Opt. Express **19**, 2193 (2011).

34. H. C. Kapteyn, M. M. Murnane, A. Szoke, and R. W. Falcone, Opt. Lett. **16**, 490 (1993).

35. D. M. Gold, Opt. Lett. **19**, 2006 (1994).

36. A. Tien, M. Nantel, G. Mourou, D. Kaplan, and M. Bouvier, Opt. Lett. **22**, 1559 (1997).

37. G. Doumy, F. Quéré, O. Gobert, M. Perdrix, Ph. Martin, P. Audebert, J. C. Gauthier, J.-P. Geindre, and T. Wittmann, Phys. Rev. E **69**, 026402 (2004).

38. Ch. Ziener, P. S. Foster, E. J. Divall, C. J. Hooker, M. H. R. Hutchinson, A. J. Langley, and D. Neely, J. Appl. Phys. **93**, 768 (2003).

26

Demonstration of laser pulse amplification by stimulated Brillouin scattering

E. Guillaume[1,2], K. Humphrey[3], H. Nakamura[4], R. M. G. M. Trines[2], R. Heathcote[2], M. Galimberti[2], Y. Amano[5], D. Doria[6], G. Hicks[4], E. Higson[7], S. Kar[6], G. Sarri[6], M. Skramic[8], J. Swain[7], K. Tang[7], J. Weston[7], P. Zak[7], E. P. Alves[9], R. A. Fonseca[9], F. Fiúza[9], H. Habara[5], K. A. Tanaka[5], R. Bingham[2,3], M. Borghesi[6], Z. Najmudin[4], L. O. Silva[9], and P. A. Norreys[7,2]

[1] *Laboratoire d'Optique Appliquée, Ecole Polytechnique, Palaiseau, 91128, France*

[2] *STFC Rutherford Appleton Laboratory, Didcot, Oxon OX11 0QX, United Kingdom*

[3] *University of Strathclyde, Glasgow G1 1XQ, United Kingdom*

[4] *Blackett Laboratory, Imperial College London, Prince Consort Road, London SW7 2BZ, United Kingdom*

[5] *Graduate School of Engineering, Osaka University, Japan*

[6] *Queens University Belfast, Belfast BT7 1NN, United Kingdom*

[7] *University of Oxford, Parks Road, Oxford OX1 3PU, United Kingdom*

[8] *University of Cambridge, Cambridge CB2 1TQ, United Kingdom*

[9] *GoLP/Instituto de Plasmas e Fusão Nuclear - Laboratorio Associado, Instituto Superior Técnico, 1049-001 Lisbon, Portugal*

Abstract

The energy transfer by stimulated Brillouin backscatter from a long pump pulse (15 ps) to a short seed pulse (1 ps) has been investigated in a proof-of-principle demonstration experiment. The two pulses were both amplified in different beamlines of a Nd:glass laser system, had a central wavelength of 1054 nm and a spectral bandwidth of 2 nm, and crossed each other in an underdense plasma in a counter-propagating geometry, off-set by $10°$. It is shown that the energy transfer and the wavelength of the generated Brillouin peak depend on the plasma density, the intensity of the laser pulses, and the competition between two-plasmon decay and stimulated Raman scatter instabilities. The highest obtained energy transfer from pump to probe pulse is 2.5%, at a plasma density of $0.17 n_{cr}$, and this energy transfer increases significantly with plasma density. Therefore, our results suggest that much higher efficiencies can be obtained when higher densities (above $0.25 n_{cr}$) are used.

Keywords: laser–plasma interactions; optical pulse generation and compression; stimulated Brillouin and Raman scattering; ultra-fast optical processes

1. Introduction

Exploration of the intensity frontier is an exciting challenge for physicists. Advances in laser technologies, particularly those associated with increased power and decreased pulse duration, are of great interest due to their application to many fields in science and engineering, for example in laser-driven inertial fusion energy, in laser- and beam-driven particle accelerators, and in next-generation light sources. Currently, most high power lasers rely on the chirped pulse amplification (CPA) technique, in which a laser pulse is stretched before going to an amplifying medium, then expanded to large area (1 m or more) and recompressed, in order to avoid optical damage that occurs at intensities close to 10^{12} W cm^{-2}. Present-day high power lasers typically reach around 1 PW peak powers. Next-generation laser systems have been designed to reach powers of 10 PW or more, by the employment of the optical parametric CPA (OPCPA) technique. However, the achievement of intensities beyond this level is still uncertain, mainly due to the requirement for precise wavefront delivery at the final focusing optic of multiple large area laser beams.

Pulse compression methods using plasmas have been promoted as a way of overcoming these obstacles. The enormous energy densities associated with focused high power

Correspondence to: P. A. Norreys, Clarendon Laboratory, University of Oxford (& STFC Rutherford Appleton Laboratory), Parks Road, Oxford OX1 3PU, UK.

lasers excite nonlinear wave amplification in a medium that is already ionized. The plasma can support intensities of up to 10^{17} W cm^{-2}, i.e., 5 orders of magnitude larger than solid-state systems, before disruption to the medium occurs[1]. Laser pulse amplification in plasma rests upon an energy transfer between a relatively long duration pump pulse and a shorter seed pulse through the generation of either an electron plasma wave, known as stimulated Raman scattering (SRS), or an ion-acoustic wave, known as stimulated Brillouin scattering (SBS). Experimental[2] and numerical[3, 4] results have already been demonstrated in the case corresponding to SRS excitation.

As SBS produces a frequency shift in the scattered wave spectra, it is necessary for the seed laser to be downshifted by an amount equal to the ion-acoustic frequency in order for coupling between the laser beams to be realized. When utilizing long duration beams, which naturally have a very narrow bandwidth, an adjustment to the seed laser is essential for coupling between the laser beams to ensure that the necessary frequency component for scattering is present in the seed. This creates an additional technical complexity to the achievement of Brillouin scattering in plasma. However, when the seed beam is sufficiently short, and its bandwidth is sufficiently wide, the necessary downshifted frequency to trigger Brillouin scattering of the pump pulse will already be available in the seed pulse, and no additional frequency modification will be needed.

In this paper, we report on experimental observations of Brillouin scattering using two beams incident from the same laser system, one long (15 ps) pump beam and one short (1 ps) seed beam counter-propagating with respect to one another through a volume of plasma, with no modifications made to the frequency of either pulse. These findings are corroborated by 1D numerical simulations using the particle-in-cell code OSIRIS[5], confirming that for sufficiently short pulses the necessary Brillouin downshifted frequency is presented in the laser bandwidth, therefore negating the requirement for a frequency downshift in the seed pulse to be performed before Brillouin scattering can be obtained. A pump-to-probe energy transfer of up to 2.5% has been obtained, which confirms earlier results by Lancia et al.[6], who obtained 2.25% in a similar experiment.

In addition, our results extend the results by Lancia et al.[6] in two important ways. First, we found that the energy transfer efficiency increases consistently with density for $0.017 < n_e/n_{cr} < 0.17$. This may indicate that our plasma density showed fewer spatial fluctuations, which were suspected to reduce the efficiency in the earlier experiment. Second, we found in our numerical simulations that there is significant competition between SRS and SBS. This both reduces the efficiency and indicates that much better results may be obtained for $n_e > 0.25 n_{cr}$, where Raman scattering is no longer possible. These new results reveal how Brillouin amplification depends on the experimental parameters, and show how future experiments should be set up for enhanced performance.

2. Theory

SBS in plasmas can be characterized as the scattering of a high frequency transverse electromagnetic wave by a low frequency ion-acoustic wave into a second transverse electromagnetic wave. This corresponds to the decay of an incident photon in the laser beam, with frequency ω_0 and wavenumber k_0, into a phonon (ion-acoustic quantum) with frequency ω_{IAW} and wavenumber k_{IAW}, and a scattered photon, with frequency ω_s and wavenumber k_s, which travels in approximately the opposite direction to the incoming laser photon. Following directly from linear theory[7], the frequency and wavenumber matching conditions, often invoked when studying the Brillouin instability, are

$$\omega_0 = \omega_s + \omega_{IAW}, \tag{1}$$
$$k_0 = k_s + k_{IAW}, \tag{2}$$

where ω_{IAW} and k_{IAW} are the frequency and wavenumber of the ion-acoustic wave, respectively. The maximum growth occurs when (for pure backscattering)

$$k_{IAW} = 2k_0 - \frac{2\omega_0}{c}\frac{c_s}{c}, \tag{3}$$
$$\gamma = \frac{1}{2\sqrt{2}}\frac{ck_0 a_0 \omega_{pi}}{\sqrt{\omega_0 k_0 c_s}}. \tag{4}$$

Here, $\sqrt{\omega_0^2 - \omega_{pe}^2}$ is the wavenumber of the pump laser wave in plasma, ω_0 and ω_{pe} are the laser and electron plasma frequencies, respectively, $c_s = \sqrt{T_e/m_i}$ is the ion sound speed, T_e is the electron temperature, m_i is the ion mass and $\omega_{pi} = \omega_{pe}\sqrt{m_e/m_i}$ is the ion plasma frequency.

In the case of Raman amplification, the minimum frequency shift that can occur is equal to the plasma frequency, meaning that the maximum density where Raman amplification techniques can be utilized is one quarter of the critical density. However, in the case of Brillouin amplification, the minimum frequency shift is equal to zero, therefore allowing Brillouin scattering to operate at densities up to the critical density[8]. In addition to this, more energy can be transferred into the scattered wave via Brillouin scattering than via Raman scattering as less energy is coupled into the ion-acoustic wave in Brillouin scattering than the Langmuir wave associated with Raman scattering. This makes the Brillouin mechanism particularly useful for applications such as laser amplification techniques[6, 9] and induced energy transfer between adjacent laser beams at facilities such as the National Ignition Facility[10, 11].

3. Experimental setup

The experiment was conducted on the Vulcan Nd:glass laser facility at the Rutherford Appleton Laboratory[12]. This facility provided two linearly polarized laser pulses of $\lambda_0 = 1054$ nm central wavelength with a $\Delta\lambda_0 = 2$ nm bandwidth. The two laser beam diameters were reduced to 20 mm using pierced plastic plates, in order to have the correct spot size on the target. Each laser pulse was focused onto the target using $f/30$ off-axis parabolic mirrors, with $f = 612$ mm focal length, giving focal spots of 130 μm diameter. The pump beam contained between 570 and 860 mJ of energy, with a pulse duration $\tau_{\text{pump}} = 15$ ps, giving a pump intensity on the target of around 3×10^{14} W cm^{-2}. The seed beam contained between 38 and 477 mJ, with a pulse duration $\tau_{\text{seed}} = 1$ ps, giving a seed intensity on the target of between 2.5×10^{14} and 3.3×10^{15} W cm^{-2}. The laser pulses were injected into the target from opposite directions, with an angle of $10°$ between the two counter-propagating beams. This angle was used for safety reasons; while it led to a small reduction in pulse growth, this was deemed acceptable. A 1.65 mm long overlap distance was achieved in this geometrical setup. The temporal delay between the pump and the seed was adjusted so that the two ascending edges of the pulses crossed in the center of the gas target in order to maximize the duration of the interaction. This was achieved by using a streak camera looking at the overlap region. The laser pulses were focused in the center of a 5 mm long supersonic gas jet target, using either argon or deuterium. The gas target produced uniform plasmas when ionized, with background electron densities n_e varying between 1.7×10^{17} and 1.7×10^{20} cm^{-3}. The plasma density was controlled by adjusting the backing pressure of the supersonic gas jet. The plasma is created by the interaction pulses themselves – without any ionization pulse needed – triggering multiphoton ionization of the gas and collisions between electrons and atoms.

The light transmitted through the plasma in the direction of propagation of the seed beam was collected and collimated using a 600 mm focal length lens. The collimated beam was then steered out of the target chamber using flat silver mirrors, and focused onto the entrance slit of an optical spectrometer, equipped with a 150 lines/mm diffraction grating coupled with an Andor 16-bit CCD camera recording the spectra with a 0.1 nm resolution. A schematic diagram of the experiment can be seen in Figure 1. It should be noted that the transmitted seed was measured only for the case of the seed and the pump interacting with parallel horizontal polarizations.

4. Experimental results

The results of the experiment are shown in Figure 2. Figure 2 shows three optical spectra of the transmitted laser light taken at different plasma densities, with different laser configurations. Figure 2(a) represents the transmitted spectrum passing through the gas jet, with the pump beam turned off. Figures 2(b) and (c) show the transmitted spectra through the gas jet in the presence of the pump beam, for two

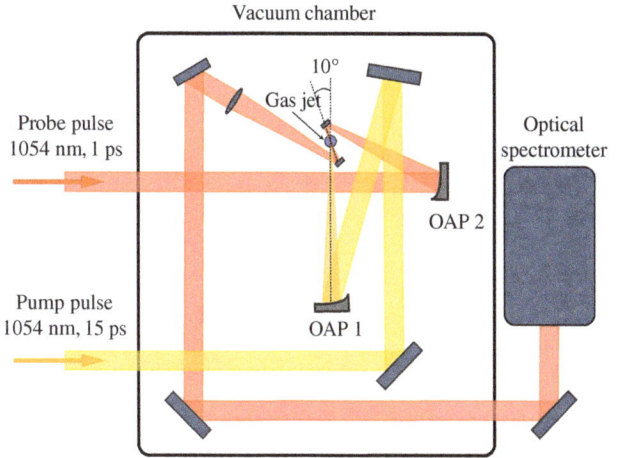

Figure 1. Schematic diagram of the experimental setup.

different plasma densities (1.7×10^{19} and 1.7×10^{20} cm^{-3}, respectively). The interaction with the gas jet and the pump beam clearly modifies the transmission spectrum of the seed. In the shot without a pump beam (Figure 2(a)), the propagation of the seed pulse through the plasma causes the formation of a small secondary peak, at $\omega/\omega_0 = 0.9975$. Since the separation between the peaks is much smaller than the plasma frequency for this configuration, the peak cannot correspond to SRS, and is presumed to correspond to spontaneous SBS by the seed pulse. The addition of an energetic pump beam, while keeping the plasma density the same, significantly enhances this downshifted spectral peak, proving that we have obtained pump-to-seed energy transfer via stimulated Brillouin backscattering. By increasing the plasma density by an order of magnitude, one can observe a significant increase of the relative intensity and energy content of the peaks: the secondary peak is now only 3.5 times smaller than the fundamental. It should be noted that for experimental shots with similar laser parameters, but with plasma densities between 1.7×10^{17} and 1.8×10^{18} cm^{-3}, no secondary peaks could be observed.

The energy transfer efficiency is calculated as follows. For the laser shot depicted in Figure 2(c), the pump energy was 675 mJ after passage through the 20 mm diameter aperture, while the seed pulse energy was 86 mJ. From the vacuum shot and the height of the Brillouin peak in Figure 2(c), it can be deduced that the Brillouin peak contains about 20% of the original seed energy, or 17 mJ. This corresponds to a 2.5% energy transfer efficiency from pump to seed.

5. Numerical simulations

The numerical simulations were conducted in 1D using the fully relativistic particle-in-cell (PIC) code OSIRIS[5] and were constructed to mirror the experimental parameters as closely as possible. For these simulations, the plasma

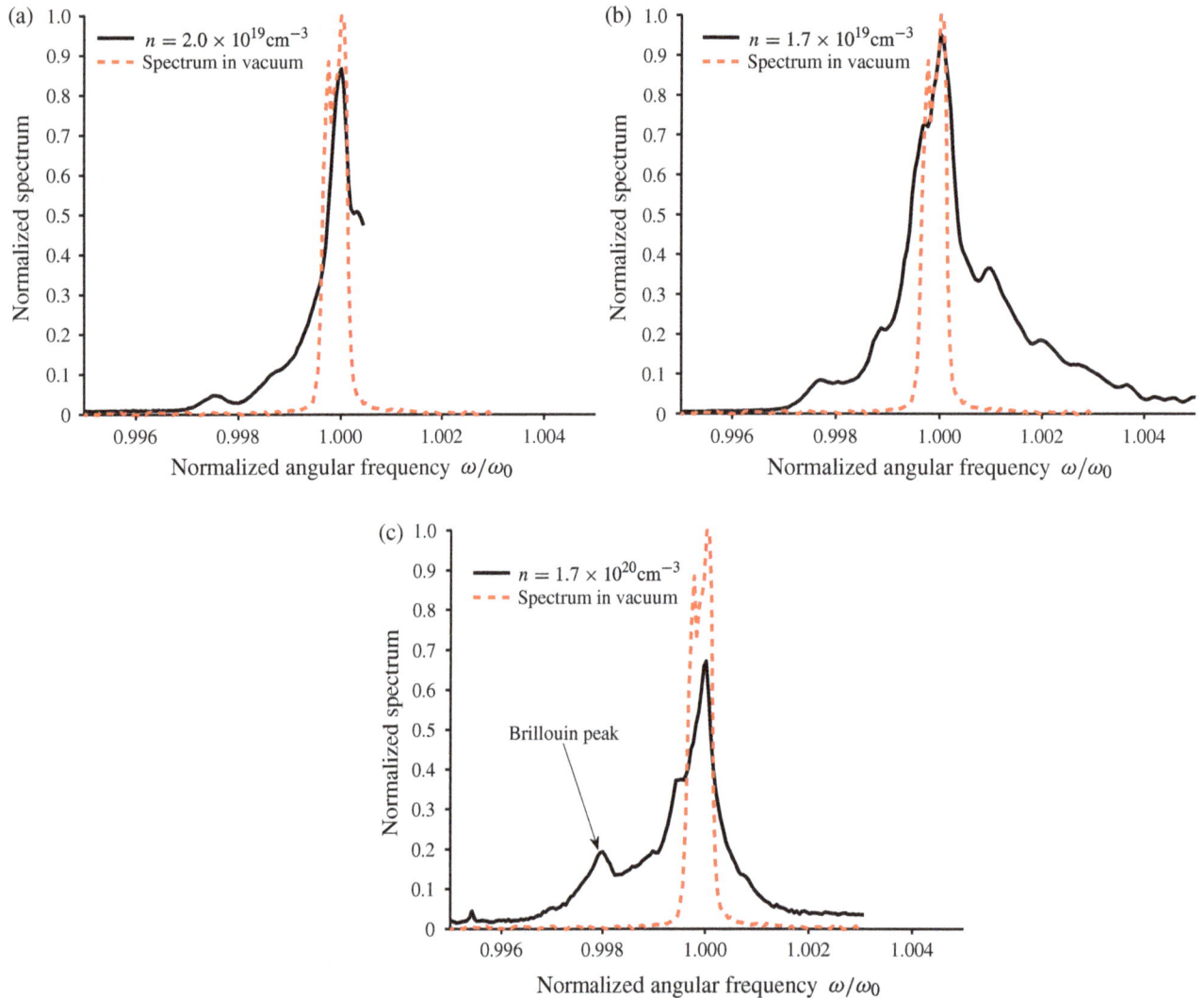

Figure 2. Experimental frequency spectra of a 1 ps laser pulse recorded after propagation through a supersonic gas jet (normalized intensity versus normalized angular frequency). A reference spectrum, recorded with the gas jet turned off, has been included in each plot. Graph (a) is the spectrum recorded with only the seed beam at an intensity of 4.0×10^{14} W cm^{-2} interacting with the gas jet at $n_e = 2.0 \times 10^{19}$ cm^{-3}, without a counter-propagating pump beam. Graph (b) was recorded with the two counter-propagating beams interacting, the seed at an intensity of 3.6×10^{15} W cm^{-2} and the pump at 4.2×10^{14} W cm^{-2}, at $n_e = 1.7 \times 10^{19}$ cm^{-3}. Graph (c) was recorded with the seed at an intensity of 3.2×10^{14} W cm^{-2} and the pump at 3.2×10^{14} W cm^{-2}, at $n_e = 1.7 \times 10^{20}$ cm^{-3}. The generation of a downshifted peak can be observed through the interaction of the laser pulses and the gas jet, with its relative intensity compared to the fundamental peak strongly depending on the plasma density and the presence of a counter-propagating pump pulse.

electron temperature and the effective ionization degree of the atoms are needed. These quantities were calculated using the laser plasma simulation code MEDUSA, in 1D planar geometry, using a corrected Thomas–Fermi equation-of-state model and an average-atom model, and assuming collisional ionization. For the shots shown in Figure 2, this yields $T_e = 130$, 120, and 5 eV, respectively, and $Z^* = 5.5$, 1.0, and 1.1. These values were used to estimate the plasma electron density for each case, which was used in the OSIRIS simulations.

Three sets of simulation results corresponding to each of the three experimental regimes examined are presented, and were set up as follows. In simulation (a) a single laser of intensity 6×10^{14} W cm^{-2} was injected into an argon plasma of density $0.018 n_c$ with a mass ratio of ions to electrons of $m_i/m_e = 14{,}688$. The plasma temperature ratio was set such that $ZT_e/T_i = 25$, where $Z = 5$ and $T_e = 20$ eV, assuming neon-like argon with the majority of the outer shell of electrons depleted. For simulation (b) two counter-propagating pulses were launched into a plasma of density $0.015 \times n_c$, in this case comprising deuterium, with a mass ratio of ions to electrons of $m_i/m_e = 3672$, with the plasma ion and electron temperatures kept constant at 20 eV for the ions and 120 eV for the electron species. Laser intensities of 6.2×10^{14} and 5.4×10^{14} W cm^{-2} for the pump and seed, respectively, were used, where the seed pulse was launched at the instant the pump laser had traversed the length of the plasma. In the case of simulation (c), two counter-

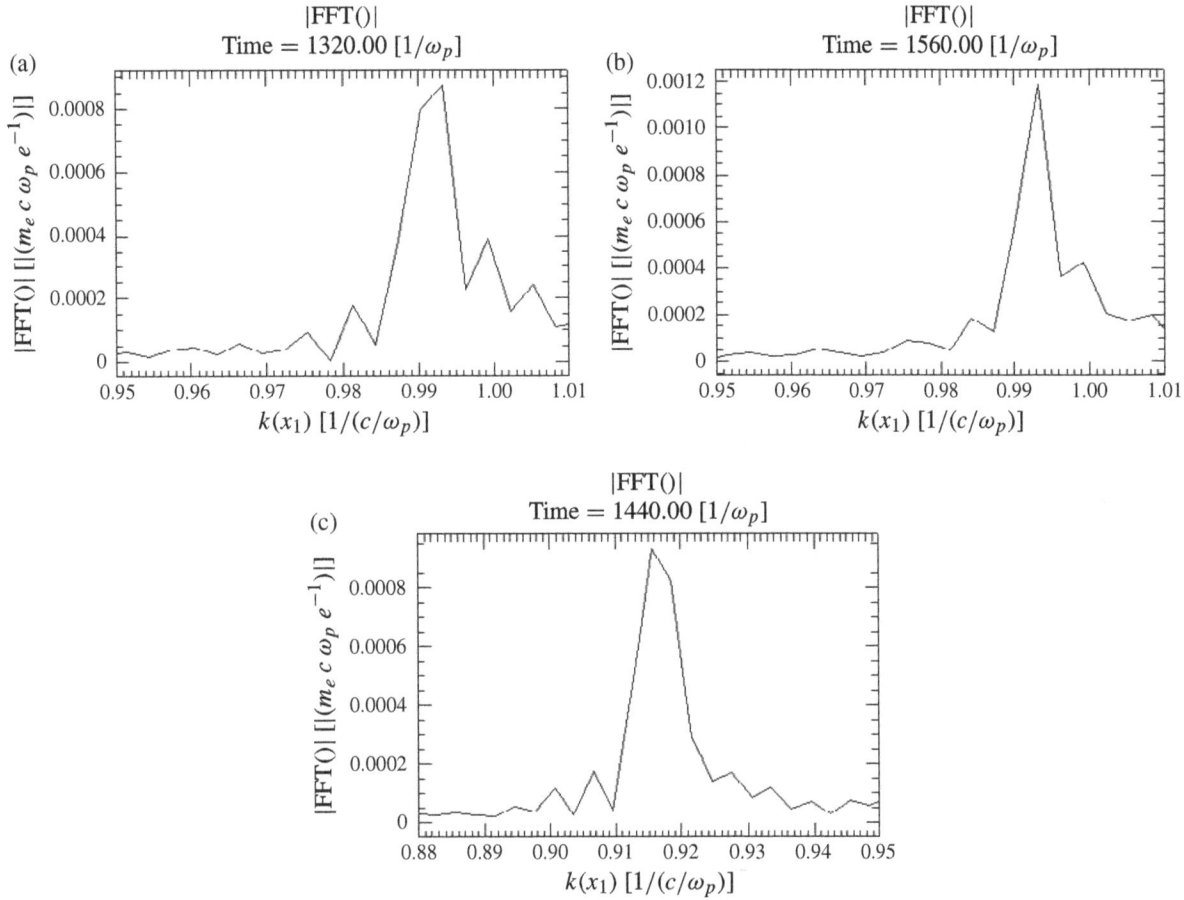

Figure 3. Simulated spectra corresponding to each of the experimental regimes presented in Figure 2 (normalized intensity versus normalized wavevector). The electron density varied from 1.7×10^{19} to 1.7×10^{20} cm^{-3}. Graph (a) is the spectrum simulated with a single laser of intensity 6×10^{14} W cm^{-2} interacting in a neon-like argon plasma with an electron temperature of about 20 eV and density of $0.018n_c$. Graph (b) was calculated with the two counter-propagating beams interacting in a deuterium plasma of density $0.015 \times n_c$, the seed at intensity 5.4×10^{15} W cm^{-2} and the pump at 6.2×10^{14} W cm^{-2}, with an electron temperature of 120 eV. Graph (c) was simulated with the seed at intensity 4.9×10^{14} W cm^{-2} and the pump at 4.9×10^{14} W cm^{-2}, in an argon plasma with an electron temperature of 5 eV and a density of $0.16n_c$.

propagating beams were used and their intensities were both set to 4.9×10^{14} W cm^{-2} and propagated through an argon plasma with a configuration such that $m_i/m_e = 73{,}440$, $ZT_e/T_i = 5$, where $Z = 1$ and $T_e = 5$ eV, and a density of $0.16n_c$. The following parameters are consistent throughout each of the three simulations presented: the pulses propagate through a plasma column of length $1410c/\omega_0$, with the pump pulse traveling from right to left through the simulation box; the pump pulse has a duration of 1.5 ps and the seed pulse a duration of 100 fs; each of the pulses is from a laser of wavelength 1 μm; the time step for integration is $\Delta t = 0.04\omega_p^{-1}$, where ω_p is the plasma electron frequency; the spatial resolution of the simulations is of the order of the Debye length, with 100 particles per cell. Due to computational limitations, the pulse lengths and plasma column were scaled down by a factor of ten from the parameters used to obtain the experimental results.

Upon examination of the spectra presented in Figure 3, it can be seen that the results obtained by OSIRIS closely

match the results obtained from the experimental observations of Brillouin scattering shown in Figure 2. In each of the three cases examined numerically, however, it can be seen that the Fourier spectrum obtained is slightly broader than that of the experimental results. This slight variation in the spectra is attributed to the fact that the simulations have no transverse dimensions as they were performed in 1D, hence putting numerical constraints on the solutions obtained as there can be no transverse variation of the laser intensity. Therefore, the amplitude of any plasma wave driven by the laser will be overestimated, which leads to an overestimation of spectral drifts and also of the temperature recorded.

6. Conclusions

These experimental observations of Brillouin scattering using two beams at the same wavelength are a promising confirmation of the observations by Lancia et al.[6] that energy transfer by SBS can be achieved with a single laser system, with no frequency downshift required in the seed pulse, as is mandatory to perform Raman amplification. In addition, we have revealed that an increase in the plasma density leads to an increase in the energy transfer efficiency, because of the increased Brillouin growth rate and the disappearance of Raman scattering above $0.25n_{cr}$. The generation of a Brillouin peak using the natural bandwidth of the laser is confirmed by the 1D PIC simulation results from OSIRIS. The same PIC simulations also revealed significant competition between SRS and SBS, for densities between $0.017n_{cr}$ and $0.17n_{cr}$. The experiments revealed a substantial increase in the SBS signal with increasing plasma density, in line with the theoretically predicted increase of the growth rate. In light of these results, it is recommended that future Brillouin amplification experiments are carried out at plasma densities above $0.25n_{cr}$ to eliminate SRS altogether and benefit from the higher Brillouin scattering growth rate.

Acknowledgements

The authors would like to gratefully acknowledge the support of the staff of the Central Laser Facility for their help. They would also like to thank R. Kirkwood, S. Wilks, D. Meyerhofer, S. Craxton, D. Froula, J. Myatt and A. Solodov for useful discussions. This work was supported by the UK Science and Technology Facilities Council and by the Engineering and Physical Sciences Research Council. H.N. was supported by the Royal Society Newton International Fellowship and the Japan Society for the Promotion of Science.

References

1. N. J. Fisch and V. M. Malkin, Phys. Plasmas **10**, 2056 (2003).
2. J. Ren, S. Li, A. Morozov, S. Suckewer, N. A. Yampolsky, V. M. Malkin, and N. J. Fisch, Phys. Plasmas **15**, 056702 (2008).
3. R. M. G. M. Trines, F. Fiúza, R. Bingham, R. A. Fonseca, L. O. Silva, R. A. Cairns, and P. A. Norreys, Nature Phys. **7**, 87 (2010).
4. R. M. G. M. Trines, F. Fiúza, R. Bingham, R. A. Fonseca, L. O. Silva, R. A. Cairns, and P. A. Norreys, Phys. Rev. Lett. **107**, 105002 (2011).
5. R. A. Fonseca, L. O. Silva, F. S. Tsung, V. K. Decyk, W. Lu, C. Ren, W. B. Mori, S. Deng, S. Lee, T. Katsouleas, and J. C. Adam, Lect. Notes Comput. Sci. **2331**, 342 (2002).
6. L. Lancia, J.-R. Marquès, M. Nakatsutsumi, C. Riconda, S. Weber, S. Hüller, A. Mančić, P. Antici, V. T. Tikhonchuk, A. Héron, P. Audebert, and J. Fuchs, Phys. Rev. Lett. **104**, 025001 (2010).
7. D. W. Forslund, J. M. Kindel, and E. L. Lindman, Phys. Fluids **18**, 1002 (1975).
8. W. L. Kruer, *The Physics of Laser Plasma Interactions* (Westview Press, 2003).
9. A. A. Andreev, C. Riconda, V. T. Tikhonchuk, and S. Weber, Phys. Plasmas **13**, 053110 (2006).
10. P. Michel, L. Divol, E. A. Williams, C. A. Thomas, D. A. Callahan, S. Weber, S. W. Haan, J. D. Salmonson, N. B. Meezan, O. L. Landen, S. Dixit, D. E. Hinkel, M. J. Edwards, B. J. MacGowan, J. D. Lindl, S. H. Glenzer, and L. J. Suter, Phys. Plasmas **16**, 042702 (2009).
11. P. Michel, S. H. Glenzer, L. Divol, D. K. Bradley, D. Callahan, S. Dixit, S. Glenn, D. Hinkel, R. K. Kirkwood, J. L. Kline, W. L. Kruer, G. A. Kyrala, S. Le Pape, N. B. Meezan, R. Town, K. Widmann, E. A. Williams, B. J. MacGowan, J. Lindl, and L. J. Suter, Phys. Plasmas **17**, 056305 (2010).
12. C. N. Danson, P. A. Brummitt, R. J. Clarke, J. L. Collier, B. Fell, A. J. Frackiewicz, S. Hancock, S. Hawkes, C. Hernandez-Gomez, P. Holligan, M. H. R. Hutchinson, A. Kidd, W. J. Lester, I. O. Musgrave, D. Neely, D. R. Neville, P. A. Norreys, D. A. Pepler, C. J. Reason, W. Shaikh, T. B. Winstone, R. W. W. Wyatt, and B. E. Wyborn, Nucl. Fusion **44**, S239 (2004).

27

A high energy nanosecond cryogenic cooled Yb:YAG active-mirror amplifier system

Xiaojin Cheng[1], Jianlei Wang[1], Zhongguo Yang[1], Jin Liu[1], Lei Li[1], Xiangchun Shi[1], Wenfa Huang[2], Jiangfeng Wang[2], and Weibiao Chen[1]

[1]*Shanghai Key Laboratory of All Solid-state Laser and Applied Techniques, Shanghai Institute of Optics and Fine Mechanics, Chinese Academy of Sciences, Shanghai 201800, China*

[2]*The Joint Laboratory for High Power Laser Physics, Shanghai Institute of Optics and Fine Mechanics, Chinese Academy of Sciences, Shanghai 201800, China*

Abstract

A diode-pumped master oscillator power amplifier system based on a cryogenic Yb:YAG active-mirror laser has been developed. The performances of the laser amplifier at low temperature and room temperature have been investigated theoretically and experimentally. A maximum output energy of 3.05 J with an optical-to-optical efficiency of 14.7% has been achieved by using the master amplifier system.

Keywords: diode-pumped; solid-state laser; Yb:YAG

1. Introduction

Solid-state lasers with high energy have been widely reported in many application fields, ranging from materials processing to remote sensing to laser-driven inertial fusion[1]. For high energy laser oscillator and amplifier design, the amplified spontaneous emission (ASE), the thermal effects of the laser material, and the laser-induced damage threshold of the optics are the most important considerations. In addition to relations to the physical character, such as the quantum defect, concentration quenching, up-conversion, etc., the thermal effects of laser materials have great relations to the pump and cooling structure[2, 3]. Prevention of damage to the optics can be achieved by scaling the laser gain medium and laser spot size. In order to suppress the ASE, reasonable design of the amplifier gain, the size and the shape of the gain medium is necessary.

The development of laser diodes has promoted the interest in Yb^{3+} doped laser gain media, such as YAG, CaF_2, Y_2O_3, and S-FAP[4–7]. In particular, Yb:YAG, with a long fluorescence lifetime, broad emission band, low quantum defect, and excellent thermo-mechanical properties, has been considered to have great potential as a material to obtain high energy, high efficiency, and short duration pulse laser output using diode-pumped solid-state laser (DPSSL) systems[8, 9].

Moreover, to obtain a much larger laser gain medium, Yb:YAG ceramic has already attracted attention in the field of high power lasers. However, the performance of an Yb:YAG laser strongly depends on the temperature because of its quasi-three-level nature. At room temperature, the laser operation threshold of Yb:YAG crystal is high for the reabsorption of a lower energy level. Imposition of a high intensity pump or decrease of the laser material temperature will allow this problem to be overcome[10–12].

Researchers have studied high energy DPSSL systems at several hertz repetition rates during the last decade, for example Mercury, Lucia, Halna, and Polaris. Mercury laser systems delivered 61 J (10 Hz) at the Lawrence Livermore National Laboratory (LLNL) with an Yb:S-FAP crystal and cooled by high pressure helium flow[13]. With Yb:YAG crystal, the Lucia laser system delivered 14 J (2 Hz) at the LULI laboratory[14, 15]. The Halna laser system at the Institute for Laser Engineering (ILE) delivered 20 J (10 Hz) with a zigzag Nd:glass slab[16]. For the Polaris laser system, 12 J (0.05 Hz) was obtained from Yb:glass at the Institute of Optics and Quantum Electronics of the Friedrich-Schiller University (IOQ, Jena, Germany)[17–20].

In this paper, in order to apply a high energy pump source with a Ti:sapphire laser to obtain ultrashort pulses, a diode-pumped master oscillator power amplifier system based on a cryogenic Yb:YAG/YAG active-mirror laser has been set up for the first step. With a doping concentration of 4 at.% for the Yb:YAG/YAG crystal and cooled by liquid nitrogen,

Correspondence to: Xiao-jin Cheng, Shanghai Key Laboratory of All Solid-state Laser and Applied Techniques, Shanghai Institute of Optics and Fine Mechanics, Chinese Academy of Sciences, 201800, No. 390, Qinghe Road, Jiading District, Shanghai, China. Email: xjcheng@siom.ac.cn

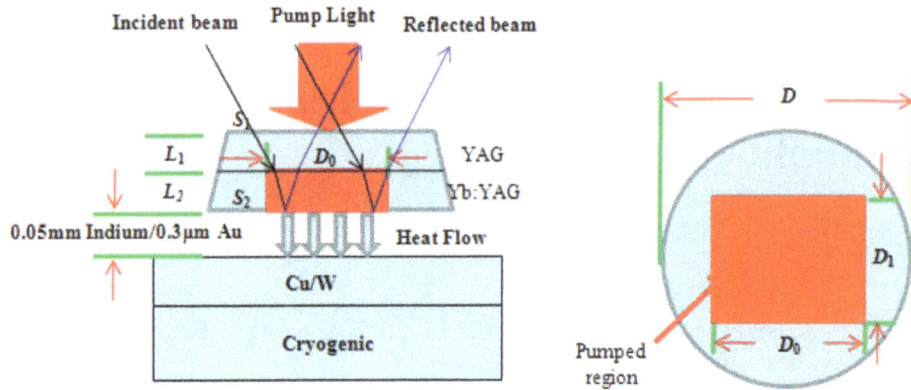

Figure 1. Illustration of the pump and cooling structure of the active-mirror amplifier with a cryogenic cooled composite Yb:YAG/YAG crystal.

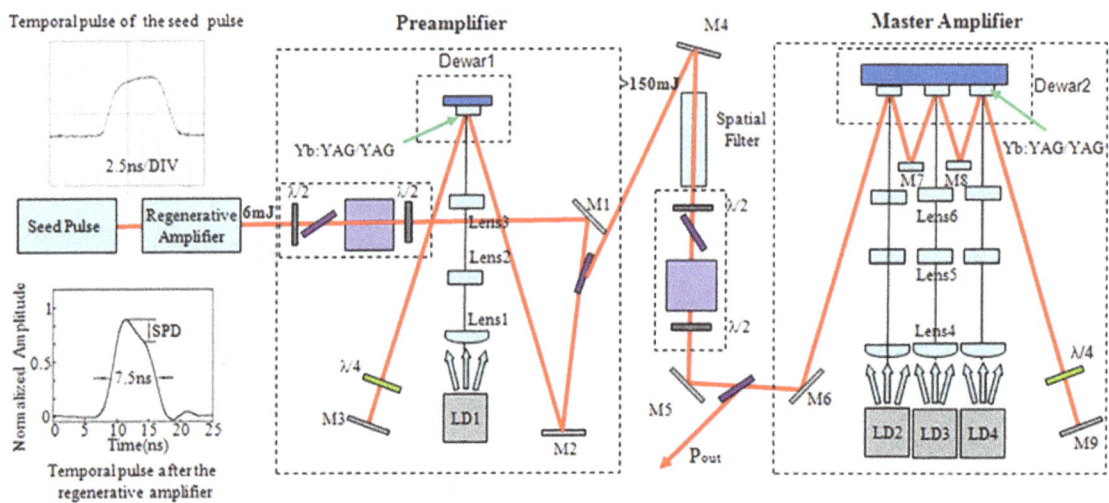

Figure 2. Scheme of the active-mirror amplifier setup.

3.05 J at an optical-to-optical efficiency of 14.7% has been achieved at a cooling temperature of 155 K.

2. Experimental setup

Figure 1 shows the pump and cooling structure of the active-mirror amplifier with cryogenic cooled composite Yb:YAG/YAG crystal. Facet S_1 is AR-coated at both 940 ± 15 nm ($0°$) and 1030 ± 15 nm ($15°$). The other end facet, S_2, is HR-coated at both 940 ± 15 nm ($0°$) and 1030 ± 15 nm ($15°$) to reflect pump light and signal light. Meanwhile, facet S_2 is also used as the cooling surface and is wrapped in indium foil ($50 \ \mu$m) and thin gold foil ($0.3 \ \mu$m). The crystal is fixed in a copper heat sink which is cooled by liquid nitrogen and sealed in a vacuum Dewar.

A diode-pumped master oscillator power amplifier system based on a cryogenic Yb:YAG/YAG active-mirror structure has also been constructed for further study of the laser performance of Yb:YAG at low temperature (Figure 2). The seed pulses with a pulse duration of 10 ns were generated

in a CW distributed-feedback Yb fiber laser following an acoustic-optic chopper and optical fiber amplifiers which produced nanosecond pulses with a 10 Hz repetition rate and amplified up to 6 mJ in a regenerative amplifier[21]. To prevent unnecessary back reflection from the amplifier stage, the regenerative amplifier was followed by an optical isolator system which consisted of two Brewster polarizers, one $45°$ rotator and two half-wave plates (HWPs). The size of the laser beam was shaped to 5.5×5.5 (mm) in order to ensure the best matching with the pump light. A four-pass preamplifier was realized using reflection mirrors (M2, M3) and a quarter-wave plate (QWP). The size of the composite Yb:YAG/YAG was $\Phi 25 \times 6.5$ (mm). In addition, the thickness of the doping part was 3.5 mm with a doping concentration of 4 at.%. The laser diode stack (LD1) with a maximum output power of 2.95 kW at a center wavelength of 940 nm was available as the pump source. Lens 1, made up of one plano-convex lens with a curvature radius of 165 mm and thickness of 5 mm, was designed to compress the fast axis beam. Meanwhile, a plano-convex group (lens 2 and lens 3), with a curvature radius of 70 mm and thickness of 15 mm, was used to shape the slow axis beam. All of these three lenses were AR-coated at 940 ± 15 and 1030 ± 15 nm. After four-pass amplification, the laser beam was output from the polarizer and expander two times. At the same time, a spatial filter system was used to filter the higher order mode laser. The same isolator system was loaded between the preamplifier and the master amplifier.

For the master amplifier, three Yb:YAG/YAG modules were sealed in one Dewar with the same pump and cooling structure. The laser diode stacks (LD3, LD4, and LD5) could deliver 6.5 kW output power. Lens 4, made up of one plano-convex lens with a curvature radius of 220 mm and thickness of 7 mm, was designed to compress the fast axis beam. Meanwhile, a plano-convex group (lens 5 and lens 6), with a curvature radius of 82 mm and thickness of 12 mm, was used to shape the slow axis beam. All of these lenses were AR-coated at 940 ± 15 and 1030 ± 15 nm.

3. Laser experiments and results

Figure 3 shows the output energy versus the pump energy at different cooling temperatures for an injection energy of 6 mJ. The spot size of the injection signal light is 5.5×5.5 (mm). As shown in Figure 3, the gain is not obvious because of the reabsorption at room temperature and will increase with decrease of the cooling temperature until 155 K. Below 155 K, the gain no longer increases because of the lateral ASE. At 155 K, a maximum output energy of 196 mJ is obtained with a pump energy of 2.95 J at 10 Hz. In order to evaluate the results of the experiment, a simulation of the preamplifier was made according to the laser rate equations. In order to reduce the complexity, the lateral ASE was not considered in our calculation. As shown in Figure 3, the simulation results match the experimental results excellently for temperatures of 200, 250, and 300 K. However, at 155 K, the experimental result deviates further from the simulation result with increase of the pump energy, which indicates that the lateral ASE is serious at 155 K.

Figure 4 shows the output energy of the master amplifier after four-pass amplification. In order to reduce the influence of lateral ASE and parasitic oscillation, we roughened the side of the Yb:YAG crystal at the master amplifier. Before the signal pulse was injected into the master amplifier, the laser beam was expanded to 11×11 (mm). At 155 K, a maximum output energy of 3.05 J at an optical-to-optical efficiency of 14.7% was achieved for an injected pulse energy of 180 mJ. To evaluate the scaling amplification of the active-mirror structure amplifier, a simulation was made for the master amplifier based on the laser rate equations. As described in Figure 4, for the same cooling temperature, the experimental result is slightly lower than the simulation result, the same growth can be obtained and the result shows that the output energy can increase linearly with a much more powerful pump.

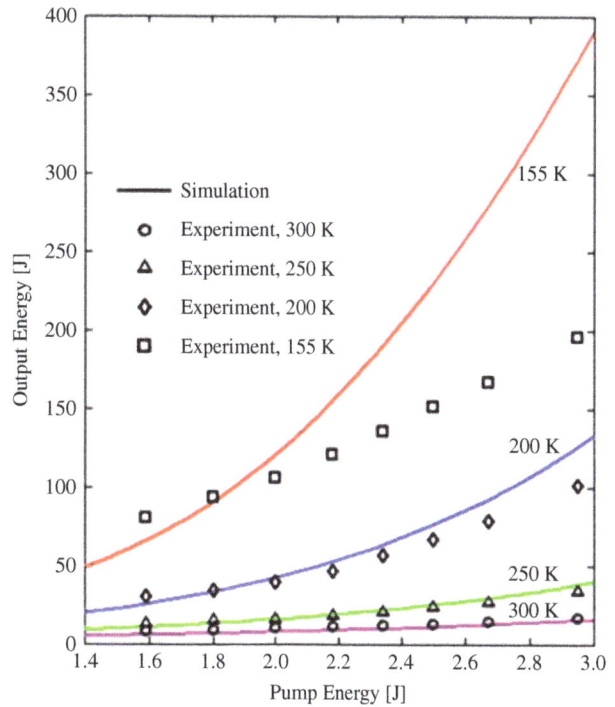

Figure 3. The output pulse energy of the preamplifier after four-pass amplification independent of the pump energy at a 10 Hz repetition rate at different cooling temperatures.

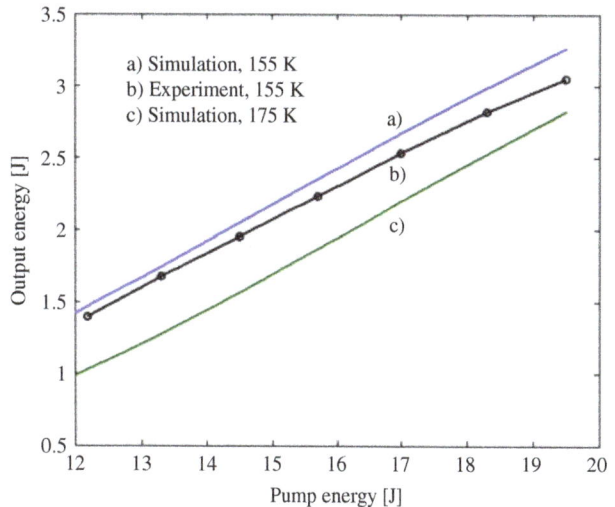

Figure 4. The measured and simulated pulse energy of the master amplifier after four-pass amplification independent of the pump energy at a 10 Hz repetition rate and injected pulse energy of 180 mJ.

4. Conclusion and outlook

In conclusion, we have shown nanosecond pulse amplification to the 3.05 J level at a repetition rate of 10 Hz. The seed pulses with a pulse duration of 10 ns were generated in an Yb fiber laser and amplified up to 6 mJ in a regenerative amplifier. The four-pass amplification system cooled by liquid nitrogen boosted the energy up to 180 mJ (preamplifier)

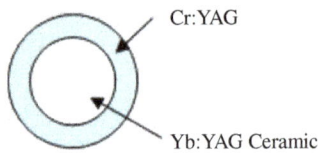

Figure 5. Edge-cladding Yb:YAG ceramic to improve the ASE suppression.

and 3.05 J (master amplifier) respectively. In future work, the optical transmission, pump and cooling structure will be further optimized to obtain a much higher energy output. In particular, an edge-cladding Yb:YAG ceramic (Figure 5) will be designed to improved the ASE suppression. Cr:YAG, which is commonly used for passive Q-switching, will be used as the edge-cladding material[22].

Acknowledgements

The authors gratefully acknowledge the support of the Knowledge Innovation Project of the Chinese Academy of Sciences and the National Natural Science Foundation of China (No. 61008020).

References

1. J. C. Chanteloup and D. Albach, IEEE Photonics J. **3**, 245 (2011).
2. T. Y. Fan, IEEE J. Quantum Electron. **29**, 1457 (1993).
3. W. Koechner, *Solid-state Laser Engineering* (Springer-Verlag, Berlin, 1999).
4. M. Siebold, M. Hornung, R. Boedefeld, S. Podleska, S. Klingebiel, C. Wandt, F. Krausz, S. Karsch, R. Uecker, A. Jochmann, J. Hein, and M. C. Kaluza, Opt. Lett. **33**, 2770 (2008).
5. F. Friebel, F. Druon, J. Boudeile, D. N. Papadopoulos, M. Hanna, P. Georges, P. Camy, J. L. Doualan, A. Benayad, R. Moncorgé, C. Cassagne, and G. Boudebs, Opt. Lett. **34**, 1474 (2009).
6. G. Q. Xie, D. Y. Tang, L. M. Zhao, L. J. Qian, and K. Ueda, Opt. Lett. **32**, 2741 (2007).
7. J. Kong, D. Y. Tang, C. C. Chan, J. Lu, K. Ueda, H. Yagi, and T. Yanagitani, Opt. Lett. **32**, 247 (2007).
8. T. S. Rutherford, W. M. Tulloch, S. Sinha, and R. L. Byer, Opt. Lett. **26**, 986 (2001).
9. Y. Akahane, M. Aoyama, K. Ogawa, K. Tsuji, S. Tokita, J. Kawanaka, H. Nishioka, and K. Yamakawa, Opt. Lett. **32**, 1899 (2007).
10. D. S. Sumida and T. Y. Fan, Opt. Lett. **20**, 2384 (1995).
11. D. J. Ripin, J. R. Ochoa, R. L. Aggarwal, and T. Y. Fan, Opt. Lett. **29**, 2154 (2004).
12. K. H. Hong, J. T. Gopinath, D. Rand, A. M. Siddiqui, S. W. Huang, E. Li, B. J. Eggleton, J. D. Hybl, T. Y. Fan, and F. X. Kärtner, Opt. Lett. **35**, 1752 (2010).
13. A. Bayramian, J. Armstrong, G. Beer, R. Campbell, B. Chai, R. Cross, A. Erlandson, Y. Fei, B. Freitas, R. Kent, J. Menapace, W. Molander, K. Schaffers, C. Siders, S. Sutton, J. Tassano, S. Telford, C. Ebbers, J. Caird, and C. Barty, J. Opt. Soc. Am. B **25**, B57 (2008).
14. A. Lucianetti, T. Novo, B. Vincent, D. Albach, and J. C. Chanteloup, Proc. SPIE **8080** (2011).
15. T. Gonçalvès-Novo, D. Albach, B. Vincent, M. Arzakantsyan, and J.-C. Chanteloup, Opt. Express **21**, 855 (2013).
16. R. Yasuhara, T. Kawashima, T. Sekine, T. Kurita, T. Ikegawa, O. Matsumoto, M. Miyamoto, H. Kan, H. Yoshida, J. Kawanaka, M. Nakatsuka, N. Miyanaga, Y. Izawa, and T. Kanabe, Opt. Lett. **33**, 1711 (2008).
17. C. Wandt, S. Klingebiel, M. Siebold, Z. Major, J. Hein, F. Krausz, and S. Karsch, Opt. Lett. **33**, 1111 (2007).
18. M. Hornung, R. Bödefeld, M. Siebold, A. Kessler, M. Schnepp, R. Wachs, A. Sävert, S. Podleska, S. Keppler, J. Hein, and M. C. Kaluza, Appl. Phys. B **101**, 93 (2010).
19. R. L. Aggarwal, D. J. Ripin, J. R. Ochoa, and T. Y. Fan, J. Appl. Phys. **98**, 103514 (2005).
20. T. Y. Fan, D. J. Ripin, R. L. Aggarwal, J. R. Ochoa, B. Chann, M. Tilleman, and J. Spitzberg, IEEE J. Sel. Top. Quantum Electron. **13**, 448 (2007).
21. X. H. Lu, J. F. Wang, X. Li, Y. E. Jiang, W. Fan, and X. Ch. Li, Opt. Lett. **9**, 111401 (2011).
22. J. Dong, K.-i. Ueda, A. Shirakawa, H Yagi, T. Yanagitani, and A. A. Kaminskii, Opt. Express **15**, 14516 (2007).

Permissions

All chapters in this book were first published in HPLSE, by Cambridge Journals; hereby published with permission under the Creative Commons Attribution License or equivalent. Every chapter published in this book has been scrutinized by our experts. Their significance has been extensively debated. The topics covered herein carry significant findings which will fuel the growth of the discipline. They may even be implemented as practical applications or may be referred to as a beginning point for another development.

The contributors of this book come from diverse backgrounds, making this book a truly international effort. This book will bring forth new frontiers with its revolutionizing research information and detailed analysis of the nascent developments around the world.

We would like to thank all the contributing authors for lending their expertise to make the book truly unique. They have played a crucial role in the development of this book. Without their invaluable contributions this book wouldn't have been possible. They have made vital efforts to compile up to date information on the varied aspects of this subject to make this book a valuable addition to the collection of many professionals and students.

This book was conceptualized with the vision of imparting up-to-date information and advanced data in this field. To ensure the same, a matchless editorial board was set up. Every individual on the board went through rigorous rounds of assessment to prove their worth. After which they invested a large part of their time researching and compiling the most relevant data for our readers.

The editorial board has been involved in producing this book since its inception. They have spent rigorous hours researching and exploring the diverse topics which have resulted in the successful publishing of this book. They have passed on their knowledge of decades through this book. To expedite this challenging task, the publisher supported the team at every step. A small team of assistant editors was also appointed to further simplify the editing procedure and attain best results for the readers.

Apart from the editorial board, the designing team has also invested a significant amount of their time in understanding the subject and creating the most relevant covers. They scrutinized every image to scout for the most suitable representation of the subject and create an appropriate cover for the book.

The publishing team has been an ardent support to the editorial, designing and production team. Their endless efforts to recruit the best for this project, has resulted in the accomplishment of this book. They are a veteran in the field of academics and their pool of knowledge is as vast as their experience in printing. Their expertise and guidance has proved useful at every step. Their uncompromising quality standards have made this book an exceptional effort. Their encouragement from time to time has been an inspiration for everyone.

The publisher and the editorial board hope that this book will prove to be a valuable piece of knowledge for researchers, students, practitioners and scholars across the globe.

List of Contributors

E. Schleifer
Racah Institute Of Physics, Hebrew University, Jerusalem,
Israel

M. Botton
Racah Institute Of Physics, Hebrew University, Jerusalem,
Israel

E. Nahum
Racah Institute Of Physics, Hebrew University, Jerusalem,
Israel

S. Eisenman
Racah Institute Of Physics, Hebrew University, Jerusalem,
Israel

A.Zigler
Racah Institute Of Physics, Hebrew University, Jerusalem,
Israel

Z. Henis
Racah Institute Of Physics, Hebrew University, Jerusalem,
Israel

Lili Hu
Shanghai Institute of Optics and Fine Mechanics, Chinese
Academy of Sciences, Shanghai 201800, China

Shubin Chen
Shanghai Institute of Optics and Fine Mechanics, Chinese
Academy of Sciences, Shanghai 201800, China

Jingping Tang
Shanghai Institute of Optics and Fine Mechanics, Chinese
Academy of Sciences, Shanghai 201800, China

Biao Wang
Shanghai Institute of Optics and Fine Mechanics, Chinese
Academy of Sciences, Shanghai 201800, China

Tao Meng
Shanghai Institute of Optics and Fine Mechanics, Chinese
Academy of Sciences, Shanghai 201800, China

Wei Chen
Shanghai Institute of Optics and Fine Mechanics, Chinese
Academy of Sciences, Shanghai 201800, China

Lei Wen
Shanghai Institute of Optics and Fine Mechanics, Chinese
Academy of Sciences, Shanghai 201800, China

Junjiang Hu
Shanghai Institute of Optics and Fine Mechanics, Chinese
Academy of Sciences, Shanghai 201800, China

Shunguang Li
Shanghai Institute of Optics and Fine Mechanics, Chinese
Academy of Sciences, Shanghai 201800, China

Yongchun Xu
Shanghai Institute of Optics and Fine Mechanics, Chinese
Academy of Sciences, Shanghai 201800, China

Yasi Jiang
Shanghai Institute of Optics and Fine Mechanics, Chinese
Academy of Sciences, Shanghai 201800, China

Junzhou Zhang
Shanghai Institute of Optics and Fine Mechanics, Chinese
Academy of Sciences, Shanghai 201800, China

Zhonghong Jiang
Shanghai Institute of Optics and Fine Mechanics, Chinese
Academy of Sciences, Shanghai 201800, China

Antonio Lucianetti
HiLASE Project, Institute of Physics AS CR, Na Slovance
2, 18221, Prague, Czech Republic

Magdalena Sawicka
HiLASE Project, Institute of Physics AS CR, Na Slovance
2, 18221, Prague, Czech Republic

Ondrej Slezak
HiLASE Project, Institute of Physics AS CR, Na Slovance
2, 18221, Prague, Czech Republic

Martin Divoky
HiLASE Project, Institute of Physics AS CR, Na Slovance
2, 18221, Prague, Czech Republic

Jan Pilar
HiLASE Project, Institute of Physics AS CR, Na Slovance
2, 18221, Prague, Czech Republic

Venkatesan Jambunathan
HiLASE Project, Institute of Physics AS CR, Na Slovance
2, 18221, Prague, Czech Republic

Stefano Bonora
HiLASE Project, Institute of Physics AS CR, Na Slovance
2, 18221, Prague, Czech Republic
CNR-IFN, Via Trasea 7, 35131, Padova, Italy

Roman Antipenkov
ELI Beamlines Project, Institute of Physics AS CR, Na Slovance 2, 18221, Prague, Czech Republic

Tomas Mocek
HiLASE Project, Institute of Physics AS CR, Na Slovance 2, 18221, Prague, Czech Republic

Guohui Li
Fujian Institute of Research on the Structure of Matter, Chinese Academy of Sciences, Fuzhou 350002, China

Guozong Zheng
Fujian Institute of Research on the Structure of Matter, Chinese Academy of Sciences, Fuzhou 350002, China

Yingkun Qi
Fujian Institute of Research on the Structure of Matter, Chinese Academy of Sciences, Fuzhou 350002, China

Peixiu Yin
Fujian Institute of Research on the Structure of Matter, Chinese Academy of Sciences, Fuzhou 350002, China

Fei Li En Tang
Fujian Institute of Research on the Structure of Matter, Chinese Academy of Sciences, Fuzhou 350002, China

Jing Xu
Fujian Institute of Research on the Structure of Matter, Chinese Academy of Sciences, Fuzhou 350002, China

Taiming Lei
Fujian Institute of Research on the Structure of Matter, Chinese Academy of Sciences, Fuzhou 350002, China

Xiuqin Lin
Fujian Institute of Research on the Structure of Matter, Chinese Academy of Sciences, Fuzhou 350002, China

Min Zhang
Fujian Institute of Research on the Structure of Matter, Chinese Academy of Sciences, Fuzhou 350002, China

Junye Lu
Fujian Institute of Research on the Structure of Matter, Chinese Academy of Sciences, Fuzhou 350002, China

Jinbo Ma
Fujian Institute of Research on the Structure of Matter, Chinese Academy of Sciences, Fuzhou 350002, China

Youping He
Fujian Institute of Research on the Structure of Matter, Chinese Academy of Sciences, Fuzhou 350002, China

Yuangen Yao
Fujian Institute of Research on the Structure of Matter, Chinese Academy of Sciences, Fuzhou 350002, China

Adi Hanuka
Department of Electrical Engineering, Technion – Israel Institute of Technology, Haifa 32000, Israel

Levi Schächter
Department of Electrical Engineering, Technion – Israel Institute of Technology, Haifa 32000, Israel

HaiyanWang
Shanghai Institute of Optics and Fine Mechanics, Chinese Academy of Sciences, Shanghai 201800, China

Cheng Liu
Shanghai Institute of Optics and Fine Mechanics, Chinese Academy of Sciences, Shanghai 201800, China

Xiaoliang He
College of Sciences, Jiangnan University, Wuxi 214122, China

Xingchen Pan
Shanghai Institute of Optics and Fine Mechanics, Chinese Academy of Sciences, Shanghai 201800, China

Shenlei Zhou
Shanghai Institute of Optics and Fine Mechanics, Chinese Academy of Sciences, Shanghai 201800, China

RongWu
Shanghai Institute of Optics and Fine Mechanics, Chinese Academy of Sciences, Shanghai 201800, China

Jianqiang Zhu
Shanghai Institute of Optics and Fine Mechanics, Chinese Academy of Sciences, Shanghai 201800, China

Yajing Guo
Joint Laboratory on High Power Laser and Physics, Shanghai Institute of Optics and Fine Mechanics, Chinese Academy of Sciences, Shanghai 201800, PR China
University of Chinese Academy of Sciences, Beijing 100039, PR China

Shunxing Tang
Joint Laboratory on High Power Laser and Physics, Shanghai Institute of Optics and Fine Mechanics, Chinese Academy of Sciences, Shanghai 201800, PR China

Xiuqing Jiang
Joint Laboratory on High Power Laser and Physics, Shanghai Institute of Optics and Fine Mechanics, Chinese Academy of Sciences, Shanghai 201800, PR China
University of Chinese Academy of Sciences, Beijing 100039, PR China

Yujie Peng
Joint Laboratory on High Power Laser and Physics, Shanghai Institute of Optics and Fine Mechanics, Chinese Academy of Sciences, Shanghai 201800, PR China
University of Chinese Academy of Sciences, Beijing 100039, PR China

Baoqiang Zhu
Joint Laboratory on High Power Laser and Physics, Shanghai Institute of Optics and Fine Mechanics, Chinese Academy of Sciences, Shanghai 201800, PR China

Zunqi Lin
Joint Laboratory on High Power Laser and Physics, Shanghai Institute of Optics and Fine Mechanics, Chinese Academy of Sciences, Shanghai 201800, PR China

M. Divoky
HiLASE, Institute of Physics, AS CR, v.v.i., Na Slovance 2, 182 21 Prague, Czech Republic

M. Smrz
HiLASE, Institute of Physics, AS CR, v.v.i., Na Slovance 2, 182 21 Prague, Czech Republic

M. Chyla
HiLASE, Institute of Physics, AS CR, v.v.i., Na Slovance 2, 182 21 Prague, Czech Republic

P. Sikocinski
HiLASE, Institute of Physics, AS CR, v.v.i., Na Slovance 2, 182 21 Prague, Czech Republic

P. Severova
HiLASE, Institute of Physics, AS CR, v.v.i., Na Slovance 2, 182 21 Prague, Czech Republic

O. Novak
HiLASE, Institute of Physics, AS CR, v.v.i., Na Slovance 2, 182 21 Prague, Czech Republic

J. Huynh
HiLASE, Institute of Physics, AS CR, v.v.i., Na Slovance 2, 182 21 Prague, Czech Republic

S.S. Nagisetty
HiLASE, Institute of Physics, AS CR, v.v.i., Na Slovance 2, 182 21 Prague, Czech Republic

T. Miura
HiLASE, Institute of Physics, AS CR, v.v.i., Na Slovance 2, 182 21 Prague, Czech Republic

J. Pilär
HiLASE, Institute of Physics, AS CR, v.v.i., Na Slovance 2, 182 21 Prague, Czech Republic

O. Slezak
HiLASE, Institute of Physics, AS CR, v.v.i., Na Slovance 2, 182 21 Prague, Czech Republic

M. Sawicka
HiLASE, Institute of Physics, AS CR, v.v.i., Na Slovance 2, 182 21 Prague, Czech Republic

V. Jambunathan
HiLASE, Institute of Physics, AS CR, v.v.i., Na Slovance 2, 182 21 Prague, Czech Republic

J. Vanda
HiLASE, Institute of Physics, AS CR, v.v.i., Na Slovance 2, 182 21 Prague, Czech Republic

A. Endo
HiLASE, Institute of Physics, AS CR, v.v.i., Na Slovance 2, 182 21 Prague, Czech Republic

A. Lucianetti
HiLASE, Institute of Physics, AS CR, v.v.i., Na Slovance 2, 182 21 Prague, Czech Republic

D. Rostohar
HiLASE, Institute of Physics, AS CR, v.v.i., Na Slovance 2, 182 21 Prague, Czech Republic

P.D. Mason
STFC Rutherford Appleton Laboratory, Didcot OX11 0QX, United Kingdom

P.J. Phillips
STFC Rutherford Appleton Laboratory, Didcot OX11 0QX, United Kingdom

K. Ertel
STFC Rutherford Appleton Laboratory, Didcot OX11 0QX, United Kingdom

S. Banerjee
STFC Rutherford Appleton Laboratory, Didcot OX11 0QX, United Kingdom

A.Hernandez-Gomez
STFC Rutherford Appleton Laboratory, Didcot OX11 0QX, United Kingdom

J.L. Collier
STFC Rutherford Appleton Laboratory, Didcot OX11 0QX, United Kingdom

T. Mocek
HiLASE, Institute of Physics, AS CR, v.v.i., Na Slovance 2, 182 21 Prague, Czech Republic

I.M. Aladi
Wigner Research Centre for Physics of the Hungarian Academy of Sciences, Association EURATOM HAS, H-1121 Budapest, Konkoly-Thege u. 29-33, Hungary

Márton
Wigner Research Centre for Physics of the Hungarian Academy of Sciences, Association EURATOM HAS, H-1121 Budapest, Konkoly-Thege u. 29-33, Hungary

P. Rácz
Wigner Research Centre for Physics of the Hungarian Academy of Sciences, Association EURATOM HAS, H-1121 Budapest, Konkoly-Thege u. 29-33, Hungary

P. Dombi
Wigner Research Centre for Physics of the Hungarian Academy of Sciences, Association EURATOM HAS, H-1121 Budapest, Konkoly-Thege u. 29-33, Hungary

I.B. FÖldes
Wigner Research Centre for Physics of the Hungarian Academy of Sciences, Association EURATOM HAS, H-1121 Budapest, Konkoly-Thege u. 29-33, Hungary

Jian-Gang Zheng
Research Center of Laser Fusion (RCLF), CAEP, P.O. Box 919-988, Mianyang, Sichuan 621900, China

Xin-Ying Jiang
Research Center of Laser Fusion (RCLF), CAEP, P.O. Box 919-988, Mianyang, Sichuan 621900, China

Xiong-Wei Yan
Research Center of Laser Fusion (RCLF), CAEP, P.O. Box 919-988, Mianyang, Sichuan 621900, China

Jun Zhang
Research Center of Laser Fusion (RCLF), CAEP, P.O. Box 919-988, Mianyang, Sichuan 621900, China

Zhen-Guo Wang
Research Center of Laser Fusion (RCLF), CAEP, P.O. Box 919-988, Mianyang, Sichuan 621900, China

Deng-Sheng Wu
Research Center of Laser Fusion (RCLF), CAEP, P.O. Box 919-988, Mianyang, Sichuan 621900, China

Xiao-Lin Tian
Research Center of Laser Fusion (RCLF), CAEP, P.O. Box 919-988, Mianyang, Sichuan 621900, China

Xiong-Jun Zhang
Research Center of Laser Fusion (RCLF), CAEP, P.O. Box 919-988, Mianyang, Sichuan 621900, China

Ming-Zhong Li
Research Center of Laser Fusion (RCLF), CAEP, P.O. Box 919-988, Mianyang, Sichuan 621900, China

Qi-Hua Zhu
Research Center of Laser Fusion (RCLF), CAEP, P.O. Box 919-988, Mianyang, Sichuan 621900, China

Jing-Qin Su
Research Center of Laser Fusion (RCLF), CAEP, P.O. Box 919-988, Mianyang, Sichuan 621900, China

Feng Jing
Research Center of Laser Fusion (RCLF), CAEP, P.O. Box 919-988, Mianyang, Sichuan 621900, China

Wan-Guo Zheng
Research Center of Laser Fusion (RCLF), CAEP, P.O. Box 919-988, Mianyang, Sichuan 621900, China

Andrea Macchi
National Institute of Optics, National Research Council (CNR/INO), Research Unit 'Adriano Gozzini', Department of Physics 'Enrico Fermi', University of Pisa, largo Bruno Pontecorvo 3, I-56127 Pisa, Italy

Yongtao Zhao
Institute of Modern Physics, Chinese Academy of Sciences, Lanzhou 730000, China

Rui Cheng
Institute of Modern Physics, Chinese Academy of Sciences, Lanzhou 730000, China

YuyuWang
Institute of Modern Physics, Chinese Academy of Sciences, Lanzhou 730000, China

Xianming Zhou
Institute of Modern Physics, Chinese Academy of Sciences, Lanzhou 730000, China

Yu Lei
Institute of Modern Physics, Chinese Academy of Sciences, Lanzhou 730000, China

Yuanbo Sun
Institute of Modern Physics, Chinese Academy of Sciences, Lanzhou 730000, China

Ge Xu
Institute of Modern Physics, Chinese Academy of Sciences, Lanzhou 730000, China

Jieru Ren
Institute of Modern Physics, Chinese Academy of Sciences, Lanzhou 730000, China

Lina Sheng
Institute of Modern Physics, Chinese Academy of Sciences, Lanzhou 730000, China

Zimin Zhang
Institute of Modern Physics, Chinese Academy of Sciences, Lanzhou 730000, China

Guoqing Xiao
Institute of Modern Physics, Chinese Academy of Sciences, Lanzhou 730000, China

Marco Hornung
Helmholtz-Institute Jena, Germany
Institute of Optics and Quantum Electronics, Jena, Germany

Hartmut Liebetrau
Institute of Optics and Quantum Electronics, Jena, Germany

Andreas Seidel
Institute of Optics and Quantum Electronics, Jena, Germany

Sebastian Keppler
Institute of Optics and Quantum Electronics, Jena, Germany

Alexander Kessler
Helmholtz-Institute Jena, Germany

Jörg Körner
Institute of Optics and Quantum Electronics, Jena, Germany

Marco Hellwing
Institute of Optics and Quantum Electronics, Jena, Germany

Frank Schorcht
Helmholtz-Institute Jena, Germany

Diethard Klöpfel
Institute of Optics and Quantum Electronics, Jena, Germany

Ajay K. Arunachalam
Helmholtz-Institute Jena, Germany

Georg A.Becker
Institute of Optics and Quantum Electronics, Jena, Germany

Alexander Säver
Helmholtz-Institute Jena, Germany
Institute of Optics and Quantum Electronics, Jena, Germany

Jens Polz
Institute of Optics and Quantum Electronics, Jena, Germany

Joachim Hein
Helmholtz-Institute Jena, Germany
Institute of Optics and Quantum Electronics, Jena, Germany

Malte C. Kaluza
Helmholtz-Institute Jena, Germany
Institute of Optics and Quantum Electronics, Jena, Germany

Qingwei Yang
National Laboratory on High Power Laser and Physics, Shanghai Institute of Optics and Fine Mechanics, Chinese Academy of Sciences, No. 390, Qinghe Road, Jiading District, Shanghai 201800, China

Xinglong Xie
National Laboratory on High Power Laser and Physics, Shanghai Institute of Optics and Fine Mechanics, Chinese Academy of Sciences, No. 390, Qinghe Road, Jiading District, Shanghai 201800, China

Jun Kang
National Laboratory on High Power Laser and Physics, Shanghai Institute of Optics and Fine Mechanics, Chinese Academy of Sciences, No. 390, Qinghe Road, Jiading District, Shanghai 201800, China

Haidong Zhu
National Laboratory on High Power Laser and Physics, Shanghai Institute of Optics and Fine Mechanics, Chinese Academy of Sciences, No. 390, Qinghe Road, Jiading District, Shanghai 201800, China

Ailin Guo
National Laboratory on High Power Laser and Physics, Shanghai Institute of Optics and Fine Mechanics, Chinese Academy of Sciences, No. 390, Qinghe Road, Jiading District, Shanghai 201800, China

Qi Gao
National Laboratory on High Power Laser and Physics, Shanghai Institute of Optics and Fine Mechanics, Chinese Academy of Sciences, No. 390, Qinghe Road, Jiading District, Shanghai 201800, China

Chao Wang
State Key Laboratory of Precision Spectroscopy, East China Normal University, Shanghai 200062, China

Wenxue Li
State Key Laboratory of Precision Spectroscopy, East China Normal University, Shanghai 200062, China

Xianghui Yang
State Key Laboratory of Precision Spectroscopy, East China Normal University, Shanghai 200062, China

Dongbi Bai
State Key Laboratory of Precision Spectroscopy, East China Normal University, Shanghai 200062, China

Kangwen Yang
State Key Laboratory of Precision Spectroscopy, East China Normal University, Shanghai 200062, China

Xuewei Ba
Key Laboratory of Transparent Opt-functional Inorganic Materials, Shanghai Institute of Ceramics, Chinese Academy of Sciences, Shanghai 200050, China

Jiang Li
Key Laboratory of Transparent Opt-functional Inorganic Materials, Shanghai Institute of Ceramics, Chinese Academy of Sciences, Shanghai 200050, China

Yubai Pan
Key Laboratory of Transparent Opt-functional Inorganic Materials, Shanghai Institute of Ceramics, Chinese Academy of Sciences, Shanghai 200050, China

Heping Zeng
State Key Laboratory of Precision Spectroscopy, East China Normal University, Shanghai 200062, China

Yudong Yao
Shanghai Institute of Optics and Fine Mechanics, Chinese Academy of Science, Shanghai 201800, China

Junyong Zhang
Shanghai Institute of Optics and Fine Mechanics, Chinese Academy of Science, Shanghai 201800, China

Yanli Zhang
Shanghai Institute of Optics and Fine Mechanics, Chinese Academy of Science, Shanghai 201800, China

Qunyu Bi
Shanghai Institute of Optics and Fine Mechanics, Chinese Academy of Science, Shanghai 201800, China

Jianqiang Zhu
Shanghai Institute of Optics and Fine Mechanics, Chinese Academy of Science, Shanghai 201800, China

Hong-bo Cai
Institute of Applied Physics and Computational Mathematics, Beijing 100094, People's Republic of China Center for Applied Physics and Technology, Peking University, Beijing 100871, People's Republic of China

Si-zhong Wu
Institute of Applied Physics and Computational Mathematics, Beijing 100094, People's Republic of China

Jun-feng Wu
Institute of Applied Physics and Computational Mathematics, Beijing 100094, People's Republic of China

Mo Chen
Institute of Applied Physics and Computational Mathematics, Beijing 100094, People's Republic of China

Hua Zhang
Institute of Applied Physics and Computational Mathematics, Beijing 100094, People's Republic of China

Min-qing He
Institute of Applied Physics and Computational Mathematics, Beijing 100094, People's Republic of China

Li-hua Cao
Institute of Applied Physics and Computational Mathematics, Beijing 100094, People's Republic of China Center for Applied Physics and Technology, Peking University, Beijing 100871, People's Republic of China

Cang-tao Zhou
Institute of Applied Physics and Computational Mathematics, Beijing 100094, People's Republic of China Center for Applied Physics and Technology, Peking University, Beijing 100871, People's Republic of China

Shao-ping Zhu
Institute of Applied Physics and Computational Mathematics, Beijing 100094, People's Republic of China

Xian-tu He
Institute of Applied Physics and Computational Mathematics, Beijing 100094, People's Republic of China Center for Applied Physics and Technology, Peking University, Beijing 100871, People's Republic of China

Mauro Temporal
Centre de Mathématiques et de Leurs Applications, ENS Cachan and CNRS, 61 Av. du President Wilson, Cachan Cedex, France

Benoit Canaud
CEA, DIF, Arpajon Cedex, France

Warren J. Garbett
AWE plc, Aldermaston, Reading, Berkshire, United Kingdom

Rafael Ramis
ETSI Aeronáuticos, Universidad Polit'ecnica de Madrid, Madrid, Spain

Stefan Weber
ELI-Beamlines, Institute of Physics, Academy of Sciences of the Czech Republic, Prague, Czech Republic

Shalom Eliezer
Nuclear Fusion Institute, Polytechnic University of Madrid, Madrid, Spain

Noaz Nissim
Applied Physics Division, Soreq NRC, Yavne, Israel

Shirly Vinikman Pinhasi
Applied Physics Division, Soreq NRC, Yavne, Israel

Erez Raicher
Applied Physics Division, Soreq NRC, Yavne, Israel
Hebrew University of Jerusalem, Jerusalem, Israel

José Maria Martinez Val
Nuclear Fusion Institute, Polytechnic University of Madrid, Madrid, Spain

W. P. Wang
State Key Laboratory of High Field Laser Physics, Shanghai Institute of Optics and Fine Mechanics, Chinese Academy of Sciences, P. O. Box 800-211, Shanghai 201800, China

X. M. Zhang
State Key Laboratory of High Field Laser Physics, Shanghai Institute of Optics and Fine Mechanics, Chinese Academy of Sciences, P. O. Box 800-211, Shanghai 201800, China

X. F. Wang
State Key Laboratory of High Field Laser Physics, Shanghai Institute of Optics and Fine Mechanics, Chinese Academy of Sciences, P. O. Box 800-211, Shanghai 201800, China

X. Y. Zhao
State Key Laboratory of High Field Laser Physics, Shanghai Institute of Optics and Fine Mechanics, Chinese Academy of Sciences, P. O. Box 800-211, Shanghai 201800, China

J. C. Xu
State Key Laboratory of High Field Laser Physics, Shanghai Institute of Optics and Fine Mechanics, Chinese Academy of Sciences, P. O. Box 800-211, Shanghai 201800, China

Y. H. Yu
State Key Laboratory of High Field Laser Physics, Shanghai Institute of Optics and Fine Mechanics, Chinese Academy of Sciences, P. O. Box 800-211, Shanghai 201800, China

China

L. Q. Yi
State Key Laboratory of High Field Laser Physics, Shanghai Institute of Optics and Fine Mechanics, Chinese Academy of Sciences, P. O. Box 800-211, Shanghai 201800, China

Y. Shi
State Key Laboratory of High Field Laser Physics, Shanghai Institute of Optics and Fine Mechanics, Chinese Academy of Sciences, P. O. Box 800-211, Shanghai 201800, China

L. G. Zhang
State Key Laboratory of High Field Laser Physics, Shanghai Institute of Optics and Fine Mechanics, Chinese Academy of Sciences, P. O. Box 800-211, Shanghai 201800, China

T. J. Xu
State Key Laboratory of High Field Laser Physics, Shanghai Institute of Optics and Fine Mechanics, Chinese Academy of Sciences, P. O. Box 800-211, Shanghai 201800, China

B.Liu
State Key Laboratory of High Field Laser Physics, Shanghai Institute of Optics and Fine Mechanics, Chinese Academy of Sciences, P. O. Box 800-211, Shanghai 201800, China

Z. K. Pei
State Key Laboratory of High Field Laser Physics, Shanghai Institute of Optics and Fine Mechanics, Chinese Academy of Sciences, P. O. Box 800-211, Shanghai 201800, China

B. F. Shen
State Key Laboratory of High Field Laser Physics, Shanghai Institute of Optics and Fine Mechanics, Chinese Academy of Sciences, P. O. Box 800-211, Shanghai 201800, China

Kazuhisa Nakajima
Center for Relativistic Laser Science, Institute for Basic Science (IBS), Gwangju 500-712, Republic of Korea

Zeev Toroker
Department of Electrical Engineering, Technion, Israel Institute of Technology, Haifa 32000, Israel

Miron Voin
Department of Electrical Engineering, Technion, Israel Institute of Technology, Haifa 32000, Israel

Levi Schächter
Department of Electrical Engineering, Technion, Israel Institute of Technology, Haifa 32000, Israel

Mauro Temporal
Centre de Mathématiques et de Leurs Applications, ENS Cachan and CNRS, 61 Av. du President Wilson, F-94235 Cachan Cedex, France

Benoit Canaud
CEA, DIF, F-91297, Arpajon Cedex, France

Warren J. Garbett
AWE plc, Aldermaston, Reading, Berkshire RG7 4PR, United Kingdom

Rafael Ramis
ETSI Aeron´auticos, Universidad Polit´ecnica de Madrid, 28040 Madrid, Spain

J. Krása
Institute of Physics, AS CR, 182 21 Prague 8, Czech Republic

D. Klír
zech Technical University in Prague, FEE, 166 27 Prague, Czech Republic

A. Velyhan
Institute of Physics, AS CR, 182 21 Prague 8, Czech Republic

E. Krousky
Institute of Physics, AS CR, 182 21 Prague 8, Czech Republic

M. Pfeifer
Institute of Physics, AS CR, 182 21 Prague 8, Czech Republic

K. Řezáč
Czech Technical University in Prague, FEE, 166 27 Prague, Czech Republic

J. Cikhardt
Czech Technical University in Prague, FEE, 166 27 Prague, Czech Republic

K. Turek
Nuclear Physics Institute, AS CR, 180 00 Prague 8, Czech Republic

J. Ullschmied
Institute of Plasma Physics, AS CR, 182 00 Prague 8, Czech Republic

K. Jungwirth
Institute of Physics, AS CR, 182 21 Prague 8, Czech Republic

Colin Danson
Centre for Inertial Fusion Studies (CIFS), Imperial College London, UK
AWE plc, Aldermaston, UK

David Neely
Central Laser Facility, STFC Rutherford Appleton Laboratory, UK\

David Hillier
AWE plc, Aldermaston, UK

B.Guillaume
Laboratoire d'Optique Appliquée, Ecole Polytechnique, Palaiseau, 91128, France
STFC Rutherford Appleton Laboratory, Didcot, Oxon OX11 0QX, United Kingdom

K. Humphrey
University of Strathclyde, Glasgow G1 1XQ, United Kingdom

H. Nakamura
Blackett Laboratory, Imperial College London, Prince Consort Road, London SW7 2BZ, United Kingdom

R. M. G. M. Trines
STFC Rutherford Appleton Laboratory, Didcot, Oxon OX11 0QX, United Kingdom

R. Heathcote
STFC Rutherford Appleton Laboratory, Didcot, Oxon OX11 0QX, United Kingdom

M. Galimberti
STFC Rutherford Appleton Laboratory, Didcot, Oxon OX11 0QX, United Kingdom

Y. Amano
Graduate School of Engineering, Osaka University, Japan

D. Doria
Queens University Belfast, Belfast BT7 1NN, United Kingdom

G. Hicks
Blackett Laboratory, Imperial College London, Prince Consort Road, London SW7 2BZ, United Kingdom

E. Higson
University of Oxford, Parks Road, Oxford OX1 3PU, United Kingdom

S. Kar
Queens University Belfast, Belfast BT7 1NN, United Kingdom

G. Sarri
Queens University Belfast, Belfast BT7 1NN, United Kingdom

M. Skramic
University of Cambridge, Cambridge CB2 1TQ, United Kingdom

J. Swain
University of Oxford, Parks Road, Oxford OX1 3PU, United Kingdom

K. Tang
University of Oxford, Parks Road, Oxford OX1 3PU, United Kingdom

J. Weston
University of Oxford, Parks Road, Oxford OX1 3PU, United Kingdom

P. Zak
University of Oxford, Parks Road, Oxford OX1 3PU, United Kingdom

E. P. Alves
GoLP/Instituto de Plasmas e Fusão Nuclear - Laboratorio Associado, Instituto Superior Técnico, 1049-001 Lisbon, Portugal

R. A. Fonseca
GoLP/Instituto de Plasmas e Fusão Nuclear - Laboratorio Associado, Instituto Superior Técnico, 1049-001 Lisbon, Portugal

F. Fiúza
GoLP/Instituto de Plasmas e Fusão Nuclear - Laboratorio Associado, Instituto Superior Técnico, 1049-001 Lisbon, Portugal

H. Habara
Graduate School of Engineering, Osaka University, Japan

K. A. Tanaka
Graduate School of Engineering, Osaka University, Japan

R. Bingham
STFC Rutherford Appleton Laboratory, Didcot, Oxon OX11 0QX, United Kingdom

M. Borghesi
Queens University Belfast, Belfast BT7 1NN, United Kingdom

Z. Najmudin
Blackett Laboratory, Imperial College London, Prince Consort Road, London SW7 2BZ, United Kingdom

L. O. Silva
GoLP/Instituto de Plasmas e Fusão Nuclear - Laboratorio Associado, Instituto Superior Técnico, 1049-001 Lisbon, Portugal

P. A. Norreys
University of Oxford, Parks Road, Oxford OX1 3PU, United Kingdom
STFC Rutherford Appleton Laboratory, Didcot, Oxon OX11 0QX, United Kingdom

Xiaojin Cheng
Shanghai Key Laboratory of All Solid-state Laser and Applied Techniques, Shanghai Institute of Optics and Fine Mechanics, Chinese Academy of Sciences, Shanghai 201800, China

Jianlei Wang
Shanghai Key Laboratory of All Solid-state Laser and Applied Techniques, Shanghai Institute of Optics and Fine Mechanics, Chinese Academy of Sciences, Shanghai 201800, China

Zhongguo Yang
Shanghai Key Laboratory of All Solid-state Laser and Applied Techniques, Shanghai Institute of Optics and Fine Mechanics, Chinese Academy of Sciences, Shanghai 201800, China

Jin Liu
Shanghai Key Laboratory of All Solid-state Laser and Applied Techniques, Shanghai Institute of Optics and Fine Mechanics, Chinese Academy of Sciences, Shanghai 201800, China

Lei Li
Shanghai Key Laboratory of All Solid-state Laser and Applied Techniques, Shanghai Institute of Optics and Fine Mechanics, Chinese Academy of Sciences, Shanghai 201800, China

Xiangchun Shi
Shanghai Key Laboratory of All Solid-state Laser and Applied Techniques, Shanghai Institute of Optics and Fine Mechanics, Chinese Academy of Sciences, Shanghai 201800, China

Wenfa Huang
The Joint Laboratory for High Power Laser Physics, Shanghai Institute of Optics and Fine Mechanics, Chinese Academy of Sciences, Shanghai 201800, China

Jiangfeng Wang
The Joint Laboratory for High Power Laser Physics, Shanghai Institute of Optics and Fine Mechanics, Chinese Academy of Sciences, Shanghai 201800, China

Weibiao Chen
Shanghai Key Laboratory of All Solid-state Laser and Applied Techniques, Shanghai Institute of Optics and Fine Mechanics, Chinese Academy of Sciences, Shanghai 201800, China